International
REVIEW OF
Neurobiology
Volume 85

International

REVIEW OF

Neurobiology

Volume 85

SERIES EDITORS

RONALD J. BRADLEY
Department of Psychiatry, College of Medicine
The University of Tennessee Health Science Center
Memphis, Tennessee, USA

R. ADRON HARRIS
Waggoner Center for Alcohol and Drug Addiction Research
The University of Texas at Austin
Austin, Texas, USA

PETER JENNER
Division of Pharmacology and Therapeutics
GKT School of Biomedical Sciences
King's College, London, UK

EDITORIAL BOARD

Advances in Neuropharmacology

EDITED BY

BAGETTA, G.
Department of Pharmacobiology and University Centre for Adaptive Disorders and Headache (UCADH), Section of Neuropharmacology of Normal and Pathological Neuronal Plasticity
University of Calabria, Arcavacata di Rende, Cosenza, Italy

CORASANITI, M. T.
Department of Pharmacobiological Sciences
University of Catanzaro "Magna Graecia", Catanzaro, Italy

SAKURADA, T.
First Department of Pharmacology
Daiichi College of Pharmaceutical Sciences, Fukuoka, Japan

SAKURADA, S.
Department of Anatomy and Physiology
Tohoku Pharmaceutical Sciences, Sendai, Japan

ELSEVIER

AMSTERDAM • BOSTON • HEIDELBERG • LONDON
NEW YORK • OXFORD • PARIS • SAN DIEGO
SAN FRANCISCO • SINGAPORE • SYDNEY • TOKYO
Academic Press is an imprint of Elsevier

AP

Academic Press is an imprint of Elsevier
360 Park Avenue South, New York, NY 10010-1700
525 B Street, Suite 1900, San Diego, California 92101-4495, USA
32 Jamestown Road, London NW1 7BY, UK

This book is printed on acid-free paper.

For information on all Academic Press publications
visit our Web site at www.elsevierdirect.com

ISBN-13: 978-0-12-374893-5

Printed and bound by CPI Group (UK) Ltd, Croydon, CR0 4YY

Transferred to Digital Print 2011

CONTENTS

Targeted Lipidomics as a Tool to Investigate Endocannabinoid Function

Giuseppe Astarita, Jennifer Geaga, Faizy Ahmed, and Daniele Piomelli

The Endocannabinoid System as a Target for Novel Anxiolytic and Antidepressant Drugs

Silvana Gaetani, Pasqua Dipasquale, Adele Romano, Laura Righetti, Tommaso Cassano, Daniele Piomelli, and Vincenzo Cuomo

$GABA_A$ Receptor Function and Gene Expression During Pregnancy and Postpartum

Giovanni Biggio, Maria Cristina Mostallino, Paolo Follesa, Alessandra Concas, and Enrico Sanna

Early Postnatal Stress and Neural Circuit Underlying Emotional Regulation

MACHIKO MATSUMOTO, MITSUHIRO YOSHIOKA, AND HIROKO TOGASHI

Roles of the Histaminergic Neurotransmission on Methamphetamine-Induced Locomotor Sensitization and Reward: A Study of Receptors Gene Knockout Mice

NAOKO TAKINO, EIKO SAKURAI, ATSUO KURAMASU, NOBUYUKI OKAMURA, AND KAZUHIKO YANAI

Developmental Exposure to Cannabinoids Causes Subtle and Enduring Neurofunctional Alterations

PATRIZIA CAMPOLONGO, VIVIANA TREZZA, MAURA PALMERY, LUIGIA TRABACE, AND VINCENZO CUOMO

Neuronal Mechanisms for Pain-Induced Aversion: Behavioral Studies Using a Conditioned Place Aversion Test

MASABUMI MINAMI

Bv8/Prokineticins and their Receptors: A New Pronociceptive System

LUCIA NEGRI, ROBERTA LATTANZI, ELISA GIANNINI, MICHELA CANESTRELLI, ANNALISA NICOTRA, AND PIETRO MELCHIORRI

$P2Y_6$-Evoked Microglial Phagocytosis

KAZUHIDE INOUE, SCHUICHI KOIZUMI, AYAKO KATAOKA, HIDETOSHI TOZAKI-SAITOH, AND MAKOTO TSUDA

PPAR and Pain

TAKEHIKO MAEDA AND SHIROH KISHIOKA

Involvement of Inflammatory Mediators in Neuropathic Pain Caused by Vincristine

Norikazu Kiguchi, Takehiko Maeda, Yuka Kobayashi, Fumihiro Saika, and Shiroh Kishioka

Nociceptive Behavior Induced by the Endogenous Opioid Peptides Dynorphins in Uninjured Mice: Evidence with Intrathecal *N*-Ethylmaleimide Inhibiting Dynorphin Degradation

Koichi Tan-No, Hiroaki Takahashi, Osamu Nakagawasai, Fukie Niijima, Shinobu Sakurada, Georgy Bakalkin, Lars Terenius, and Takeshi Tadano

Mechanism of Allodynia Evoked by Intrathecal Morphine-3-Glucuronide in Mice

Takaaki Komatsu, Shinobu Sakurada, Sou Katsuyama, Kengo Sanai, and Tsukasa Sakurada

(–)-Linalool Attenuates Allodynia in Neuropathic Pain Induced by Spinal Nerve Ligation in C57/Bl6 Mice

Laura Berliocchi, Rossella Russo, Alessandra Levato, Vincenza Fratto, Giacinto Bagetta, Shinobu Sakurada, Tsukasa Sakurada, Nicola Biagio Mercuri, and Maria Tiziana Corasaniti

Intraplantar Injection of Bergamot Essential Oil into the Mouse Hindpaw: Effects on Capsaicin-Induced Nociceptive Behaviors

Tsukasa Sakurada, Hikari Kuwahata, Soh Katsuyama, Takaaki Komatsu, Luigi Antonio Morrone, Maria Tiziana Corasaniti, Giacinto Bagetta, and Shinobu Sakurada

New Therapy for Neuropathic Pain

Hirokazu Mizoguchi, Chizuko Watanabe, Akihiko Yonezawa, and Shinobu Sakurada

Regulated Exocytosis from Astrocytes: Physiological and Pathological Related Aspects

CORRADO CALÌ, JULIE MARCHALAND, PAOLA SPAGNUOLO, JULIEN GREMION, AND PAOLA BEZZI

Glutamate Release from Astrocytic Gliosomes Under Physiological and Pathological Conditions

MARCO MILANESE, TIZIANA BONIFACINO, SIMONA ZAPPETTINI, CESARE USAI, CARLO TACCHETTI, MARIO NOBILE, AND GIAMBATTISTA BONANNO

Neurotrophic and Neuroprotective Actions of an Enhancer of Ganglioside Biosynthesis

JIN-ICHI INOKUCHI

Involvement of Endocannabinoid Signaling in the Neuroprotective Effects of Subtype 1 Metabotropic Glutamate Receptor Antagonists in Models of Cerebral Ischemia

Elisa Landucci, Francesca Boscia, Elisabetta Gerace, Tania Scartabelli, Andrea Cozzi, Flavio Moroni, Guido Mannaioni, and Domenico E. Pellegrini-Giampietro

NF-kappaB Dimers in the Regulation of Neuronal Survival

Ilenia Sarnico, Annamaria Lanzillotta, Marina Benarese, Manuela Alghisi, Cristina Baiguera, Leontino Battistin, PierFranco Spano, and Marina Pizzi

Oxidative Stress in Stroke Pathophysiology: Validation of Hydrogen Peroxide Metabolism as a Pharmacological Target to Afford Neuroprotection

Diana Amantea, Maria Cristina Marrone, Robert Nisticò, Mauro Federici, Giacinto Bagetta, Giorgio Bernardi, and Nicola Biagio Mercuri

Role of Akt and Erk Signaling in the Neurogenesis following Brain Ischemia

Norifumi Shioda, Feng Han, and Kohji Fukunaga

Prevention of Glutamate Accumulation and Upregulation of Phospho-Akt may Account for Neuroprotection Afforded by Bergamot Essential Oil against Brain Injury Induced by Focal Cerebral Ischemia in Rat

Diana Amantea, Vincenza Fratto, Simona Maida, Domenicantonio Rotiroti, Salvatore Ragusa, Giuseppe Nappi, Giacinto Bagetta, and Maria Tiziana Corasaniti

Identification of Novel Pharmacological Targets to Minimize Excitotoxic Retinal Damage

Rossella Russo, Domenicantonio Rotiroti, Cristina Tassorelli, Carlo Nucci, Giacinto Bagetta, Massimo Gilberto Bucci, Mari Tiziana Corasaniti, and Luigi Antonio Morrone

CONTRIBUTORS

Numbers in parentheses indicate the pages on which the author's contributions begin.

Faizy Ahmed (35), Department of Pharmacology, University of California, Irvine, California 92967, USA

Kazuyuki Aihara (1), Institute of Industrial Science, University of Tokyo, 4-6-1 Komaba, Meguro-ku, Tokyo 153-8505, Japan; and ERATO Aihara Complexity Modelling Project, JST, 45–18 Oyama, Shibuya-ku, Tokyo 151-0065, Japan

Manuela Alghisi (351), Division of Pharmacology and Experimental Therapeutics, Department of Biomedical Sciences and Biotechnologies, School of Medicine, University of Brescia, Brescia 25123, Italy

Diana Amantea (363, 389), Department of Pharmacobiology and Center of Neuropharmacology of Normal and Pathological Neuronal Plasticity, UCADH, University of Calabria, 87036 Cosenza, Italy

Yosefu Arime (29), Department of Biological Psychiatry, Tohoku University Graduate School of Medicine, Sendai 980-8574, Japan

Giuseppe Astarita (35), Department of Pharmacology, University of California, Irvine, California 92967, USA

Giacinto Bagetta (221, 237, 363, 389, 407), Department of Pharmacobiology and University Centre for Adaptive Disorders and Headache (UCADH), Section of Neuropharmacology of Normal and Pathological Neuronal Plasticity, University of Calabria, Arcavacata di Rende, Cosenza 87036, Italy

Cristina Baiguera (351), Division of Pharmacology and Experimental Therapeutics, Department of Biomedical Sciences and Biotechnologies, School of Medicine, University of Brescia, Brescia 25123, Italy

Georgy Bakalkin (191), Division of Biological Research on Drug Dependence, Department of Pharmaceutical Biosciences, Uppsala University, Uppsala S-751 24, Terenius, Sweden

Leontino Battistin (351), IRCCS San Camillo Hospital, 30100 Venice, Italy

Marina Benarese (351), Division of Pharmacology and Experimental Therapeutics, Department of Biomedical Sciences and Biotechnologies, School of Medicine, University of Brescia, Brescia 25123, Italy

Laura Berliocchi (221), Department of Pharmacobiological Sciences, University of Catanzaro "Magna Graecia", 88100 Catanzaro, Italy

Girogio Bernardi (363), Neurological Clinic, Department of Neuroscience, "Tor Vergata" University; and C.E.R.C.—Santa Lucia Foundation IRCCS, Rome, Italy

Paola Bezzi (261), Department of Cellular Biology and Morphology (DBCM), Faculty of Medicine, University of Lausanne, rue du Bugnon 9, 1005 Lausanne, Switzerland

Giovanni Biggio (73), C.N.R. Institute of Neuroscience, Section of Cagliari, Cagliari, Italy; and Department of Experimental Biology, Center of Excellence for the Neurobiology of Dependence, University of Cagliari, 09100 Cagliari, Italy

Giambattista Bonanno (295), National Institute for Neuroscience (INN), 10125 Turin; and Center of Excellence for Biomedical Research; and Department of Experimental Medicine, Section of Pharmacology and Toxicology, University of Genoa, 16148 Genoa, Italy

Tiziana Bonifacino (295), Department of Experimental Medicine, Section of Pharmacology and Toxicology, University of Genoa, 16148 Genoa, Italy

Zuner A. Bortolotto (13), Department of Anatomy, MRC Centre for Synaptic Plasticity, School of Medical Sciences, University of Bristol, Bristol BS8 1TD, United Kingdom

Francesca Boscia (337), Dipartimento di Farmacologia Preclinica e Clinica, Università di Firenze, Firenze 50139; and Dipartimento di Neuroscienze, Università di Napoli "Federico II," Napoli 80131, Italy

Massimo Gilberto Bucci (407), Fondazione G. B. Bietti, IRCCS, 00199 Roma, Italy

Corrado Calì (261), Department of Cellular Biology and Morphology (DBCM), Faculty of Medicine, University of Lausanne, rue du Bugnon 9, 1005 Lausanne, Switzerland

Patrizia Campolongo (117), Department of Physiology and Pharmacology, "Vittorio Erspamer", Sapienza University of Rome, P.le A. Moro 5, 00185 Rome, Italy

Michela Canestrelli (145), Department of Physiology and Pharmacology, University "Sapienza" of Roma, P.le Aldo Moro 5, 00185 Rome, Italy

Tommaso Cassano (57), Department of Biomedical Sciences, Medical School, University of Foggia, V.le Luigi Pinto 1, 71100 Foggia, Italy

Graham L. Collingridge (13), Department of Anatomy, MRC Centre for Synaptic Plasticity, School of Medical Sciences, University of Bristol, Bristol BS8 1TD, United Kingdom

Alessandra Concas (73), C.N.R. Institute of Neuroscience, Section of Cagliari, Cagliari, Italy; and Department of Experimental Biology, Center of Excellence for the Neurobiology of Dependence, University of Cagliari, 09100 Cagliari, Italy

Maria Tiziana Corasaniti (221, 237, 389, 407), Department of Pharmacobiological Sciences, University of Catanzaro "Magna Graecia", 88100 Catanzaro, Italy

Andrea Cozzi (337), Dipartimento di Farmacologia Preclinica e Clinica, Università di Firenze, Firenze 50139, Italy

Vincenzo Cuomo (57, 117), Department of Physiology and Pharmacology, "Vittorio Erspamer", Sapienza University of Rome, P.le A. Moro 5, 00185 Rome, Italy

Sheila Dargan (13), Department of Physiology and Pharmacology; and Department of Anatomy, MRC Centre for Synaptic Plasticity, School of Medical Sciences, University of Bristol, Bristol BS8 1TD, United Kingdom

Pasqua Dipasquale (57), Department of Physiology and Pharmacology, Sapienza University of Rome, P.le Aldo Moro 5, 00185 Rome, Italy

Mauro Federici (363), C.E.R.C.—Santa Lucia Foundation IRCCS, Rome, Italy

Stephen M. Fitzjohn (13), Department of Anatomy, MRC Centre for Synaptic Plasticity, School of Medical Sciences, University of Bristol, Bristol BS8 1TD, United Kingdom

Paolo Follesa (73), Department of Experimental Biology, Center of Excellence for the Neurobiology of Dependence, University of Cagliari, 09100 Cagliari, Italy

Vincenza Fratto (221, 389), Department of Pharmacobiological Sciences, University of Catanzaro "Magna Graecia", 88100 Catanzaro, Italy

Asami Fukui (29), Department of Biological Psychiatry, Tohoku University Graduate School of Medicine, Sendai 980-8574, Japan

Kohji Fukunaga (375), Department of Pharmacology, Graduate School of Pharmaceutical Sciences, Tohoku University, Sendai 980-8578, Japan

Setsu Fumushima (29), Department of Biological Psychiatry, Tohoku University Graduate School of Medicine, Sendai 980-8574, Japan

Silvana Gaetani (57), Department of Physiology and Pharmacology, Sapienza University of Rome, P.le Aldo Moro 5, 00185 Rome, Italy

Jennifer Geaga (35), Department of Pharmacology, University of California, Irvine, California 92967, USA

Elisabetta Gerace (337), Dipartimento di Farmacologia Preclinica e Clinica, Università di Firenze, Firenze 50139, Italy

Elisa Giannini (145), Department of Physiology and Pharmacology, University "Sapienza" of Roma, P.le Aldo Moro 5, 00185 Rome, Italy

Julien Gremion (261), Department of Cellular Biology and Morphology (DBCM), Faculty of Medicine, University of Lausanne, rue du Bugnon 9, 1005 Lausanne, Switzerland

Feng Han (375), Institute of Pharmacology & Toxicology and Biochemical Pharmaceutics, College of Pharmaceutical Sciences, Zhejiang University, Hangzhou 310058, China

Kazutaka Ikeda (29), Molecular Psychiatry Research, Tokyo Institute of Psychiatry, Tokyo 156-8505, Japan

Jin-ichi Inokuchi (319), Division of Glycopathology, Institute of Molecular Biomembranes and Glycobiology, Tohoku Pharmaceutical University, 4-4-1, komatsushima, Aoba-ku, Sendai 981-8558, Miyagi, Japan

Kazuhide Inoue (159), Department of Molecular and System Pharmacology, Graduate School of Pharmaceutical Sciences, Kyushu University, 3-1-1 Maidashi, Higashi, Fukuoka 812-8582, Japan

Toshiro Inui (1), ERATO Asada Project, Japan Science and Technology Agency, Yamadaoka, Suita, Osaka 565-0871, Japan; and Department of Intelligence Science and Technology, Graduate School of Informatics, Kyoto University, Kyoto 606-8501, Japan

David E. Jane (13), Department of Physiology and Pharmacology, MRC Centre for Synaptic Plasticity, School of Medical Sciences, University of Bristol, Bristol BS8 1TD, United Kingdom

Yoshiyuki Kasahara (29), Department of Biological Psychiatry, Tohoku University Graduate School of Medicine, Sendai 980-8574, Japan

Ayako Kataoka (159), Department of Molecular and System Pharmacology, Graduate School of Pharmaceutical Sciences, Kyushu University, 3-1-1 Maidashi, Higashi, Fukuoka 812-8582, Japan

Sou Katsuyama (207), First Department of Pharmacology, Daiichi College of Pharmaceutical Sciences, 22-1 Tamagawa-cho, Minami-ku, Fukuoka 815-8511, Japan

Soh Katsuyama (237), First Department of Pharmacology, Daiichi College of Pharmaceutical Sciences, Fukuoka, Japan

Norikazu Kiguchi (179), Department of Pharmacology, Wakayama Medical University, 811–1 Kimiidera, Wakayama 641-0012, Japan

Shiroh Kishioka (165, 179), Department of Pharmacology, Wakayama Medical University, 811–1 Kimiidera, Wakayama 641-0012, Japan

Yuka Kobayashi (179), Department of Pharmacology, Wakayama Medical University, 811–1 Kimiidera, Wakayama 641-0012, Japan

Schuichi Koizumi (159), Department of Pharmacology, Interdisciplinary Graduate School of Medicine and Engineering, University of Yamanashi, 1110 Shimokato, Chuo, Yamanashi 409-3893, Japan

Takaaki Komatsu (207, 237), First Department of Pharmacology, Daiichi College of Pharmaceutical Sciences, 22-1 Tamagawa-cho, Minami-ku, Fukuoka 815-8511, Japan

Atsuo Kuramasu (109), Department of Pharmacology, Tohoku University Graduate School of Medicine, Sendai 980-8575, Japan

Hikari Kuwahata (237), Department of Physiology and Anatomy, Tohoku Pharmaceutical University, Sendai, Japan

Elisa Landucci (337), Dipartimento di Farmacologia Preclinica e Clinica, Università di Firenze, Firenze 50139, Italy

Annamaria Lanzillotta (351), Division of Pharmacology and Experimental Therapeutics, Department of Biomedical Sciences and Biotechnologies, School of Medicine, University of Brescia, Brescia 25123, Italy

Roberta Lattanzi (145), Department of Physiology and Pharmacology, University "Sapienza" of Roma, P.le Aldo Moro 5, 00185 Rome, Italy

Alessandra Levato (221), Department of Pharmacobiological Sciences, University of Catanzaro "Magna Graecia", 88100 Catanzaro, Italy

BingJin Li (29), Department of Biological Psychiatry, Tohoku University Graduate School of Medicine, Sendai 980-8574, Japan

David Lodge (13), Department of Anatomy, MRC Centre for Synaptic Plasticity, School of Medical Sciences, University of Bristol, Bristol BS8 1TD, United Kingdom

Takehiko Maeda (165, 179), Department of Pharmacology, Wakayama Medical University, 811–1 Kimiidera, Wakayama 641–0012, Japan

Simona Maida (389), Department of Pharmacobiological Sciences, University of Catanzaro "Magna Graecia", 88100 Catanzaro, Italy

Guido Mannaioni (337), Dipartimento di Farmacologia Preclinica e Clinica, Università di Firenze, Firenze 50139, Italy

Julie Marchaland (261), Department of Cellular Biology and Morphology (DBCM), Faculty of Medicine, University of Lausanne, rue du Bugnon 9, 1005 Lausanne, Switzerland

Maria Cristina Marrone (363), C.E.R.C.—Santa Lucia Foundation IRCCS, Rome, Italy

Machiko Matsumoto (95), Department of Pharmacology, School of Pharmaceutical Science, Health Sciences University of Hokkaido, Ishikari-Tobetsu, Japan

Pietro Melchiorri (145), Department of Physiology and Pharmacology, University "Sapienza" of Roma, P.le Aldo Moro 5, 00185 Rome, Italy

Nicola Biagio Mercuri (221, 363), Department of Neuroscience, University "Tor Vergata", and Laboratory of Experimental Neurology, CERC-IRCCS Santa Lucia, Rome, Italy

Marco Milanese (295), Department of Experimental Medicine, Section of Pharmacology and Toxicology, University of Genoa, 16148 Genoa, Italy

Masabumi Minami (135), Department of Pharmacology, Graduate School of Pharmaceutical Sciences, Hokkaido University, Sapporo 060–0812, Japan

Hirokazu Mizoguchi (249), Department of Physiology and Anatomy, Tohoku Pharmaceutical University, Sendai 981-8558, Japan

Flavio Moroni (337), Dipartimento di Farmacologia Preclinica e Clinica, Università di Firenze, Firenze 50139, Italy

Luigi Antonio Morrone (237, 407), Department of Pharmacobiology and University Center for Adaptive Disorders and Headache (UCADH), Section of Neuropharmacology of Normal and Pathological Neuronal Plasticity, University of Calabria, 87036 Arcavacata di Rende, Italy

Maria Cristina Mostallino (73), C.N.R. Institute of Neuroscience, Section of Cagliari, Cagliari, Italy

Hajime Mushiake (1), Department of Physiology, Tohoku University School of Medicine, 2-1 Seiryo-cho, Aoba-ku, Sendai 980-8575, Japan

Osamu Nakagawasai (191), Department of Pharmacology, Tohoku Pharmaceutical University, 4-4-1 Komatsushima, Aoba-ku, Sendai 981-8558, Japan

Giuseppe Nappi (389), Chair of Neurology, University "La Sapienza" of 00161 Rome and IRCCS "C Mondino Institute of Neurology" Foundation, 27100 Pavia, Italy

Lucia Negri (145), Department of Physiology and Pharmacology, University "Sapienza" of Roma, P.le Aldo Moro 5, 00185 Rome, Italy

Annalisa Nicotra (145), Department of Physiology and Pharmacology, University "Sapienza" of Roma, P.le Aldo Moro 5, 00185 Rome, Italy

Fukie Niijima (191), Department of Pharmacology, Tohoku Pharmaceutical University, 4-4-1 Komatsushima, Aoba-ku, Sendai 981-8558, Japan

Robert Nisticò (13, 363), Department of Anatomy, MRC Centre for Synaptic Plasticity, School of Medical Sciences, University of Bristol, Bristol BS8 1TD, United Kingdom; and Department of Pharmacobiology, University Center for Adaptive Disorders and Headache (UCHAD), Section of Neuropharmacology of Normal and Pathological Neuronal Plasticity, University of Calabria, Arcavacata di Rende, Italy

Mario Nobile (295), Institute of Biophysics, National Research Council, 16149 Genoa, Italy

Carlo Nucci (407), "Mondino-Tor Vergata" Center for Experimental Neurobiology; and Physiopathological Optics, Department of Biopathology, University of Rome "Tor Vergata", 00133 Rome, Italy

Nobuyuki Okamura (109), Department of Pharmacology, Tohoku University Graduate School of Medicine, Sendai 980-8575, Japan

Maura Palmery (117), Department of Physiology and Pharmacology, "Vittorio Erspamer", Sapienza University of Rome, P.le A. Moro 5, 00185 Rome, Italy

Domenico E. Pellegrini-Giampietro (337), Dipartimento di Farmacologia Preclinica e Clinica, Università di Firenze, Firenze 50139, Italy

Daniele Piomelli (35, 57), Department of Pharmacology, University of California, 3101 Gillespie NRF Irvine, 92697-4625 Irvine, California USA

Marina Pizzi (351), National Institute of Neuroscience, 10125 Turin; and Division of Pharmacology and Experimental Therapeutics, Department of

Biomedical Sciences and Biotechnologies, School of Medicine, University of Brescia, Brescia 25123, Italy

Salvatore Ragusa (389), Department of Pharmacobiological Sciences, University of Catanzaro "Magna Graecia", 88100 Catanzaro, Italy

Laura Righetti (57), Department of Physiology and Pharmacology, Sapienza University of Rome, P.le Aldo Moro 5, 00185 Rome, Italy

Adele Romano (57), Department of Physiology and Pharmacology, Sapienza University of Rome, P.le Aldo Moro 5, 00185 Rome, Italy

Domenicantonio Rotiroti (389, 407), Department of Pharmacobiological Sciences, University of Catanzaro "Magna Graecia", 88100 Catanzaro, Italy

Rossella Russo (221, 407), Department of Pharmacobiology, University of Calabria, Arcavacata di Rende, Cosenza 87036, Italy

Fumihiro Saika (179), Department of Pharmacology, Wakayama Medical University, 811–1 Kimiidera, Wakayama 641–0012, Japan

Naohiro Saito (1), Department of Physiology, Tohoku University School of Medicine, 2-1 Seiryo-cho, Aoba-ku, Sendai 980-8575, Japan

Kazuhiro Sakamoto (1), Research Institute of Electrical Communication, Tohoku University, 2-1-1 Katahira, Aoba-ku, Sendai 980-8577, Japan

Shinobu Sakurada (191, 207, 221, 249), Department of Physiology and Anatomy, Tohoku Pharmaceutical University, 4-4-1 Kamatsushima, Aoba-ku, Sendai 981-8588, Japan

Tsukasa Sakurada (207, 221, 237), First Department of Pharmacology, Daiichi College of Pharmaceutical Sciences, 22-1 Tamagawa-cho, Minami-ku, Fukuoka 815-8511, Japan

Eiko Sakurai (109), Department of Pharmacology, Tohoku University Graduate School of Medicine, Sendai 980-8575, Japan

Kengo Sanai (207), Department of Physiology and Anatomy, Tohoku Pharmaceutical University, 4-4-1 Kamatsushima, Sendai 981-8558, Japan

Enrico Sanna (73), Department of Experimental Biology, Center of Excellence for the Neurobiology of Dependence, University of Cagliari, 09100 Cagliari, Italy

Ilenia Sarnico (351), Division of Pharmacology and Experimental Therapeutics, Department of Biomedical Sciences and Biotechnologies, School of Medicine, University of Brescia, Brescia 25123, Italy

Tania Scartabelli (337), Dipartimento di Farmacologia Preclinica e Clinica, Università di Firenze, Firenze 50139, Italy

Norifumi Shioda (375), Department of Pharmacology, Graduate School of Pharmaceutical Sciences, Tohoku University, Sendai 980-8578, Japan

Ichiro Sora (29), Department of Biological Psychiatry, Tohoku University Graduate School of Medicine, Sendai 980-8574, Japan

Paola Spagnuolo (261), Department of Cellular Biology and Morphology (DBCM), Faculty of Medicine, University of Lausanne, rue du Bugnon 9, 1005 Lausanne, Switzerland

PierFranco Spano (351), National Institute of Neuroscience, 10125 Turin; IRCCS San Camillo Hospital, 30100 Venice; and Division of Pharmacology and Experimental Therapeutics, Department of Biomedical Sciences and Biotechnologies, School of Medicine, University of Brescia, Brescia 25123, Italy

Takeshi Tadano (191), Department of Pharmacology, Tohoku Pharmaceutical University, 4-4-1 Komatsushima, Aoba-ku, Sendai 981-8558, Japan

Carlo Tacchetti (295), FIRC Institute of Molecular Oncology (IFOM), 20139 Milan; and Department of Experimental Medicine, Section of Human Anatomy, University of Genoa, 16132 Genoa, Italy

Hiroaki Takahashi (191), Department of Pharmacology, Tohoku Pharmaceutical University, 4-4-1 Kamatsushima, Aoba-ku, Sendai 981-8558, Japan

Naoko Takino (109), Department of Pharmacology, Tohoku University Graduate School of Medicine, Sendai 980-8575, Japan

Koichi Tan-No (191), Department of Pharmacology, Tohoku Pharmaceutical University, 4-4-1 Komatsushima, Aoba-ku, Sendai 981-8558, Japan

Jun Tanji (1), Brain Research Institute, Tamagawa University, Tamagawa Gakuen 6-1-1, Machida, Tokyo 194-8610, Japan

Cristina Tassorelli (407), IRCCS "C. Mondino Institute of Neurology" Foundation Department of Neurological Sciences, University of Pavia, 27100 Pavia, Italy

Lars Terenius (191), Department of Clinical Neuroscience, Section of Alcohol and Drug Dependence Research, Karolinska Institute, Stockholm S-171 76, Sweden

Hiroko Togashi (95), Department of Pharmacology, School of Pharmaceutical Science, Health Sciences University of Hokkaido, Ishikari-Tobetsu, Japan

Hiroaki Tomita (29), Department of Biological Psychiatry, Tohoku University Graduate School of Medicine, Sendai 980-8574, Japan

Hidetoshi Tozaki-Saitoh (159), Department of Molecular and System Pharmacology, Graduate School of Pharmaceutical Sciences, Kyushu University, 3-1-1 Maidashi, Higashi, Fukuoka 812-8582, Japan

Luigia Trabace (117), Department of Biomedical Sciences, University of Foggia, V.le L. Pinto, 71100 Foggia, Italy

Viviana Trezza (117), Department of Physiology and Pharmacology, "Vittorio Erspamer", Sapienza University of Rome, P.le A. Moro 5, 00185 Rome, Italy

Makoto Tsuda (159), Department of Molecular and System Pharmacology, Graduate School of Pharmaceutical Sciences, Kyushu University, 3-1-1 Maidashi, Higashi, Fukuoka 812-8582, Japan

Cesare Usai (295), Institute of Biophysics, National Research Council, 16149 Genoa, Italy

Chizuko Watanabe (249), Department of Physiology and Anatomy, Tohoku Pharmaceutical University, Sendai 981-8558, Japan

Kazuhiko Yanai (109), Department of Pharmacology, Tohoku University Graduate School of Medicine, Sendai 980-8575, Japan

Akihiko Yonezawa (249), Department of Physiology and Anatomy, Tohoku Pharmaceutical University, Sendai 981-8558, Japan

Mitsuhiro Yoshioka (95), Department of Neuropharmacology, Hokkaido University Graduate School of Medicine, Sapporo, Japan

Simona Zappettini (295), Department of Experimental Medicine, Section of Pharmacology and Toxicology, University of Genoa, 16148 Genoa, Italy

PREFACE

A great deal of information has been accumulated during the last three decades and this has improved our understanding of neuronal transmission and plasticity under normal and pathological conditions. For instance, the discovery at the synaptic level of the molecular mechanisms regulating the expression and transduction mechanisms of receptors for glutamate, the main excitatory neurotransmitter in the mammalian central nervous system (CNS), has been essential for understanding how events related to normal neuronal communication, to neuronal plasticity (e.g., LTP and LTD) under normal and pathological conditions, as well as to abusive, for example, excitotoxic, NMDA receptor subtype-mediated events, are tuned. These advances have been made possible by the multidisciplinary approach used by neuroscientists in their researches. The concept has then emerged that redistribution between synaptic and extrasynaptic membranes of specific receptor subunits together with anomalous potentiation of NMDA receptor function, operated via interaction with neuroinflammatory mediators (e.g., cytokines) or kinases (e.g., Src kinase), causes disturbance of the glutamatergic synapse and this, for instance, may form the basis for disorders of learning and memory processes as well as for the development of pain sensitization and excitotoxicity. Accordingly, pathological glutamate transmission has been implicated in several diseases including Alzheimer's disease, a chronic neurodegenerative malady paradigmatic for learning and memory deficits, neuropathic pain and stroke, clinical conditions in great demand for effective therapies.

These and other subjects are the main topics discussed at the Workshop on Apoptosis in Biology and Medicine held in Sendai (Japan) on September 12–15, 2008, at the Tohoku Pharmaceutical University that have generated peer-reviewed papers now incorporated in this "Advances in Neuropharmacology" issue of *International Review on Neurobiology.* The workshop has been organized in the frame of the Scientific and Academic Cooperation Agreement signed by the Tohoku Pharmaceutical University and the University of Calabria. The latter is the XI in a series of workshops yielded by the activity of the PhD Course on Pharmacology and Biochemistry of Cell Death run at the University of Calabria at Cosenza in Consortium with the University "Magna Graecia" at Catanzaro and the University of Rome "Tor Vergata." The Guest Editors of this book would like to express their gratitude to all the authors for their contribution to the volume that has been made possible by the precious and skillful collaboration of Susan Lee and Narmada Thangavelu from Elsevier. The workshop has been organized with the important financial support of the Tohoku Pharmaceutical University (Sendai) and of the

University Magna Graecia of Catanzaro (Catanzaro). Also, we would like to acknowledge the financial support from the Federation of Pharmaceutical Manufactures' Associations of Japan (Tokyo); the Sankyo Foundation of Life Science (Tokyo); the Sendai Tourism and Convention Bureau (Sendai); the Research Foundation of Intelligent Cosmos (Sendai); the Italian Ministry for Instruction, University and Research (MIUR, Rome); the University of Calabria (Cosenza); the Department of Pharmacobiology, University of Calabria (Cosenza); and the Department of Physiology and Anatomy, Tohoku Pharmaceutical University (Sendai).

BAGETTA, G.
CORASANITI, M. T.
SAKURADA, T.
SAKURADA, S.

INVOLVEMENT OF THE PREFRONTAL CORTEX IN PROBLEM SOLVING

Hajime Mushiake,* Kazuhiro Sakamoto,† Naohiro Saito,* Toshiro Inui,‡,§ Kazuyuki Aihara,¶,|| and Jun Tanji**

*Department of Physiology, Tohoku University School of Medicine, 2-1 Seiryo-cho, Aoba-ku, Sendai 980–8575, Japan

†Research Institute of Electrical Communication, Tohoku University, 2-1-1 Katahira, Aoba-ku, Sendai 980–8577, Japan

‡Department of Intelligence Science and Technology, Graduate School of Informatics, Kyoto University, Kyoto 606–8501, Japan

§ERATO Asada Project, Japan Science and Technology Agency, Yamadaoka, Suita, Osaka 565–0871, Japan

¶Institute of Industrial Science, University of Tokyo, 4-6-1 Komaba, Meguro-ku, Tokyo 153–8505, Japan

||ERATO Aihara Complexity Modelling Project, JST, 45–18 Oyama, Shibuya-ku, Tokyo 151–0065, Japan

**Brain Research Institute, Tamagawa University, Tamagawa Gakuen 6-1-1, Machida, Tokyo 194–8610, Japan

To achieve a behavioral goal in a complex environment, such as problem-solving situations, we must plan multiple steps of action. On planning a series of actions, we anticipate future events that will occur as a result of each action, and mentally organize the temporal sequence of events. To investigate the involvement of the lateral prefrontal cortex (PFC) in such multistep planning, we examined neuronal activity in the PFC while monkeys performed a maze path-finding task. In this task, we set monkeys the job of capturing a goal in the maze by moving a cursor on the screen. Cursor movement was linked to movements of each wrist. To dissociate the outcomes of the intended action from the motor commands, we trained the monkeys to use three different hand–cursor assignments. We found that monkeys were able to perform this task in a flexible manner.

INTERNATIONAL REVIEW OF NEUROBIOLOGY, VOL. 85
DOI: 10.1016/S0074-7742(09)85001-0

This report first introduces a problem-solving framework for studying the function of the PFC, from the view point of cognitive science. Then, this chapter will cover the neuronal representation of a series of actions, goal subgoal transformation, and synchrony of PFC neurons. We reported PFC neurons reflected final goals and immediate goals during the preparatory period. We also found some PFC neurons reflected each of *all* forthcoming steps of actions during the preparatory period and increased their activity step by step during the execution period. Recently, we found that the transient increase in synchronous activity of PFC neurons was involved in goal subgoal transformations. Our data suggest that the PFC is involved primarily in the dynamic representation of multiple future events that occur as a consequence of behavioral actions in problem-solving situations.

I. Introduction

The prefrontal cortex (PFC) is the anterior part of the frontal lobes of the brain. It lies in front of the motor and premotor areas. The PFC is divided into the lateral, orbitofrontal, and medial prefrontal areas (Barbas and Pandya, 1987, 1989). Comprehensive reviews of this structure and functions of the PFC have been published in various forms (Fuster, 1997; Goldman-Rakic, 1987; Miller and Cohen, 2001; Passingham, 1993; Tanji and Hoshi, 2008). The PFC possesses a wealth of anatomical connectivity with multiple cortical and subcortical areas, and is involved in broad aspects of behavioral control. The PFC has been implicated in complex cognitive behaviors, social behaviors, and personality expression. Recent studies of this area have revealed its role in the control of a much broader spectrum of functions, such as cross-modal and cross-temporal association of information, in the executive control of behavior, and in the top-down control of neural networks involving the cortical and subcortical areas. Among them, the executive control of action was a term coined to capture various aspects of PFC function.

The term "executive" can be understood both in the sense of one who issues commands from the top rank of a hierarchy within an organization, and in the sense of one who carries out those commands. According to Shallice, the PFC plays a supervisory, reflective role, and controls multiple functional modules (Shallice, 1982). Stuss and Beson (1986) described executive functions as a list of abilities such as anticipation, goal establishment, planning, response trials, monitoring of results, and the use of feedback. What do these executive functions involve? The executive functions constitute a set of processes that underlie flexible goal-directed behavior, such as planning, inhibitory control, attentional flexibility, and working memory (Baddeley, 1986; Stuss and Knight, 2002). What is the best way to understand such diversity of function in the PFC? One approach is to consider the common outcome of executive function. We can ask the question: What is the purpose of the set of executive functions? One answer might be problem solving. Problem solving requires the integration of all of the processes involved in executive function.

Problem solving has been defined as the higher order cognitive process used to generate a specific series of actions aimed at changing the current situation in order to achieve a goal state. Problem solving has been intensely studied in cognitive science since the 1960s (Anzai and Simon, 1979; Miller *et al.*, 1960; Newell and Simon, 1972). According to a recent view of cognitive science, any organism, such as a human or animal, or even a robot, is considered an embodied agent that interacts with its environment. The environment is considered to give the agent a problem-solving situation. In general, when agents are confronted with problems, they attempt to solve problems in four temporally distinct phases of executive function: goal representation, planning, execution, and evaluation (Fig. 1A). The purpose of this study was to investigate executive function in the PFC in a problem-solving framework. Our findings demonstrate that the PFC is involved in each phase of problem solving.

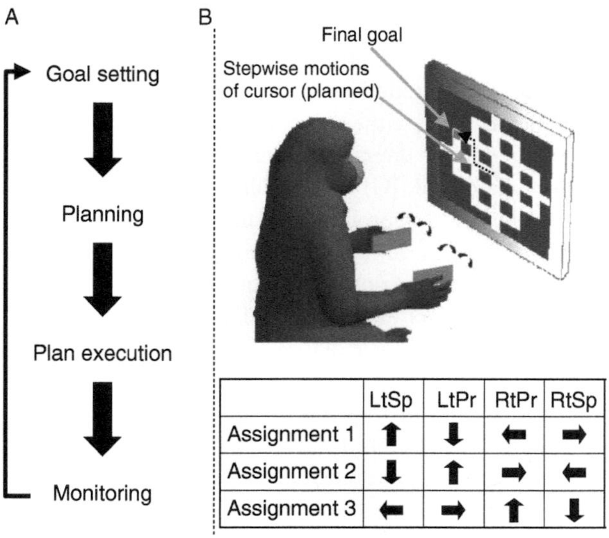

FIG. 1. (A) Cognitive processes in problem solving. (B) Experimental set up. Each monkey, faced a maze on a screen, was required to plan a stepwise motion of the cursor from the starting position to the final goal using two hand-operated controllers. The direction of cursor movement on the screen was linked to the movements of the controllers in three assignments in the lower panel.

II. Problem-Solving Behavior, Using a Path-Planning Task

To investigate executive function in the PFC in a problem-solving framework, we trained monkeys to perform a path-planning task and recorded neuronal activity from the PFC (Mushiake *et al.*, 2001). We illustrated the experimental setting in Fig. 1 for the path-planning task (Fig. 1B). In this task, the animals were required to use two manipulanda to move the cursor from an initial position (the "Start") to reach a goal. Multiple movements of the cursor were required en route to the goal. The animal planned each step of its actions to achieve the final goal (FG). In this task, it is important to dissociate the outcome of an intended action from the intended motor action. If the relationship between arm movement and cursor movement is fixed, one cannot tell whether the neuronal activity reflects the outcome of the intended action or the intended motor action. To dissociate the outcomes of the intended action from the motor commands, we trained monkeys to use three different hand–cursor assignments (Fig. 1B). The animal was trained to control two manipulanda, one with each arm, to move the cursor up, down, left, and right. For example, upward movement was assigned to left supination in cursor assignment 1, while it was assigned to left pronation in cursor assignment 2, and to right pronation in cursor assignment 3.

To begin the trial, the animal was required to hold the two manipulanda in the neutral position for 1 s. On doing so, a cursor was presented at the center of the maze to indicate the starting position. One second later, a goal cursor position was presented for 1 s. After a delay, the color of the cursor changed from green to yellow, which served as an initiation signal. The animal was then required to move the cursor within 1 s by either supination or pronation of the right or left arm. The cursor moved in the direction specified by the animal; this direction was defined as the immediate goal (IG) to be reached in the first step.

To investigate whether this problem-solving behavior was flexible, obstacles were introduced in the mazes. In this situation, the animals were required to take a detour to reach the goal. We constructed the obstacles in such a way that they blocked multiple pathways simultaneously. The obstacles always required the cursor to detour around the obstacle before reaching the goal. The starting points were now positioned either at the center or periphery of the maze. We designed this path-finding task so that the shortest paths required 3, 4, or 5 steps to reach the goal. Because there was a large number of possible combinations of starting points, goals, and obstacles, the animals encountered a new path-finding task in each session. Two monkeys reached the goal by taking the shortest path in more than 90% of the trials (Mushiake *et al.*, 2001).

So, these findings demonstrate that the animals solved the path-finding task in a flexible manner. The animals could take a detour to avoid an obstacle.

According to this analysis, the animals' behavior did not result from the direct association of visual stimuli with a set of arm movements. Rather, the animal planned actions involving multiple steps to reach the FG.

III. Goal Representation and Planning in the Prefrontal Cortex

In discussing the concept of a behavioral goal, the term "goal" has different meanings for different researchers (Duncan, 2000; Watanabe, 2007). There are at least two different types of goal: (1) cognitive goals and (2) motivational goals. A "cognitive goal" is a desired state in problem solving, such as the situation in the path-planning task. In problem solving, a problem is defined as the difference between the current state and the desired state in a problem space. The desired state is the goal to be achieved. An agent is required to plan a series of actions to change the current state into the goal state. The second type of goal is a motivational goal. Rewards, such as juice or a piece of fruit, are used as motivational goals for monkeys. The motivational goal is related to the evaluation of a performed action. It is important for the agent to predict the future reward and to select an action that will maximize the reward. Thus, this type of goal acts as a positive reinforcer. The frequency and intensity of contingent behaviors increase, depending on the amount of the reward.

Our path-planning task required sequential information processing, During a delay period in which the animal prepared to start the first of three cursor movements in order to approach the preinstructed goal, we identified two types of neuronal activity: the first type reflected the position within the maze to which the animal intended to move the cursor, as an initial step to an IG, and the second type reflected the position within the maze that was to be captured, as a FG. Representative IG neurons discharged markedly during delay periods only when the animal prepared to move the cursor to a particular IG, regardless of the location of the FG. By contrast, representative FG neurons discharged markedly during delay periods only when the animal planned to move the cursor to a particular FG, regardless of the location of IGs (Saito *et al.*, 2005). We also found prefrontal neurons reflecting combinations of the FGs and IGs. The PFC represents goal information useful to guide behavior. We made population analyses of goal-selective neurons during the delay period. During the early delay period, the representation of a FG was dominant; however, during the late delay period, the representation of an IG was dominant. In other words, the representation of behavioral goals in the PFC was transformed from the FG to the IG gradually (Fig. 2A). In the next section, we focus on the relationship of this goal transformation and discharge synchrony.

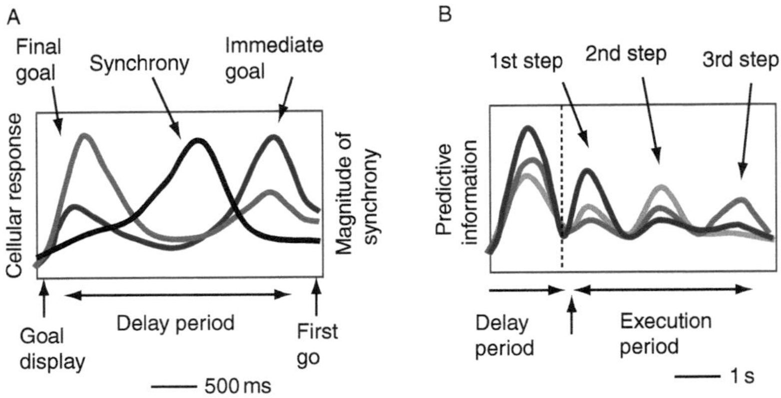

FIG. 2. (A) Plots of cellular responses representing the positions of the final goal (red) and the immediate goal (blue) on the screen and the time course of magnitude of synchrony (black). Based on Saito *et al.* (2005) and Sakamoto *et al.* (2008). (B) Plots of cellular responses representing the first, second, and third steps of cursor movements during the late delay period and execution period. Based on Mushiake *et al.* (2006). (See Color Insert.)

IV. Synchrony as a State Transition and Goal Transformation

PFC neurons represent FGs and IGs. At the population level, we found that goal representation was transformed from the FG to the IG. We made statistical analyses of activity in the prefrontal neurons in terms of these two types of goals and found that even a single prefrontal neuron showed the transition from FGs to IGs. We called this type of neuron an "F–I neuron." Typical F–I neurons showed biphasic activity: neuronal activity selective to the FG during the early delay period, and neuronal activity selective to the IG during the late delay period. A question arises as to what kind of mechanism is involved in this representational change. We hypothesized that interactions among task-related PF neurons somehow enabled such a representational change. How can we evaluate such neuronal interactions? One possible approach is a dynamic system approach (Haken, 2006). In this regard, we try to describe the phenomena that involve change over time. The broadest definition of a dynamic system is simply any system that changes over time. According to dynamic system theory, a dynamic system is a manifold, M, called the phase (or state) space and a smooth evolution function that maps a point of the phase space back into the state space for any element of *t*, the time. The evolution rule of the dynamic system describes what future states follow from the current state. Dynamic system approaches consider the target system, in terms of possible trajectories and stability. A dynamic system often shows bifurcation phenomena and allows multiple attractor states. Dynamic system approaches provide a set of conceptual tools that help to elucidate how a system changes over time.

A recent theoretical study has suggested that persistent activity in the PFC is considered to be an attractor state, in that relatively small amounts of variation in this state lead it back to the same state. This idea has been examined in detail theoretically, especially by Amit, who described persistent activity in terms of "dynamical attractors" (Amit and Brunel, 1997; Rolls *et al.*, 2008). The spontaneous state and stimulus-selective memory states are assumed to represent multiple attractors, such that a memory state can be switched on or off by transient inputs. This formulation is plausible, insomuch as stimulus-selective persistent firing patterns are dynamically stable in time. These properties of attractors result from interactions in neuronal circuits. Neural synchrony is a general mechanism for dynamically linking together cells coding task-relevant information (Salinas and Sejnowski, 2001). The dynamics of neuronal activities and the representations they reflect are two sides of a coin.

In the path-planning task, we can distinguish two states: the FG-dominant state, during the early delay period, and the IG-dominant state, during the late delay period. The transition from the FG to an IG in a neural representation can be considered a state transition of neural circuits. How synchronous is the discharge of the PF neurons in a state transition? We used the time-resolved cross-correlation method to assess the changes in synchrony of neuron pairs independently of changes in the firing rate of individual neurons during a behavioral task (Baker *et al.*, 2001). At the population level, we found that synchrony was transiently enhanced at the period of state transition from final-goal dominant state to immediate-goal dominant state (Fig. 2A). In each pair of PF neurons, we found significant correlation between the time of state transition and time of synchrony peak (Sakamoto *et al.*, 2008). These results suggested that the transient increase in discharge synchrony is an indication of a process that facilitates dynamic changes in the prefrontal neural circuits in order to undergo profound state changes.

V. Involvement of the Prefrontal Cortex in Planning and Execution

The PFC is involved not only in goal setting, but also in planning multiple steps of actions. In our task, animals took at least three steps of cursor movements to reach the FG. During the preparatory delay period, PFC neurons reflected each of all forthcoming cursor movements, rather than arm movements. We found that many PF neurons showed distinct activity during the delay period, reflecting intended future action, or cursor movements of the first, second, and third steps (Mushiake *et al.*, 2006). Furthermore, some of these PF neurons were reactivated immediately before each step of action during the execution period.

Based on these findings, the PFC was involved in planning the consequences of actions and the execution of the planned sequence of actions.

Can we predict future actions to be performed by the monkeys based on neuronal activity? We calculated the mutual information of the PF neurons; the behavior of the PFC neurons did not reflect future arm movements. To evaluate the extent to which PFC neuronal activity predicted information associated with cursor movements, we calculated the predictive information (Is) carried by the occurrence of spikes by quantifying the decrease in entropy in the cursor directions. The predictive information was defined as a decrease in uncertainty of path selection after observing neuronal activity, that is, a difference between the entropy of path selection and the conditional entropy of path selection given a particular neuronal activity. In the population analyses of PFC neurons, we illustrated the time course of information that predicted the first (plotted in blue), second (green), and third (red) cursor movements (Fig. 2B). Each of the three cursor movements was represented during the delay period, while information that reflected the first, second, and third cursor movements also appeared during and immediately before the first, second, and third movements, in that order. The properties of PFC neurons that we observed in the present study are compatible with behavioral planning, based on future events. These findings suggested that PFC neurons in the monkey brain process information for future events in a prospective manner to generate action plans, based on a series of events during the course of reaching a behavioral goal.

VI. Monitoring Action

In our path-planning task, the monkeys were required to plan multistep cursor movements according to hand–cursor assignments (rules). Trials were blocked according to rules and rules were changed without any external instructions. Thus, the animal was required to find appropriate rules based on trial and error. The question arises as to how it is possible for the animals to update the rules without any instructions. First, we conducted a behavioral analysis. We found that the performance rate, the ratio of the number of successful trials to the total number of trials, was lowered to about 0.8 when behavioral rules were changed. In successful trials after the rule changes, the animal tended to take more steps to reach the FG. However, this deterioration in performance was observed for only a few trials after a rule change. Behavioral analysis suggested that animals were able to find an appropriate rule when the cursor–arm relationship was changed, without instructions. It took only a few trials to change from one rule to anther, suggesting that the animals were able to change all of the cursor–arm assignments, even when animals experienced only a few cases of cursor–arm relationships. These results further suggest that this rule finding was a result of behavioral switching, based on monitoring behavioral outcomes rather than learning new rules by trial and error. The animal knew how many rules existed and what type of relationship between hands and cursor movements each rule determined. Based on this prior knowledge of the rules and actual consequences of intended actions, the animal was able to compare the actual outcome and intended outcome of actions. If these two matched, the animal stayed with the current rule. However, if they did not match, the animal detected the change in the rule. To find out the appropriate current rule within a few trials, detection of rule change is not enough; updating by a kind of inference, based on acquired knowledge of task and active monitoring of behavioral outcome is important. Bayesian inference is a statistical inference in which evidence or observations are used to update the subjective probability that a hypothesis is true. The subjective probability can be defined as the degree to which a person believes a proposition. The basic assumption in this inference is a causal model between the hypothesis and data. As evidence accumulates, the degree of belief in the hypothesis changes. We hypothesized that the monkey was able to switch one rule to another by monitoring the outcomes of their actions, using Bayesian inference. We calculated the subjective probability of each rule and difference between posterior

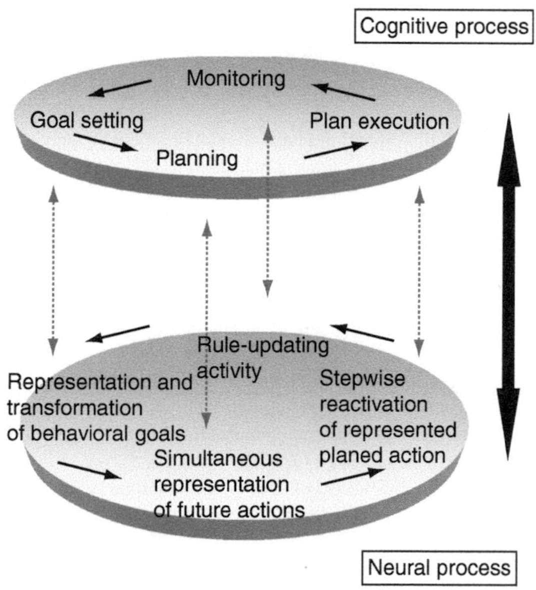

FIG. 3. Summary scheme illustrating the neural mechanisms underlying cognitive processing in problem solving.

probability and prior probability of each rule. We called this difference the "updating signal," because a large value in this difference suggests a rule change. We made a regression analysis of neuronal activity with factors for the "updating signal." In our preliminary study, we found many neurons reflected this updating signal in the medial frontal areas.

VII. Summary and Conclusions

We have demonstrated that the PFC is involved in each phase of problem solving: goal setting, planning, execution, and monitoring actions (Fig. 3). We also reported orderly activations of human cortical areas including the PFC during path-planning task (Mushiake *et al.*, 2002). In the problem-solving framework, the PFC is involved in guiding actions based on various cognitive resources, such as behavioral goals, rules, and categories of actions (Shima *et al.*, 2007; Tanji *et al.*, 2007). To accomplish these functions, the PFC adaptively encodes abstract forms of representations of task-relevant information (Duncan, 2001). These representations are dynamically transformed from one representation into another to cope with various task demands.

References

Amit, D. J., and Brunel, N. (1997). Model of global spontaneous activity and local structured activity during delay periods in the cerebral cortex. *Cereb. Cortex* **7,** 237–252.

Anzai, K., and Simon, H. A. (1979). The theory of learning by doing. *Psychol. Rev.* **86,** 124–140.

Baddeley, A. D. (1986). "Working Memory." Clarendon Press, Oxford, UK.

Barbas, H., and Pandya, D. N. (1987). Architecture and frontal cortical connections of the premotor cortex (area 6) in the rhesus monkey. *J. Comp. Neurol.* **256,** 211–228.

Baker, S. N., Spinks, R., Jackson, A., and Lemon, R. N. (2001). Synchronization in monkey cortex during a precision grip task. I. Task-dependent modulation in single-unit synchrony. *J. Neurophysiol.* **85,** 869–885.

Barbas, H., and Pandya, D. N. (1989). Architecture and intrinsic connections of the prefrontal cortex in the rhesus monkey. *J. Comp. Neurol.* **286,** 353–375.

Duncan, J., and Owen, A. M. (2000). Common regions of the human frontal lobe recruited by diverse cognitive demands. *Trends Neurosci.* **23,** 475–483.

Duncan, J. (2001). An adaptive coding model of neural function in prefrontal cortex. *Nat. Rev. Neurosci.* **2,** 820–829.

Fuster, J. M. (1997). "The Prefrontal Cortex: Anatomy, Physiology, Neuropsychology of the Frontal Lobe." Lippincott-Raven, Philadelphia, PA.

Goldman-Rakic, P. S. (1987). Circuitry of primate prefrontal cortex and regulation of behavior by representational memory. *In* "Handbook of Physiology: The Nervous System" (F. Plum, Ed.), pp. 373–417. American Physiological Society, Bethesda, MD.

Haken, H. (2006). Synergetics of brain function. *Int. J. Psychophysiol.* **60,** 110–124.

Miller, E. K., and Cohen, J. D. (2001). An integrative theory of prefrontal cortex function. *Annu. Rev. Neurosci.* **24,** 167–202.

Miller, G., Galanter, E., and Pribram, K. H. (1960). "Plans and the Structure of Behavior." Henry Holt, New York.

Mushiake, H., Saito, N., Sakamoto, K., Sato, Y., and Tanji, J. (2001). Visually based path-planning by Japanese monkeys. *Brain Res. Cogn. Brain Res.* **11,** 165–169.

Mushiake, H., Saito, N., Furusawa, Y., Izumiyama, M., Sakamoto, K., Shamoto, H., Shimizu, H., and Yoshimoto, T. (2002). Orderly activations of human cortical areas during path-planning task. *Neuroreport* **13,** 423–426.

Mushiake, H., Saito, N., Sakamoto, K., Itoyama, Y., and Tanji, J. (2006). Activity in the lateral prefrontal cortex reflects multiple steps of future events in action plans. *Neuron* **50,** 631–641.

Newell, A., and Simon, H. A. (1972). "Human Problem Solving." Prentice-Hall, Englewood Cliffs, NJ.

Passingham, R. E. (1993). "The Frontal Lobes and Voluntary Action." Oxford University Press, Oxford, UK.

Rolls, E. T., Loh, M., Deco, G., and Winterer, G. (2008). Computational models of schizophrenia and dopamine modulation in the prefrontal cortex. *Nat. Rev. Neurosci.* **9,** 696–709.

Saito, N., Mushiake, H., Sakamoto, K., Itoyama, Y., and Tanji, J. (2005). Representation of immediate and final behavioral goals in the monkey prefrontal cortex during an instructed delay period. *Cereb. Cortex* **15,** 1535–1546.

Sakamoto, K., Mushiake, H., Saito, N., Aihara, K., Yano, M., and Tanji, J. (2008). Discharge synchrony during the transition of behavioral goal representations encoded by discharge rates of prefrontal neurons. *Cereb. Cortex* **18,** 2036–2045.

Salinas, E., and Sejnowski, T. J. (2001). Correlated neuronal activity and the flow of neuronal information. *Nat. Rev. Neurosci.* **2,** 539–550.

Shallice, T. (1982). Specific impairments of planning. *Philos. Trans. R. Soc. Lond., B, Biol. Sci.* **298,** 199–209.

Shima, K., Isoda, M., Mushiake, H., and Tanji, J. (2007). Categorization of behavioral sequences in the prefrontal cortex. *Nature* **445,** 315–318.

Stuss, D. T., and Benson, D. S. (1986). "The Frontal Lobes." Raven Press, New York.

Stuss, D. T., and Knight, R. T. (2002). "Principles of Frontal Lobe Function." Oxford University Press, New York, NY.

Tanji, J., and Hoshi, E. (2008). Role of the lateral prefrontal cortex in executive behavioral control. *Physiol. Rev.* **88,** 37–57.

Tanji, J., Shima, K., and Mushiake, H. (2007). Concept based behavioural planning and lateral prefrontal cortex. *Trends Cogn. Sci.* **12,** 528–534.

Watanabe, M. (2007). Role of anticipated reward in cognitive behavioral control. *Curr. Opin. Neurobiol.* **17,** 213–219.

GLUK1 RECEPTOR ANTAGONISTS AND HIPPOCAMPAL MOSSY FIBER FUNCTION

Robert Nisticò,[*,†] Sheila Dargan,[*,‡] Stephen M. Fitzjohn,[*] David Lodge,[*] David E. Jane,[‡] Graham L. Collingridge,[*] and Zuner A. Bortolotto[*]

[*]Department of Anatomy, MRC Centre for Synaptic Plasticity, School of Medical Sciences, University of Bristol, Bristol BS8 1TD, United Kingdom

[†]Department of Pharmacobiology, University Center for Adaptive Disorders and Headache (UCHAD), Section of Neuropharmacology of Normal and Pathological Neuronal Plasticity, University of Calabria, Arcavacata di Rende, Italy

[‡]Department of Physiology and Pharmacology, MRC Centre for Synaptic Plasticity, School of Medical Sciences, University of Bristol, Bristol BS8 1TD, United Kingdom

Kainate receptors, one of the three subtypes of ionotropic receptors for the excitatory transmitter L-glutamate, play a variety of functions in the regulation of synaptic activity. Their physiological properties and functional roles have been identified only recently, following the discovery of selective pharmacological tools that allow for isolation of kainate receptor-mediated events. A considerable amount of data indicates that this class of glutamate receptors is located both at the pre- and postsynaptic site, playing a special role in regulating transmission and controlling short- and long-term plasticity. In this review, we summarize some data obtained in our laboratory over the last decade illustrating how various ligands have contributed to our understanding of the physiological role for neuronal kainate receptors. In particular, we show that the GluK1-containing KARs are important for regulating synaptic facilitation and LTP induction at hippocampal mossy fiber synapses.

DOI: 10.1016/S0074-7742(09)85002-2

0074-7742/09 $35.00

I. Introduction

Kainate receptors (KARs) were originally identified by Watkins and coworkers (Davies *et al.*, 1979; Watkins and Evans, 1981) based on pharmacological experiments describing neuronal responses to excitatory amino acids. Since then, numerous studies confirmed the existence of three different receptor subtypes (Dingledine *et al.*, 1999; Hollmann and Heinemann, 1994), although it also has been recognized that many excitatory amino acids, including kainate and AMPA, are not entirely selective for only one receptor class. Thus, AMPA can activate at least some types of KAR (Herb *et al.*, 1992) and kainate activates AMPA receptors to produce large sustained currents (Keinänen *et al.*, 1990; Kiskin *et al.*, 1986; Patneau and Mayer, 1991).

Among the ionotropic glutamate receptors, the kainate subtype has, for a long time, been the poorly understood relative of its famous brothers, the AMPA and NMDA receptors (AMPARs and NMDARs, respectively). In fact, the lack of pharmacological specificity has hindered our understanding of KAR function for several years (Lerma *et al.*, 2001). KARs are tetrameric assemblies, comprising various combinations of the five different subunits GluK1–5, which are derived from the genes *GRIK*1–5, respectively. These subunits are named according to the new IUPHAR nomenclature (Collingridge *et al.*, 2009): GluK1, GluK2, GluK3, GluK4, and GluK5, formerly known as GluR5, GluR6, GluR7, KA1, and KA2. KARs may exist as certain homomeric assemblies, although native receptors are most likely to be organized in heteromeric combinations. In the early 1980s, Monaghan and Cotman (1982), using radioactive kainate (a glutamate receptor agonist), showed the presence of high-affinity binding sites restricted to stratum lucidum, which is the mossy fiber termination zone. Since then, in an attempt to understand KAR synaptic function, many studies focused on hippocampal neurons. It is well known now that these heteromeric receptors are located both at the presynaptic and postsynaptic site (reviewed by Huettner, 2003; Lerma, 2003). What functional role might these KARs have at the hippocampal mossy fiber synapse? Despite several contradictory results over the recent years, concerning the KAR subunit(s) involved, all parties agree that KARs do have a role in mossy fiber LTP, although they are not essential for its induction under all circumstances. In the present work, we describe the presence of facilitating presynaptic KARs that enhance transmission during repetitive pattern of stimuli. This would enhance Ca^{2+} entry into the presynaptic terminal, which, in our *in vitro* models is the trigger that initiates short- and long-term plasticity.

II. Pharmacological Tools to Investigate the Roles of KARs

More in depth studies concerning the role played by KARs in synaptic transmission have only been possible since the discovery of selective AMPAR ligands (Bleakman *et al.*, 1996a,b; Paternain *et al.*, 1995), which allowed researchers to selectively block AMPAR-mediated responses to examine KARs in isolation. Previously, interesting effects on excitability (Robinson and Deadwyler, 1981; Westbrook and Lothman, 1983) and synaptic transmission (Collingridge *et al.*, 1983; Fisher and Alger, 1984; Kehl *et al.*, 1984) have been described following application of low doses of kainate. Seemingly, KARs mediated many of these effects, particularly when kainate was applied at submicromolar concentrations; yet, the possibility that also AMPARs could partially be involved in mediating its action still remained. The development of more selective ligands which affect KARs has deepened our knowledge on the roles played by these receptors in a variety of physiological functions. Figure 1 summarizes the structures of AMPA/KAR selective orthosteric antagonists. These studies have also been supported by the availability of KAR knockout mice (Contractor *et al.*, 2003).

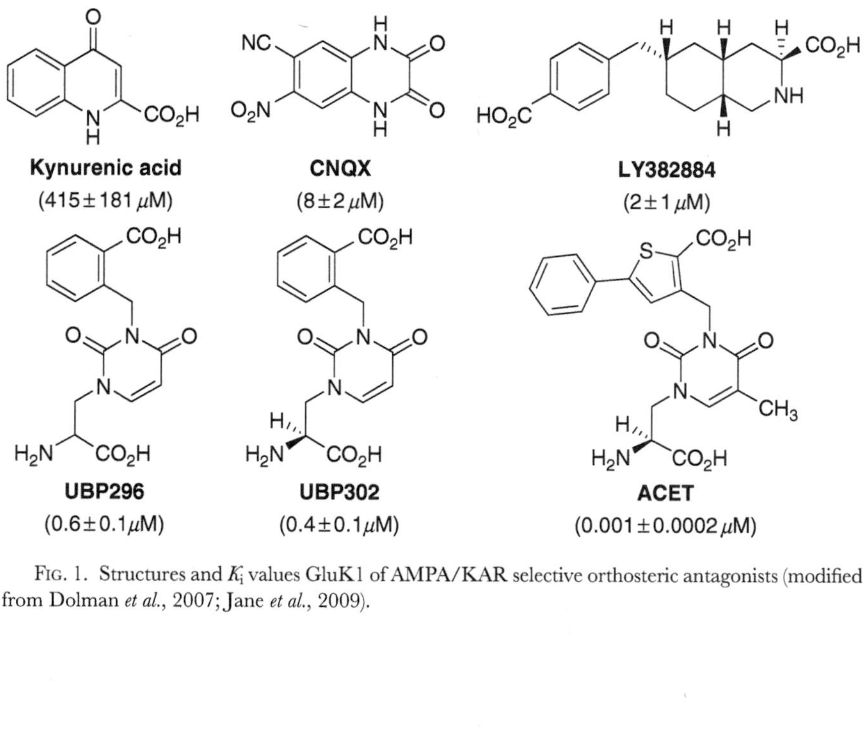

FIG. 1. Structures and K_i values GluK1 of AMPA/KAR selective orthosteric antagonists (modified from Dolman *et al.*, 2007; Jane *et al.*, 2009).

The use of both pharmacology and knockouts to understand KAR function has been the subject of several recent reviews (Jane *et al.*, 2009; Lerma, 2006; Vincent and Mulle, 2008).

A. Quinoxalinediones and 2,3-Benzodiazepines

Until recently, the discrimination of non-NMDAR-mediated responses into those mediated by AMPARs or KARs has been challenging due to a lack of selective compounds. The agonists used to activate KARs—usually kainate and domoate—are also potent AMPAR agonists. Moreover, quinoxalinediones, such as CNQX and NBQX, act as competitive antagonists at native and recombinant KARs; whereas CNQX exhibits little selectivity for KARs over AMPARs, NBQX is far more potent at AMPARs (Mayer *et al.*, 2006). Another factor that challenges the study of the pharmacological properties within the KAR family is the numerous heteromeric assemblies formed by different subunits. With the advent of the 2,3-benzodiazepines, in particular GYKI52466 and GYKI53655 which are more potent antagonists of AMPARs than KARs (Paternain *et al.*, 1995; Wilding and Huettner, 1995), it was possible to select KAR-mediated responses. Accordingly, in the presence of GYKI, compounds such as kainate and domoate could be applied as selective KAR agonists. Thus, it was possible to identify functional KARs distributed in various neuronal populations with the use of KAR agonists coapplied with AMPAR antagonists. These studies led us to demonstrate presynaptic and postsynaptic actions of exogenous KAR agonists. In fact, submicromolar concentration of kainate elicited a facilitation of transmission, whereas micromolar concentrations caused a depression. The interpretation of these data led us to hypothesize a presynaptic locus for KARs on excitatory terminals that would regulate l-glutamate release (Kamiya and Ozawa, 2000; Lauri *et al.*, 2001; Schmitz *et al.*, 2000). In addition, the first demonstration of postsynaptic KARs came more recently studying the synapses between mossy fibers and CA3 pyramidal cells in the hippocampus (Castillo *et al.*, 1997; Vignes and Collingridge, 1997). Accordingly, in the presence of antagonists for both AMPARs (GYKI53655) and NMDARs, a slow EPSC was still obtained in response to the stimulation of the mossy fiber pathway. Since this slow EPSC was inhibited by CNQX, it was assumed to be mediated by KARs. This synaptic response was slower than the AMPAR-mediated component but faster than the NMDAR-mediated component.

Subsequent work using these ligands has demonstrated that KARs are involved in controlling synaptic plasticity at mossy fiber synapses. Hence, it was shown that mossy fiber LTP was fully blocked by broad spectrum antagonists kynurenic acid and CNQX (Bortolotto *et al.*, 1999), although both caused a depression of AMPAR-mediated EPSPs (Fig. 2A and B). The block was

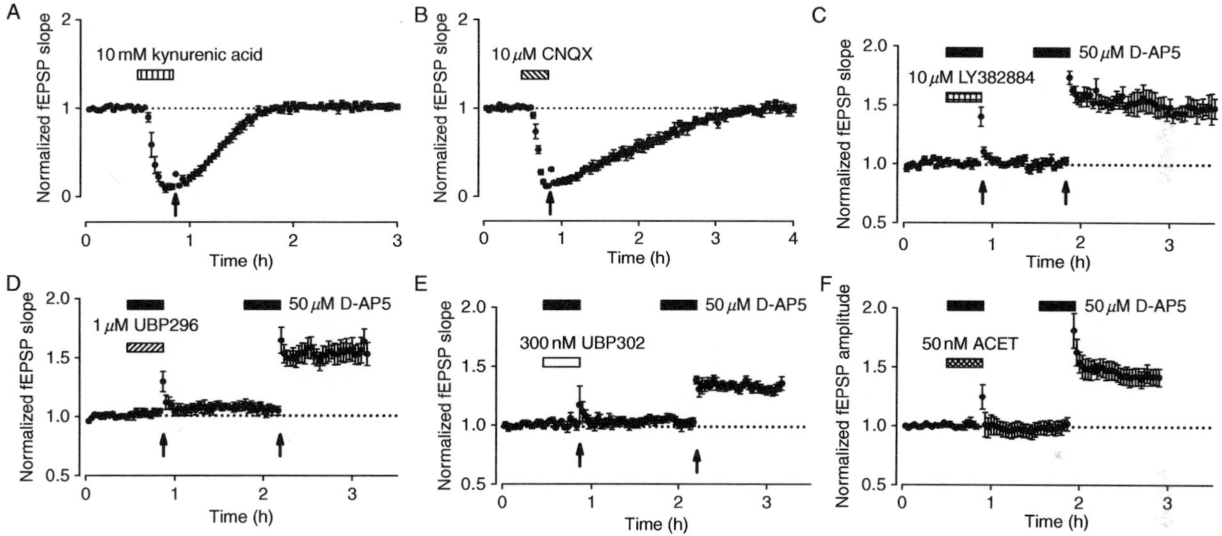

FIG. 2. AMPA/KAR antagonists and selective KAR orthosteric antagonists block the induction of mossy fiber LTP. (A–C) Modified from Bortolotto *et al.* (1999), Lauri *et al.* (2001), (D) Modified from More *et al.* (2004), (E) Modified from Dolman *et al.* (2005), (F) Modified from Dargan *et al.* (2009).

reversible, since AMPAR-mediated synaptic transmission slowly recovered following washout of the antagonist. That AMPARs are not required for induction of mossy fiber LTP is also confirmed by the evidence that GYKI53655, a more selective AMPAR antagonist, was not able to prevent its induction. Although these results have been disputed by other studies showing no effects for these antagonists on LTP induction (Nicoll *et al.*, 2000) they paved the way for elucidating the involvement of KARs in regulating mossy fiber LTP. Moreover the use of these pharmacological tools has concurred in delineating the short-term plasticity features that MF-CA3 synapses classically exhibit. Accordingly, when monitoring mossy fiber NMDAR-mediated EPSCs (recorded in the presence of 20 μM GYKI53655), CNQX used as a KAR antagonist is able to partially inhibit frequency facilitation elicited by repetitive stimulation at 25 or 100 Hz (Schmitz *et al.*, 2001). Taken together the combined use of these compounds, although nonselective, has represented a useful tool in the understanding of the role played by KARs in this brain region. Interestingly, in addition to regulating synaptic function at CA3 synapses, they revealed a role for these KARs in other brain regions such as in regulating the activity of GABAergic interneurons of the CA1 region (Cossart *et al.*, 1998; Frerking *et al.*, 1998), in the basolateral amygdala (Li and Rogawski, 1998), in the neocortex (Eder *et al.*, 2003; Wu *et al.*, 2005), and in the superficial dorsal horn of the spinal cord (Li *et al.*, 1999).

B. LY Compounds

The first selectivity towards KARs was obtained when a series of decahydroisoquinolines, synthesized at Eli Lilly, were screened against AMPA and KAR subunits in HEK293 cell lines. Surprisingly, these compounds, besides their activity on AMPARs, also antagonized homomeric GluK1 receptors. The first compound, LY293558, was almost equipotent at GluK1 and AMPAR subunits (Bleakman *et al.*, 1996a,b) but subsequent compounds, such as LY294486 (Clarke *et al.*, 1997) and its active isomer LY377770 (O'Neill *et al.*, 2000; Smolders *et al.*, 2002) displayed improved selectivity for GluK1-containing KARs. However, the use of these compounds has been limited given that at slightly higher doses these compounds act as AMPAR antagonists. Nevertheless, both compounds have been important in describing a role for GluK1 subunits in excitatory synaptic transmission in the hippocampus (Cossart *et al.*, 1998) and amygdala (Li and Rogawski, 1998). In particular, LY294486 inhibited mossy fiber synaptic transmission and, like LY293558, also depressed the inward current induced in CA3 neurons by the activation of KARs (Vignes *et al.*, 1997).

A major aid in the study of KARs has been the development of LY382884 (Bortolotto *et al.*, 1999), which displays greater selectivity for the GluK1 receptor subunit (K_i value of ~4 μM), compared to AMPA or NMDA receptors. Thus, for some years LY382884 was the most useful antagonist with which to explore the functions of the GluK1 subunit. Unlike its predecessors, LY382884 can block GluK1-containing KAR-mediated responses at concentrations subthreshold for antagonizing AMPARs. For example, LY382884 was able to block the depression of AMPAR-mediated EPSPs induced by ATPA, a selective GluK1 receptor agonist (Bortolotto *et al.*, 1999) at concentrations subthreshold for depressing the EPSP directly. Other studies carried out in the spinal cord isolated from KAR subunit-deficient mice further suggested the selectivity of LY382884 for GluK1 receptors (Kerchner *et al.*, 2002). Interestingly, this report also shows that GluK1-dependent functions in wild-type mice can be compensated for in GluK1$^{-/-}$ mice, an observation with important implications when using KAR knockout mice. Having established that LY382884 is a selective KAR antagonist, the next step in our work was to study whether GluK1-containing KARs are involved in synaptic facilitation and the induction of LTP at mossy fiber synapses.

1. *A Role for GluK1-Containing KARs in Regulating Synaptic Facilitation and Mossy Fiber LTP*

The use of LY382884 enabled several new principles of synaptic function to be discovered. The selective KAR antagonist LY382884, applied at a concentration (10 μM) that did not affect the baseline synaptic transmission, was capable of blocking LTP induction at mossy fiber synapses in a reversible manner (Fig. 2C). In contrast, LY382884 had no effect on the induction of NMDAR-dependent LTP by a second input stimulating the associational/commissural synapses onto CA3 cells. This provided the first evidence that GluK1-containing KARs act as a trigger of mossy fiber LTP (Bortolotto *et al.*, 1999). LTP was invariably blocked even when applying different stimulation protocols (Bortolotto *et al.*, 2003), which can evoke different forms of mossy fiber LTP (Urban and Barrionuevo, 1996). The nature of the synaptic response during tetanic stimulation was most likely explained by a rapid, facilitatory autoreceptor mechanism resulting in enhanced glutamate release (Bortolotto *et al.*, 1999). Indeed subsequent work revealed a role for KARs as facilitatory autoreceptors, rapidly augmenting synaptic transmission during high-frequency stimulation of the mossy fiber pathway (Lauri *et al.*, 2001). A facilitatory autoreceptor displaying positive feedback mechanism was an unusual mechanism in the mammalian CNS, which might partly explain the extreme sensitivity of the CA3 region to excitotoxicity (Meldrum, 1991). We therefore wanted to deepen our knowledge of the properties of such a mechanism.

One of the first observations we made was that Ca^{2+}-induced Ca^{2+} release from intracellular stores mediates an important component of mossy fiber LTP and synaptic facilitation (Lauri *et al.*, 2003). This was achieved by treatments that inhibited Ca^{2+} release from intracellular stores and by philanthotoxin, which blocks Ca^{2+} permeable receptors. These results suggest that the necessary trigger for Ca^{2+} release from internal stores can be provided by Ca^{2+} entry through Ca^{2+}-permeable KARs. The magnification of the synaptic Ca^{2+} transient by release of Ca^{2+} from intracellular stores has previously been described in hippocampal CA1 neurons (Alford *et al.*, 1993; Emptage *et al.*, 1999) and is the element that triggers an intracellular cascade involved in the induction of LTP, at least under certain experimental conditions. Thus, there are parallels between NMDAR- and KAR-dependent LTP at the level of Ca^{2+} release from intracellular stores.

In these studies (Lauri *et al.*, 2003), it was also found that the role of KARs could be compensated for by Ca^{2+} entry via L-type voltage-gated calcium channels, with entry through voltage-gated channels being the dominant factor when extracellular calcium levels are high (4 mM). This study helps to resolve earlier disagreements over the ability of KAR antagonists to block LTP induction at this synapse (Nicoll *et al.*, 2000), because experiments that appeared to be in conflict were performed under different ambient calcium levels.

Based on the extensive pharmacological characterization of LY382884, it seems extremely likely that its effects on synaptic facilitation and LTP induction are due to antagonism of GluK1-containing KARs. Nevertheless the role of GluK1-containing KARs in mossy fiber LTP and synaptic facilitation remains still controversial following an earlier paper that failed to observe any effect of LY382884 on these synaptic processes under similar conditions (Breustedt and Schmitz, 2004). This report supports previous evidence showing that mossy fiber LTP and synaptic facilitation are preserved in GluK1 knockout mice but deficient in GluK2 knockout mice (Contractor *et al.*, 2001) and that the expression levels of GluK2 are higher than those of GluK1 (Bahn *et al.*, 1994). Other more recent data suggest a role for GluK3-containing KARs (Pinheiro *et al.*, 2007), thus claiming that the KAR involved might be a heteromeric combination of GluK2/3 subunits. Nevertheless, we believe that the differences with the knockouts can be resolved since the critical heteromeric KAR could contain GluK1 and there could be compensation in the $GluK1^{-/-}$ knockout mouse (Kerchner *et al.*, 2002). Of course, GluK2-subtype-selective antagonists would be useful to address this issue. Another issue that has been disputed is the possibility that another KAR subunit combination could be the target for the actions of LY382884 (Nicoll *et al.*, 2000). It is impossible to exclude the possibility, however remote, that at least part of the effects mediated by this compound could be nonspecific in nature. One way to circumvent this possible doubt was the need for more selective GluK1 antagonists derived from structurally unrelated series.

C. The University of Bristol Pharmaceuticals Series: UBP296, UBP302, and ACET

The availability of a series of compounds based on the willardiine structure allowed further investigation on the functions of KARs in the mossy fiber function. These compounds, part of the University of Bristol Pharmaceuticals (UBP) series, have been described as the most potent and selective GluK1 antagonists available so far. Three drugs from the UBP series have been applied to study LTP at mossy fiber synapses—UBP296 (More *et al.*, 2004), UBP302 (Dolman *et al.*, 2005), and more recently ACET (UBP316; Dargan *et al.*, 2009). All three compounds were able to block LTP induction (Fig. 2D–F). Accordingly, UBP296 was found to reversibly block its induction at a concentration of 5 μM. As for LY382884, its effect depended on the Ca^{2+} concentration in the bathing solution, consistent with both compounds working via the same mechanism. Next, we showed that mossy fiber LTP was blocked by UBP302 at a concentration of 300 nM (Dolman *et al.*, 2005). This effect was obtained by the S-enantiomer and LTP could still be evoked in the presence of much higher concentrations of the R-enantiomer, suggesting that these compounds act in a stereoselective fashion. Noteworthy, the potency of inhibition of all GluK1 antagonists described here correlates with the ability to block MF-LTP (Fig. 3).

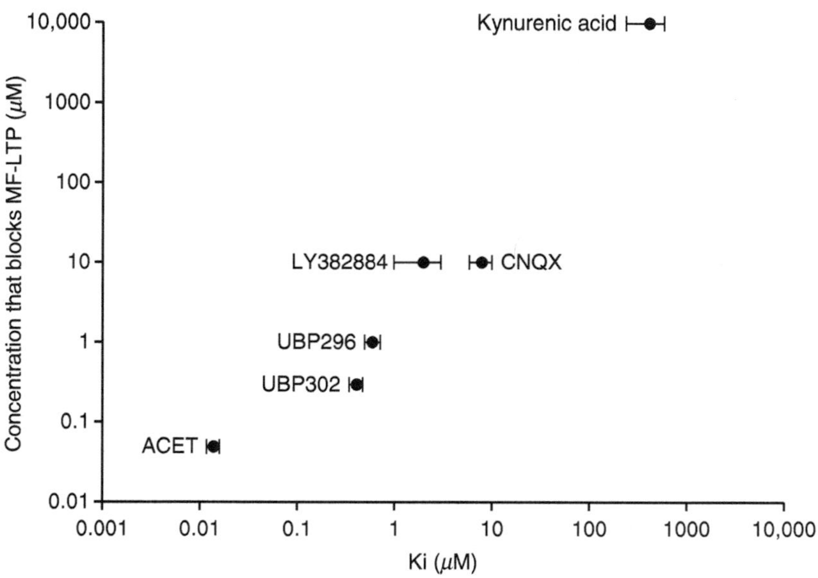

Fig. 3. Correlation between potency of inhibition of GluK1 and the ability to block MF-LTP.

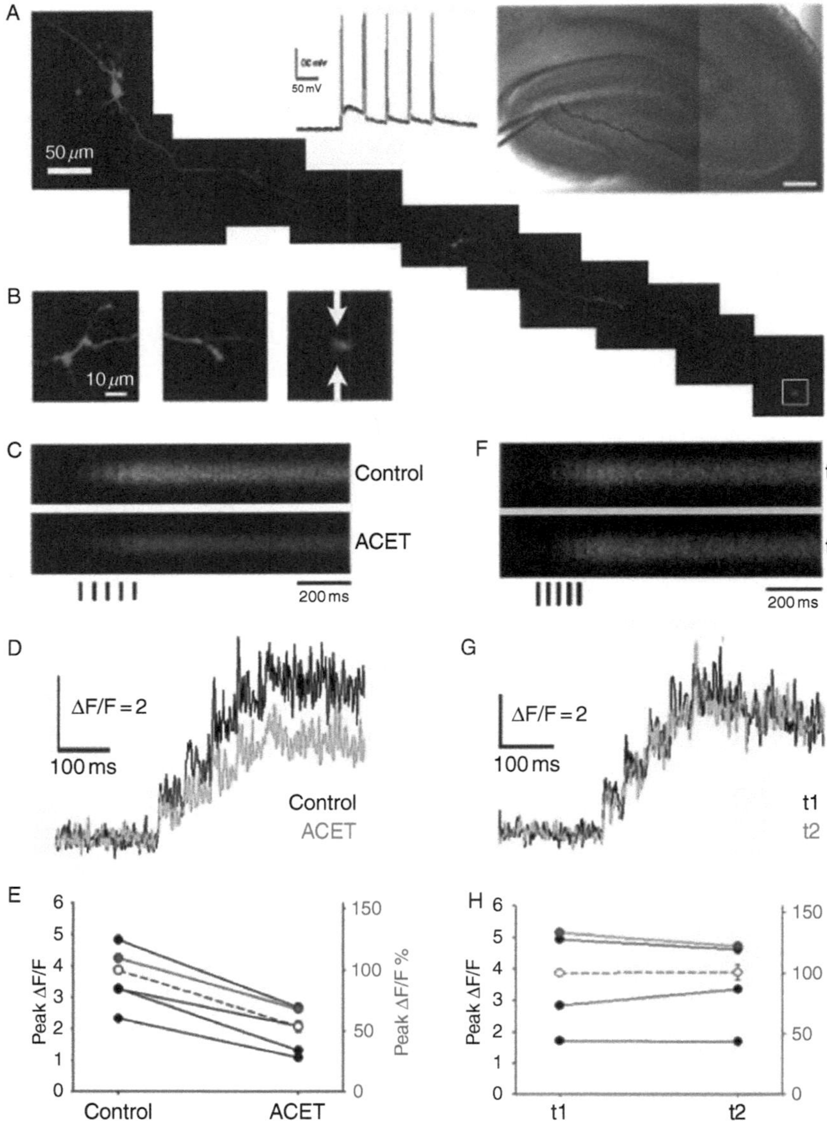

FIG. 4. ACET reduced short-term facilitation of presynaptic calcium transients, evoked by multiple spikes, in individual giant mossy fiber boutons (modified from Dargan *et al.*, 2009). (A) Axon of a single dentate granule cell traced into CA3 region of hippocampus. Inset shows five action potentials at 20 Hz recorded before (black) and after (grey) application of 200 nM ACET, together with the axon trajectory (grey arrow indicates location of giant bouton indicated by white square in A; Scale bar = 200 µm). (B) Enlarged images of the three giant boutons shown in (A). White arrows indicate position of line scan. (C) Fluorescence transients (averages of five successive scans) recorded before and after

The X-ray crystal structure of UBP310 in complex with the ligand binding core of GluK1 (Mayer *et al.*, 2006) has been used to design new willardiine-based antagonists with improved selectivity and potency (Dolman *et al.*, 2007). The most potent of these compounds to date is ACET (UBP316), which has no activity at NMDARs or group I mGluRs, yet shows improved potency and selectivity for GluK1-containing KARs versus AMPARs, and is highly selective for homomeric GluK1 over GluK2, whilst showing weaker antagonist activity at GluK3 (Dargan *et al.*, 2009; Dolman *et al.*, 2007). ACET has similar antagonist potency at heteromeric GluK1-containing KARs (GluK1/GluK2 and GluK1/GluK5) but does not affect heteromeric GluK1-lacking KARs (GluK2/GluK5) (Jane *et al.*, 2009). The X-ray crystal structure of ACET in complex with the ligand binding core of GluK1 has recently been published (Dargan *et al.*, 2009), and the improved selectivity of ACET is thought to be attributed to the presence of the phenyl group, which increases hydrophobic interactions in the GluK1 ligand binding core but causes a steric clash in the binding core of AMPARs (Jane *et al.*, 2009).

We have recently employed a relatively new technique that combines patch clamp electrophysiology and 2-photon microscopy to image calcium dynamics in individual giant mossy fiber boutons (Scott and Rusakov, 2006) to directly investigate the role of presynaptic GluK1-containing KARs. We found that ACET consistently reduced short-term facilitation of presynaptic calcium transients induced by repetitive spikes (Fig. 4, modified from Dargan *et al.*, 2009), which is in agreement with data from another group using NBQX (Scott *et al.*, 2006b) and more recently UBP302 (Scott *et al.*, 2008). These data support the idea that there are presynaptic KARs that regulate Ca^{2+} dynamics in mossy fiber giant terminals.

Although ACET showed poor antagonist activity at GluK3-containing KARs in our hands (Dargan *et al.*, 2009) this is in apparent disagreement with recent data from another group showing that ACET has potent antagonist activity at native rat GluK3-contining KARs (Perrais *et al.*, 2009). The most obvious difference between our study and theirs is the species of receptor used, human versus rat, respectively. Although there is high (98%) homology between rat and human, it remains a possibility that a single amino acid difference could account for differences in receptor pharmacology. Another possible reason underlying the apparent discrepancy could be KAR desensitization state, since we used slow application of glutamate in the presence of Concanavalin A (Dargan *et al.*, 2009)

application of 200 nM ACET. Action potentials are indicated by vertical black bars. (D) Profiles of fluorescence transients before (black) and after (grey) application of ACET. (E) Peak fluorescence for individual cells (representative cell shown in doted line) and pooled data showing percentage change (open circles, $n = 5$). (F, G) Fluorescence transients and profiles under control conditions at different time points: t_1 corresponds to the control time point in C and t_2 corresponds to the time point where recordings were made in the presence of ACET. (H) Peak fluorescence for individual cells (representative cell shown in doted line) and pooled data showing percentage change (open circles, $n = 4$). (See Color Insert.)

whereas the other group use fast glutamate application (Perrais *et al.*, 2009). However, previous studies on dorsal root C-fibers showed that the potency of ACET for antagonizing GluK1-containing receptors was similar in the presence and absence of Concanavalin A, suggesting that the desensitization state of the receptor does not alter the affinity of ACET for the GluK1 binding site (Jane *et al.*, 2009). In summary, ACET is a useful and potent tool for investigating the physiological function of GluK1-containing KARs, but some care should be taken when interpreting data in light of the fact that this compound may also act as an antagonist of GluK3-containing KARs.

Overall, the data obtained with UBP compounds are in accordance with their different potencies for antagonizing GluK1-containing receptors, thus providing stronger evidence for a role for GluK1-containing KARs in mossy fiber function.

III. Conclusions

The pharmacological profile and subunit composition of presynaptic receptors that affect excitatory transmission in the CA3 region of the hippocampus still remain controversial. Results obtained in our laboratory over the past years have suggested that presynaptic receptors, involved both in short- and long-term synaptic plasticity, include a GluK1 subunit. This finding was achieved by the use of various antagonists, with an unrelated chemical structure, over a 200,000 fold concentration range (between 10 mM for kynurenic acid and 50 nM for ACET), which correlates with their potency as GluK1 antagonists. In summary, despite the number of various conflicting results concerning the KAR subunit involved, we strongly believe that GluK1-containing KARs mediate a significant component of short-term frequency-dependent facilitation and regulate LTP in the mossy fiber pathway.

Acknowledgments

We thank our colleagues who have participated in the studies described in this review that have helped to elucidate the roles of KARs.

References

Alford, S., Frenguelli, B. G., Schofield, J. G., and Collingridge, G. L. (1993). Characterization of Ca^{2+} signals induced in hippocampal CA1 neurones by the synaptic activation of NMDA receptors. *J. Physiol.* **469,** 693–716.

Bahn, S., Volk, B., and Wisden, W. (1994). Kainate receptor gene expression in the developing rat brain. *J. Neurosci.* **14,** 5525–5547.

Bleakman, D., Ballyk, B. A., Schoepp, D. D., Palmer, A. J., Bath, C. P., Sharpe, E. F., Woolley, M. L., Bufton, H. R., Kamboj, R. K., Tarnawa, I., and Lodge, D. (1996a). Activity of 2,3-benzodiazepines at native rat and recombinant human glutamate receptors *in vitro*: Stereospecificity and selectivity profiles. *Neuropharmacology* **35,** 1689–1702.

Bleakman, R., Schoepp, D. D., Ballyk, B., Bufton, H., Sharpe, E. F., Thomas, K., Ornstein, P. L., and Kamboj, R. K. (1996b). Pharmacological discrimination of GluR5 and GluR6 kainate receptor subtypes by (3S,4aR,6R,8aR)-6-[2-(1(2)H-tetrazole-5-yl)ethyl]decahyd roisdoquinoline-3 carboxylic-acid. *Mol. Pharmacol.* **49,** 581–585.

Bortolotto, Z. A., Clarke, V. R., Delany, C. M., Parry, M. C., Smolders, I., Vignes, M., Ho, K. H., Miu, P., Brinton, B. T., Fantaske, R., Ogden, A., Gates, M., *et al.* (1999). Kainate receptors are involved in synaptic plasticity. *Nature* **402,** 297–301.

Bortolotto, Z. A., Lauri, S., Isaac, J. T., and Collingridge, G. L. (2003). Kainate receptors and the induction of mossy fibre long-term potentiation. *Trans. R. Soc. Lond. B Biol. Sci.* **358,** 657–666.

Breustedt, J., and Schmitz, D. (2004). Assessing the role of GLUK5 and GLUK6 at hippocampal mossy fiber synapses. *J. Neurosci.* **24,** 10093–10098.

Castillo, P. E., Malenka, R. C., and Nicoll, R. A. (1997). Kainate receptors mediate a slow postsynaptic current in hippocampal CA3 neurons. *Nature* **388,** 182–186.

Clarke, V. R., Ballyk, B. A., Hoo, K. H., Mandelzys, A., Pellizzari, A., Bath, C. P., Thomas, J., Sharpe, E. F., Davies, C. H., Ornstein, P. L., Schoepp, D. D., Kamboj, R. K., *et al.* (1997). A hippocampal GluR5 kainate receptor regulating inhibitory synaptic transmission. *Nature* **389,** 599–603.

Collingridge, G. L., Kehl, S. J., Loo, R., and McLennan, H. (1983). Effects of kainic and other amino acids on synaptic excitation in rat hippocampal slices: 1. Extracellular analysis. *Exp. Brain Res.* **1952,** 170–178.

Collingridge, G. L., Olsen, R. W., Peters, J., and Spedding, M. (2009). A nomenclature for ligand-gated ion channels. *Neuropharmacology* **56,** 2–5.

Contractor, A., Swanson, G., and Heinemann, S. F. (2001). Kainate receptors are involved in short- and long-term plasticity at mossy fiber synapses in the hippocampus. *Neuron* **29,** 209–216.

Contractor, A., Sailer, A. W., Darstein, M., Maron, C., Xu, J., Swanson, G. T., and Heinemann, S. F. (2003). Loss of kainate receptor-mediated heterosynaptic facilitation of mossy-fiber synapses in KA2S/S mice. *J. Neurosci.* **23,** 422–429.

Cossart, R., Esclapez, M., Hirsch, J. C., Bernard, C., and Ben-Ari, Y. (1998). GluR5 kainate receptor activation in interneurons increases tonic inhibition of pyramidal cells. *Nat. Neurosci.* **1,** 470–478.

Dargan, S. L., Clarke, V. R., Alushin, G. M., Sherwood, J. L., Nisticò, R., Bortolotto, Z. A., Ogden, A. M., Bleakman, D., Doherty, A. J., Lodge, D., Mayer, M. L., Fitzjohn, S. M., *et al.* (2009). ACET is a highly potent and specific kainate receptor antagonist: Characterisation and effects on hippocampal mossy fibre function. *Neuropharmacology* **56,** 121.

Davies, J., Evans, R. H., Francis, A. A., and Watkins, J. C. (1979). Excitatory amino acid receptors and synaptic excitation in the mammalian central nervous system. *J. Physiol. (Paris)* **75,** 641–654.

Dingledine, R., Borges, K., Bowie, D., and Traynelis, S. F. (1999). The glutamate receptor ion channels. *Pharmacol. Rev.* **51,** 7–61.

Dolman, N. P., Troop, H. M., More, J. C., Alt, A., Knauss, J. L., Nistico, R., Jack, S., Morley, R. M., Bortolotto, Z. A., Roberts, P. J., Bleakman, D., Collingridge, G. L., *et al.* (2005). Synthesis and pharmacology of willardiine derivatives acting as antagonists of kainate receptors. *J. Med. Chem.* **48,** 7867–7881.

Dolman, N. P., More, J. C., Alt, A., Knauss, J. L., Pentikäinen, O. T., Glasser, C. R., Bleakman, D., Mayer, M. L., Collingridge, G. L., and Jane, D. E. (2007). Synthesis and pharmacological characterization of N3-substituted willardiine derivatives: Role of the substituent at the

5-position of the uracil ring in the development of highly potent and selective GLUK5 kainate receptor antagonists. *J. Med. Chem.* **50,** 1558–1570.

Eder, M., Becker, K., Rammes, G., Schierloh, A., Azad, S. C., Zieglgänsberger, W., and Dodt, H. U. (2003). Distribution and properties of functional postsynaptic kainate receptors on neocortical layer V pyramidal neurons. *J. Neurosci.* **23,** 6660–6670.

Emptage, N., Bliss, T. V., and Fine, A. (1999). Single synaptic events evoke NMDA receptor-mediated release of calcium from internal stores in hippocampal dendritic spines. *Neuron* **22,** 115–124.

Fisher, R. S., and Alger, B. E. (1984). Electrophysiological mechanisms of kainic acid-induced epileptiform activity in the rat hippocampal slice. *J. Neurosci.* **4,** 1312–1323.

Frerking, M., Malenka, R. C., and Nicoll, R. A. (1998). Synaptic activation of kainate receptors on hippocampal interneurons. *Nat. Neurosci.* **1,** 479–486.

Herb, A., Burnashev, N., Werner, P., Sakmann, B., Wisden, W., and Seeburg, P. H. (1992). The KA-2 subunit of excitatory amino acid receptors shows widespread expression in brain and forms ion channels with distantly related subunits. *Neuron* **8,** 775–785.

Hollmann, M., and Heinemann, S. (1994). Cloned glutamate receptors. *Annu. Rev. Neurosci.* **17,** 31–108.

Huettner, J. E. (2003). Kainate receptors and synaptic transmission. *Prog. Neurobiol.* **70,** 387–407.

Jane, D. E., Lodge, D., and Collingridge, G. L. (2009). Kainate receptors: Pharmacology, function and therapeutic potential. *Neuropharmacology* **56,** 90–113.

Kamiya, H., and Ozawa, S. (2000). Kainate receptor-mediated presynaptic inhibition at the mouse hippocampal mossy fibre synapse. *J. Physiol.* **523,** 653–665.

Kehl, S. J., McLennan, H., and Collingridge, G. L. (1984). Effects of folic and kainic acids on synaptic responses of hippocampal neurones. *Neuroscience* **11,** 111–124.

Keinänen, K., Wisden, W., Sommer, B., Werner, P., Herb, A., Verdoorn, T. A., Sakmann, B., and Seeburg, P. H. (1990). A family of AMPA-selective glutamate receptors. *Science* **249,** 556–560.

Kerchner, G. A., Wilding, T. J., Huettner, J. E., and Zhuo, M. (2002). Kainate receptor subunits underlying presynaptic regulation of transmitter release in the dorsal horn. *J. Neurosci.* **22,** 8010–8017.

Kiskin, N. I., Krishtal, O. A., and Tsyndrenko, A. Y. (1986). Excitatory amino acid receptors in hippocampal neurons: Kainate fails to desensitize them. *Neurosci. Lett.* **63,** 225–230.

Lauri, S. E., Bortolotto, Z. A., Bleakman, D., Ornstein, P. L., Lodge, D., Isaac, J. T., and Collingridge, G. L. (2001). A critical role of a facilitatory presynaptic kainate receptor in mossy fiber LTP. *Neuron* **32,** 697–709.

Lauri, S. E., Bortolotto, Z. A., Nistico, R., Bleakman, D., Ornstein, P. L., Lodge, D., Isaac, J. T., and Collingridge, G. L. (2003). A role for Ca^{2+} stores in kainate receptor-dependent synaptic facilitation and LTP at mossy fiber synapses in the hippocampus. *Neuron* **39,** 327–341.

Lerma, J. (2003). Roles and rules of kainate receptors in synaptic transmission. *Nat. Rev. Neurosci.* **4,** 481–495.

Lerma, J. (2006). Kainate receptor physiology. *Curr. Opin. Pharmacol.* **6,** 89–97.

Lerma, J., Paternain, A. V., Rodríguez-Moreno, A., and López-García, J. C. (2001). Molecular physiology of kainate receptors. *Physiol. Rev.* **81,** 971–998.

Li, H., and Rogawski, M. A. (1998). GluR5 kainate receptor mediated synaptic transmission in rat basolateral amygdale *in vitro*. *Neuropharmacology* **37,** 1279–1286.

Li, P., Wilding, T. J., Kim, S. J., Calejesan, A. A., Huettner, J. E., and Zhuo, M. (1999). Kainate-receptor-mediated sensory synaptic transmission in mammalian spinal cord. *Nature* **397,** 161–164.

Mayer, M. L., Ghosal, A., Dolman, N. P., and Jane, D. E. (2006). Crystal structures of the kainate receptor GluR5 ligand binding core dimer with novel GluR5-selective antagonists. *J. Neurosci.* **26,** 2852–2861.

Meldrum, B. (1991). Excitotoxicity and epileptic brain damage. *Epilepsy Res.* **10,** 55–61.

Monaghan, D. T., and Cotman, C. W. (1982). The distribution of [3 H]kainic acid binding sites in rat CNS as determined by autoradiography. *Brain Res.* **252,** 91–100.

More, J. C., Nistico, R., Dolman, N. P., Clarke, V. R., Alt, A. J., Ogden, A. M., Buelens, F. P., Troop, H. M., Kelland, E. E., Pilato, F., Bleakman, D., Bortolotto, Z. A., *et al.* (2004). Characterisation of UBP296: A novel, potent and selective kainate receptor antagonist. *Neuropharmacology* **47,** 46–64.

Nicoll, R. A., Mellor, J., Frerking, M., and Schmitz, D. (2000). Kainate receptors and synaptic plasticity. *Nature* **406,** 957.

O'Neill, M. J., Bogaert, L., Hicks, C. A., Bond, A., Ward, M. A., Ebinger, G., Ornstein, P. L., Michotte, Y., and Lodge, D. (2000). LY377770, a novel iGlu5 kainate receptor antagonist with neuroprotective effects in global and focal cerebral ischaemia. *Neuropharmacology* **39,** 1575–1588.

Paternain, A. V., Morales, M., and Lerma, J. (1995). Selective antagonism of AMPA receptors unmasks kainate receptor-mediated responses in hippocampal neurons. *Neuron* **14,** 185–189.

Patneau, D. K., and Mayer, M. L. (1991). Kinetic analysis of interactions between kainate and AMPA: Evidence for activation of a single receptor in mouse hippocampal neurons. *Neuron* **6,** 785–798.

Perrais, D., Pinheiro, P. S., Jane, D. E., and Mulle, C. (2009). Antagonism of recombinant and native GluK3-containing kainate receptors. *Neuropharmacology* **56,** 131–140.

Pinheiro, P. S., Perrais, D., Coussen, F., Barhanin, J., Bettler, B., Mann, J. R., Malva, J. O., Heinemann, S. F., and Mulle, C. (2007). GluR7 is an essential subunit of presynaptic kainate autoreceptors at hippocampal mossy fiber synapses. *Proc. Natl. Acad. Sci. USA* **104,** 12181–12186.

Robinson, J. H., and Deadwyler, S. A. (1981). Kainic acid produces depolarization of CA3 pyramidal cells in the vitro hippocampal slice. *Brain Res.* **221,** 117–127.

Schmitz, D., Frerking, M., and Nicoll, R. A. (2000). Synaptic activation of presynaptic kainate receptors on hippocampal mossy fiber synapses. *Neuron* **27,** 327–338.

Schmitz, D., Mellor, J., and Nicoll, R. A. (2001). Presynaptic kainate receptor mediation of frequency facilitation at hippocampal mossy fiber synapses. *Science* **291,** 1972–1976.

Scott, R., and Rusakov, D. A. (2006). Main determinants of presynaptic Ca^{2+} dynamics at individual mossy fiber-CA3 pyramidal cell synapses. *J. Neurosci.* **26,** 7071–7081.

Scott, R., Kullmann, D. M., and Rusakov, D. A. (2006). Cell target specific presynaptic actions of kainate receptors at individual mossy fibre synapses. *Proc. Physiol. Soc.* **3,** PC168.

Scott, R., Lalic, T., Kullmann, D. M., Capogna, M., and Rusakov, D. A. (2008). Target-cell specificity of kainate autoreceptor and Ca^{2+}-store-dependent short term plasticity at hippocampal mossy fiber synapses. *J. Neurosci.* **28,** 13139–13149.

Smolders, I., Bortolotto, Z. A., Clarke, V. R., Warre, R., Khan, G. M., O'Neill, M. J., Ornstein, P. L., Bleakman, D., Ogden, A., Weiss, B., Stables, J. P., Ho, K. H., *et al.* (2002). Antagonists of GLU(K5)-containing kainate receptors prevent pilocarpine-induced limbic seizures. *Nat. Neurosci.* **5,** 796–804.

Urban, N. N., and Barrionuevo, G. (1996). Induction of hebbian and non-hebbian mossy fiber long-term potentiation by distinct patterns of high-frequency stimulation. *J. Neurosci.* **16,** 4293–4299.

Vignes, M., and Collingridge, G. L. (1997). The synaptic activation of kainate receptors. *Nature* **388,** 179–182.

Vignes, M., Bleakman, D., Lodge, D., and Collingridge, G. L. (1997). The synaptic activation of the GluR5 subtype of kainate receptor in area CA3 of the rat hippocampus. *Neuropharmacology* **36,** 1477–1481.

Vincent, P., and Mulle, C. (2009). Kainate receptors in epilepsy and excitotoxicity. *Neuroscience* **158,** 309–323.

Watkins, J. C., and Evans, R. H. (1981). Excitatory amino acid transmitters. *Annu. Rev. Pharmacol. Toxicol.* **21,** 165–204.

Westbrook, G. L., and Lothman, E. W. (1983). Cellular and synaptic basis of kainic acid-induced hippocampal epileptiform activity. *Brain Res.* **273,** 97–109.

Wilding, T. J., and Huettner, J. E. (1995). Differential antagonism of alpha-amino-3-hydroxy-5-methyl-4-isoxazolepropionic acid-preferring and kainate-preferring receptors by 2,3-benzodiazepines. *Mol. Pharmacol.* **47,** 582–587.

Wu, L. J., Zhao, M. G., Toyoda, H., Ko, S. W., and Zhuo, M. (2005). Kainate receptor-mediated synaptic transmission in the adult anterior cingulated cortex. *J. Neurophysiol.* **94,** 1805–1813.

MONOAMINE TRANSPORTER AS A TARGET MOLECULE FOR PSYCHOSTIMULANTS

Ichiro Sora,* BingJin Li,* Setsu Fumushima,* Asami Fukui,* Yosefu Arime,* Yoshiyuki Kasahara,* Hiroaki Tomita,* and Kazutaka Ikeda[†]

*Department of Biological Psychiatry, Tohoku University Graduate School of Medicine, Sendai 980–8574, Japan

[†]Molecular Psychiatry Research, Tokyo Institute of Psychiatry, Tokyo 156–8585, Japan

Methamphetamine (MAP), a drug of abuse known worldwide for its addictive effects and neurotoxicity, causes somatic and psychiatric disorders. MAP enters terminals/neurons via monoamine transporters, displaces both vesicular and intracellular monoamines, and facilitates the release of monoamines into the extraneuronal space through synaptic transport via the monoamine transporters. Chronic psychostimulant abusers exhibit psychotic features, including delusions and auditory hallucinations. The dopamine transporter (DAT) and the vesicular monoamine transporter 2 (VMAT2) play pivotal roles in the action of MAP, including locomotor effects. The deletion of DAT attenuates the locomotor effects of MAP and may play larger role in behavioral responses to MAP compared to the deletion of VMAT2. MAP produces hyperthermia and/or neuronal toxicity in most species. The effects of MAP in DAT or serotonin transporter (SERT) single knockout (KO) mice and DAT/SERT double KO mice suggested that DAT and SERT are key molecules for hyperthermia and neuronal toxicity of MAP.

I. Introduction

Methamphetamine (MAP) is a psychostimulant that induces enhanced arousal and euphoria acutely, and psychosis and addiction chronically. MAP enters the terminals/neuron via the monoamine transporters (dopamine transporter: DAT, serotonin transporter: SERT, or norepinephrine transporter: NET), displaces

INTERNATIONAL REVIEW OF NEUROBIOLOGY, VOL. 85
DOI: 10.1016/S0074-7742(09)85003-4

0074-7742/09 $35.00

both vesicular and intracellular monoamines, and facilitates release of monoamines into the extraneuronal space by synaptic transport in the monoamine transporters (Seiden *et al.*, 1993). The large release of monoamine produced by psychostimulant is thought to contribute to the drug's effects in the brain.

II. MAP-Induced Behavioral Sensitization

The acute and chronic pharmacological consequences of MAP in human users have been observed in behavioral experiments in animals, including both hyperactivity and sensitization of locomotor responses (Segal and Schuckit, 1983). Behavioral sensitization is a phenomenon whereby repeated intermittent exposure to MAP-like psychostimulant elicits a progressive enhancement of those responses, which persists for extended time periods following withdrawal from the drug and are easily reinstated by exposure to the drug or psychosocial stress (Robinson and Becker, 1986). This process closely resembles the course of the relapse in MAP-induced psychosis or schizophrenia, thus sensitization in animals has been suggested to model these psychoses (Sato *et al.*, 1983). Behavioral sensitization is thought to be an early and enduring manifestation of neuronal plasticity associated with changes in mesolimbic dopamine neurotransmission (Kalivas *et al.*, 1993). MAP induces dopamine release through exchange diffusion of plasma membrane DAT (Seiden *et al.*, 1993), and release of vesicular dopamine into the cytosol by acting on the vesicular monoamine transporter 2 (VMAT2) (Sulzer *et al.*, 2005). The dopamine releasing effect of MAP has been postulated to mediate its locomotor stimulant and rewarding effects (White and Kalivas, 1998). Therefore, DAT and VMAT2 should play pivotal roles in the mechanisms underlying the actions of MAP.

DAT knockout (KO) mice and VMAT2 KO mice have been used to investigate the roles of DAT and VMAT2 in dopamine neurotransmission and pharmacological mechanisms underlying the actions of psychostimulants. Homozygous deletion of the DAT gene has been reported to produce a 10-fold increase (Shen *et al.*, 2004) or fivefold elevation (Jones *et al.*, 1998) of extracellular dopamine concentrations in the striatum measured by *in vivo* microdialysis, while heterozygous deletion of DAT was not found to significantly increase extracellular dopamine (Shen *et al.*, 2004) or to produce a smaller twofold elevation (Jones *et al.*, 1998) of dopamine in the striatum. Homozygous DAT KO mice show growth retardation and hyperactivity, whereas heterozygous DAT KO mice did not show gross abnormalities in either development or baseline behavioral parameters (Sora *et al.*, 1998). Habituated homozygous DAT KO mice do not show any significant cocaine-induced increase in locomotion (Sora *et al.*, 1998, 2001; Uhl *et al.*, 2002).

We examined locomotor activity and sensitization in heterozygous DAT KO ($DAT^{+/-}$), heterozygous VMAT2 KO ($VMAT2^{+/-}$), double heterozygous DAT/VMAT2 KO ($DAT^{+/-}$ $VMAT2^{+/-}$), and wild-type (WT) mice to evaluate the roles of DAT and VMAT2 in MAP-induced locomotor behavior (Fukushima *et al.*, 2007). In $DAT^{+/-}$ $VMAT2^{+/-}$ mice, all of MAP-induced behavioral responses were similar to those in $DAT^{+/-}$, but not $VMAT2^{+/-}$ mice. The behavioral effects of both acute and chronic MAP administration were suppressed in heterozygous DAT KO mice, whether or not it was combined with heterozygous VMAT2 KO. Contrary to the effect observed in heterozygous DAT KO mice, acute MAP administration produced greater locomotor responses in heterozygous VMAT2 KO mice. These findings indicate that the half deletion of DAT plays a major role in both acute and chronic behavioral responses to MAP, while the effect of the half deletion of VMAT2 is less prominent.

III. MAP-Induced Hyperthermia and Neuronal Toxicity

MAP abuse causes serious health hazards including irreversible neuronal degeneration, seizures, hyperthermia, and death in human and experimental animals (Davidson *et al.*, 2001). Among these side effects, MAP produces hyperthermia and/or dopaminergic neurotoxicity in most species. Clinical reports and animal studies indicate that lethality by MAP closely correlates with hyperthermia, which may be the primary cause of death. Animal studies suggest that dopamine receptor activation is crucial for MAP-induced hyperthermia (Broening *et al.*, 2005) and lethality (Bronstein and Hong, 1995). There has also been an assumption that the hyperthermia that follows MAP administration is serotonin receptor-mediated (Green *et al.*, 2003).

We examined hyperthermic and lethal toxic effects of MAP in DAT, SERT, and DAT/SERT double KO mice to elucidate the role of these two transporters in MAP-induced hyperthermia and lethality (Numachi *et al.*, 2007). MAP caused significant hyperthermia even in the mice with a single DAT gene copy and no SERT copies ($DAT^{+/-}$ $SERT^{-/-}$ mice). Mice with no DAT copies and a single SERT gene copy ($DAT^{-/-}$ $SERT^{+/-}$ mice) showed significant but reduced hyperthermia when compared to WT mice after MAP. These results demonstrate that MAP exerts a hyperthermic effect via DAT, or via SERT, in the absence of DAT. DAT gene deletion in mice strikingly increased LD_{50} of MAP by 1.7–1.8 times that of WT mice, suggesting that the lethal toxic effect of MAP is mainly dependent on DAT. Although DAT and SERT were shown here to be involved in both the effects of MAP on temperature as well as MAP lethal toxicity, the mechanisms are nonetheless different; DAT single KO mice exhibited hyperthermia but greatly reduced MAP lethality, and the lethality was no different from

DAT/SERT double KO mice that had hypothermic responses to MAP. Thus, although the lethal toxic effect of MAP is mainly dependent on DAT, with some contribution from SERT, hyperthermia is not prerequisite for MAP-induced lethality.

In conclusion, these findings lead us to hypothesize that DAT variants may have more profound effects than VMAT2 or SERT variants on the clinically important consequences of acute and chronic MAP abuse in humans.

Acknowledgments

This study was supported in part by Grant-in-Aid for Health and Labor Science Research (Research on Pharmaceutical and Medical Safety) from the Ministry of Health, Labor and Welfare of Japan; by Grants-in-Aid for Scientific Research (B), Scientific Research on Priority Areas—System study on higher order brain functions and Research on Pathomechanisms of Brain Disorders, Core Research for Evolutional Science and Technology (CREST), from the Ministry of Education, Culture, Sports, Science, and Technology of Japan.

References

Broening, H. W., Morford, L. L., and Vorhees, C. V. (2005). Interactions of dopamine D1 and D2 receptor antagonists with D-methamphetamine-induced hyperthermia and striatal dopamine and serotonin reductions. *Synapse.* **56,** 84–93.

Bronstein, D. M., and Hong, J. S. (1995). Effects of sulpiride and SCH 23390 on methamphetamine-induced changes in body temperature and lethality. *J. Pharmacol. Exp. Ther.* **274,** 943–950.

Davidson, C., Gow, A. J., Lee, T. H., and Ellinwood, E. H. (2001). Methamphetamine neurotoxicity: Necrotic and apoptotic mechanisms and relevance to human abuse and treatment. *Brain Res. Brain Res. Rev.* **36,** 1–22.

Fukushima, S., Shen, H., Hata, H., Ohara, A., Ohmi, K., Ikeda, K., Numachi, Y., Kobayashi, H., Hall, F. S., Uhl, G. R., and Sora, I. (2007). Methamphetamine-induced locomotor activity and sensitization in dopamine transporter and vesicular monoamine transporter 2 double mutant mice. *Psychopharmacology (Berl.)* **193,** 55–62.

Green, A. R., Mechan, A. O., Elliott, J. M., O'Shea, E., and Colado, M. I. (2003). The pharmacology and clinical pharmacology of 3,4-methylenedioxymethamphetamine (MDMA, "ecstasy"). *Pharmacol. Rev.* **55,** 463–508.

Jones, S. R., Gainetdinov, R. R., Jaber, M., Giros, B., Wightman, R. M., and Caron, M. G. (1998). Profound neuronal plasticity in response to inactivation of the dopamine transporter. *Proc. Natl. Acad. Sci. USA* **95,** 4029–4034.

Kalivas, P. W., Sorg, B. A., and Hooks, M. S. (1993). The pharmacology and neural circuitry of sensitization to psychostimulants. *Behav. Pharmacol.* **4,** 315–334.

Numachi, Y., Ohara, A., Yamashita, M., Fukushima, S., Kobayashi, H., Hata, H., Watanabe, H., Hall, F. S., Lesch, K. P., Murphy, D. L., Uhl, G. R., and Sora, I. (2007). Methamphetamine-induced hyperthermia and lethal toxicity: Role of the dopamine and serotonin transporters. *Eur. J. Pharmacol.* **572,** 120–128.

Robinson, T. E., and Becker, J. B. (1986). Enduring changes in brain and behavior produced by chronic amphetamine administration: A review and evaluation of animal models of amphetamine psychosis. *Brain Res.* **396,** 157–198.

Sato, M., Chen, C. C., Akiyama, K., and Otsuki, S. (1983). Acute exacerbation of paranoid psychotic state after long-term abstinence in patients with previous methamphetamine psychosis. *Biol. Psychiatry* **18,** 429–440.

Segal, D. S., and Schuckit, M. A. (1983). Animal models of stimulant-induced psychosis. *In* "Stimulants: Neurochemical, Behavioral and Clinical Perspectives" (I. Grees, ed.), pp. 131–167. Reven, New York.

Seiden, L. S., Sabol, K. E., and Ricaurte, G. A. (1993). Amphetamine: Effects on catecholamine systems and behavior. *Annu. Rev. Pharmacol. Toxicol.* **33,** 639–677.

Shen, H. W., Hagino, Y., Kobayashi, H., Shinohara-Tanaka, K., Ikeda, K., Yamamoto, H., Yamamoto, T., Lesch, K. P., Murphy, D. L., Hall, F. S., Uhl, G. R., and Sora, I. (2004). Regional differences in extracellular dopamine and serotonin assessed by *in vivo* microdialysis in mice lacking dopamine and/or serotonin transporters. *Neuropsychopharmacology* **29,** 1790–1799.

Sora, I., Wichems, C., Takahashi, N., Li, X. F., Zeng, Z., Revay, R., Lesch, K. P., Murphy, D. L., and Uhl, G. R. (1998). Cocaine reward models: Conditioned place preference can be established in dopamine- and in serotonin-transporter knockout mice. *Proc. Natl. Acad. Sci. USA* **95,** 7699–7704.

Sora, I., Hall, F. S., Andrews, A. M., Itokawa, M., Li, X. F., Wei, H. B., Wichems, C., Lesch, K. P., Murphy, D. L., and Uhl, G. R. (2001). Molecular mechanisms of cocaine reward: Combined dopamine and serotonin transporter knockouts eliminate cocaine place preference. *Proc. Natl. Acad. Sci. USA* **98,** 5300–5305.

Sulzer, D., Sonders, M. S., Poulsen, N. W., and Galli, A. (2005). Mechanisms of neurotransmitter release by amphetamines: A review. *Prog. Neurobiol.* **75,** 406–433.

Uhl, G. R., Hall, F. S., and Sora, I. (2002). Cocaine, reward, movement and monoamine transporters. *Mol. Psychiatry* **7,** 21–26.

White, F. J., and Kalivas, P. W. (1998). Neuroadaptations involved in amphetamine and cocaine addiction. *Drug Alcohol Depend.* **51,** 141–153.

TARGETED LIPIDOMICS AS A TOOL TO INVESTIGATE ENDOCANNABINOID FUNCTION

Giuseppe Astarita, Jennifer Geaga, Faizy Ahmed, and Daniele Piomelli

Department of Pharmacology, University of California, Irvine, California 92967, USA

Endocannabinoids are a family of lipid messengers present in a wide range of living organisms. They bind and activate the membrane receptors that are targeted by Δ^9-tetrahydrocannabinol, the main psychoactive principle in marijuana (*Cannabis*). In the brain, they regulate ion-channel activity and neurotransmitter release critical to biological processes such as synaptic plasticity and learning and memory. Endocannabinoids are embedded within an intricate network of lipid pathways, the regulation of which controls the strength and duration of their signaling. Therefore, physiological, pathological, or pharmacological perturbations of these interconnected lipid pathways have a profound effect on the regulation of endocannabinoid signaling. The recent development of high-sensitivity and high-throughput analytical tools affords a broader view of the endocannabinoid system, allowing researchers to place individual endocannabinoid molecules in the context of the interconnected network of their precursors and derivatives. Targeted lipidomics provides new opportunities for understanding endocannabinoid metabolism.

INTERNATIONAL REVIEW OF NEUROBIOLOGY, VOL. 85
DOI: 10.1016/S0074-7742(09)85004-6

I. Introduction

Lipidomics has been defined as "the full characterization of lipid molecular species and of their biological roles with respect to expression of proteins involved in lipid metabolism and function, including gene regulation" (Piomelli *et al.*, 2007; Spener *et al.*, 2003). This definition implies (1) the identification and quantification of lipids present in biological samples, and (2) the interpretation of the biological significance of such analytical data. While unbiased lipidomics is used for the large-scale analysis of as many lipids as possible, targeted lipidomics aims to look at one specific lipid pathway. In this chapter, we describe targeted lipidomics as a tool for the analysis of endocannabinoid metabolism (Astarita *et al.*, 2008; Jung *et al.*, 2007; Piomelli *et al.*, 2007).

II. Endocannabinoids

Endocannabinoids are endogenous lipids derived from polyunsaturated fatty acids (PUFAs) (Di Marzo, 2008; Katona and Freund, 2008). The two most intensively studied endocannabinoids, arachidonoylethanolamide (anandamide) (Devane *et al.*, 1992) and 2-arachidonoyl-*sn*-glycerol (2-AG) (Sugiura *et al.*, 1995), are the derivatives of arachidonic acid. These molecules are present in a wide range of vertebrate animals (Schmid *et al.*, 1990)—including fish, reptiles (Astarita *et al.*, 2006), and mammals—plants (Chapman, 2000), and recently have been found in nematodes (Lehtonen *et al.*, 2008), suggesting that the endocannabinoid system developed very early in evolution. In the mammalian nervous system, endocannabinoids are produced "on demand" from neural cells and bind to specific G-protein-coupled receptors. These receptors have been named cannabinoid (CB) receptors because they also recognize Δ^9-tetrahydrocannabinol, the psychoactive component of *Cannabis*. The activation of CB receptors plays critical roles in various physiological processes including appetite, pain, immunity, inflammation, and cognition. In the last 20 years, considerable progress has been made in understanding the physiological and pathological roles for individual component of endocannabinoid signaling (i.e., anandamide and 2-AG) (Di Marzo, 2008; Katona and Freund, 2008). In this chapter, we will consider the endocannabinoid system as an ensemble (i.e., all the interconnected networks of precursors and derivatives).

A. Endocannabinoid Lipid Network

The endocannabinoid system is constituted of various lipid species, which include precursors and metabolites of anandamide and 2-AG (Figs. 1 and 2). Each of these lipids appears to have functionally distinct biosynthetic pathways and

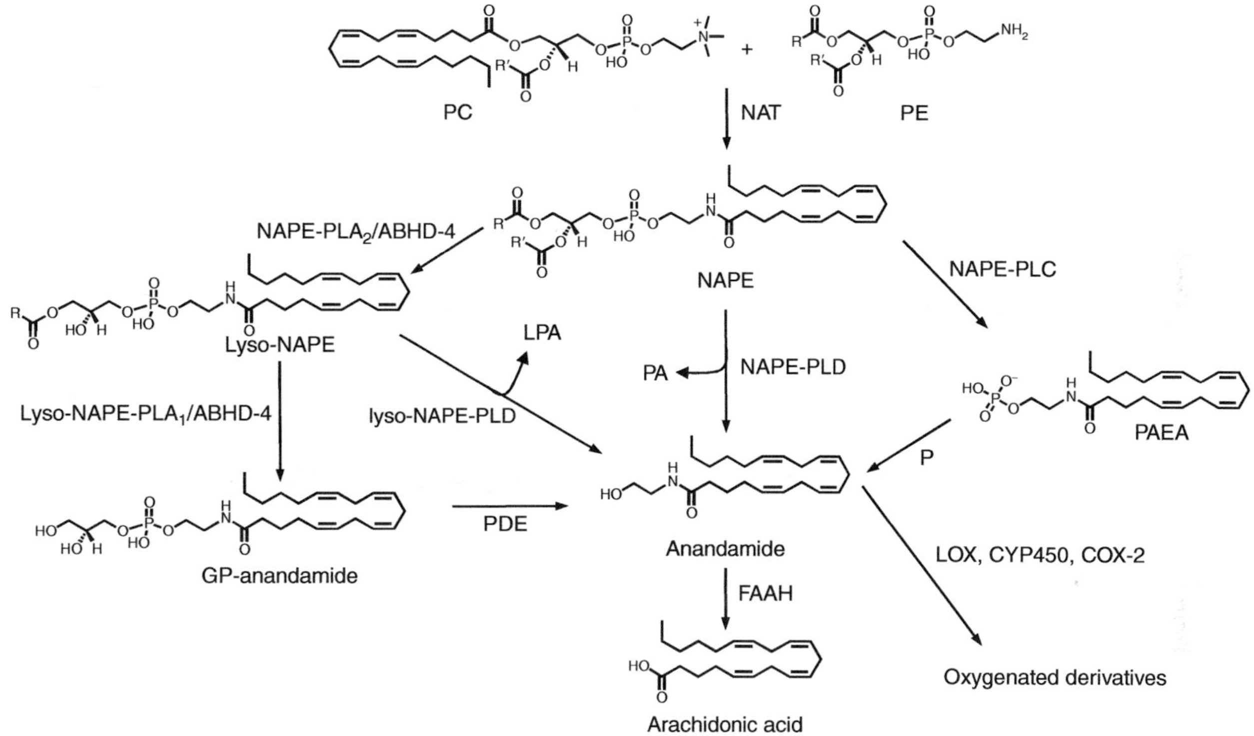

FIG. 1. Targeted lipidomics of anandamide metabolism. Postulated pathways of anandamide metabolism. *Abbreviations*: PC, phosphatidylcholine; PE, phosphatidylethanolamine; NAT, *N*-acyl transferase; LPA, lysophosphatidic acid; PA, phosphatidic acid; NAPE, *N*-acyl-phosphatidylethanolamine; Lyso-NAPE, 1-lyso,2-acyl-*sn*-glycero-3-phosphoethanolamine-*N*-acyl; ABHD-4, α/β hydrolase-4; GP-anandamide, glycerophospho-anandamide; PAEA, phospho-anandamide; PLA, phospholipase A; NAPE-PLD, NAPE phospholipase D; PLC, phospholipase C; FAAH, fatty acid amide hydrolase; P, phosphatase; COX, cyclooxygenase; LOX, lipoxygenase; CYP450, cytochrome P450; PDE, phosphodiesterase.

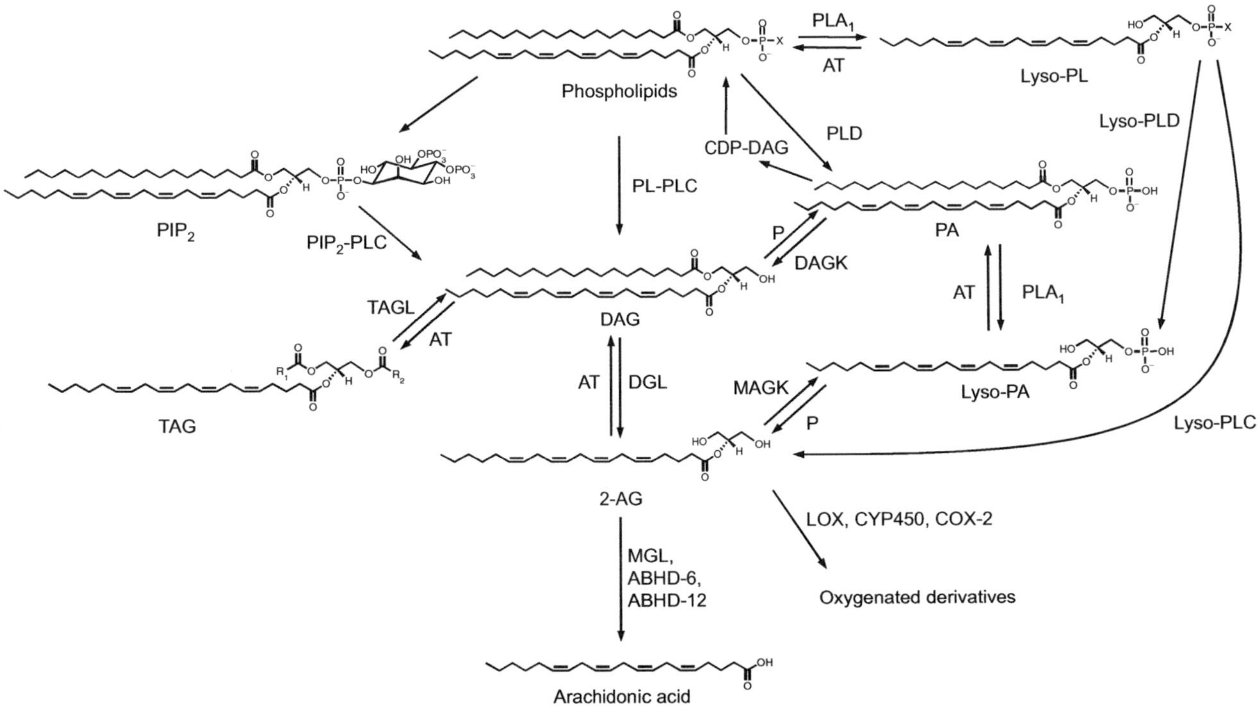

FIG. 2. Targeted lipidomics of 2-AG metabolism. Postulated pathways for 2-AG metabolism. *Abbreviations*: PLC, phospholipase C; DAG, diacylglycerol; DGL, diacylglycerol lipase; MGL, monoacylglycerol lipase; PLA, phospholipase A; AT, acyltransferase; TAGL, triacylglycerol lipase; PIP_2, phosphatidylinositol bisphosphate; ABHD-6/12 hydrolase; lyso-PL, lysophospholipid; lyso-PA, lysophosphatidic acid; PA, phosphatidic acid; P, phosphatase; COX, cyclooxygenase; LOX, lipoxygenase; CYP450, cytochrome P450; CDP, cytidine diphosphate.

biological roles, placing the endocannabinoid system in the context of an intricate lipid network. Physiological and pathological processes that modulate these related lipid pathways might deeply affect endocannabinoid signaling. Targeted lipidomic analysis can help us to elucidate how each lipid component is involved in the regulation of the endocannabinoid metabolism.

B. Lipidomic Analysis of Endocannabinoids by LC/MS

In recent years the rapid development of high-sensitivity analytical techniques such as mass spectrometry (MS) and liquid chromatography (LC) supported the investigation of the endocannabinoids as part of a complex lipid network. The identification of lipid components of the endocannabinoid system can be achieved in a single analytical step by state-of-the-art platforms such as tandem mass spectrometry (MS/MS), which provides the detailed structural information necessary for characterization of lipids and increases specificity in complex biological matrices. Furthermore, the implementation of ionization techniques such as electrospray ionization (ESI) and atmospheric pressure chemical ionization (APCI) allow the coupling of LC to MS, and permits the separation and analysis of endocannabinoids with greater speed and accuracy.

C. Application of Targeted Lipidomics for the Study of Endocannabinoid Metabolism

Targeted lipidomic approaches may be used to reveal several aspects of endocannabinoid metabolism that are still incompletely understood. For example, we do not fully understand the specific contribution of the multiple biosynthetic and degradative pathways to the regulation of endocannabinoid signaling. In addition, it is unclear how changes in endocannabinoid signaling affect distal lipid pathways, and, vice versa, how alterations in cellular lipid metabolism influence endocannabinoid signaling. Lipidomics can help answer these questions by offering a large-scale view of lipid alterations accompanying perturbations in endocannabinoid metabolism. For example, analyses of lipid profiles associated with genetic manipulations (i.e., gene overexpression or knock down) or pharmacological treatments can help elucidate the biochemical mechanisms underlying the regulation of endocannabinoid metabolism in biological systems (Astarita *et al.*, 2008; Jung *et al.*, 2007; Saghatelian *et al.*, 2004). Furthermore, lipidomics-based information on the endocannabinoid system can be integrated into a multidisciplinary set of data deriving from genomics, transcriptomics, and proteomics. Such a systems biology approach may offer novel insights on the effects of genetic diversity (e.g., genotype, epigenetic regulation, mutation, and polymorphism), messenger RNA expression profiles, or protein diversity (e.g., isoforms, posttranslational modifications, cofactors) on endocannabinoid metabolism.

In the following sections we illustrate the complexity of endocannabinoid metabolism, highlighting the many questions that still remain unanswered.

III. Targeted Lipidomics of the Anandamide Pathway

Multiple lipid pathways take part in anandamide biosynthesis and degradation (Fig. 1). Here, we review the lipid components that constitute such networks and how they might interact to regulate anandamide metabolism.

A. Anandamide Precursors

1. *N-Acyl-Phosphatidylethanolamine (NAPE)*

The most probable physiological route for the formation of anandamide requires the generation of NAPE. This lipid species is formed through the transfer of arachidonic acid from the *sn*-1 position of phospholipids (e.g., phosphatidylcholine, PC) to the primary amine group of phosphatidylethanolamine (PE, Fig. 1), which is catalyzed by an *N*-acyltransferase (NAT) (Astarita *et al.*, 2008; Cadas *et al.*, 1996a,b, 1997) (Fig. 1). In cultures of rat cortical neurons, two intracellular second messengers control NAT activity: Ca^{2+} and cyclic AMP (cAMP). Ca^{2+} is required to engage NAT activity, which is inactive in its absence, whereas cAMP works through protein kinase A-dependent phosphorylation to enhance NAT activity. Recently, a targeted lipidomic analysis has shown that Ca^{2+} specifically stimulates the formation of PUFA-containing NAPE species (Fig. 3). Such a preference might derive either from a substrate selectivity of NAT toward polyunsaturated PE species or from a localization of this enzyme in membrane domains (e.g., lipid rafts) enriched in PUFA-containing PE (Dainese *et al.*, 2007; Terova *et al.*, 2005). Furthermore, NAPE composition may affect the physical properties of membrane domains (Brites *et al.*, 2004) and the ratio of diacyl-NAPEs over alkenyl-NAPEs may influence signaling processes in the brain (Terova *et al.*, 2005). We recently reported that diacyl-NAPE species constitute almost the totality of NAPE species in intestinal mucosal cells, while intestinal serosa contains comparable amounts of alkenyl-NAPEs and diacyl-NAPE (Fu *et al.*, 2007), suggesting that NAPE composition may be linked to as-yet-uncharacterized biological functions. Considering that plasmalogens have a larger dipole moment and form hexagonal phases at lower temperatures than diacyl-PE analogs, it is likely that different composition in NAPE species could be involved in membrane fluidity and fusion processes.

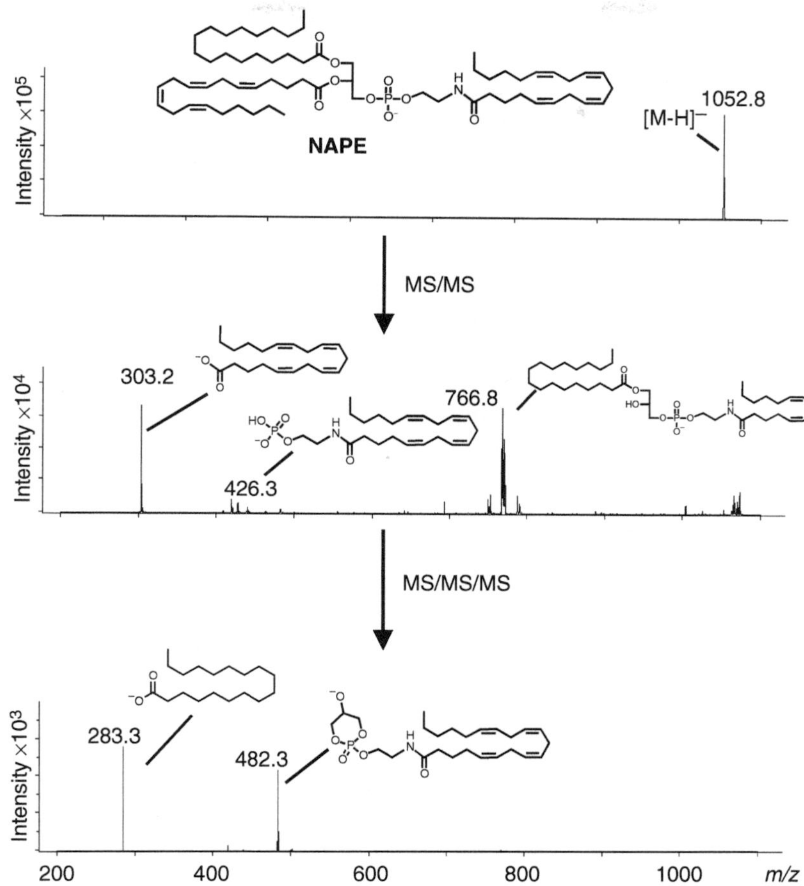

FIG. 3. Identification of anandamide precursors. Representative MS^n spectra of anandamide precursor, NAPE extracted from biological samples.

Besides serving as biological precursors for fatty acid ethanolamides (FAEs), NAPEs have been recently shown to play a physiological role on their own, reducing food intake and arousal after a fat-containing meal (Gillum *et al.*, 2008).

B. NAPE Derivatives

Alternative biosynthetic pathways can interconvert NAPE precursors into other anandamide-containing lipids whose biological activities remain unknown. We recently observed that the combination of soft ionization techniques such as

ESI and the gentle process of fragmentation produced by the ion trap mass spectrometer (MS^n) generates a series of fragment ions that may be found physiologically as neutral molecular species (Astarita and Piomelli, 2009). For example, fragments generated by the ESI/MS^n analysis of complex lipid species are also common products of enzymatic hydrolysis (e.g., lysophospholipids and fatty acids) or chemical rearrangement (e.g., products of intramolecular cyclization of phosphate groups) (Fig. 3) (Astarita *et al.*, 2008). This suggests that a gentle fragmentation approach can be used as a discovery tool in biochemistry for the study of catabolic products of known biomolecules and the identification of novel biologically relevant components of the endocannabinoid system. Our observations have been recently confirmed by *in vivo* studies showing that NAPE can generate alternative precursors for anandamide through the action of two different enzymes: (1) a NAPE-specific phospholipases A_1/A_2 (e.g., α/β hydrolase-4, ABHD-4), which forms the intermediates lyso-NAPE and glycerophospho-anandamide (Simon and Cravatt, 2006); and (2) a NAPE-specific phospholipase C (PLC), which forms the intermediate phospho-anandamide (Liu *et al.*, 2006, 2007). The physiopathological relevance of these alternative pathways for anandamide biosynthesis is the focus of current research.

C. Anandamide

Anandamide was discovered as a high-affinity ligand for CB receptors in 1992 (Devane *et al.*, 1992). Anandamide is produced in response to an increase in intracellular Ca^{2+} levels or activation of G-protein-coupled receptors (e.g., D_2 receptors) (Piomelli, 2003). In mammalian tissues, anandamide is present at concentrations of 1–50 pmol/g (Fig. 4), and can be formed through three distinct biochemical pathways: (1) the direct hydrolysis of NAPE by a NAPE-specific phospholipase D (NAPE-PLD) (Okamoto *et al.*, 2005; Wang *et al.*, 2006); (2) the hydrolysis of lyso-NAPE or glycerophospho-anandamide by a specific PLD (Leung *et al.*, 2006; Simon and Cravatt, 2008; Sun *et al.*, 2004); and (3) the hydrolysis of phospho-anandamide by a lipid phosphatase (Liu *et al.*, 2006, 2007) (Fig. 1).

D. Anandamide Derivatives

Anandamide can be transported inside neural cells (neurons and glia) by a carrier-mediated facilitate diffusion mechanism (Beltramo *et al.*, 1997) and transformed through two main pathways: (1) hydrolysis to arachidonic acid and ethanolamine, and (2) oxidation to various oxygenated derivatives.

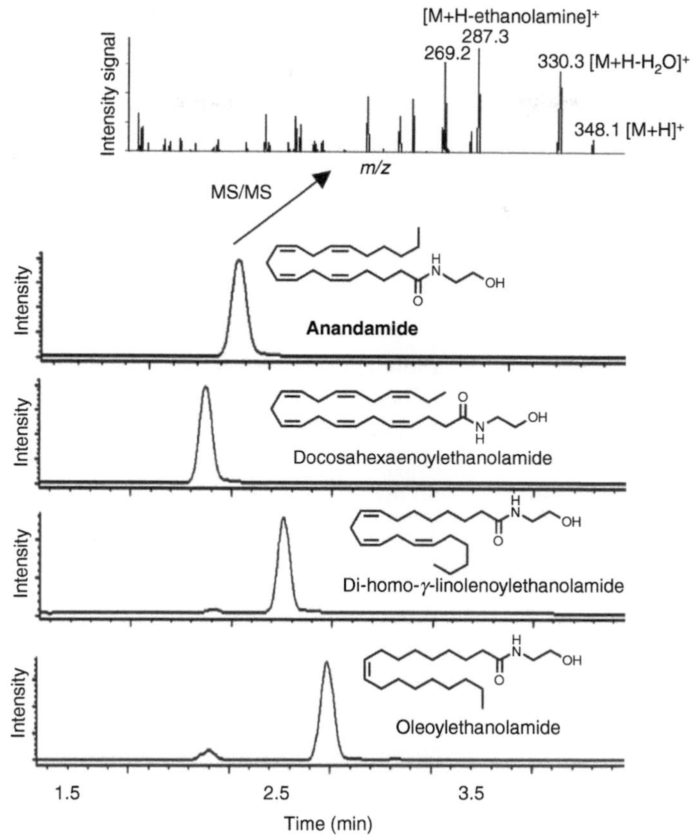

FIG. 4. Representative LC/MS^n chromatogram and MS/MS spectra of anandamide and its analogs extracted from human brain tissue.

1. *Arachidonic Acid*

Anandamide can be hydrolyzed to arachidonic acid and ethanolamine by fatty acid amide hydrolase (FAAH) (Cravatt *et al.*, 1996; Wei *et al.*, 2006) (Fig. 1). FAAH is highly expressed in the brain, where it is expressed at high concentrations in neuronal cell bodies and dendrites that are juxtaposed to axon terminals containing CB receptors. This suggests that anandamide hydrolysis occurs postsynaptically (Piomelli, 2003).

2. *Oxygenated Derivatives*

Anandamide can be metabolized by cyclooxygenase-2 (COX-2), 12- and 15-lipoxygenase (LOX), and cytochrome P450 (CYP450) to generate prostaglandin ethanolamides (prostamides), hydroperoxide, hydroxide, and epoxide metabolites (Fig. 1) (Fowler, 2007; Hampson *et al.*, 1995; Kozak and Marnett, 2002; Kozak *et al.*, 2001, 2004; Snider *et al.*, 2007, 2008; Ueda *et al.*, 1995). Most of these molecules have low affinity for CB receptors, though they display moderate activity as FAAH inhibitors, which may result in a potentiation of anandamide signaling. However, because of their extremely low concentration, the physiological significance of these oxygenated metabolites of anandamide remains to be further elucidated.

E. Analogs of Anandamide

Highly sensitive targeted lipidomic approaches are rapidly leading to the identification of new analogs of anandamide (Tan *et al.*, 2006). The two major families of lipids that share common chemical structure with anandamide are FAEs and fatty acid amides. Although many of these lipids show no activity at CB receptors, they are known to bind and activate other receptors, such as transient receptor potential vanilloid type-1 (TRPV-1) and the nuclear receptor peroxisome proliferator-activated (PPAR-α).

1. *Fatty Acid Ethanolamides*

Anandamide is a member of the FAE family (also called *N*-acylethanolamides, NAEs) (Fig. 4), which include both cannabimimetic and noncannabimimetic compounds present in most mammalian tissues. These molecules, which only differ in length and degree of unsaturation of their acyl chains, share common biosynthetic and degradative enzymes.

a. Cannabimimetic FAEs. Two polyunsaturated FAEs, docosatetraenoylethanolamide (K_i for CB_1 receptors = 34 nM) and dihomo-γ-linolenoylethanolamide (K_i for CB_1 = 53 nM) (Felder *et al.*, 1995; Hanus *et al.*, 1993), bind with high affinity to CB receptors and produce pharmacological effects similar to those of anandamide. Notably, these two lipids are present in human brains in levels comparable to those of anandamide (Fig. 4) (Astarita and Piomelli, 2009). However, their specific role and relative importance as cannabinergic messengers have not yet been determined.

b. Noncannabimimetic FAEs. In mammalian tissues, most of the endogenous FAEs are present at levels ranging from 1 to 10 times higher than anandamide, but they do not bind CB receptors (Farrell and Merkler, 2008). Yet, FAEs such as palmitoylethanolamide (C16:0) and oleoylethanolamide (C18:1) (Fig. 4), bind to PPAR-α, mediating important physiological processes such as analgesia, anti-inflammation, and anorexia (Fu *et al.*, 2003; LoVerme *et al.*, 2006; Rodríguez de Fonseca *et al.*, 2001). Stearoylethanolamide (C18:0) has also been shown to produce anti-inflammatory (Dalle Carbonare *et al.*, 2008) and anorexic effects (Terrazzino *et al.*, 2004). Docosahexaenoylethanolamide, whose structure closely resembles those of CB agonists, is present in brain tissue at levels comparable to anandamide but its biological role remains unknown (Astarita and Piomelli, 2009) (Fig. 4). Notably, although FAEs share similar metabolic pathways, they have distinct metabolic regulation that still are awaiting characterization (Astarita *et al.*, 2006, 2008; Fu *et al.*, 2007, 2008; Schwartz *et al.*, 2008).

2. *Fatty Acid Amides*

Fatty acyl amides analogs of anandamide comprise two classes of endogenous molecules: (1) fatty acid primary amides (e.g., oleamide) (Cravatt *et al.*, 1995; McDonald *et al.*, 2008), and (2) *N*-acyl amino acids (also called elmiric acids), which derive from the conjugation of fatty acids with amino acids or amino acid derivatives (e.g., *N*-acyl glycine, *N*-acyl taurine, *N*-acyl serotonins, and *N*-acyl dopamines) (Farrell and Merkler, 2008; Vuong *et al.*, 2008). With the exception of *N*-arachidonoyl dopamine, none of these lipids activate CB receptors and their physiological roles are not yet known. However, it is important to consider that many of these amides (e.g., oleamide) are commonly used in the production of plasticware and may thus leach in biological samples during sample preparation, producing false positives (McDonald *et al.*, 2008).

IV. Targeted Lipidomics of the 2-AG Pathway

The regulation of 2-AG levels is interconnected with lipid pathways distinct from those of anandamide. Here, we illustrate the metabolism of 2-AG and present unresolved questions that could be answered using lipidomic strategies.

A. 2-AG Precursors

As in the case of anandamide, 2-AG is derived from the hydrolysis of complex membrane lipids. The direct precursors of 2-AG can either be 1-acyl, 2-arachidonoyl-diacylglycerols (DAGs) or 2-arachidonoyl-lysoglycerophospholipids (e.g., lysophosphatidic acid, LPA).

1. *DAG*

DAG species are derived from three main routes: (1) PLC-mediated hydrolysis of phospholipids; (2) phosphatase-mediated hydrolysis of phosphatidic acid (PA); and (3) lipase-mediated hydrolysis of triacylglycerol (TAG) species (Fig. 2). Targeted lipidomic analyses show that the fatty acid compositions of the DAGs formed by these various routes reflect the composition of the parent lipid (Fig. 5). In particular, those derived from inositol phospholipids are highly enriched in

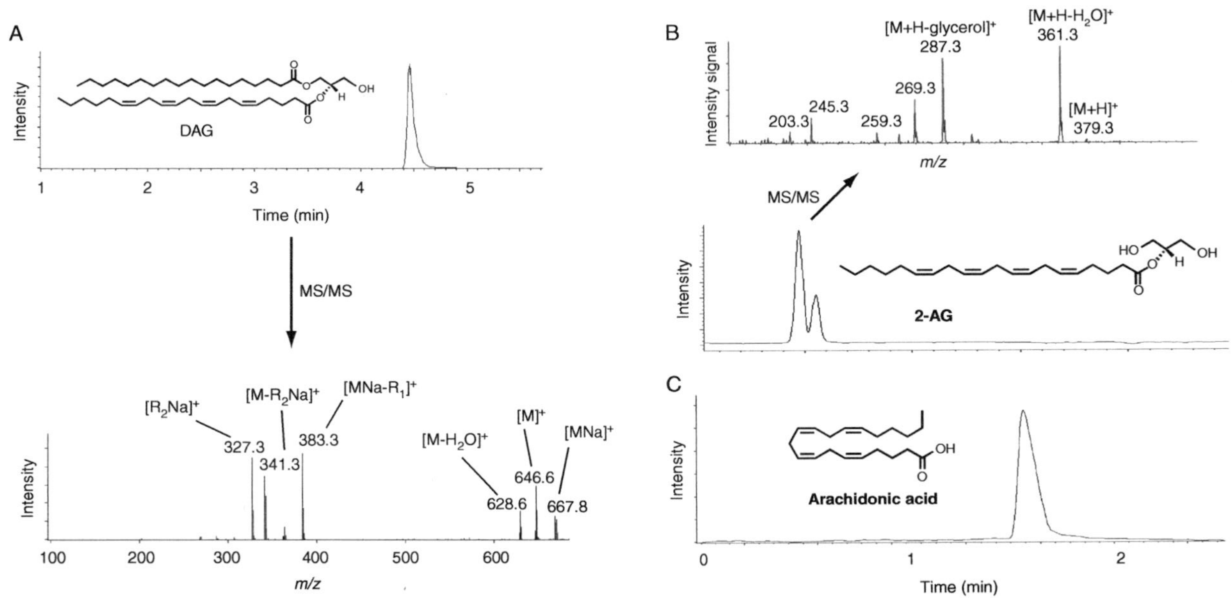

FIG. 5. (A) Representative LC/MS^n chromatogram and MS/MS spectra of 2-AG precursor DAG (1-stearoyl,2-arachidonoyl-*sn*-glycerol). (B, C) Representative LC/MS^n chromatogram of 2-AG and arachidonic acid from human brain tissue.

molecular species containing arachidonic acid in position *sn*-2 (Jung *et al.*, 2007). However, PC and PA may also serve as precursors for 1-acyl, 2-arachidonoyl-DAG (Bisogno *et al.*, 1999; Nakane *et al.*, 2002; Oka *et al.*, 2005). Furthermore, our recent lipidomic analysis revealed that several arachidonoyl-containing TAG species occur in human brain and may contribute to 2-AG formation (Astarita and Piomelli, 2009) (Fig. 2). The existence of a stimulus-induced formation of bioactive DAG species deriving from TAG is under investigation. Finally, targeted lipidomic analyses suggest the cellular existence of preformed DAG pools in neuronal cells that could directly be hydrolyzed to 2-AG in response to stimuli (Jung *et al.*, 2007).

2. *Lysophospholipids*

Arachidonoyl-bound lysophospholipids such as LPA may contribute to 2-AG formation (Sugiura *et al.*, 2006). LPA can derive either from PLA_1-mediated hydrolysis of PA or by lyso-PLD-mediated hydrolysis of lysophospholipids (Fig. 2).

B. 2-AG

In 1995, 2-AG was discovered to activate CB receptors with an affinity 20-fold weaker than anandamide (Sugiura *et al.*, 1995) (Fig. 5). However, 2-AG concentrations in brain tissue range from 0.5 to 20 nmol/g of tissue, which is about 100-fold greater than anandamide levels. A significant fraction of brain 2-AG might be engaged in housekeeping functions rather than in signaling, and the extracellular concentrations of 2-AG and anandamide are nearly equivalent (Ahn *et al.*, 2008). 2-AG biosynthesis inside cell membranes is driven in response to a rise in intracellular Ca^{2+} levels and/or the activation of $G_{q/11}$-coupled receptors (Di Marzo, 2008). Three main biosynthetic pathways are ascribed to 2-AG formation: (1) diacylglycerol lipase (DGL)-mediated hydrolysis of 1-acyl, 2-arachidonoyl-DAGs (Fig. 2) (Bisogno *et al.*, 2003); (2) lyso-PLC-mediated hydrolysis of lyso-phospholipids such as 2-arachidonoyl-lyso-phosphocholine (Sugiura *et al.*, 2006); and (3) phosphatase-mediated hydrolysis of LPA (Nakane *et al.*, 2002).

Using a combination of genetic and targeted lipidomic approaches we recently determined the role of DGL in metabotropic glutamate receptor-dependent 2-AG mobilization (Jung *et al.*, 2007). However, the relative importance of other biosynthetic routes for the biosynthesis of 2-AG is still under investigation.

1. *Topological Distribution of 2-AG in Brain Microstructures*

Most of the analytical approaches do not take into account the topological distribution of 2-AG in tissues, often missing out on critical information about the endocannabinoid signaling. For example, a physiological analysis of the

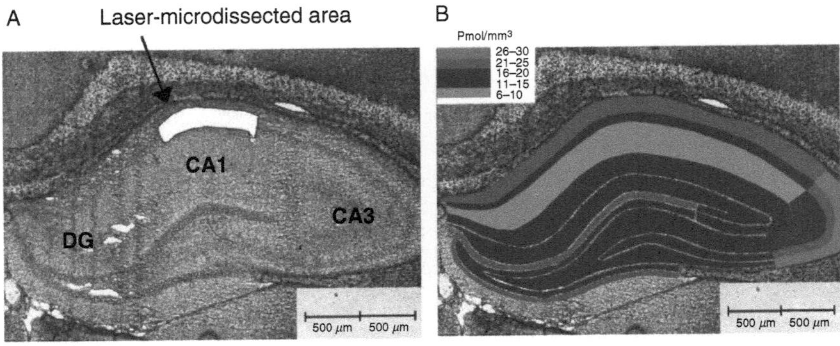

FIG. 6. Topological analysis of 2-AG distribution in different layers of rat hippocampus. Selected microstructure from hippocampus was laser-cut under the microscope (A) and analyzed using a state-of-the-art microflow LC/MS system. A heat map was generated to show topological differences in 2-AG levels (B).

hippocampus shows that neural information flows through a series of subregions—from the dentate gyrus to CA3 and finally to CA1. Lipid alterations in each of these microstructures may contribute to hippocampal function in health and disease (e.g., Alzheimer's disease). Recently, high-spatial resolution and high-sensitive analytical techniques, such as laser microdissection in combination with microflow LC/MS, have been used by our laboratory to determine the topological distribution of 2-AG in microstructures from rat hippocampus (Fig. 6).

C. 2-AG Derivatives

2-AG is internalized inside the cells through either passive diffusion or carrier-mediated transport. Once inside the cell, 2-AG can be transformed through three main mechanisms: (1) hydrolysis to arachidonic acid and glycerol; (2) oxidation to a series of oxygenated derivatives; and (3) anabolic metabolism (Fig. 2).

1. *Arachidonic Acid*

2-AG is hydrolyzed by a series of serine hydrolases to generate free arachidonic acid and glycerol. Between 50% and 85% of 2-AG-hydrolyzing activity can be ascribed to monoacylglycerol lipase (MGL) (Dinh *et al.*, 2002, 2004). The remaining is likely due to the activities of α/β hydrolase-6 (ABHD-6) and α/β hydrolase-12 (ABHD-12) (Blankman *et al.*, 2007). Less than 1% of 2-AG is hydrolyzed by FAAH, neuropathy-target esterase and hormone-sensitive lipase (Blankman *et al.*, 2007). Notably, because 2-AG levels are on the order of nanomoles per gram of wet tissue, the hydrolysis of this compound can provide a significant source of arachidonic acid and eicosanoids inside the cells (Jung *et al.*, 2007; Nomura *et al.*, 2008) (Fig. 5). We recently showed that upregulation of the biosynthetic pathway for 2-AG also leads to an increase in the levels of arachidonic acid in neural cells (Jung *et al.*, 2007).

2. *Oxygenated Derivatives*

As in the case of anandamide, 2-AG can also be metabolized by oxidative enzymes as effectively as arachidonic acid. 12- and 15-LOX-, COX-2-, and CYP450-mediated 2-AG oxygenation provides hydroxyperoxy, hydroxy, prostaglandin, and epoxide metabolites of the glycerol ester (Kozak *et al.*, 2002a,b, 2004; Moody *et al.*, 2001). Because of the higher endogenous concentration of 2-AG compared to anandamide, its oxidative metabolism is probably physiologically relevant. Oxygenated metabolites of 2-AG may represent a novel class of signal mediators with distinct biological activities (Kozak and Marnett, 2002; Woodward *et al.*, 2008) and could contribute to the formation of eicosanoids (i.e., prostaglandins, hydroxy and epoxide derivatives of arachidonic acid) (Kozak and Marnett, 2002). Notably, 2-(14,15-epoxyeicosatrienoyl)-glycerol, which is a 2-AG metabolite produced through the action of CYP450, activates CB receptors with high affinity and exerts mitogenic activity (Chen *et al.*, 2008).

3. *Anabolic Metabolism*

2-AG can be converted back into complex lipid molecules by anabolic enzymes. This recycling process is mediated by MAG kinases (MAGK) and acyltransferases (AT) to generate 2-arachidonoyl-lysophosphatidic acid and DAG, respectively. In turn, these lipids can be then converted into glycerophospholipids or TAG (Fig. 2).

D. Analogs of 2-AG

1. *Esters and Ether of Glycerol*

2-AG analogs include both cannabimimetic and noncannabimimetic molecules.

a. Cannabimimetic Analogs. Noladin ether is a putative endocannabinoid discovered in 2001 that binds CB receptors with high affinity (K_i = 20–450 nM) (Hanus *et al.*, 2001) and is metabolically more stable than 2-AG. The biosynthetic pathway leading to the formation of this unusual lipid is still unknown and its existence in mammalian tissues is still controversial (Fezza *et al.*, 2002; Oka *et al.*, 2003; Richardson *et al.*, 2007; Thomas *et al.*, 2009).

b. Noncannabimimetic Analogs. Targeted lipidomic analysis showed that the biosynthesis of 2-AG is accompanied by the formation of various 2-acylglycerol esters such as 2-oleoyl-*sn*-glycerol and 2-linoleyl-*sn*-glycerol (Jung *et al.*, 2007). Although these molecules show no activity at CB receptors, they are known to potentiate the binding of 2-AG to the CB receptors, probably through inhibition of lipase-mediated inactivation (Ben-Shabat *et al.*, 1998).

V. Conclusions

Targeted lipidomics is leading to a better understanding of endocannabinoid metabolism. Recent innovations in LC/MS analyses have led to the rapid identification and quantification of numerous lipids, and these data are creating new opportunities to enhance our knowledge of endocannabinoid function in the context of lipid metabolism.

Acknowledgments

The contribution of the Agilent Technologies/University of California Irvine Analytical Discovery Facility, Center for Drug Discovery, the Agilent Technologies Foundation, Advanced Chemistry Development, Inc., the Institute for Brain Aging & Dementia, and the UC Irvine Alzheimer's Disease Research Center is gratefully acknowledged. This work was supported by grants from the National Institute of Health (R21DA-022702, R01DK-073955, R01DA-012413, R01DA-012447, RR274-297/3504008, RR274-305/3505998, 1RL1AA017538 to D.P. and UCI ADRC P50 AG16573).

References

Ahn, K., McKinney, M., and Cravatt, B. (2008). Enzymatic pathways that regulate endocannabinoid signaling in the nervous system. *Chem. Rev.* **108,** 1687–1707.

Astarita, G., and Piomelli, D. (2009). Lipidomic analysis of endocannabinoid metabolism in biological samples. *J. Chromatogr. B Analyt. Technol. Biomed. Life Sci.* [Epub ahead of print].

Astarita, G., Rourke, B. C., Andersen, J. B., Fu, J., Kim, J. H., Bennett, A. F., Hicks, J. W., and Piomelli, D. (2006). Postprandial increase of oleoylethanolamide mobilization in small intestine of the Burmese python (Python molurus). *Am. J. Physiol. Regul. Integr. Comp. Physiol.* **290,** R1407–R1412.

Astarita, G., Ahmed, F., and Piomelli, D. (2008). Identification of biosynthetic precursors for the endocannabinoid anandamide in the rat brain. *J. Lipid Res.* **49,** 48–57.

Beltramo, M., Stella, N., Calignano, A., Lin, S. Y., Makriyannis, A., and Piomelli, D. (1997). Functional role of high-affinity anandamide transport, as revealed by selective inhibition. *Science* **277,** 1094–1097.

Ben-Shabat, S., Fride, E., Sheskin, T., Tamiri, T., Rhee, M. H., Vogel, Z., Bisogno, T., De Petrocellis, L., Di Marzo, V., and Mechoulam, R. (1998). An entourage effect: Inactive endogenous fatty acid glycerol esters enhance 2-arachidonoyl-glycerol cannabinoid activity. *Eur. J. Pharmacol.* **353,** 23–31.

Bisogno, T., Melck, D., De Petrocellis, L., and Di Marzo, V. (1999). Phosphatidic acid as the biosynthetic precursor of the endocannabinoid 2-arachidonoylglycerol in intact mouse neuroblastoma cells stimulated with ionomycin. *J. Neurochem.* **72,** 2113–2119.

Bisogno, T., Howell, F., Williams, G., Minassi, A., Cascio, M. G., Ligresti, A., Matias, I., Schiano-Moriello, A., Paul, P., Williams, E. J., Gangadharan, U., Hobbs, C., *et al.* (2003). Cloning of the first sn1-DAG lipases points to the spatial and temporal regulation of endocannabinoid signaling in the brain. *J. Cell Biol.* **163,** 463–468.

Blankman, J. L., Simon, G. M., and Cravatt, B. F. (2007). A comprehensive profile of brain enzymes that hydrolyze the endocannabinoid 2-arachidonoylglycerol. *Chem. Biol.* **14,** 1347–1356.

Brites, P., Waterham, H. R., and Wanders, R. J. (2004). Functions and biosynthesis of plasmalogens in health and disease. *Biochim. Biophys. Acta* **1636,** 219–231.

Cadas, H., Gaillet, S., Beltramo, M., Venance, L., and Piomelli, D. (1996a). Biosynthesis of an endogenous cannabinoid precursor in neurons and its control by calcium and cAMP. *J. Neurosci.* **16,** 3934–3942.

Cadas, H., Schinelli, S., and Piomelli, D. (1996b). Membrane localization of *N*-acylphosphatidylethanolamine in central neurons: Studies with exogenous phospholipases. *J. Lipid Mediat. Cell Signal* **14,** 63–70.

Cadas, H., di Tomaso, E., and Piomelli, D. (1997). Occurrence and biosynthesis of endogenous cannabinoid precursor, *N*-arachidonoyl phosphatidylethanolamine, in rat brain. *J. Neurosci.* **17,** 1226–1242.

Chapman, K. D. (2000). Emerging physiological roles for *N*-acylphosphatidylethanolamine metabolism in plants: Signal transduction and membrane protection. *Chem. Phys. Lipids* **108,** 221–229.

Chen, J. K., Chen, J., Imig, J. D., Wei, S., Hachey, D. L., Guthi, J. S., Falck, J. R., Capdevila, J. H., and Harris, R. C. (2008). Identification of novel endogenous cytochrome p450 arachidonate metabolites with high affinity for cannabinoid receptors. *J. Biol. Chem.* **283,** 24514–24524.

Cravatt, B. F., Próspero-García, O., Siuzdak, G., Gilula, N. B., Heriksen, S. J., Boger, D. L., and Lerner, R. A. (1995). Chemical characterization of a family brain lipids that induce sleep. *Science* **268,** 1506–1509.

Cravatt, B. F., Giang, D. K., Mayfield, S. P., Boger, D. L., Lerner, R. A., and Gilula, N. B. (1996). Molecular characterization of an enzyme that degrades neuromodulatory fatty-acid amides. *Nature* **384,** 83–87.

Dainese, E., Oddi, S., Bari, M., and Maccarrone, M. (2007). Modulation of the endocannabinoid system by lipid rafts. *Curr. Med. Chem.* **14,** 2702–2715.

Dalle Carbonare, M., Del Giudice, E., Stecca, A., Colavito, D., Fabris, M., D'Arrigo, A., Bernardini, D., Dam, M., and Leon, A. (2008). A saturated *N*-acylethanolamine other than *N*-palmitoyl ethanolamine with anti-inflammatory properties: A neglected story.... *J. Neuroendocrinol.* **20,** 26–34.

Devane, W. A., Hanus, L., Breuer, A., Pertwee, R. G., Stevenson, L. A., Griffin, G., Gibson, D., Mandelbaum, A., Etinger, A., and Mechoulam, R. (1992). Isolation and structure of a brain constituent that binds to the cannabinoid receptor. *Science* **258,** 1946–1949.

Di Marzo, V. (2008). Targeting the endocannabinoid system: To enhance or reduce? *Nat. Rev. Drug Discov.* **7,** 438–455.

Dinh, T. P., Carpenter, D., Leslie, F. M., Freund, T. F., Katona, I., Sensi, S. L., Kathuria, S., and Piomelli, D. (2002). Brain monoglyceride lipase participating in endocannabinoid inactivation. *Proc. Natl. Acad. Sci. USA* **99,** 10819–10824.

Dinh, T. P., Kathuria, S., and Piomelli, D. (2004). RNA interference suggests a primary role for monoacylglycerol lipase in the degradation of the endocannabinoid 2-arachidonoylglycerol. *Mol. Pharmacol.* **66,** 1260–1264.

Farrell, E. K., and Merkler, D. J. (2008). Biosynthesis, degradation and pharmacological importance of the fatty acid amides. *Drug Discov. Today* **13,** 558–568.

Felder, C. C., Joyce, K. E., Briley, E. M., Mansouri, J., Mackie, K., Blond, O., Lai, Y., Ma, A. L., and Mitchell, R. L. (1995). Comparison of the pharmacology and signal transduction of the human cannabinoid CB1 and CB2 receptors. *Mol. Pharmacol.* **48,** 443–450.

Fezza, F., Bisogno, T., Minassi, A., Appendino, G., Mechoulam, R., and Di Marzo, V. (2002). Noladin ether, a putative novel endocannabinoid: Inactivation mechanisms and a sensitive method for its quantification in rat tissues. *FEBS Lett.* **513,** 294–298.

Fowler, C. J. (2007). The contribution of cyclooxygenase-2 to endocannabinoid metabolism and action. *Br. J. Pharmacol.* **152,** 594–601.

Fu, J., Gaetani, S., Oveisi, F., Lo Verme, J., Serrano, A., Rodriguez de Fonseca, F., Rosengarth, A., Luecke, H., Di Giacomo, B., Tarzia, G., and Piomelli, D. (2003). Oleylethanolamide regulates feeding and body weight through activation of the nuclear receptor PPAR-alpha. *Nature* **425,** 90–93.

Fu, J., Astarita, G., Gaetani, S., Kim, J., Cravatt, B. F., Mackie, K., and Piomelli, D. (2007). Food intake regulates oleoylethanolamide formation and degradation in the proximal small intestine. *J. Biol. Chem.* **282,** 1518–1528.

Fu, J., Kim, J., Oveisi, F., Astarita, G., and Piomelli, D. (2008). Targeted enhancement of oleoylethanolamide production in proximal small intestine induces across-meal satiety in rats. *Am. J. Physiol. Regul. Integr. Comp. Physiol.* **295,** R45–R50.

Gillum, M., Zhang, D., Zhang, X., Erion, D., Jamison, R., Choi, C., Dong, J., Shanabrough, M., Duenas, H., and Frederick, D. (2008). *N*-Acylphosphatidylethanolamine, a gut-derived circulating factor induced by fat ingestion, inhibits food intake. *Cell* **135,** 813–824.

Hampson, A. J., Hill, W. A. G., Zan-Phillips, M., Makriyannis, A., Leung, E., Eglen, R. M., and Bornheim, L. M. (1995). Anandamide hydroxylation by brain lipoxygenase: Metabolite structures and potencies at the cannabinoid receptor. *Biochim. Biophys. Acta* **1259,** 173–179.

Hanus, L., Gopher, A., Almog, S., and Mechoulam, R. (1993). Two new unsaturated fatty acid ethanolamides in brain that bind to the cannabinoid receptor. *J. Med. Chem.* **36,** 3032–3034.

Hanus, L., Abu-Lafi, S., Fride, E., Breuer, A., Vogel, Z., Shalev, D. E., Kustanovich, I., and Mechoulam, R. (2001). 2-Arachidonyl glyceryl ether, an endogenous agonist of the cannabinoid CB_1 receptor. *Proc. Natl. Acad. Sci. USA* **98,** 3662–3665.

Jung, K. M., Astarita, G., Zhu, C., Wallace, M., Mackie, K., and Piomelli, D. (2007). A key role for diacylglycerol lipase-alpha in metabotropic glutamate receptor-dependent endocannabinoid mobilization. *Mol. Pharmacol.* **72,** 612–621.

Katona, I., and Freund, T. F. (2008). Endocannabinoid signaling as a synaptic circuit breaker in neurological disease. *Nat. Med.* **14,** 923–930.

Kozak, K. R., and Marnett, L. J. (2002). Oxidative metabolism of endocannabinoids. *Prostaglandins Leukot. Essent. Fatty Acids* **66,** 211–220.

Kozak, K. R., Crews, B. C., Ray, J. L., Tai, H. H., Morrow, J. D., and Marnett, L. J. (2001). Metabolism of prostaglandin glycerol esters and prostaglandin ethanolamides *in vitro* and *in vivo*. *J. Biol. Chem.* **276,** 36993–36998.

Kozak, K. R., Crews, B. C., Morrow, J. D., Wang, L. H., Ma, Y. H., Weinander, R., Jakobsson, P. J., and Marnett, L. J. (2002a). Metabolism of the endocannabinoids, 2-arachidonylglycerol and anandamide, into prostaglandin, thromboxane, and prostacyclin glycerol esters and ethanolamides. *J. Biol. Chem.* **277,** 44877–44885.

Kozak, K. R., Gupta, R. A., Moody, J. S., Ji, C., Boeglin, W. E., DuBois, R. N., Brash, A. R., and Marnett, L. J. (2002b). 15-Lipoxygenase metabolism of 2-arachidonylglycerol: Generation of a PPARalpha agonist. *J. Biol. Chem.* **15,** 15.

Kozak, K. R., Prusakiewicz, J. J., and Marnett, L. J. (2004). Oxidative metabolism of endocannabinoids by COX-2. *Curr. Pharm. Des.* **10,** 659–667.

Lehtonen, M., Reisner, K., Auriola, S., Wong, G., and Callaway, J. (2008). Mass-spectrometric identification of anandamide and 2-arachidonoylglycerol in nematodes. *Chem. Biodivers.* **5,** 2431.

Leung, D., Saghatelian, A., Simon, G. M., and Cravatt, B. F. (2006). Inactivation of *N*-acyl phosphatidylethanolamine phospholipase D reveals multiple mechanisms for the biosynthesis of endocannabinoids. *Biochemistry* **45,** 4720–4726.

Liu, J., Wang, L., Harvey-White, J., Osei-Hyiaman, D., Razdan, R., Gong, Q., Chan, A. C., Zhou, Z., Huang, B. X., Kim, H. Y., and Kunos, G. (2006). A biosynthetic pathway for anandamide. *Proc. Natl. Acad. Sci. USA* **103,** 13345–13350.

Liu, J., Wang, L., Harvey-White, J., Huang, B. X., Kim, H. Y., Luquet, S., Palmiter, R. D., Krystal, G., Rai, R., Mahadevan, A., Razdan, R. K., and Kunos, G. (2007). Multiple pathways involved in the biosynthesis of anandamide. *Neuropharmacology* **54,** 1–7.

LoVerme, J., Russo, R., La Rana, G., Fu, J., Farthing, J., Mattace-Raso, G., Meli, R., Hohmann, A., Calignano, A., and Piomelli, D. (2006). Rapid broad-spectrum analgesia through activation of peroxisome proliferator-activated receptor-alpha. *J. Pharmacol. Exp. Ther.* **319,** 1051–1061.

McDonald, G. R., Hudson, A. L., Dunn, S. M., You, H., Baker, G. B., Whittal, R. M., Martin, J. W., Jha, A., Edmondson, D. E., and Holt, A. (2008). Bioactive contaminants leach from disposable laboratory plasticware. *Science* **322,** 917.

Moody, J. S., Kozak, K. R., Ji, C., and Marnett, L. J. (2001). Selective oxygenation of the endocannabinoid 2-arachidonylglycerol by leukocyte-type 12-lipoxygenase. *Biochemistry* **40,** 861–866.

Nakane, S., Oka, S., Arai, S., Waku, K., Ishima, Y., Tokumura, A., and Sugiura, T. (2002). 2-Arachidonoyl-sn-glycero-3-phosphate, an arachidonic acid-containing lysophosphatidic acid: Occurrence and rapid enzymatic conversion to 2-arachidonoyl-sn-glycerol, a cannabinoid receptor ligand, in rat brain. *Arch. Biochem. Biophys.* **402,** 51–58.

Nomura, D. K., Hudak, C. S. S., Ward, A. M., Burston, J. J., Issa, R. S., Fisher, K. J., Abood, M. E., Wiley, J. L., Lichtman, A. H., and Casida, J. E. (2008). Monoacylglycerol lipase regulates 2-arachidonoylglycerol action and arachidonic acid levels. *Bioorg. Med. Chem. Lett.* **18,** 5875–5878.

Oka, S., Tsuchie, A., Tokumura, A., Muramatsu, M., Suhara, Y., Takayama, H., Waku, K., and Sugiura, T. (2003). Ether-linked analogue of 2-arachidonoylglycerol (noladin ether) was not detected in the brains of various mammalian species. *J. Neurochem.* **85,** 1374–1381.

Oka, S., Yanagimoto, S., Ikeda, S., Gokoh, M., Kishimoto, S., Waku, K., Ishima, Y., and Sugiura, T. (2005). Evidence for the involvement of the cannabinoid CB2 receptor and its endogenous ligand 2-arachidonoylglycerol in 12-*O*-tetradecanoylphorbol-13-acetate-induced acute inflammation in mouse ear. *J. Biol. Chem.* **280,** 18488–18497.

Okamoto, Y., Morishita, J., Wang, J., Schmid, P. C., Krebsbach, R. J., Schmid, H. H., and Ueda, N. (2005). Mammalian cells stably overexpressing *N*-acylphosphatidylethanolamine-hydrolysing phospholipase D exhibit significantly decreased levels of *N*-acylphosphatidylethanolamines. *Biochem. J.* **389,** 241–247.

Piomelli, D. (2003). The molecular logic of endocannabinoid signalling. *Nat. Rev. Neurosci.* **4,** 873–884.

Piomelli, D., Astarita, G., and Rapaka, R. (2007). A neuroscientist's guide to lipidomics. *Nat. Rev. Neurosci.* **8,** 743–754.

Richardson, D., Ortori, C., Chapman, V., Kendall, D., and Barrett, D. (2007). Quantitative profiling of endocannabinoids and related compounds in rat brain using liquid chromatography–tandem electrospray ionization mass spectrometry. *Anal. Biochem.* **360,** 216–226.

Rodríguez de Fonseca, F., Navarro, M., Gómez, R., Escuredo, L., Nava, F., Fu, J., Murillo-Rodríguez, E., Giuffrida, A., LoVerme, J., Gaetani, S., Kathuria, S., Gall, C., *et al.* (2001). An anorexic lipid mediator regulated by feeding. *Nature* **414,** 209–212.

Saghatelian, A., Trauger, S. A., Want, E. J., Hawkins, E. G., Siuzdak, G., and Cravatt, B. F. (2004). Assignment of endogenous substrates to enzymes by global metabolite profiling. *Biochemistry* **43,** 14332–14339.

Schmid, H. H., Schmid, P. C., and Natarajan, V. (1990). *N*-Acylated glycerophospholipids and their derivatives. *Prog. Lipid Res.* **29,** 1–43.

Schwartz, G. J., Fu, J., Astarita, G., Li, X., Gaetani, S., Campolongo, P., Cuomo, V., and Piomelli, D. (2008). The lipid messenger OEA links dietary fat intake to satiety. *Cell Metab.* **8,** 281–288.

Simon, G. M., and Cravatt, B. F. (2006). Endocannabinoid biosynthesis proceeding through glycerophospho-*N*-acyl ethanolamine and a role for {alpha}/beta-hydrolase 4 in this pathway. *J. Biol. Chem.* **281,** 26465–26472.

Simon, G. M., and Cravatt, B. F. (2008). Anandamide biosynthesis catalyzed by the phosphodiesterase GDE1 and detection of glycerophospho-*N*-acyl ethanolamine precursors in mouse brain. *J. Biol. Chem.* **283,** 9341.

Snider, N. T., Kornilov, A. M., Kent, U. M., and Hollenberg, P. F. (2007). Anandamide metabolism by human liver and kidney microsomal cytochrome P450 enzymes to form hydroxyeicosatetraenoic and epoxyeicosatrienoic acid ethanolamides. *J. Pharmacol. Exp. Ther.* **321,** 590–597.

Snider, N. T., Sikora, M. J., Sridar, C., Feuerstein, T. J., Rae, J. M., and Hollenberg, P. F. (2008). The endocannabinoid anandamide is a substrate for the human polymorphic cytochrome P450 2D6. *J. Pharmacol. Exp. Ther.* **327,** 538–545.

Spener, F., Lagarde, M., Geloen, A., and Record, M. (2003). What is lipidomics? *Eur. J. Lipid Sci. Technol.* **105,** 481–482.

Sugiura, T., Kondo, S., Sukagawa, A., Nakane, S., Shinoda, A., Itoh, K., Yamashita, A., and Waku, K. (1995). 2-Arachidonoylglycerol: A possible endogenous cannabinoid receptor ligand in brain. *Biochem. Biophys. Res. Commun.* **215,** 89–97.

Sugiura, T., Kishimoto, S., Oka, S., and Gokoh, M. (2006). Biochemistry, pharmacology and physiology of 2-arachidonoylglycerol, an endogenous cannabinoid receptor ligand. *Prog. Lipid Res.* **45,** 405–446.

Sun, Y. X., Tsuboi, K., Okamoto, Y., Tonai, T., Murakami, M., Kudo, I., and Ueda, N. (2004). Biosynthesis of anandamide and *N*-palmitoylethanolamine by sequential actions of phospholipase A2 and lysophospholipase D. *Biochem. J.* **380,** 749–756.

Tan, B., Bradshaw, H. B., Rimmerman, N., Srinivasan, H., Yu, Y. W., Krey, J. F., Monn, M. F., Chen, J. S., Hu, S. S., Pickens, S. R., and Walker, J. M. (2006). Targeted lipidomics: Discovery of new fatty acyl amides. *AAPS J.* **8,** E461–E465.

Terova, B., Petersen, G., Hansen, H. S., and Slotte, J. P. (2005). *N*-Acyl phosphatidylethanolamines affect the lateral distribution of cholesterol in membranes. *Biochim. Biophys. Acta* **1715,** 49–56.

Terrazzino, S., Berto, F., Carbonare, M. D., Fabris, M., Guiotto, A., Bernardini, D., and Leon, A. (2004). Stearoylethanolamide exerts anorexic effects in mice via down-regulation of liver stearoyl-coenzyme A desaturase-1 mrna expression. *FASEB J.* **18,** 1580–1582.

Thomas, A., Hopfgartner, G., Giroud, C., and Staub, C. (2009). Quantitative and qualitative profiling of endocannabinoids in human plasma using a triple quadrupole linear ion trap mass spectrometer with liquid chromatography. *Rapid Commun. Mass Spectrom.* **23,** 629–638.

Ueda, N., Yamamoto, K., Yamamoto, S., Tokunaga, T., Shirakawa, E., Shinkai, H., Ogawa, M., Sato, T., Kudo, I., Inoue, K., Takizawa, H., and Nagano, T. (1995). Lipoxygenase-catalyzed oxygenation of arachidonylethanolamide, a cannabinoid receptor agonist. *Biochim. Biophys. Acta* **1254,** 127–134.

Vuong, L. A. Q., Mitchell, V. A., and Vaughan, C. W. (2008). Actions of *N*-arachidonyl-glycine in a rat neuropathic pain model. *Neuropharmacology* **54,** 189–193.

Wang, J., Okamoto, Y., Morishita, J., Tsuboi, K., Miyatake, A., and Ueda, N. (2006). Functional analysis of the purified anandamide-generating phospholipase D as a member of the metallo-beta-lactamase family. *J. Biol. Chem.* **281,** 12325–12335.

Wei, B. Q., Mikkelsen, T. S., McKinney, M. K., Lander, E. S., and Cravatt, B. F. (2006). A second fatty acid amide hydrolase with variable distribution among placental mammals. *J. Biol. Chem.* **281,** 36569–36578.

Woodward, D. F., Carling, R. W. C., Cornell, C. L., Fliri, H. G., Martos, J. L., Pettit, S. N., Liang, Y., and Wang, J. W. (2008). The pharmacology and therapeutic relevance of endocannabinoid derived cyclo-oxygenase (COX)-2 products. *Pharmacol. Ther.* **120,** 71–80.

THE ENDOCANNABINOID SYSTEM AS A TARGET FOR NOVEL ANXIOLYTIC AND ANTIDEPRESSANT DRUGS

Silvana Gaetani,* Pasqua Dipasquale,* Adele Romano,* Laura Righetti,* Tommaso Cassano,† Daniele Piomelli,‡ and Vincenzo Cuomo*

*Department of Physiology and Pharmacology, Sapienza University of Rome, P.le Aldo Moro 5, 00185 Rome, Italy

†Department of Biomedical Sciences, Medical School, University of Foggia, V.le Luigi Pinto 1, 71100 Foggia, Italy

‡Department of Pharmacology, University of California, 3101 Gillespie NRF Irvine, Irvine, California 92697–4625, USA

Observational studies in humans suggest that exposure to marijuana and other cannabis-derived drugs produces a wide range of subjective effects on mood tone and emotionality. These observations have their counterpart in animal studies, showing that cannabinoid agonists strongly affect emotional reactivity in directions that vary depending on dose and context. Based on these evidence, the activation of central CB_1 receptor has emerged as potential target for the development of antianxiety and antidepressant therapies. However, the variable effects of exogenous cannabinoid agonists have gradually shifted the interest to the alternative approach of amplifying the effects of endogenous cannabinoids (EC), namely anandamide (AEA) and 2-arachidonoylglycerol (2-AG), by preventing their deactivation. The enzyme fatty acid amide hydrolase (FAAH) has been the target of intense research efforts aimed at developing potent and selective inhibitors that might prolong AEA actions *in vivo*. Among the inhibitors developed, the compound URB597 was found to potently inhibit FAAH activity *in vivo* and cause brain AEA levels to increase. Interestingly, the enhanced AEA tone produced by URB597 does not result in the behavioral effects typical of a direct-acting cannabinoid agonist. Though URB597 does not elicit a full-fledged cannabinoid profile of behavioral responses, it does elicit marked anxiolytic-like and

INTERNATIONAL REVIEW OF
NEUROBIOLOGY, VOL. 85
DOI: 10.1016/S0074-7742(09)85005-8

antidepressant-like effects in rats and mice. Such effects involve the downstream activation of CB_1 receptors, since they are attenuated by the CB_1 antagonist SR141716 (rimonabant).

Parallel to FAAH inhibition, similar results can also be observed by pharmacologically blocking the AEA transport system, which is responsible of the intracellular uptake of AEA from the synaptic cleft.

The reason why FAAH inhibition approach produces a smaller set of cannabimimetic effects might depend on the mechanism of EC synthesis and release upon neuronal activation and on the target selectivity of the drug. The mechanism of EC release is commonly referred to as "on request", since they are not synthesized and stored in synaptic vesicles, such as classical neurotransmitters, but are synthesized from membrane precursors and immediately released in the synaptic cleft following neuronal activation. The neural stimulation in specific brain areas, for example, those involved in the regulation of mood tone and/or emotional reactivity, would result in an increased EC tone in these same areas, but not necessarily in others. Therefore, inhibition of AEA metabolism activity could amplify CB_1 activation mainly where AEA release is higher. Furthermore, the inhibition of FAAH causes an accumulation of AEA but not 2-AG, which, being 200-fold more abundant than AEA in the brain, might differently modulate CB_1-mediated behavioral responses.

The evidence outlined above supports the hypothesis that the EC system plays an important role in anxiety and mood disorders and suggests that modulation of FAAH activity might be a pharmacological target for novel anxiolytic and antidepressant therapies.

I. The Endogenous Cannabinoid System

Derivatives of *Cannabis sativa* have been used for thousands of years, but only in 1964 the isolation of the active ingredient delta-9-tetrahydrocannabinol (Gaoni and Mechoulam, 1964) triggered the discovery of the endogenous cannabinoid system, constituted by the cannabinoid receptors, CB_1 (the most abundant G-coupled receptors in the brain) and CB_2 (expressed mostly in immune cells) and their attending family of endogenous ligands, called endogenous cannabinoids (ECs) (for review, see Freund *et al.*, 2003; Piomelli, 2003).

Best studied, among these ligands, are the lipid derivatives anandamide (AEA) (Devane *et al.*, 1992) and 2-arachidonoylglycerol (2-AG) (Mechoulam *et al.*, 1995; Sugiura *et al.*, 1995), which differ from classical and peptide neurotransmitters in several ways. The first difference concerns their synthesis and release. In fact, the release of EC transmitters from neurons is not mediated by exocytosis of storage vesicles; rather the compounds are produced through cleavage of membrane lipid precursors and immediately released in response to depolarization (Di Marzo *et al.*, 1994; Stella *et al.*, 1997). AEA is produced in two concerted steps. The first consists in the enzymatic transfer of arachidonic acid from phosphatidylcholine to the primary amino group of phosphatidylethanolamine (PE). This reaction is catalyzed by an *N*-acyltransferase, which remains molecularly uncharacterized, and yields a diverse group of *N*-arachidonoyl-substituted PE species (NAPE) (Di Marzo, 2008; Piomelli, 2003). The second step consists in the cleavage of NAPE to produce AEA and can be mediated by three distinct enzyme pathways. One involves the phospholipase D (PLD)-mediated hydrolysis of NAPE to yield AEA, which is likely catalyzed by a specific NAPE-PLD (Okamoto *et al.*, 2005). A second pathway proceeds through the sequential deacylation of NAPE by α,β-hydrolase 4 (Abhd4) and the subsequent cleavage of glycerophosphate to yield AEA (Simon and Cravatt, 2006). Finally, phospholipase C-mediated cleavage of NAPE may yield phosphoanandamide, which may then be dephosphorylated by phosphatases such as the tyrosine phosphatase PTPN22 and inositol 5′ phosphatase SHIP1 (Liu *et al.*, 2006, 2008).

The best characterized route of 2-AG production involves a two-step process in which phospholipase C-β (PLC-β) hydrolyses phosphatidylinositol-4,5-bisphosphate (PIP_2) to generate 1,2-diacylglycerol (DAG), which is then hydrolyzed by diacylglycerol lipase (DGL) to yield 2-AG (Stella *et al.*, 1997). In brain slices and neurons in culture, 2-AG formation is stimulated by electrical activity (Stella *et al.*, 1997), membrane depolarizing agents (Stella and Piomelli, 2001), or activation of G-protein-coupled receptors such as group I metabotropic glutamate receptors (Jung *et al.*, 2007). Two DGL isoforms have been molecularly cloned, DGL-α and DGL-β (Bisogno *et al.*, 2003). Anatomical and biochemical evidence points to an important role of DGL-α in 2-AG-mediated signaling (Hashimotodani *et al.*, 2008; Jung *et al.*, 2007; Katona *et al.*, 2006).

EC actions are rapidly terminated through a two-step elimination process, transport into cells and intracellular hydrolysis. Intracellular uptake is mediated by a putative "EC transporter" not cloned yet, but characterized both biochemically and pharmacologically. Internalization of ECs in neural cells is a rapid, temperature sensitive, saturable process that is independent of EC hydrolysis and sodium gradients (Beltramo *et al.*, 1997; Hillard *et al.*, 1997; Kathuria *et al.*, 2003), but is susceptible to selective pharmacological inhibition by drugs such as AM404 and VDM11 (Fegley *et al.*, 2004; Piomelli *et al.*, 1999). Once inside the cell, AEA is cleaved by fatty acid amide hydrolase (FAAH), a membrane-bound intracellular serine hydrolase (Cravatt *et al.*, 1996; Schmid *et al.*, 1985), whereas 2-AG is mostly metabolized by monoglyceride lipase (MGL), a cytosolic serine hydrolase (Dinh *et al.*, 2002). FAAH is widely distributed in the rat brain, particularly in cell bodies juxtaposed to axon terminals that contain CB_1 receptors suggesting that FAAH activity has an important role in the postsynaptic inactivation of AEA. On the other hand, MGL, the expression of which partially overlaps with that of FAAH in the central nervous system, is found in presynaptic boutons (Dinh *et al.*, 2002). Genetic deletion of the *faah* gene or pharmacological inhibition of FAAH activity impairs AEA hydrolysis, resulting in elevated brain levels of this lipid mediator (Cravatt *et al.*, 2001; Fegley *et al.*, 2004). Potent and selective FAAH inhibitors include the compounds URB597 (Kathuria *et al.*, 2003; Mor *et al.*, 2004; Tarzia *et al.*, 2003) and PF-750 (Ahn *et al.*, 2007). Another AEA-hydrolyzing amidase, called *N*-acylethanolamide-hydrolyzing acid amidase (NAAA), has been identified in inflammatory cells (Tsuboi *et al.*, 2007).

Adenovirus-mediated overexpression of MGL enhances hydrolysis of endogenous 2-AG, while RNA interference-mediated silencing of MGL elevates 2-AG levels, suggesting that this enzyme is an important determinant of 2-AG levels in cells (Dinh *et al.*, 2002, 2004). Additionally, administration of the MGL inhibitor URB602 to organotypic brain slices or microinjection into selected brain regions raises the levels of 2-AG without affecting those of AEA (Hohmann *et al.*, 2005). Similarly, the novel selective and potent MGL inhibitor called JZL184 upon administration to mice, raises brain 2-AG by eightfold without altering AEA (Long *et al.*, 2008).

Recently, Stella and colleagues identified a novel 2-AG-hydrolyzing activity in the mouse microglial cell line, BV-2, which does not express MGL. This activity is pharmacologically distinct from that of cloned MGL, but remains to be molecularly characterized (Muccioli *et al.*, 2007). In a separate study, two additional 2-AG-hydrolyzing lipases, Abhd6 and Abhd12, were identified by functional proteomics (Blankman *et al.*, 2007). In agreement with previous work (Dinh *et al.*, 2004), these enzymes were shown to account for approximately 15% of the total 2-AG-hydrolyzing activity in the brain, while MGL accounted for the remaining 85% (Blankman *et al.*, 2007).

II. Endocannabinoid Role in Emotional Reactivity and Mood Tone

At the synapse, ECs are thought to act as retrograde messengers, inhibiting neurotransmitter release (Llano *et al.*, 1991; Pitler and Alger, 1992). This phenomenon was firstly observed at synapses between GABA interneurons and pyramidal cells in the CA1 field of the hippocampus and at parallel fiber–Purkinje cell synapses in the cerebellum. ECs, released upon postsynaptic depolarization, activate presynaptic CB_1 receptors. In hippocampal neurons such activation engages G-protein $\beta\gamma$-subunits, leading to closure of Ca^{2+} channels and inhibition of GABA release. In the cerebellar synapses, CB_1 activation involves

G-protein α-subunits, opening of K^+ channels, membrane hyperpolarization, and consequent fall of Ca^{2+} entry that, in turn, decreases glutamate release. Similar mechanisms underlie cannabinoid-mediated inhibition of neurotransmitter release in other brain regions, such as the striatum (Gerdeman and Lovinger, 2001; Huang *et al.*, 2001), the nucleus accumbens (Robbe *et al.*, 2001) and lateral amygdala (Azad *et al.*, 2003).

This peculiar cellular mechanism has a wide range of functional consequences, such as those on memory, cognition, pain perception, motor behavior, and emotional states. Several lines of evidence support the hypothesis of an important role of cannabinoid modulation in anxiety.

In fact, cannabinoids modulate the release of transmitters implicated in the control of anxious states. They suppress the outflow of glutamate in the hippocampus and periacqueductal gray; inhibit the corticolimbic release of noradrenaline, dopamine, serotonin and the anxiogenic neuropeptides, cholecystokinin and corticotropin releasing factor (CRF) (for review see Millan, 2003). On the other hand, by presynaptic mechanisms, they interfere with GABAergic transmission in the amygdala, hippocampus, frontal cortex, and other regions (Piomelli, 2003). The latter effect of GABAergic outflow may underlie their indirect disinhibition of cortical glutamatergic and dopaminergic transmission pathways in the frontal cortex and nucleus accumbens (Millan, 2003).

III. Effects of Exogenously Administered Cannabinoid Agonists and Antagonists

Probably as a consequence of this complex pattern of influence upon different neurotransmitter systems that divergently modulate emotional behavior and mood states, cannabinoid agonists can produce both anxiolytic and anxiogenic effects.

The main feature of the recreational use of cannabis is that it produces an euphoriant effect. This "high" can be accompanied by decreased anxiety and increased sociability. However, retrospective studies in cannabis users (Hall and Solowij, 1998; Tournier *et al.*, 2003) and small clinical trials (Wachtel *et al.*, 2002) showed that exposure to cannabis produces, in humans, a wide range of subjective emotional effects, as a function of personality and the precise circumstances of consumption. Such effects can range from relaxation and euphoria to feeling of anxiety, paranoia, acute panic attacks, and psychosis (Zuardi *et al.*, 1982). Typically higher incidence of adverse effects is reported after consumption of large doses of marijuana, or in naïve users or in the case of stressful or novel environmental conditions (Abel, 1997; Gregg *et al.*, 1976; Naliboff *et al.*, 1976). These clinical observations have their counterpart in animal studies, showing that cannabinoids elicit dose-dependent and environment-dependent anxiolytic and anxiogenic-like responses in rodent models (Onaivi *et al.*, 1995). Relatively low doses of cannabinoids produce anxiolytic-like effects in animals, whereas higher doses produce an anxiogenic profile (for review, see Viveros *et al.*, 2005), while preexposure to stressful environment makes animals more prone to the anxiogenic effects of cannabinoids.

Emotional distress—including anxiety, depressed mood, and irritability—is a pervasive feature of cannabis withdrawal in human subjects (Budney *et al.*, 2003, 2004; Vandrey *et al.*, 2005, 2008). Likewise, pharmacologically induced cannabinoid withdrawal in rats is associated with marked anxiety-related behavioral and physiological responses, as well as with activation of CRF-mediated transmission in the limbic system (Rodriguez de Fonseca *et al.*, 1997).

Moreover, genetic or pharmacological interruption of CB_1 receptor activity results in enhanced anxiety- and depressive-like states in rats and mice (Haller *et al.*, 2002, 2004a,b; Marsicano *et al.*, 2002; Martin *et al.*, 2002; Navarro *et al.*, 1997; Uriguen *et al.*, 2004). Severe adverse psychiatric events, mainly depression and anxiety, were observed in obese patients treated with SR141716 (rimonabant), a CB_1-receptor antagonist previously in the market as an antiobesity agent (Christensen *et al.*, 2007).

Furthermore, a single nucleotide polymorphism of the CB_1 receptor gene (rs1049353) was recently found to confer an increased risk of antidepressant treatment resistance in female patients with high comorbid anxiety (Domschke *et al.*, 2008).

IV. Enhancement of the Endogenous Cannabinoid Tone

The distribution of the CB_1 receptors in the brain and the molecular mechanism of EC signaling might explain in part this contradictory scenario. High densities of CB_1 are found in the basolateral amygdala, the anterior cingulate cortex, the prefrontal cortex, and the hippocampus, brain regions that serve different functions in the regulation of emotions (Cahill and McGaugh, 1998; McDonald and Mascagni, 2001). Moreover, in the forebrain, CB_1 is primarily localized to axon terminals of GABAergic interneurons (Freund *et al.*, 2003). In some of these areas, however, CB_1 (or a functionally related isoform) is also found on glutamatergic nerve terminals (Hájos *et al.*, 2001). As a result of such distribution, CB_1 activation by a cannabinoid agonist or CB_1 blockade by a cannabinoid antagonist might have dramatically different behavioral consequences, depending on the balance of its effects on GABAergic and glutamatergic transmission.

Although these evidences point at the CB_1 receptors as an attractive target for the development of antianxiety therapies, the variable effects of exogenous cannabinoid agonists have shifted the interest of pharmacologists to an alternative approach that could rather amplify the effects of ECs by preventing their inactivation (Gaetani *et al.*, 2003a). The EC transporter and the enzyme FAAH have been the targets of intense research efforts aimed at developing potent and selective inhibitors that might prolong AEA and 2-AG actions *in vivo*. These efforts produced the compound AM404, the first transporter inhibitor shown to inhibit AEA transport in rat neurons ($IC_{50} = 1$ M) and in rat astrocytes ($IC_{50} = 5$ M) and to amplify *in vivo* the analgesic effects of exogenous AEA (Beltramo *et al.*, 1997; Di Marzo *et al.*, 1994; Hillard *et al.*, 1995). Among FAAH inhibitors a new potent, selective, and systemically active inhibitor is URB597, which inhibits FAAH activity both *in vitro* (IC_{50} = 4.6 nM in rat brain membranes and IC_{50} = 0.5 nM in intact neurons) and *in vivo* (ID_{50} = 0.15 mg/kg, intraperitoneally, in the rat) (Kathuria *et al.*, 2003; Tarzia *et al.*, 2003).

It is now clear that pharmacological blockade of AEA deactivation elicits anxiolytic- and antidepressant-like effects in rodents. For example, administration of the FAAH inhibitor URB597 lowered isolation-induced ultrasonic vocalizations in rat pups and increased the time spent on the open arms of the elevated zero maze (Kathuria *et al.*, 2003), the elevated plus maze (Naidu *et al.*, 2007; Patel and Hillard, 2006), and the light–dark test (Moreira *et al.*, 2008) in rats, mice, and hamsters. Similarly, the AEA transport inhibitor AM404 reduced isolation-induced ultrasonic vocalizations in rat pups, and increased the time adult rats spent in the open arms of the elevated plus maze or in the open field during defensive withdrawal (Bortolato *et al.*, 2006). The anxiolytic-like effects of URB597 and AM404 were accompanied by elevations of brain AEA and were blocked by the CB_1 antagonist rimonabant, providing evidence that they were due to enhanced AEA activity at CB_1 receptors. Interestingly, several investigators found that URB597 produced linear, dose-dependent anxiolytic effects, while AM404 was found in some cases to produce, like direct-acting CB_1 agonists, a biphasic effect on anxiety, with low doses producing anxiolytic effects and the highest dose having no effect (Patel and Hillard, 2006). Similar profiles of the two drugs were observed with regard to inhibition of stress-induced corticosterone release (Patel *et al.*, 2004). AM404 is thought to increase both AEA and 2-AG tone at the CB_1 receptors, by inhibiting their intracellular uptake, whereas URB597 directly modulates AEA levels, without affecting MGL activity. URB597 not only does not induce 2-AG accumulation, but also it seems, at least in the striatum, that through the elevation of AEA tone it can reduce the levels, metabolism, and physiological effects of 2-AG. Based on these observations, we might speculate that selective augmentation of AEA signaling may be preferred as a novel approach to developing antianxiety pharmacotherapeutics (Patel and Hillard, 2006).

In addition to the anxiolytic-like actions, EC deactivation inhibitors may also produce antidepressant-like effects (Patel and Hillard, 2006). Thus, AM404 decreased immobility time in the rat forced-swim test (Hill *et al.*, 2005).

URB597 exerted a similar effect and also increased struggling behavior in the mouse tail-suspension test (Gobbi *et al.*, 2005). Notably, in the experiments of Gobbi *et al.* (2005), the enhancement of stress-coping behaviors caused by URB597 was accompanied by elevations of AEA levels in the prefrontal cortex, hippocampus, and midbrain—regions that are implicated in the regulation of mood and the processing of emotional information (Berton and Nestler, 2006). Moreover, the antidepressant-like activities of AM404 and URB597 were prevented by CB_1 receptor blockade, again suggesting that they were mediated by AEA acting at CB_1 (Gobbi *et al.*, 2005). Compelling evidence for an antidepressant-like action of the EC system was recently provided by a study, which reported that prolonged URB597 treatment (5 weeks) elevates AEA levels in selected regions of the brain (thalamus, midbrain, striatum) and reverses stress-induced reductions in sucrose consumption and body-weight gain in rats subjected to chronic mild stress (Bortolato *et al.*, 2007). Consistent with these data, experiments by Rademacher and Hillard have shown that URB597 attenuates the reductions in sucrose preference and consumption produced by restraint stress in mice (Rademacher and Hillard, 2007).

Although FAAH inhibition *in vivo* is accompanied by increased brain AEA levels, URB597 does not mimic the spectrum of pharmacological responses produced by typical CB_1 agonists, except for a modest analgesic effect. In fact, systemic doses of URB597 that maximally block FAAH activity *in vivo* were not found to produce catalepsy (rigid immobility), hypothermia, or hyperphagia, three typical effects produced by CB_1 activation (Kathuria *et al.*, 2003). The behavioral selectivity of URB597 correlates with its target specificity, as it was found not significantly interacting with other cannabinoid-related targets, including cannabinoid receptors and AEA transport, or with a broad panel of receptors, ion channels, transporters, and enzymes (Piomelli *et al.*, 2006).

Even more importantly, URB597 seems to lack hedonic properties, as it had no effect on two rat models of abuse potential, the conditioned place preference test and the drug discrimination test (Kathuria *et al.*, 2003). Similar results have been recently obtained also in experiments with monkeys (Justinova *et al.*, 2008).

The evidence reviewed above supports the idea that attenuating EC deactivation might represent a better pharmacological tool to manipulate selectively the neuromodulatory effects of CB_1 activation *in vivo*. The reason why this approach should produce a smaller set of side effects might depend on the mechanism of EC release and on the target selectivity of the FAAH inhibitor. An anxiety arousal relates to neuronal activity and, in turn, increased AEA release, in specific brain areas. In fact, increased levels of ECs were observed during aversive stimuli in key regions related to fear and anxiety responses, where this system may act on-demand to oppose the impact of stress (Marsicano *et al.*, 2002). Blocking FAAH will amplify CB_1 activation mainly where AEA release is occurring, explaining the restricted spectrum of action of URB597 (Gaetani *et al.*, 2003a). Conversely, the administration of a direct cannabinoid agonist will produce an activation of CB_1 receptors throughout the brain, regardless of the EC signaling. The selective effect of URB597 on AEA but not 2-AG levels might also play a role in its behavioral specificity. In fact, the administration of the MGL inhibitor JZL184 to mice induced a broad array of CB_1-dependent behavioral effects, including analgesia, hypothermia, and hypomotility, which is not observed after URB597 treatment, thus suggesting that 2-AG modulates, likely better than AEA, different behavioral processes classically associated with the pharmacology of cannabinoids (Long *et al.*, 2008).

V. Faah-Knockout Phenotype

Cravatt *et al.* (2001) generated mutant mice lacking the *faah* gene ($FAAH^{-/-}$ mice) and characterized by altered nociceptive threshold, enhanced memory extinction, and increased sensitivity to the effects of exogenously administered ECs (Cravatt *et al.*, 2001; Varvel *et al.*, 2007). Such behavioral phenotype is compatible with the higher AEA levels measured in these mice (Cravatt *et al.*, 2001).

In accordance with the observation reported from pharmacological studies, $FAAH^{-/-}$ mice exhibit reduced anxiety-like behavior in two experimental models, the elevated plus maze and the light–dark test (Moreira *et al.*, 2008). Their behavioral profile in both tests is reversed by systemic administration of rimonabant, thus suggesting that an enhancement of CB_1-receptor-mediated signaling, likely due to increased AEA tone, might be responsible for the decreased anxiety in these animals.

Discrepant observation on these mice has been reported by Naidu *et al.* (2007), who did not detect a phenotype of $FAAH^{-/-}$ mice in the elevated plus maze, with respect to wild-type controls. Differences in the genetic background of mice as well as in the animal housing and experimental context have been proposed as key factors determining such discrepancies (Moreira *et al.*, 2008).

VI. Conclusions

FAAH appears as a promising target for the development of novel antianxiety and antidepressant therapies. However, several issues need to be considered.

The first one concerns the role of other endogenous compound, such as other *N*-acylethanolamides, which are also cleaved by this enzyme and can, therefore, accumulate following its pharmacological inhibition. Important biological roles have been described for some of these compounds, such as palmitoylethanolamide (LoVerme *et al.*, 2005, 2006), which controls pain threshold, and oleoylethanolamide, which acts as an endogenous regulator of satiety (Fu *et al.*, 2003; Gaetani *et al.*, 2003b; Rodriguez de Fonseca *et al.*, 2001; Schwartz *et al.*, 2008). Moreover, not only AEA, but also oleoylethanolamide tone have been found altered in the blood and in the cerebrospinal fluid of patients suffering from different psychiatric disorders, such as schizophrenia and eating disorders (Gaetani *et al.*, 2008; Giuffrida *et al.*, 2004; Leweke *et al.*, 2007), thus suggesting that other effects or other therapeutic applications might arise with a chronic inhibition of FAAH.

The second concern is related to the compensatory reduction of 2-AG signaling following AEA chronic accumulation. If such reduction might contribute, on one hand, to the behavioral selectivity of URB597 or similarly acting novel FAAH inhibitors, on the other hand, it might limit its potential clinical efficacy, since alterations in 2-AG levels are observed during stress-related states in both rodents and humans (Hill *et al.*, 2005; Hungund *et al.*, 2004).

Finally, a potential problem in human is the finding of a second *faah* gene, which is not present in rodents, although URB597 appears to be effective in inhibiting also this human FAAH isoform (Wei *et al.*, 2006).

Irrespective of these potential problems, the pharmacological manipulation of the EC system, accomplished by specifically inhibiting their metabolism, represents a highly innovative approach that arises from the common observation of the effects of cannabis derivatives on mood and emotionality and goes far beyond the problematic debate on the therapeutic potential of such derivatives. The discovery of such pharmacological target allows, indeed, to selectively focus on the benefits deriving from the enhancement of CB_1 activation in discrete brain areas, avoiding the whole set of psychotropic and systemic effects of direct-acting cannabinoid agonists.

References

Abel, E. L. (1997). Changes in anxiety feelings following marihuana smoking. The alternation in feelings of anxiety resulting from the smoking of marihuana (*Cannabis sativa* L.). *Br. J. Addict. Alcohol Other Drugs* **66,** 185–187.

Ahn, K., Johnson, D. S., Fitzgerald, L. R., Liimatta, M., Arendse, A., Stevenson, T., Lund, E. T., Nugent, R. A., Nomanbhoy, T. K., Alexander, J. P., and Cravatt, B. F. (2007). Novel mechanistic class of fatty acid amide hydrolase inhibitors with remarkable selectivity. *Biochemistry* **46,** 13019–13030.

Azad, S. C., Eder, M., Marsicano, G., Lutz, B., Zieglgänsberger, W., and Rammes, G. (2003). Activation of the cannabinoid receptor type 1 decreases glutamatergic and gabaergic synaptic transmission in the lateral amygdala of the mouse. *Learn. Mem.* **10,** 116–128.

Beltramo, M., Stella, N., Calignano, A., Lin, S. Y., Makriyannis, A., and Piomelli, D. (1997). Functional role of high-affinity anandamide transport, as revealed by selective inhibition. *Science* **277,** 1094–1097.

Berton, O., and Nestler, E. J. (2006). New approaches to antidepressant drug discovery: Beyond monoamines. *Nat. Rev. Neurosci.* **7,** 137–151.

Bisogno, T., Howell, F., Williams, G., Minassi, A., Cascio, M. G., Ligresti, A., Matias, I., Schiano-Moriello, A., Paul, P., Williams, E. J., Gangadharan, U., Hobbs, C., *et al.* (2003). Cloning of the first sn1-DAG lipases points to the spatial and temporal regulation of endocannabinoid signaling in the brain. *J. Cell Biol.* **163,** 463–468.

Blankman, J. L., Simon, G. M., and Cravatt, B. F. (2007). A comprehensive profile of brain enzymes that hydrolyze the endocannabinoid 2-arachidonoylglycerol. *Chem. Biol.* **14,** 1347–1356.

Bortolato, M., Campolongo, P., Mangieri, R. A., Scattoni, M. L., Frau, R., Trezza, V., La Rana, G., Russo, R., Calignano, A., Gessa, G. L., Cuomo, V., and Piomelli, D. (2006). Anxiolytic-like properties of the anandamide transport inhibitor AM404. *Neuropsychopharmacology* **31,** 2652–2659.

Bortolato, M., Mangieri, R. A., Fu, J., Kim, J. H., Arguello, O., Duranti, A., Tontini, A., Mor, M., Tarzia, G., and Piomelli, D. (2007). Antidepressant-like activity of the fatty acid amide hydrolase inhibitor URB597 in a rat model of chronic mild stress. *Biol. Psychiatry* **62,** 1103–1110.

Budney, A. J., Moore, B. A., Vandrey, R. G., and Hughes, J. R. (2003). The time course and significance of cannabis withdrawal. *J. Abnorm. Psychol.* **112,** 393–402.

Budney, A. J., Hughes, J. R., Moore, B. A., and Vandrey, R. (2004). Review of the validity and significance of cannabis withdrawal syndrome. *Am. J. Psychiatry* **161,** 1967–1977.

Cahill, L., and McGaugh, J. L. (1998). Mechanisms of emotional arousal and lasting declarative memory. *Trends Neurosci.* **21,** 294–299.

Christensen, R., Kristensen, P. K., Bartels, E. M., Bliddal, H., and Astrup, A. (2007). Efficacy and safety of the weight-loss drug rimonabant: A meta-analysis of randomised trials. *Lancet* **370,** 1706–1713.

Cravatt, B. F., Giang, D. K., Mayfield, S. P., Boger, D. L., Lerner, R. A., and Gilula, N. B. (1996). Molecular characterization of an enzyme that degrades neuromodulatory fatty-acid amides. *Nature* **384,** 83–87.

Cravatt, B. F., Demarest, K., Patricelli, M. P., Bracey, M. H., Giang, D. K., Martin, B. R., and Lichtman, A. H. (2001). Supersensitivity to anandamide and enhanced endogenous cannabinoid signaling in mice lacking fatty acid amide hydrolase. *Proc. Natl. Acad. Sci. USA* **98,** 9371–9376.

Devane, W. A., Hanus, L., Breuer, A., Pertwee, R. G., Stevenson, L. A., Griffin, G., Gibson, D., Mandelbaum, A., Etinger, A., and Mechoulam, R. (1992). Isolation and structure of a brain constituent that binds to the cannabinoid receptor. *Science* **258,** 1946–1949.

Di Marzo, V. (2008). Targeting the endocannabinoid system: To enhance or reduce? *Nat. Rev. Drug Discov.* **7,** 438–455.

Di Marzo, V., Fontana, A., Cadas, H., Schinelli, S., Cimino, G., Schwartz, J. C., and Pomelli, D. (1994). Formation and inactivation of endogenous cannabinoid anandamide in central neurons. *Nature* **372,** 686–691.

Dinh, T. P., Carpenter, D., Leslie, F. M., Freund, T. F., Katona, I., Sensi, S. L., Kathuria, S., and Piomelli, D. (2002). Brain monoglyceride lipase participating in EC inactivation. *Proc. Natl. Acad. Sci. USA* **99,** 10819–10824.

Dinh, T. P., Kathuria, S., and Piomelli, D. (2004). RNA interference suggests a primary role for monoacylglycerol lipase in the degradation of the endocannabinoid 2-arachidonoylglycerol. *Mol. Pharmacol.* **66,** 1260–1264.

Domschke, K., Dannlowski, U., Ohrmann, P., Lawford, B., Bauer, J., Kugel, H., Heindel, W., Young, R., Morris, P., Arolt, V., Deckert, J., Suslow, T., *et al.* (2008). Cannabinoid receptor 1 (CNR1) gene: Impact on antidepressant treatment response and emotion processing in major depression. *Eur. Neuropsychopharmacol.* **18,** 751–759.

Fegley, D., Kathuria, S., Mercier, R., Li, C., Goutopoulos, A., Makriyannis, A., and Piomelli, D. (2004). Anandamide transport is independent of fatty-acid amide hydrolase activity and is blocked by the hydrolysis-resistant inhibitor AM1172. *Proc. Natl. Acad. Sci. USA* **101,** 8756–8761.

Freund, T. F., Katona, I., and Piomelli, D. (2003). Role of endogenous cannabinoids in synaptic signaling. *Physiol. Rev.* **83,** 1017–1066.

Fu, J., Gaetani, S., Oveisi, F., Lo Verme, J., Serrano, A., Rodríguez De Fonseca, F., Rosengarth, A., Luecke, H., Di Giacomo, B., Tarzia, G., and Piomelli, D. (2003). Oleylethanolamide regulates feeding and body weight through activation of the nuclear receptor PPAR-alpha. *Nature* **425,** 90–93.

Gaetani, S., Cuomo, V., and Pomelli, D. (2003a). Anandamide hydrolysis: A new target for anti-anxiety drugs? *Trends Mol. Med.* **9,** 474–478.

Gaetani, S., Oveisi, F., and Piomelli, D. (2003b). Modulation of meal pattern in the rat by the anorexic lipid mediator oleoylethanolamide. *Neuropsychopharmacology* **28,** 1311–1316.

Gaetani, S., Kaye, W. H., Cuomo, V., and Piomelli, D. (2008). Role of endocannabinoids and their analogues in obesity and eating disorders. *Eat Weight Disord.* **13,** 42–48.

Gaoni, Y., and Mechoulam, R. (1964). Isolation, structure and partial synthesis of an active constituent of hashish. *J. Am. Chem. Soc.* **86,** 1646–1647.

Gerdeman, G., and Lovinger, D. M. (2001). CB1 cannabinoid receptor inhibits synaptic release of glutamate in rat dorsolateral striatum. *J. Neurophysiol.* **85,** 468–471.

Giuffrida, A., Leweke, F. M., Gerth, C. W., Schreiber, D., Koethe, D., Faulhaber, J., Klosterkötter, J., and Piomelli, D. (2004). Cerebrospinal anandamide levels are elevated in acute schizophrenia and are inversely correlated with psychotic symptoms. *Neuropsychopharmacology* **29,** 2108–2114.

Gobbi, G., Bambico, F. R., Mangieri, R., Bortolato, M., Campolongo, P., Solinas, M., Cassano, T., Morgese, M. G., Debonnel, G., Duranti, A., Tontini, A., Tarzia, G., *et al.* (2005). Antidepressant-like activity and modulation of brain monoaminergic transmission by blockade of anandamide hydrolysis. *Proc. Natl. Acad. Sci. USA* **102,** 18620–18625.

Gregg, J. M., Small, E. W., Moore, R., Raft, D., and Toomey, T. C. (1976). Emotional response to intravenous delta9tetrahydrocannabinol during oral surgery. *J. Oral Surg.* **34,** 301–313.

Hájos, N., Ledent, C., and Freund, T. F. (2001). Novel cannabinoid-sensitive receptor mediates inhibition of glutamatergic synaptic transmission in the hippocampus. *Neuroscience* **106,** 1–4.

Hall, W., and Solowij, N. (1998). Adverse effects of cannabis. *Lancet* **352,** 1611–1616.

Haller, J., Bakos, N., Szirmay, M., Ledent, C., and Freund, T. F. (2002). The effects of genetic and pharmacological blockade of the CB1 cannabinoid receptor on anxiety. *Eur. J. Neurosci.* **16,** 1395–1398.

Haller, J., Varga, B., Ledent, C., Barna, I., and Freund, T. F. (2004a). Context-dependent effects of CB1 cannabinoid gene disruption on anxiety-like and social behaviour in mice. *Eur. J. Neurosci.* **19,** 1906–1912.

Haller, J., Varga, B., Ledent, C., and Freund, T. F. (2004b). CB1 cannabinoid receptors mediate anxiolytic effects: Convergent genetic and pharmacological evidence with CB1-specific agents. *Behav. Pharmacol.* **15,** 299–304.

Hashimotodani, Y., Ohno-Shosaku, T., Maejima, T., Fukami, K., and Kano, M. (2008). Pharmacological evidence for the involvement of diacylglycerol lipase in depolarization-induced endocanabinoid release. *Neuropharmacology* **54,** 58–67.

Hill, M. N., Patel, S., Carrier, E. J., Rademacher, D. J., Ormerod, B. K., Hillard, C. J., and Gorzalka, B. B. (2005). Downregulation of endocannabinoid signaling in the hippocampus following chronic unpredictable stress. *Neuropsychopharmacology* **30,** 508–515.

Hillard, C. J., Wilkison, D. M., Edgemond, W. S., and Campbell, W. B. (1995). Characterization of the kinetics and distribution of *N*-arachidonylethanolamine (anandamide) hydrolysis by rat brain. *Biochim. Biophys. Acta* **1257,** 249–256.

Hillard, C. J., Edgemond, W. S., Jarrahian, A., and Campbell, W. B. (1997). Accumulation of *N*-arachidonoylethanolamine (anandamide) into cerebellar granule cells occurs via facilitated diffusion. *J. Neurochem.* **69,** 631–638.

Hohmann, A. G., Suplita, R. L., Bolton, N. M., Neely, M. H., Fegley, D., Mangieri, R., Krey, J. F., Walker, J. M., Holmes, P. V., Crystal, J. D., Duranti, A., Tontini, A., *et al.* (2005). An endocannabinoid mechanism for stress-induced analgesia. *Nature* **435,** 1108–1112.

Huang, C. C., Lo, S. W., and Hsu, K. S. (2001). Presynaptic mechanisms underlying cannabinoid inhibition of excitatory synaptic transmission in rat striatal neurons. *J. Physiol.* **532,** 731–748.

Hungund, B. L., Vinod, K. Y., Kassir, S. A., Basavarajappa, B. S., Yalamanchili, R., Cooper, T. B., Mann, J. J., and Arango, V. (2004). Upregulation of CB1 receptors and agonist-stimulated [^{35}S] GTPgammaS binding in the prefrontal cortex of depressed suicide victims. *Mol. Psychiatry* **9,** 184–190.

Jung, K. M., Astarita, G., Zhu, C., Wallace, M., Mackie, K., and Piomelli, D. (2007). A key role for diacylglycerol lipase-alpha in metabotropic glutamate receptor-dependent endocannabinoid mobilization. *Mol. Pharmacol.* **72,** 612–621.

Justinova, Z., Manieri, R. A., Bortolato, M., Chefir, S. I., Mukhin, A. G., Clipper, J. R., King, A. R., Redhi, G. H., Yasar, S., Pomelli, D., and Goldberg, S. R. (2008). Fatty acid amide hydrolase inhibition heightens anandamide signaling without producing reinforcing effects in primates. *Biol. Psychiatry* **64,** 930–937.

Kathuria, S., Gaetani, S., Fegley, D., Valiño, F., Duranti, A., Tontini, A., Mor, M., Tarzia, G., La Rana, G., Malignano, A., Giustino, A., Tattoli, M., *et al.* (2003). Modulation of anxiety through blockade of anandamide hydrolyis. *Nat. Med.* **9,** 76–81.

Katona, I., Urbán, G. M., Wallace, M., Ledent, C., Jung, K. M., Piomelli, D., Mackie, K., and Freund, T. F. (2006). Molecular composition of the endocannabinoid system at glutamatergic synapses. *J. Neurosci.* **26,** 5628–5637.

Leweke, F. M., Giuffrida, A., Koethe, D., Schreiber, D., Nolden, B. M., Kranaster, L., Neatby, M. A., Schneider, M., Gerth, C. W., Hellmich, M., Klosterkötter, J., and Piomelli, D. (2007). Anandamide levels in cerebrospinal fluid of first-episode schizophrenic patients: Impact of cannabis use. *Schizophr. Res.* **94,** 29–36.

Liu, J., Wang, L., Harvey-White, J., Osei-Hyiaman, D., Razdan, R., Gong, Q., Chan, A. C., Zhou, Z., Huang, B. X., Kim, H. Y., and Kunos, G. (2006). A biosynthetic pathway for anandamide. *Proc. Natl. Acad. Sci. USA* **103,** 13345–13350.

Liu, J., Wang, L., Harvey-White, J., Huang, B. X., Kim, H. Y., Luquet, S., Palmiter, R. D., Krystal, G., Rai, R., Mahadevan, A., Razdan, R. K., and Kunos, G. (2008). Multiple pathways involved in the biosynthesis of anandamide. *Neuropharmacology* **54,** 1–7.

Llano, I., Leresche, N., and Marty, A. (1991). Calcium entry increases the sensitivity of cerebellar Purkinje cells to applied GABA and decreases inhibitory synaptic currents. *Neuron* **6,** 565–574.

Long, J. Z., Li, W., Booker, L., Burston, J. J., Kinsey, S. G., Schlosburg, J. E., Pavón, F. J., Serrano, A. M., Selley, D. E., Parsons, L. H., Lichtman, A. H., and Cravatt, B. F. (2008). Selective blockade of 2-arachidonoylglycerol hydrolysis produces cannabinoid behavioral effects. *Nat. Chem. Biol.* **5,** 37–44.

LoVerme, J., Fu, J., Astarita, G., La Rana, G., Russo, R., Calignano, A., and Piomelli, D. (2005). The nuclear receptor peroxisome proliferator-activated receptor-alpha mediates the anti-inflammatory actions of palmitoylethanolamide. *Mol. Pharmacol.* **67,** 15–29.

LoVerme, J., Russo, R., La Rana, G., Fu, J., Farthing, J., Mattace-Raso, G., Meli, R., Hohmann, A., Calignano, A., and Piomelli, D. (2006). Rapid broad-spectrum analgesia through activation of peroxisome proliferator-activated receptor-alpha. *J. Pharmacol. Exp. Ther.* **319,** 1051–1061.

Marsicano, G., Wotjak, C. T., Azad, S. C., Bisogno, T., Rammes, G., Cascio, M. G., Hermann, H., Tang, J., Hofmann, C., Zieglgänsberger, W., Di Marzo, V., and Lutz, B. (2002). The endogenous cannabinoid system controls extinction of aversive memories. *Nature* **418,** 530–534.

Martin, M., Ledent, C., Parmentier, M., Maldonado, R., and Valverde, O. (2002). Involvement of CB1 cannabinoid receptors in emotional behaviour. *Psychopharmacology* **159,** 379–387.

McDonald, A. J., and Mascagni, F. (2001). Localization of the CB1 type cannabinoid receptor in the rat basolateral amygdala: High concentrations in a subpopulation of cholecystokinin-containing interneurons. *Neuroscience* **107,** 641–652.

Mechoulam, R., Ben-Shabat, S., Hanus, L., Ligumsky, M., Kaminski, N. E., Schatz, A. R., Gopher, A., Almog, S., Martin, B. R., Compton, D. R., Pertwee, R. G., Griffin, G., *et al.* (1995). Identification of an endogenous 2-monoglyceride, present in canine gut, that binds to cannabinoid receptors. *Biochem. Pharmacol.* **50,** 83–90.

Millan, M. J. (2003). The neurobiology and control of anxious states. *Prog. Neurobiol.* **70,** 83–244.

Mor, M., Rivara, S., Lodola, A., Plazzi, P. V., Tarzia, G., Duranti, A., Tontini, A., Piersanti, G., Kathuria, S., and Piomelli, D. (2004). Cyclohexylcarbamic acid 3′- or 4′-substituted biphenyl-3-yl esters as fatty acid amide hydrolase inhibitors: Synthesis, quantitative structure-activity relationships, and molecular modeling studies. *J. Med. Chem.* **47,** 4998–5008.

Moreira, F. A., Kaiser, N., Monory, K., and Lutz, B. (2008). Reduced anxiety-like behaviour induced by genetic and pharmacological inhibition of the endocannabinoid-degrading enzyme fatty acid amide hydrolase (FAAH) is mediated by CB1 receptors. *Neuropharmacology* **54,** 141–150.

Muccioli, G. G., Xu, C., Odah, E., Cudaback, E., Cisneros, J. A., Lambert, D. M., López Rodríguez, M. L., Bajjalieh, S., and Stella, N. (2007). Identification of a novel endocannabinoid-hydrolyzing enzyme expressed by microglial cells. *J. Neurosci.* **27,** 2883–2889.

Naidu, P. S., Varvel, S. A., Ahn, K., Cravatt, B. F., Martin, B. R., and Lichtman, A. H. (2007). Evaluation of fatty acid amide hydrolase inhibition in murine models of emotionality. *Psychopharmacology* **192,** 61–70.

Naliboff, B. D., Rickles, W. H., Cohen, M. J., and Naimark, R. S. (1976). Interactions of marijuana and induced stress: Forearm blood flow, heart rate, and skin conductance. *Psychophysiology* **13,** 517–522.

Navarro, M., Hernández, E., Muñoz, R. M., del Arco, I., Villanúa, M. A., Carrera, M. R., and Rodríguez de Fonseca, F. (1997). Acute administration of the CB1 cannabinoid receptor antagonist SR 141716A induces anxiety-like responsesin the rat. *Neuroreport* **8,** 491–496.

Okamoto, Y., Morishita, J., Wang, J., Schmid, P. C., Krebsbach, R. J., Schmid, H. H., and Ueda, N. (2005). Mammalian cells stably overexpressing *N*-acylphosphatidylethanolamine-hydrolysing phospholipase D exhibit significantly decreased levels of *N*-acylphosphatidylethanolamines. *Biochem. J.* **389,** 241–247.

Onaivi, E. S., Chakrabarti, A., Gwebu, E. T., and Chaudhuri, G. (1995). Neurobehavioral effects of D9-THC and cannabinoid (CB1) receptor gene expression in mice. *Behav. Brain Res.* **72,** 115–125.

Patel, S., and Hillard, C. J. (2006). Pharmacological evaluation of cannabinoid receptor ligands in a mouse model of anxiety: Further evidence for an anxiolytic role for endogenous cannabinoid signaling. *J. Pharmacol. Exp. Ther.* **318,** 304–311.

Patel, S., Roelke, C. T., Rademacher, D. J., Cullinan, W. E., and Hillard, C. J. (2004). Endocannabinoid signaling negatively modulates stress-induced activation of the hypothalamic-pituitary-adrenal axis. *Endocrinology* **145,** 5431–5438.

Piomelli, D. (2003). The molecular logic of endocannabinoid signalling. *Nat. Rev. Neurosci.* **4,** 873–884.

Piomelli, D., Beltramo, M., Glasnapp, S., Lin, S. Y., Goutopoulos, A., Xie, X. Q., and Makriyannis, A. (1999). Structural determinants for recognition and translocation by the anandamide transporter. *Proc. Natl. Acad. Sci. USA* **96,** 5802–5807.

Piomelli, D., Tarzia, G., Duranti, A., Tontini, A., Mor, M., Compton, T. R., Dasse, O., Monaghan, E. P., Parrott, J. A., and Putman, D. (2006). Pharmacological profile of the selective FAAH inhibitor KDS-4103 (URB597). *CNS Drug Rev.* **12,** 21–38.

Pitler, T. A., and Alger, B. E. (1992). Postsynaptic spike firing reduces synaptic GABAA responses in hippocampal pyramidal cells. *J. Neurosci.* **12,** 4122–4132.

Rademacher, D. J., and Hillard, C. J. (2007). Interactions between endocannabinoids and stress-induced decreased sensitivity to natural reward. *Prog. Neuropsychopharmacol. Biol. Psychiatry* **31,** 633–641.

Robbe, D., Alonso, G., Duchamp, F., Bockaert, J., and Manzoni, O. J. (2001). Localization and mechanisms of action of cannabinoid receptors at the glutamatergic synapses of the mouse nucleus accumbens. *J. Neurosci.* **21,** 109–116.

Rodríguez de Fonseca, F., Carrera, M. R., Navarro, M., Koob, G. F., and Weiss, F. (1997). Activation of corticotropin-releasing factor in the limbic system during cannabinoid withdrawal. *Science* **276,** 2050–2054.

Rodríguez de Fonseca, F., Navarro, M., Gómez, R., Escuredo, L., Nava, F., Fu, J., Murillo-Rodríguez, E., Giuffrida, A., LoVerme, J., Gaetani, S., Kathuria, S., Gall, C., *et al.* (2001). An anorexic lipid mediator regulated by feeding. *Nature* **414,** 209–212.

Schmid, P. C., Zuzarte-Augustin, M. L., and Schmid, H. H. (1985). Properties of rat liver *N*-acylethanolamin amidohydrolase. *J. Biol. Chem.* **260,** 14145–14149.

Schwartz, G. J., Fu, J., Astarita, G., Li, X., Gaetani, S., Campolongo, P., Cuomo, V., and Piomelli, D. (2008). The lipid messenger OEA links dietary fat intake to satiety. *Cell Metab.* **8,** 281–288.

Simon, G. M., and Cravatt, B. F. (2006). Endocannabinoid biosynthesis proceeding through glycerophospho-*N*-acyl ethanolamine and a role for alpha/beta-hydrolase 4 in this pathway. *J. Biol. Chem.* **281,** 26465–26472.

Stella, N., and Piomelli, D. (2001). Receptor-dependent formation of endogenous cannabinoids in cortical neurons. *Eur. J. Pharmacol.* **425,** 189–196.

Stella, N., Schweitzer, P., and Piomelli, D. (1997). A second endogenous cannabinoid that modulates long-term potentiation. *Nature* **388,** 773–778.

Sugiura, T., Kondo, S., Sukagawa, A., Nakane, S., Shinoda, A., Itoh, K., Yamashita, A., and Waku, K. (1995). 2-Arachidonoylglycerol: A possible endogenous cannabinoid receptor ligand in brain. *Biochem. Biophys. Res. Commun.* **215,** 89–97.

Tarzia, G., Duranti, A., Tontini, A., Piersanti, G., Mor, M., Rivara, S., Plazzi, P. V., Park, C., Kathuria, S., and Piomelli, D. (2003). Design, synthesis, and structure-activity relationships of alkylcarbamic acid aryl esters, a new class of fatty acidamide hydrolase inhibitors. *J. Med. Chem.* **46,** 2352–2360.

Tournier, M., Sorbara, F., Gindre, C., Swendsen, J. D., and Verdoux, H. (2003). Cannabis use and anxiety in daily life: A naturalistic investigation in a non-clinical population. *Psychiatry Res.* **118,** 1–8.

Tsuboi, K., Takezaki, N., and Ueda, N. (2007). The *N*-acylethanolamine-hydrolyzing acid amidase (NAAA). *Chem. Biodivers.* **4,** 1914–1925.

Urigüen, L., Pérez-Rial, S., Ledent, C., Palomo, T., and Manzanares, J. (2004). Impaired action of anxiolytic drugs in mice deficient in cannabinoid CB1 receptors. *Neuropharmacology* **46,** 966–973.

Vandrey, R. G., Budney, A. J., Moore, B. A., and Hughes, J. R. (2005). A cross-study comparison of cannabis and tobacco withdrawal. *Am. J. Addict.* **14,** 54–63.

Vandrey, R. G., Budney, A. J., Hughes, J. R., and Liguori, A. (2008). A within-subject comparison of withdrawal symptoms during abstinence from cannabis, tobacco, and both substances. *Drug Alcohol Depend.* **92,** 48–54.

Varvel, S. A., Wise, L. E., Niyuhire, F., Cravatt, B. F., and Lichtman, A. H. (2007). Inhibition of fatty-acid amide hydrolase accelerates acquisition and extinction rates in a spatial memory task. *Neuropsychopharmacology* **32,** 1032–1041.

Viveros, M. P., Marco, E. M., and File, S. E. (2005). Endocannabinoid system and stress and anxiety responses. *Pharmacol. Biochem. Behav.* **81,** 331–342.

Wachtel, S. R., ElSohly, M. A., Ross, S. A., Ambre, J., and de Wit, H. (2002). Comparison of the subjective effects of D(9)-tetrahydrocannabinol and marijuana in humans. *Psychopharmacology (Berl.)* **161,** 331–339.

Wei, B. Q., Mikkelsen, T. S., McKinney, M. K., Lander, E. S., and Cravatt, B. F. (2006). A second fatty acid amide hydrolase with variable distribution among placental mammals. *J. Biol. Chem.* **281,** 36569–36578.

Zuardi, A. W., Shirakawa, I., Finkelfarb, E., and Karniol, I. G. (1982). Action of cannabidiol on the anxiety and other effects produced by D9-THC in normal subjects. *Psychopharmacology (Berl.)* **76,** 245–250.

GABA$_A$ RECEPTOR FUNCTION AND GENE EXPRESSION DURING PREGNANCY AND POSTPARTUM

Giovanni Biggio,[*,†] Maria Cristina Mostallino,[†] Paolo Follesa,[*] Alessandra Concas,[*,†] and Enrico Sanna[*]

[*]Department of Experimental Biology, Center of Excellence for the Neurobiology of Dependence, University of Cagliari, 09100 Cagliari, Italy
[†]C.N.R. Institute of Neuroscience, Section of Cagliari, Cagliari, Italy

Neuroactive steroids such as 3α,5α-THP are reduced metabolites of progesterone and are considered to play an important physiological role to locally modulate neuronal excitability by "fine-tuning" the action of GABA acting at GABA$_A$ receptors (GABA$_A$-Rs). In different brain regions, such as the hippocampus, different subpopulations of nerve cells exhibit two components of inhibitory GABAergic transmission: a phasic component mediated by synaptic GABA$_A$-Rs, and a tonic component mediated by "ambient" GABA acting at extrasynaptic GABA$_A$-Rs mainly containing the δ subunit and endowed with a higher sensitivity to neuroactive steroids compared to synaptic receptors. It is also well accepted that fluctuations in brain neuroactive steroid levels may result in plastic changes of GABA$_A$-Rs. In this article we review some of our results obtained with the model of pregnancy in rats. Pregnancy, in fact, is characterized by a marked and progressive increase in plasma and brain levels of neuroactive steroids which could contribute to changes in mood, anxiety as well as other psychiatric conditions.

INTERNATIONAL REVIEW OF NEUROBIOLOGY, VOL. 85
DOI: 10.1016/S0074-7742(09)85006-X

0074-7742/09 $35.00

Such elevation in brain neuroactive steroid concentrations during pregnancy, in turn, is accompanied by alterations in both gene expression and function of synaptic and extrasynaptic $GABA_A$-Rs in the hippocampus as well as other areas.

I. Introduction

The neurotransmitter γ-aminobutyric acid (GABA), acting through $GABA_A$ receptors ($GABA_A$-Rs), mediates the majority of "fast" inhibitory synaptic signaling in the mammalian central nervous system and consequently has a profound influence on mood and behavior. $GABA_A$-Rs are pentameric complexes of α, β and one other type of subunit (for example γ, δ, or ε) in a likely stoichiometry ratio of 2:2:1 (Barnard *et al.*, 1998). Although both synaptic and nonsynaptic receptors are likely to share this stoichiometry and pentameric structure, it is now well established that subcellular localization may be influenced by the presence of a γ (synaptic) or a δ (extrasynaptic) subunit (Glykys and Mody, 2007; Nusser *et al.*, 1995, 1998). Moreover, it is now widely accepted that extrasynaptically located $GABA_A$-Rs underpin a sustained tonic form of inhibition via which they play a significant role in brain excitability, both in normal and pathophysiological states (Farrant and Nusser, 2005; Semyanov *et al.*, 2004).

Several studies have confirmed, using pharmacological approaches, null mutant mice or both, that in neurons where δ subunit is expressed, $GABA_A$-Rs are not only responsible for the mediation of tonic inhibition (Drasbek and Jensen, 2006; Glykys and Mody, 2006; Jia *et al.*, 2005; Porcello *et al.*, 2003; Stell *et al.*, 2003; Wei *et al.*, 2003), but are also the physiological natural target of some neuroactive steroids such as the progesterone metabolite $3\alpha,5\alpha$-THP that acts by increasing the low efficacy of GABA on these δ containing receptors that show exceptionally high affinity for GABA and low degree of desensitization (Bianchi and Macdonald, 2003; Wohlfarth *et al.*, 2002). Indeed, some neuroactive steroids in the concentration range assumed to be present in the extracellular space under various physiological conditions, including pregnancy, selectively enhance the magnitude of tonic inhibition through δ subunit-containing $GABA_A$-Rs (Stell *et al.*, 2003), with the net effect being the reduction of network excitability (Stell *et al.*, 2003). Although various other sites for the action of physiological concentrations of neuroactive steroids have been postulated in the brain, these hormones appear to preferentially target the tonic inhibition mediated by δ subunit-containing $GABA_A$-Rs. Thus, fluctuations in the concentrations of neurosteroids can affect the receptor expression and function. For example it has been shown a dynamic, ovarian cycle-linked modification in specific $GABA_A$-R expression and function in the rodent brain (Maguire *et al.*, 2005). During the diestrus stage of the estrus cycle in mice when levels of progesterone and neuroactive steroids are elevated, there is an increased expression of the GABA$_A$-R δ subunit in the membranes of hippocampal neurons and an increase in GABA$_A$-R δ subunit-mediated tonic inhibition in dentate gyrus granule cells. Moreover, other investigators have shown the same receptors to parallel ovarian cycle-related changes in the periaqueductal gray matter (Griffiths and Lovick, 2005; Lovick, 2006; Lovick *et al.*, 2005) or to be upregulated in a steroid withdrawal model of premenstrual dysphoric disorder (Smith *et al.*, 2006).

In this chapter, we have summarized our most relevant findings on the differential changes in expression and function of both synaptic and extrasynaptic GABA$_A$-Rs in the rat hippocampus during pregnancy and postpartum.

II. Concentrations of 3α,5α-THP in Rat Brain and Plasma During Pregnancy and After Delivery

Pregnancy is characterized by a gradual increase in hormone levels and is followed by a dramatic and rapid fall during postpartum. Progesterone, one of the principal hormones secreted by the corpus luteum and placenta to ensure the establishment and the maintenance of pregnancy, reaches its highest levels in a woman's life during this period. Given that 3α,5α-THP is a progesterone metabolite, its circulating levels are also increased in pregnant women (Luisi *et al.*, 2000; Paoletti *et al.*, 2006). In addition, the activity of the enzymes responsible for the synthesis of 3α,5α-THP is increased in both maternal (especially the placenta) and fetal tissue (Buster, 1983; Milewich *et al.*, 1979).

We have shown in rats that the plasma and brain concentrations of progesterone and 3α,5α-THP increase during pregnancy, return to the values of estrus female immediately before delivery, and remain unchanged during postpartum (Concas *et al.*, 1998). Although the brain has the capability to synthesize 3α,5α-THP *de novo* from cholesterol, the majority of 3α,5α-THP in the brain would be derived from circulating progesterone. However, the kinetics of the cerebrocortical concentration of these neurosteroids differ from those of the plasma concentration. Whereas the concentrations of progesterone in the cerebral cortex and hippocampus, similarly to plasma, peak on P15 (approximately 12 and 10 times the values of estrus female, respectively) and remain substantially increased on P19, the concentrations of 3α,5α-THP in the same brain areas do not peak until P19 (+208% in the cerebral cortex and +194% in the hippocampus). Then, both progesterone and 3α,5α-THP concentrations return to control value immediately before delivery (P21) and do not change further during the postpartum period (days 2 and 7 after delivery) (Concas *et al.*, 1998).

These results imply that the synthesis or accumulation of 3α,5α-THP in the brain of pregnant rats is not simply a function of the circulating levels of progesterone. Accordingly, differential regulation of neurosteroid concentrations between plasma and the brain has been suggested on the basis of the effects of various pharmacological treatments (Barbaccia *et al.*, 2001; Concas *et al.*, 2000; Follesa *et al.*, 2002). Moreover, an ethanol-induced increase in *de novo* neurosteroid synthesis in the brain that is independent from periphery has been also demonstrated (Sanna *et al.*, 2004). One possible explanation for this difference in kinetics is that steroid hormones such as estrogens or other agents may regulate the activity and expression of 5α-reductase or 3α-hydroxysteroid dehydrogenase in the brain during pregnancy. Indeed, estradiol has been shown to regulate 5α-reductase expression (Maayan *et al.*, 2004) and 3α-hydroxysteroid dehydrogenase activity in rat brain (Penning *et al.*, 1985).

Given that 3α,5α-THP positively modulates $GABA_A$-Rs (Belelli and Lambert, 2005; Majewska, 1992), subsequent studies have been undertaken to determine whether the physiological fluctuations in the concentrations of these neuroactive steroids that occur during pregnancy and after delivery may also affect the gene expression of different subunits of both synaptic and extrasynaptic $GABA_A$-Rs in the rat brain.

III. Changes in Synaptic $GABA_A$-R Gene Expression During Pregnancy and After Delivery

In the rat brain the amounts of γ_2 subunit mRNA, measured by RNase protection assay, are progressively decreased during pregnancy (Concas *et al.*, 1998; Follesa *et al.*, 1998). In particular, in the hippocampus, the expression of this subunit mRNA remained unchanged at P10 but was reduced (about 30%) at P15 and P19 compared with the levels measured in estrus rats. The amounts of γ_2 subunit mRNA in the hippocampus during the postpartum period increased, compared to P19, and then returned to control estrus values (Concas *et al.*, 1998; Follesa *et al.*, 1998). Immunohistochemical (Sanna *et al.*, 2009) studies revealed that in estrus rats, the γ_2 subunit peptide was expressed in the strata oriens and radiatum of CA1 and CA3, in the stratum lacunosum-moleculare of CA1, and in granule cells of the dentate gyrus. During pregnancy the level of γ_2 subunit immunoreactivity was decreased at P15 and P19 in the dentate gyrus and the CA1 region with a time course overlapping that previously described for the mRNA (Concas *et al.*, 1998; Follesa *et al.*, 1998). It then increased immediately before delivery at P21, remained increased at 2 days after delivery, and returned to control values by 7 days after delivery (Fig. 1) (Sanna *et al.*, 2009).

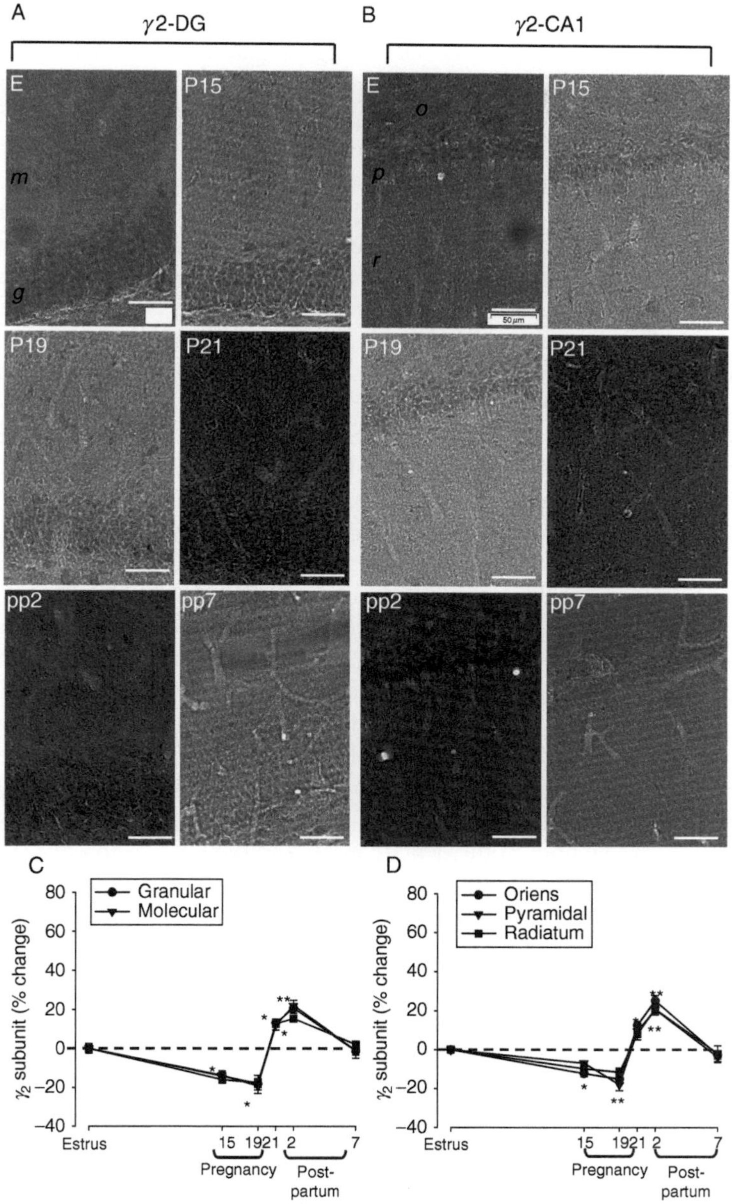

FIG. 1. Changes in immunoreactivity for the γ_2 subunit of the GABA$_A$-R in the rat hippocampus during pregnancy and after delivery. (A, B) Representative immunohistochemical images of the distribution of the γ_2 subunit in the dentate gyrus (A) and in the CA1 region (B) in hippocampal sections from rats in estrus (E); at day 15 (P15), 19 (P19), or 21 (P21) of pregnancy; or at 2 (pp2) or 7

These results are consistent with the observation that muscimol-induced $^{36}Cl^-$ uptake was markedly reduced during pregnancy compared with rats in estrus. At this same time point, the potentiating effects of diazepam and $3\alpha,5\alpha$-THP on muscimol stimulation of $^{36}Cl^-$ uptake also were reduced. In contrast, the effects of muscimol, $3\alpha,5\alpha$-THP and diazepam were significantly increased, relative to animals in estrus, after delivery (Concas *et al.*, 1998; Follesa *et al.*, 1998). The densities of [^{3}H]GABA, [^{3}H]flunitrazepam, and *t*-[^{35}S]butylbicyclophosphorotionate ([^{35}S]TBPS) binding sites in the cerebral cortex also increased during pregnancy, again peaking at P19 and returning to control values at P21; receptor density was decreased further 2 days after delivery and again returned to control values within 7 days. These changes were accompanied by a decrease in the apparent affinity of the binding sites for the corresponding radioligand on P19 (Concas *et al.*, 1998, 1999).

As opposed to what observed for the γ_2 subunit, the abundance of mRNA encoding the α_4 subunit of the $GABA_A$-R was unchanged during pregnancy in both rat hippocampus and cerebral cortex (Concas *et al.*, 1998; Follesa *et al.*, 1998) but specifically increased only in the hippocampus and only after delivery (Concas *et al.*, 1999). Moreover, no significant changes were apparent for α_1, α_2, α_3, β_1, β_2, and β_3 subunit mRNAs in both brain regions during pregnancy and after delivery (Concas *et al.*, 1998; Follesa *et al.*, 1998). Immunohistochemical studies (Sanna *et al.*, 2009) revealed that the α_4 subunit peptide of the $GABA_A$-R appeared diffuse throughout the hippocampal formation of control rats, being most abundant in the granule cell layer of the dentate gyrus and in the pyramidal cell layer of CA1 (Fig. 2). Staining was also prominent in the molecular layer of the dentate gyrus. Consistently with the mRNA data (Concas *et al.*, 1999) the amount of the α_4 subunit did not change significantly during pregnancy, but was increased markedly both in the dentate gyrus and the CA1 region after delivery only (Fig. 2) (Sanna *et al.*, 2009).

These results demonstrate that pregnancy and delivery are accompanied by marked changes in the expression of specific $GABA_A$-R subunit genes. Thus, the amount of the γ_2 subunit mRNA and corresponding peptide decreased in the hippocampus during pregnancy and returned to control values after delivery. On the other hand, both the α_4 subunit mRNA and peptide remained unchanged during pregnancy and markedly increased late in the postpartum period.

(pp7) days after delivery. (*m*) Molecular and (*g*) granule layer of the dentate gyrus; and (*o*) stratum oriens, (*p*) pyramidal, and (*r*) radians of the CA1 region. Scale bar, 50 μm. Images similar to those in (A) and (B) were subjected to semiquantitative measurement of γ_2 subunit immunoreactivity in the granular and molecular cell layers of the dentate gyrus (C); and in the stratum oriens, pyramidal cell layer, and stratum radiatum of CA1 (D). Data are expressed as the percentage change in gray-scale values relative to the corresponding value for rats in estrus (control) and are mean ± S.E.M. from six to eight animals at each time point. *$P < 0.05$, **$P < 0.01$ versus estrus (one-way ANOVA followed by Scheffe's test). Reproduced with permission from Sanna *et al.* (2009).

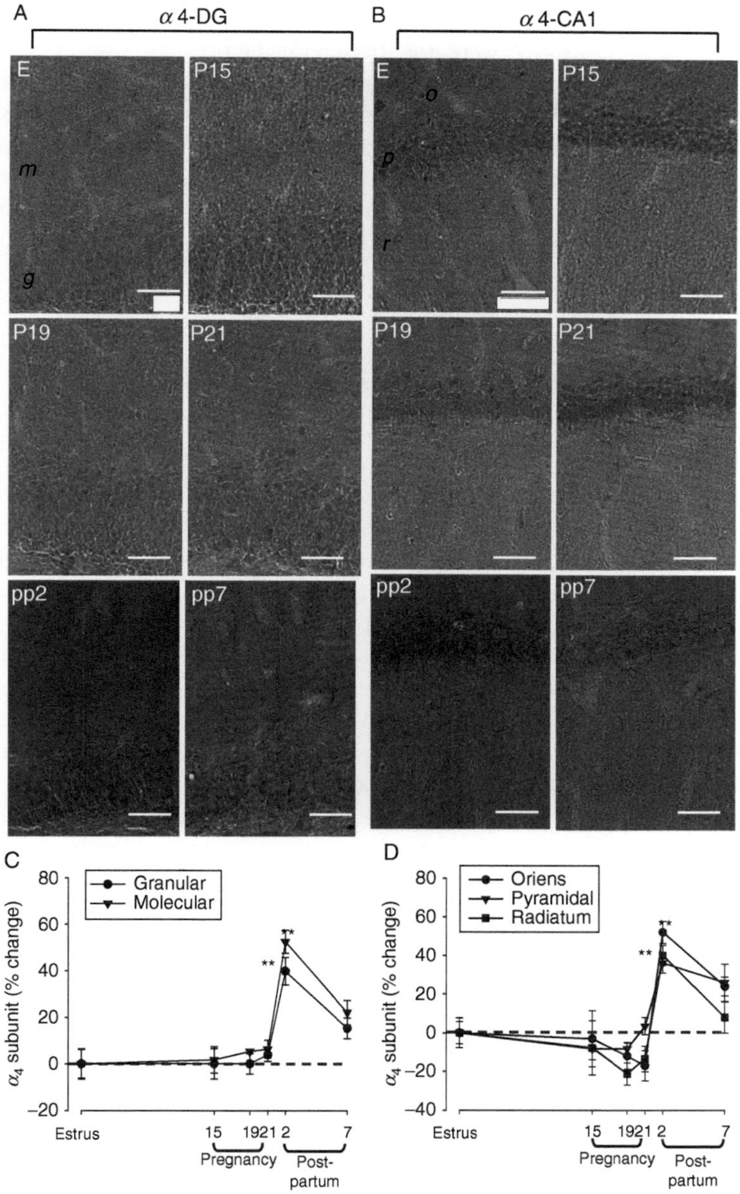

FIG. 2. Changes in immunoreactivity for the α_4 subunit of the GABA$_A$-R in the rat hippocampus during pregnancy and after delivery. (A, B) Representative immunohistochemical images of the distribution of the α_4 subunit in the dentate gyrus (A) and in the CA1 region (B) in hippocampal sections from rats in estrus (E); at day 15 (P15), 19 (P19), or 21 (P21) of pregnancy; or at 2 (pp2) or 7 (pp7) days after delivery. (*m*) Molecular and (*g*) granule layer of the dentate gyrus; and (*o*) stratum

These observations, together with the evidence that the amounts of α_1, α_2, α_3, β_1, β_2, and β_3 subunit mRNAs were not affected during pregnancy or after delivery, suggest that the changes in γ_2 and α_4 subunits are time- and region-specific and do not represent a generalized phenomenon. Accordingly, it has been previously shown (Brussaard *et al.*, 1997; Fenelon and Herbison, 1996) that the expression of α_1 subunit mRNA increases during the course of pregnancy specifically in hypothalamic magnocellular neurons, which play a critical role in promoting parturition and lactation, whereas an increase in γ_2 gene expression is apparent in the same area only after delivery.

IV. Changes in Extrasynaptic δ-Containing $GABA_A$-R Gene Expression During Pregnancy and After Delivery

In control estrus rats, moderate levels of immunoreactivity for the δ subunit of the $GABA_A$-R were apparent throughout most regions of the hippocampal formation (Sanna *et al.*, 2009). The abundance of the δ subunit appeared greatest in the granule cell layer of the dentate gyrus and in the pyramidal cell layer of region CA1 (Fig. 3). Expression of the δ subunit was also prominent in the molecular layer of the dentate gyrus. Measurement of δ subunit immunoreactivity in rats at P15, P19, and P21 revealed that specific staining increased progressively in the dentate gyrus (Fig. 3) and the CA1 region compared with that in control rats. A significant increase was already evident at P15 in the dentate gyrus and the pyramidal cell layer of CA1, with maximal upregulation in all regions being apparent at P19 or P21. In contrast, the amount of δ subunit immunoreactivity had decreased to values significantly less than those for control rats by 2 days after delivery (Fig. 3) (Sanna *et al.*, 2009). This downregulation of subunit expression was still apparent 7 days postpartum in the granular and molecular layers of the dentate gyrus and in the stratum oriens of CA1 (Fig. 3)(Sanna *et al.*, 2009).

These results provide further support to the notion that both the structure and function of $GABA_A$-Rs in the brain change during pregnancy and immediately after delivery as a result of the marked associated fluctuations in the brain levels of neuroactive steroids.

---

oriens, (*p*) pyramidal, and (*r*) radians of the CA1 region. Scale bar, 50 μm. Images similar to those in (A) and (B) were subjected to semiquantitative measurement of α_4 subunit immunoreactivity in the granular and molecular cell layers of the dentate gyrus (C); and in the stratum oriens, pyramidal cell layer, and stratum radiatum of CA1 (D). Data are expressed as the percentage change in gray-scale values relative to the corresponding value for rats in estrus (control) and are mean $\pm$ S.E.M. from six to eight animals at each time point. $^{**}P < 0.01$ versus estrus (one-way ANOVA followed by Scheffe's test). Reproduced with permission from Sanna *et al.* (2009).

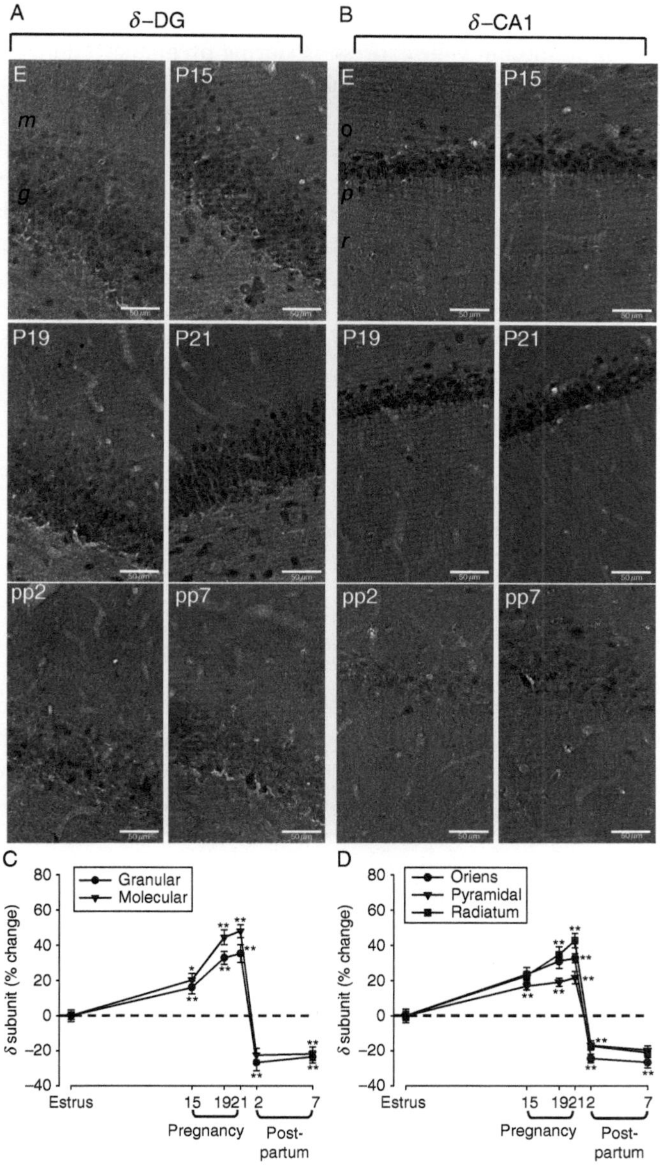

FIG. 3. Changes in immunoreactivity for the δ subunit of the GABA$_A$-R in the rat hippocampus during pregnancy and after delivery. (A, B) Representative immunohistochemical images of the distribution of the δ subunit in the dentate gyrus (A) and in the CA1 region (B) in hippocampal sections from rats in estrus (E); at day 15 (P15), 19 (P19), or 21 (P21) of pregnancy; or at 2 (pp2) or 7 (pp7) days after delivery. (*m*) Molecular and (*g*) granule layer of the dentate gyrus; and (*o*) stratum oriens, (*p*) pyramidal, and (*r*) radians of the CA1 region. Scale bar, 50 μm. Images similar to those in

The abundance of immunoreactivity for the δ subunit showed changes opposite to those for γ_2 subunit gene expression during pregnancy and after delivery. Specific expression for the δ subunit in the hippocampus increased progressively during pregnancy, reaching a peak at P19–21, and then fell markedly, being below control levels at 2 and 7 days after delivery. Consistent with the higher sensitivity of $GABA_A$-Rs containing the δ subunit to $3\alpha,5\alpha$-THP (Adkins *et al.*, 2001; Brown *et al.*, 2002; Wohlfarth *et al.*, 2002), fluctuations in the plasma and brain content of this compound associated with various pathophysiological or pharmacological conditions are accompanied by changes in the expression of the δ subunit together, in some instances, with changes in α_4 subunit expression (Griffiths and Lovick, 2005; Lovick *et al.*, 2005; Maguire *et al.*, 2005; Shen *et al.*, 2005; Smith *et al.*, 1998a,b; Sundstrom-Poromaa *et al.*, 2002) that is present in both synaptic and extrasynaptic receptors (Nusser and Mody, 2002; Pirker *et al.*, 2000; Wei *et al.*, 2003).

V. Changes in $GABA_A$-R Function in the Rat Hippocampus During Pregnancy and After Delivery

A. Extrasynaptic $GABA_A$-R-Mediated Tonic Inhibition

In the hippocampus of adult rats, granule cells of the dentate gyrus express, in addition to synaptic $GABA_A$-Rs that are responsible for the phasic inhibition, a major subpopulation of extrasynaptic receptors formed by α_4, β_n, and δ subunits (Nusser and Mody, 2002) that mediate the tonic component of the GABAergic inhibition. Moreover, in the hippocampal formation there are also, although in a significant lower proportion, extrasynaptic receptors formed by α_1, β_n, and δ which are expressed selectively in GABAergic interneurons of the dentate gyrus molecular layer (Glykys and Mody, 2007) as well as other populations of extrasynaptic $GABA_A$-Rs containing α_5, β_n, and γ subunits that are localized in the CA1/CA3 region and dentate gyrus (Caraiscos *et al.*, 2004; Glykys *et al.*, 2008; Pirker *et al.*, 2000). Based on the marked changes in the expression of the α_4, γ_2, and δ subunits in the hippocampus of rats during pregnancy and in the postpartum period, we applied the patch clamp technique in order to record the GABAergic tonic currents in dentate gyrus granule cells. In granule cells of

(A) and (B) were subjected to semiquantitative measurement of δ subunit immunoreactivity in the granular and molecular cell layers of the dentate gyrus (C); and in the stratum oriens, pyramidal cell layer, and stratum radiatum of CA1 (D). Data are expressed as the percentage change in gray-scale values relative to the corresponding value for rats in estrus (control) and are mean ± S.E.M. from six to eight animals at each time point. *$P < 0.05$, **$P < 0.01$ versus estrus (one-way ANOVA followed by Scheffe's test). Reproduced with permission from Sanna *et al.* (2009).

control (estrus) rats, bath application of exogenous GABA (5 μM) for 5 min to stimulate high affinity extrasynaptic $GABA_A$-Rs resulted in an increase in current noise variance and a negative shift in the holding current with respect to baseline (Fig. 4). In line with the increase in the expression of the δ subunit during pregnancy, we found that the effect of GABA on tonic current parameters started to increase in granule cells of P15 rats and reached a maximal enhancement at P19, and returned to control values 2 days after delivery (Fig. 4). Thus, the last portion of pregnancy in the rat hippocampus is associated with a marked increase in GABAergic tonic inhibition in dentate gyrus granule cells. This conclusion is further supported by the experiments in which we tested the sensitivity of extrasynaptic $GABA_A$-Rs to neurosteroids. Neurosteroids such as $3\alpha,5\alpha$-THP are known to exert a more robust positive modulatory action on extrasynaptic $GABA_A$-Rs compared to synaptic receptors (Belelli and Lambert, 2005). We found that bath application of $3\alpha,5\alpha$-THP (1 μM) to hippocampal slices induced a greater enhancement of tonic current noise variance and a larger negative shift of holding current in granule cells from P19 rats with respect to granule cells from estrus animals (Fig. 5). It is worthy to note that in the dentate gyrus granule cells of rat at P19, the increase in δ subunit expression, together with the enhancement in the function of extrasynaptic receptors, is paralleled by the decrease in the expression of the γ_2 subunit with no change in that of α_4 subunit. Because the α_4 subunit can form receptor constructs with both δ and γ_2 subunits (Whiting *et al.*, 1999), it can be speculated that the enhancement of δ subunit may be balanced by a decreased assembly of $\alpha_4\beta\gamma_2$ receptors without changing the expression levels of the α_4 subunit. Therefore, a mechanism involving δ/γ_2 subunit exchange in α_4-containing receptors may account for the stable expression of the α_4 subunit during pregnancy. This pattern of alterations is very similar to that described during the ovary cycle (Maguire *et al.*, 2005).

B. Synaptic $GABA_A$-R-Mediated Phasic Inhibition

To assess further the functional consequences of the changes in $GABA_A$-R subunit expression associated with pregnancy and the postpartum period, we studied the phasic GABAergic inhibition by recording both spontaneous and miniature (due to action potential-independent release of GABA) inhibitory postsynaptic currents (sIPSCs and mIPSCs, respectively) in voltage-clamped granule cells. Analysis of the basal kinetic properties (decay time, amplitude, area, and frequency) of these currents revealed no significant differences among rats at P15, P19, or 2 days after delivery compared to animals in estrus (Sanna *et al.*, 2009). We thus investigate whether granule cell synaptic $GABA_A$-Rs may have an altered sensitivity to the action of different allosteric modulators such as neurosteroids and benzodiazepines. The effects of $3\alpha,5\alpha$-THP (1 μM) on

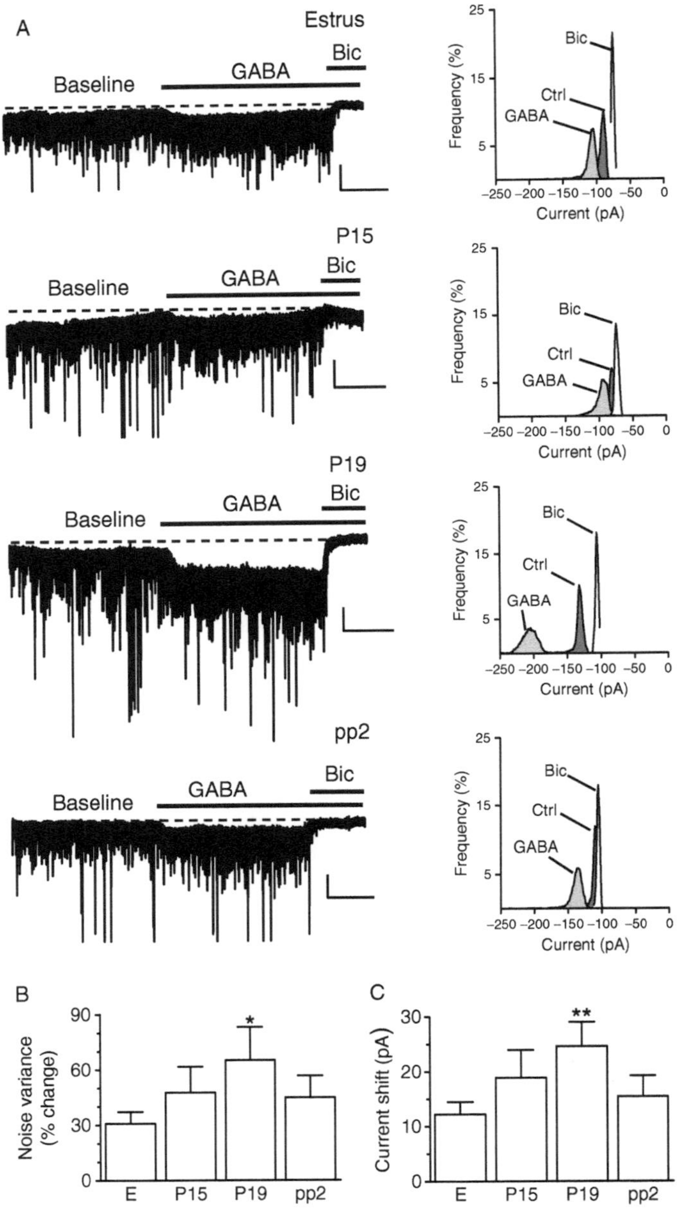

FIG. 4. Changes in tonic GABAergic current in granule cells of the dentate gyrus during pregnancy and after delivery. (A) Representative traces of GABAergic currents recorded in the whole-cell mode (holding potential, −65 mV) from granule cells of the dentate gyrus are shown (left panels) for hippocampal slices isolated from female rats in estrus (E), at day 15 (P15) or 19 (P19) of pregnancy,

GABA$_A$-R-mediated sIPSCs were examined and we found that the marked increase of decay time constant, amplitude, and area caused by this neurosteroid was not different between rats at P19 and those in estrus. Similarly, the potentiation of sIPSCs recorded in dentate gyrus granule cells induced by the benzodiazepine lorazepam (3 μM), although slightly reduced in rats at P19 compared to control, resulted not different among these animals.

To evaluate the functional relevance of the increased expression of the α_4 and γ_2 subunits in the hippocampus 2 days after delivery with respect to estrus, we tested the modulatory activity of Ro15-4513 on sIPSCs in granule cells of the dentate gyrus (Table I). Ro15-4513 is an inverse agonist at the benzodiazepine binding site of GABA$_A$-Rs that contain the α_1, α_2, α_3, or α_5 subunit together with the β and γ_2 subunit (Barnard *et al.*, 1998), but it behaves as a positive modulator at receptors formed by α_4, β, and γ_2 subunits (Knoflach *et al.*, 1996; Wafford *et al.*, 1996). In granule cells of rats in estrus and at P19, Ro15-4513 (3 μM) reduced the decay time constant of sIPSCs, but was capable of enhancing the same parameter in granule cells obtained from rats 2 days after delivery, that well correlates with our observation of an increased expression of α_4 subunit (Table I) (Sanna *et al.*, 2009).

VI. Role of Neuroactive Steroids in GABA$_A$-R Plasticity During Pregnancy: Effects of Chronic Blockade of 5α-Reductase by Finasteride

Steroid 5α-reductase, which exists in two isoforms, catalyzes the NAPDH-dependent reduction of various Δ4-3-keto steroids (progestogens and androgens) to their 5α-reduced metabolites in peripheral organ and brain (Celotti *et al.*, 1992). The 5α-reduced metabolite of testosterone, dihydrotestosterone, is essential for normal development of the male external genitalia and the prostate, whereas the 5α-reduced metabolite of progesterone is 5α-dihydroprogesterone which is then hydroxylated to 3α,5α-THP. Therefore, change in the activity of this enzyme, in addition to affecting testosterone metabolism and contributing to

---

or at 2 days after delivery (pp2). All recordings were performed in the presence of kynurenic acid (3 mM) to block glutamatergic current. After a baseline period of 5 min, application of GABA (5 μM) induced an increase in noise variance and a shift in the holding current. All GABAergic currents were blocked by the application of bicuculline (Bic) at 30 μM. Scale bars: 50 pA and 2 min. An all-point histogram for 2-min periods before (Ctrl) or during the application of GABA alone or with bicuculline is also shown for each trace (right panels). (B, C) Bar graphs summarizing the changes in noise variance (B) and holding current (C) induced by the application of exogenous GABA in experiments similar to those shown in (A). Data are mean ± S.E.M. from 32 to 67 neurons. *$P < 0.05$, **$P < 0.01$ versus estrus (ANOVA followed by Scheffe's test). Reproduced with permission from Sanna *et al.* (2009).

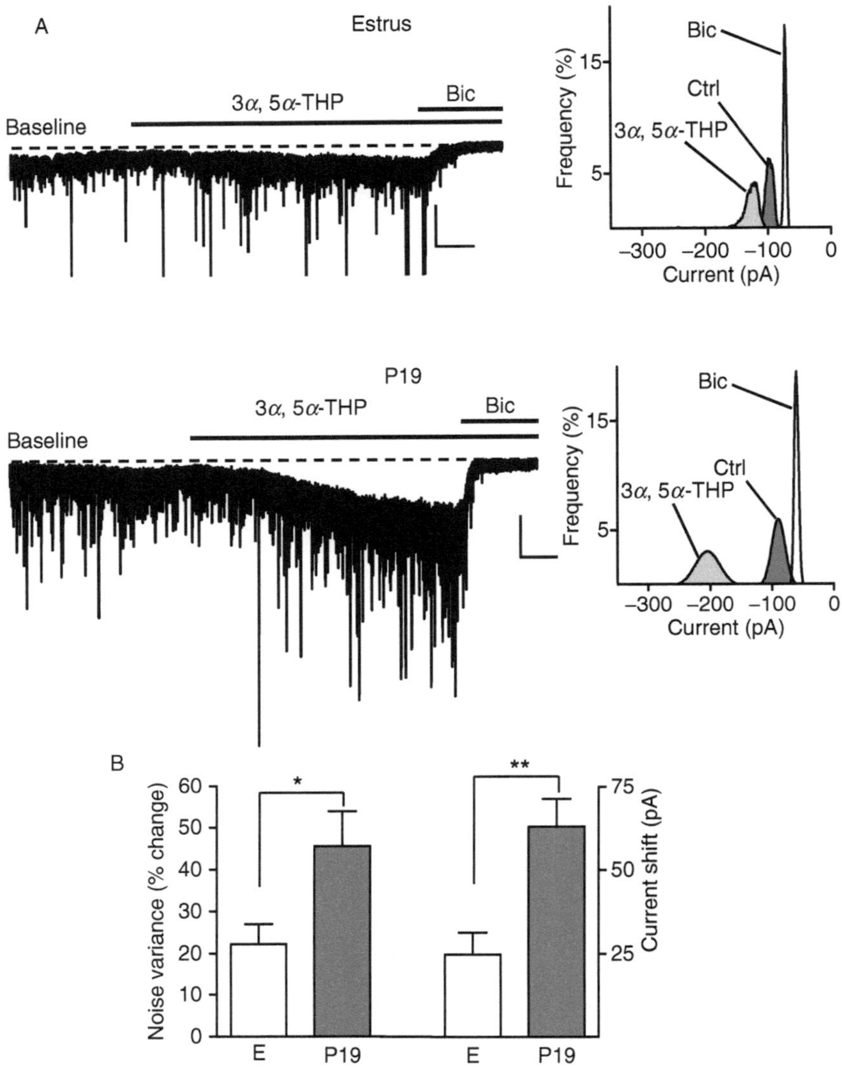

FIG. 5. Modulatory action of 3α,5α-THP on tonic GABAergic current in granule cells of the dentate gyrus during pregnancy. (A) Representative traces of GABAergic currents recorded from granule cells in hippocampal slices from rats in estrus or at day 19 (P19) of pregnancy (left panels). Recordings were obtained before (Baseline) and during bath application of 1 μM 3α,5α-THP either alone or in the presence of 30 μM bicuculline. Scale bars: 100 pA and 1 min. An all-point histogram for 2-min periods at the baseline (Ctrl) or during the application of 3α,5α-THP alone (final 2 min before the addition of bicuculline) or with bicuculline is shown for each trace (right panels). (B) Bar graph summarizing the percentage changes in noise variance (left panel) and holding current (right panel) induced by 3α,5α-THP in experiments similar to those shown in (A). Data are mean ± S.E.M. from 13 to 23 neurons. $^*P < 0.05$, $^{**}P < 0.01$ versus estrus (ANOVA followed by Scheffe's test). Reproduced with permission from Sanna *et al.* (2009).

TABLE I
EFFECT OF RO15-4513 ON GABA$_A$-R–MEDIATED sIPSCs IN GRANULE CELLS OF THE DENTATE GYRUS DURING PREGNANCY AND AFTER DELIVERY

Group (*n*)	Amplitude (% change)	Decay time (% change)	Area (% change)	Frequency (% change)
E (5)	−5.5 ± 7.2	−24.1 ± 4.1*	−31.2 ± 9.5*	−11.2 ± 7.3
P19 (8)	−10.3 ± 6.8	−19.5 ± 5.2*	−23.1 ± 10.3*	−3.6 ± 6.9
pp2 (8)	12.2 ± 10.2	28.6 ± 8.3*	25 ± 11.5	−5.8 ± 7.7

Data are expressed as the percentage change in current parameters induced by Ro15-4513 (3 μM) and are mean ± S.E.M. for the indicated numbers (*n*) of neurons in hippocampal slices isolated from rats in estrus, at day 19 of pregnancy, or at 2 days after delivery. *$P < 0.05$ versus respective control response (one-way ANOVA followed by Scheffe's test). Reproduced with permission from Sanna *et al.* (2009).

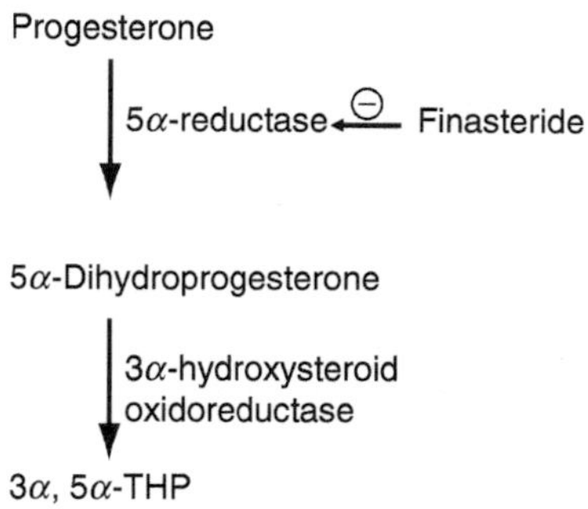

FIG. 6. Schematic mechanism of the inhibitory action of finasteride on 3α,5α-THP synthesis.

the associated disturbances, may result in changes in the concentrations of neuroactive steroids in brain and plasma, thus affecting GABA$_A$-R plasticity and function.

To clarify the role of neuroactive steroids in the plasticity of GABA$_A$-Rs during pregnancy and after delivery, we investigated the effect of blocking 5α-reductase with finasteride (Fig. 6), a 4-azasteroid developed to reduce dihydrotestosterone production in prostatic hyperplasia (Rittmaster *et al.*, 1994), on the concentrations of progesterone and 3α,5α-THP as well as on the expression and function of GABA$_A$-Rs. Subcutaneous administration of finasteride (25 mg/kg) from P12 to P18 markedly reduced the increases in the brain and plasma concentrations of 3α,5α-THP normally apparent at P19 (Concas *et al.*, 1998), while progesterone concentrations were not affected by the same treatment. Administration of finasteride also prevented both the increase in the densities of [^{3}H]flunitrazepam and [^{35}S]TBPS binding normally observed during pregnancy

(Concas *et al.*, 1998, 1999). Moreover, finasteride, but not antiestrogen clomiphene (Homburg, 2005), treatment inhibited the downregulation of γ_2 subunit expression (Concas *et al.*, 1998; Follesa *et al.*, 1998; Sanna *et al.*, 2009) and the upregulation of δ subunit expression observed in the dentate gyrus and the CA1 region at P19 (Fig. 7) (Sanna *et al.*, 2009).

Given that the enhancement of GABAergic tonic current in dentate gyrus granule cells detected in rats at P19 appeared also temporally correlated with the increase in the brain concentrations of progesterone and $3\alpha,5\alpha$-THP (Concas *et al.*, 1998), we examined whether these events might be causally related by treating the animals with finasteride. The difference in the effects of bath applied GABA to hippocampal slices on tonic current noise variance and holding current shift observed between rats at P19 and those in estrus was significantly reduced by finasteride treatment (Fig. 7). In contrast, finasteride treatment did not alter the kinetic properties of GABAergic sIPSCs in granule cells of rats at P19 compared to control.

These results demonstrate that the changes in the plasticity of $GABA_A$-Rs that occur in rat brain during pregnancy and after delivery are strictly dependent from the physiological changes in plasma and brain concentrations of neurosteroids. In fact, the finasteride-induced reversal of the increases in $3\alpha,5\alpha$-THP concentration and of the changes in the expression and function of $GABA_A$-Rs that occur during pregnancy and after delivery indicates that these latter changes are the consequence of $3\alpha,5\alpha$-THP physiological action at the steroid recognition site on the $GABA_A$-R.

VII. Conclusions

In conclusion, our data reviewed here together with those of other studies (Brussaard *et al.*, 1997; Concas *et al.*, 1998; Fenelon and Herbison, 1996; Follesa *et al.*, 1998; Maguire and Mody, 2008; Maguire *et al.*, 2005; Sanna *et al.*, 2009) support the idea of a functional correlation between the fluctuations in the plasma and brain concentrations of neuroactive steroids and $GABA_A$-R plasticity during pregnancy and immediately after delivery. We have found that the dramatic progressive increase in the brain concentrations of $3\alpha,5\alpha$-THP starting from P15 and peaking at P19 is associated causally, in a finasteride-sensitive manner, with the reduction in the expression of the γ_2 subunit of the $GABA_A$-R in the cerebral cortex and hippocampus (Concas *et al.*, 1998; Follesa *et al.*, 1998; Sanna *et al.*, 2009). The decreased expression of the γ_2 subunit was shown to correlate with a decreased activity of $GABA_A$-Rs in biochemical studies using rat brain membrane preparations (Concas *et al.*, 1998; Follesa *et al.*, 1998). However, in our more recent work (Sanna *et al.*, 2009) we did not detect a similar change when

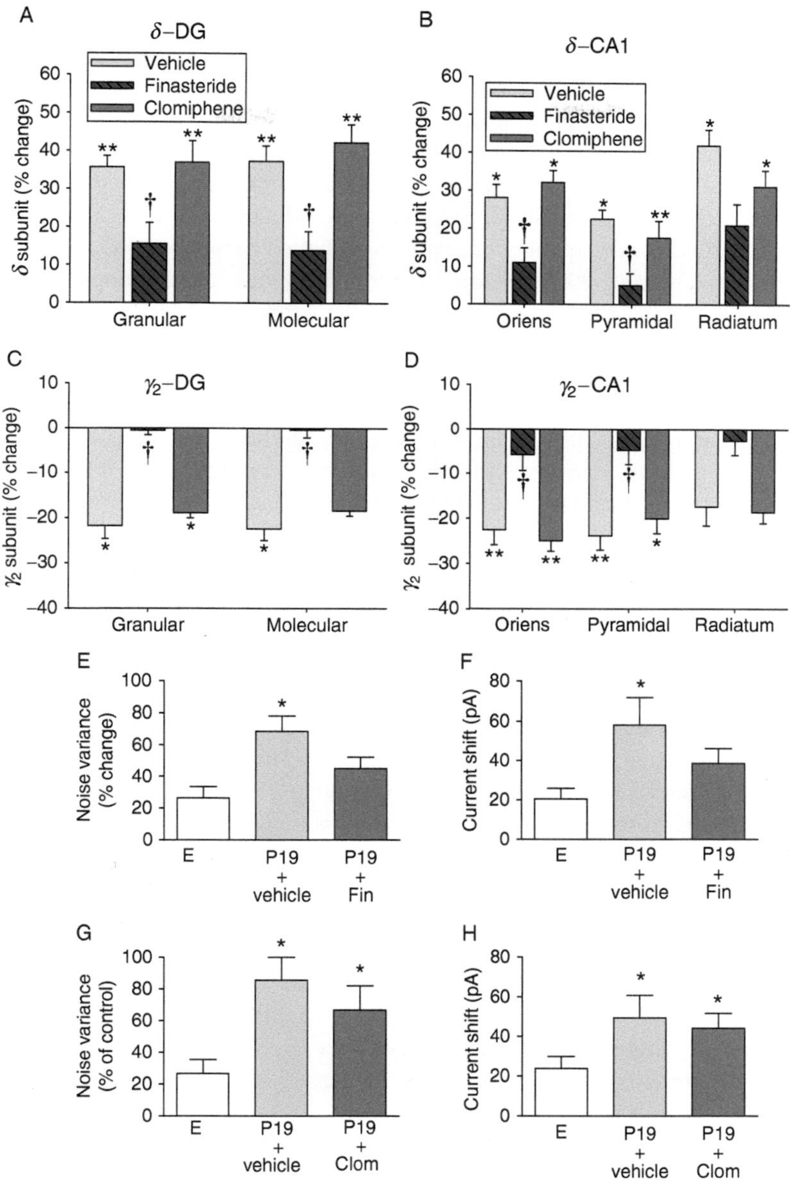

FIG. 7. Effect of treatment with finasteride or clomiphene on expression of the δ and $\gamma 2$ subunits expression and GABAergic tonic currents during pregnancy. Finasteride (25 mg/kg), clomiphene (5 mg/kg), or vehicle was administered daily from day 12 to day 18 of pregnancy, and rats were analyzed at day 19 of pregnancy for both immunohistochemistry (A–D), and electrophysiological recordings (E–H). In graphs A–D, data are expressed as percentage change relative to the

examining single dentate gyrus granule cells by whole-cell patch clamp recording. It is possible that plasticity of γ_2-containing $GABA_A$-Rs may occur preferentially in a pool of extrasynaptic receptors that are not influential on the synaptic inhibition, or it could also be that the decreased expression of the γ_2 subunit (Concas *et al.*, 1998; Sanna *et al.*, 2009) takes place at synapses that are located in the more distal portion of dendrites very distant from the cell body where the recording electrode is usually placed, so that its reduction in these conditions may be very difficult to detect functionally. It should also be considered that in another study (Maguire *et al.*, 2005), synaptic currents in dentate gyrus granule cells were not altered during the ovary cycle despite the expression of the γ_2 subunit was reduced.

We have shown also that late pregnancy is associated with a marked upregulation of δ subunit expression together with an increase in tonic currents mediated by extrasynaptic $GABA_A$-Rs in dentate gyrus granule cells. These effects are consistent with an increased density of receptors containing the δ subunit and are prevented by the treatment with finasteride, suggesting that they are mediated by the associated changes in brain levels of neuroactive steroids. Such an increase in GABAergic tonic inhibition at P19 may be relevant to further reduce the excitability and anxiety levels usually associated with the final phase of pregnancy and immediately preceding delivery (de Brito Faturi *et al.*, 2006; Skouteris *et al.*, 2009; Zuluaga *et al.*, 2005).

We found also that the expression of the α_4 subunit in the hippocampus did not change during pregnancy but rather was increased markedly after delivery, likely as a consequence of the sudden decrease in the anxiolytic neuroactive steroid concentrations. This finding is consistent with those obtained in other pharmacological studies of prolonged treatment with and subsequent withdrawal of neuroactive steroids (Biggio *et al.*, 2006; Smith *et al.*, 1998a,b; Sundstrom-Poromaa *et al.*, 2002).

Our results of an enhanced expression and function of extrasynaptic $GABA_A$-Rs in the rat dentate gyrus during pregnancy are in contrast with those reported in a similar recent study in the mouse (Maguire and Mody, 2008). These Authors found a decreased expression of both γ_2 and δ subunits in the hippocampus at P18 compared to virgin mice. These changes were accompanied by a reduction in both tonic and phasic inhibitory currents recorded in dentate gyrus granule cells

---

corresponding value for rats in estrus and are mean ± S.E.M. from at least eight animals per group. Results from rats treated with the vehicle for finasteride or with that for clomiphene were pooled. $*P <$ 0.05, $**P < 0.01$ versus rats in estrus; $^{\dagger}P < 0.05$ versus vehicle-treated rats at day 19 of pregnancy (one-way ANOVA followed by Scheffe's test). For electrophysiological analysis (E–H), data are expressed the percentage changes in noise variance (E, G) and holding current (F, H) induced by GABA in hippocampal slices from rats in estrus as well as from those at day 19 of pregnancy that had been treated with finasteride (E, F) or clomiphene (G, H). $*P < 0.05$ versus estrus (ANOVA followed by Scheffe's test). Modified with permission from Sanna *et al.* (2009).

at P18, with these effects rebounding 48 h after delivery. Although the reasons that may account for this difference in results in these two studies are presently not clear, it is possible that species difference as well as other differences related to the experimental conditions, among others the choice of the control animals (estrus vs diestrus phase of the ovary cycle), may explain the different results.

In conclusion, the opposite regulation of extrasynaptic GABA$_A$-R-mediated tonic currents in response to fluctuations in brain concentrations of neuroactive steroids may be crucial for the changes in mood and for the anxiolytic effect associated with pregnancy (de Brito Faturi *et al.*, 2006) as well as for the arousal state characteristic of the period immediately preceding delivery and of early postpartum.

Acknowledgments

This work was supported by grants from the National Research Council, the Istituto Superiore della Sanità, the Sardinian Department of Health and Welfare, and the Gioventù in Armonia (GIO I A) Foundation (Pisa, Italy).

References

Adkins, C. E., Pillai, G. V., Kerby, J., Bonnert, T. P., Haldon, C., McKernan, R. M., Gonzalez, J. E., Oades, K., Whiting, P. J., and Simpson, P. B. (2001). Alpha4beta3delta GABA(A) receptors characterized by fluorescence resonance energy transfer-derived measurements of membrane potential. *J. Biol. Chem.* **276,** 38934–38939.

Barbaccia, M. L., Affricano, D., Purdy, R. H., Maciocco, E., Spiga, F., and Biggio, G. (2001). Clozapine, but not haloperidol, increases brain concentrations of neuroactive steroids in the rat. *Neuropsychopharmacology* **25,** 489–497.

Barnard, E. A., Skolnick, P., Olsen, R. W., Mohler, H., Sieghart, W., Biggio, G., Braestrup, C., Bateson, A. N., and Langer, S. Z. (1998). International Union of Pharmacology. XV. Subtypes of gamma-aminobutyric acidA receptors: Classification on the basis of subunit structure and receptor function. *Pharmacol. Rev.* **50,** 291–313.

Belelli, D., and Lambert, J. J. (2005). Neurosteroids: Endogenous regulators of the GABA(A) receptor. *Nat. Rev. Neurosci.* **6,** 565–575.

Bianchi, M. T., and Macdonald, R. L. (2003). Neurosteroids shift partial agonist activation of GABA(A) receptor channels from low- to high-efficacy gating patterns. *J. Neurosci.* **23,** 10934–10943.

Biggio, F., Gorini, G., Caria, S., Murru, L., Mostallino, M. C., Sanna, E., and Follesa, P. (2006). Plastic neuronal changes in GABA(A) receptor gene expression induced by progesterone metabolites: *In vitro* molecular and functional studies. *Pharmacol. Biochem. Behav.* **84,** 545–554.

Brown, N., Kerby, J., Bonnert, T. P., Whiting, P. J., and Wafford, K. A. (2002). Pharmacological characterization of a novel cell line expressing human alpha(4)beta(3)delta GABA(A) receptors. *Br. J. Pharmacol.* **136,** 965–974.

Brussaard, A. B., Kits, K. S., Baker, R. E., Willems, W. P., Leyting-Vermeulen, J. W., Voorn, P., Smit, A. B., Bicknell, R. J., and Herbison, A. E. (1997). Plasticity in fast synaptic inhibition of adult oxytocin neurons caused by switch in GABA(A) receptor subunit expression. *Neuron* **19,** 1103–1114.

Buster, J. E. (1983). Gestational changes in steroid hormone biosynthesis, secretion, metabolism, and action. *Clin. Perinatol.* **10,** 527–552.

Caraiscos, V. B., Elliott, E. M., You-Ten, K. E., Cheng, V. Y., Belelli, D., Newell, J. G., Jackson, M. F., Lambert, J. J., Rosahl, T. W., Wafford, K. A., MacDonald, J. F., and Orser, B. A. (2004). Tonic inhibition in mouse hippocampal CA1 pyramidal neurons is mediated by alpha5 subunit-containing gamma-aminobutyric acid type A receptors. *Proc. Natl. Acad. Sci. USA* **101,** 3662–3667.

Celotti, F., Melcangi, R. C., and Martini, L. (1992). The 5 alpha-reductase in the brain: Molecular aspects and relation to brain function. *Front. Neuroendocrinol.* **13,** 163–215.

Concas, A., Mostallino, M. C., Porcu, P., Follesa, P., Barbaccia, M. L., Trabucchi, M., Purdy, R. H., Grisenti, P., and Biggio, G. (1998). Role of brain allopregnanolone in the plasticity of gamma-aminobutyric acid type A receptor in rat brain during pregnancy and after delivery. *Proc. Natl. Acad. Sci. USA* **95,** 13284–13289.

Concas, A., Follesa, P., Barbaccia, M. L., Purdy, R. H., and Biggio, G. (1999). Physiological modulation of GABA(A) receptor plasticity by progesterone metabolites. *Eur. J. Pharmacol.* **375,** 225–235.

Concas, A., Porcu, P., Sogliano, C., Serra, M., Purdy, R. H., and Biggio, G. (2000). Caffeine-induced increases in the brain and plasma concentrations of neuroactive steroids in the rat. *Pharmacol. Biochem. Behav.* **66,** 39–45.

de Brito Faturi, C., Teixeira-Silva, F., and Leite, J. R. (2006). The anxiolytic effect of pregnancy in rats is reversed by finasteride. *Pharmacol. Biochem. Behav.* **85,** 569–574.

Drasbek, K. R., and Jensen, K. (2006). THIP, a hypnotic and antinociceptive drug, enhances an extrasynaptic GABAA receptor-mediated conductance in mouse neocortex. *Cereb. Cortex.* **16,** 1134–1141.

Farrant, M., and Nusser, Z. (2005). Variations on an inhibitory theme: Phasic and tonic activation of GABA(A) receptors. *Nat. Rev. Neurosci.* **6,** 215–229.

Fenelon, V. S., and Herbison, A. E. (1996). Plasticity in GABAA receptor subunit mRNA expression by hypothalamic magnocellular neurons in the adult rat. *J. Neurosci.* **16,** 4872–4880.

Follesa, P., Floris, S., Tuligi, G., Mostallino, M. C., Concas, A., and Biggio, G. (1998). Molecular and functional adaptation of the GABA(A) receptor complex during pregnancy and after delivery in the rat brain. *Eur. J. Neurosci.* **10,** 2905–2912.

Follesa, P., Porcu, P., Sogliano, C., Cinus, M., Biggio, F., Mancuso, L., Mostallino, M. C., Paoletti, A. M., Purdy, R. H., Biggio, G., and Concas, A. (2002). Changes in GABAA receptor gamma 2 subunit gene expression induced by long-term administration of oral contraceptives in rats. *Neuropharmacology* **42,** 325–336.

Glykys, J., and Mody, I. (2006). Hippocampal network hyperactivity after selective reduction of tonic inhibition in GABA A receptor alpha5 subunit-deficient mice. *J. Neurophysiol.* **95,** 2796–2807.

Glykys, J., and Mody, I. (2007). Activation of GABAA receptors: Views from outside the synaptic cleft. *Neuron* **56,** 763–770.

Glykys, J., Mann, E. O., and Mody, I. (2008). Which GABA(A) receptor subunits are necessary for tonic inhibition in the hippocampus? *J. Neurosci.* **28,** 1421–1426.

Griffiths, J. L., and Lovick, T. A. (2005). GABAergic neurones in the rat periaqueductal grey matter express alpha4, beta1 and delta GABAA receptor subunits: Plasticity of expression during the estrous cycle. *Neuroscience* **136,** 457–466.

Homburg, R. (2005). Clomiphene citrate--end of an era? A mini-review. *Hum. Reprod.* **20,** 2043–2051.

Jia, F., Pignataro, L., Schofield, C. M., Yue, M., Harrison, N. L., and Goldstein, P. A. (2005). An extrasynaptic GABAA receptor mediates tonic inhibition in thalamic VB neurons. *J. Neurophysiol.* **94,** 4491–4501.

Knoflach, F., Benke, D., Wang, Y., Scheurer, L., Luddens, H., Hamilton, B. J., Carter, D. B., Mohler, H., and Benson, J. A. (1996). Pharmacological modulation of the diazepam-insensitive recombinant gamma-aminobutyric acidA receptors alpha 4 beta 2 gamma 2 and alpha 6 beta 2 gamma 2. *Mol. Pharmacol.* **50,** 1253–1261.

Lovick, T. A. (2006). Plasticity of GABAA receptor subunit expression during the oestrous cycle of the rat: Implications for premenstrual syndrome in women. *Exp. Physiol.* **91,** 655–660.

Lovick, T. A., Griffiths, J. L., Dunn, S. M., and Martin, I. L. (2005). Changes in GABA(A) receptor subunit expression in the midbrain during the oestrous cycle in Wistar rats. *Neuroscience* **131,** 397–405.

Luisi, S., Petraglia, F., Benedetto, C., Nappi, R. E., Bernardi, F., Fadalti, M., Reis, F. M., Luisi, M., and Genazzani, A. R. (2000). Serum allopregnanolone levels in pregnant women: Changes during pregnancy, at delivery, and in hypertensive patients. *J. Clin. Endocrinol. Metab.* **85,** 2429–2433.

Maayan, R., Fisch, B., Galdor, M., Kaplan, B., Shinnar, N., Kinor, N., Zeldich, E., Valevski, A., and Weizman, A. (2004). Influence of 17beta-estradiol on the synthesis of reduced neurosteroids in the brain (*in vivo*) and in glioma cells (*in vitro*): Possible relevance to mental disorders in women. *Brain Res.* **1020,** 167–172.

Maguire, J., and Mody, I. (2008). GABA(A)R plasticity during pregnancy: Relevance to postpartum depression. *Neuron* **59,** 207–213.

Maguire, J. L., Stell, B. M., Rafizadeh, M., and Mody, I. (2005). Ovarian cycle-linked changes in GABA(A) receptors mediating tonic inhibition alter seizure susceptibility and anxiety. *Nat. Neurosci.* **8,** 797–804.

Majewska, M. D. (1992). Neurosteroids: Endogenous bimodal modulators of the GABAA receptor. Mechanism of action and physiological significance. *Prog. Neurobiol.* **38,** 379–395.

Milewich, L., Gant, N. F., Schwarz, B. E., Chen, G. T., and MacDonald, P. C. (1979). 5 alpha-Reductase activity in human placenta. *Am. J. Obstet. Gynecol.* **133,** 611–617.

Nusser, Z., and Mody, I. (2002). Selective modulation of tonic and phasic inhibitions in dentate gyrus granule cells. *J. Neurophysiol.* **87,** 2624–2628.

Nusser, Z., Roberts, J. D., Baude, A., Richards, J. G., and Somogyi, P. (1995). Relative densities of synaptic and extrasynaptic GABAA receptors on cerebellar granule cells as determined by a quantitative immunogold method. *J. Neurosci.* **15,** 2948–2960.

Nusser, Z., Sieghart, W., and Somogyi, P. (1998). Segregation of different GABAA receptors to synaptic and extrasynaptic membranes of cerebellar granule cells. *J. Neurosci.* **18,** 1693–1703.

Paoletti, A. M., Romagnino, S., Contu, R., Orru, M. M., Marotto, M. F., Zedda, P., Lello, S., Biggio, G., Concas, A., and Melis, G. B. (2006). Observational study on the stability of the psychological status during normal pregnancy and increased blood levels of neuroactive steroids with GABA-A receptor agonist activity. *Psychoneuroendocrinology* **31,** 485–492.

Penning, T. M., Sharp, R. B., and Krieger, N. R. (1985). Purification and properties of 3 alpha-hydroxysteroid dehydrogenase from rat brain cytosol. Inhibition by nonsteroidal anti-inflammatory drugs and progestins. *J. Biol. Chem.* **260,** 15266–15272.

Pirker, S., Schwarzer, C., Wieselthaler, A., Sieghart, W., and Sperk, G. (2000). GABA(A) receptors: Immunocytochemical distribution of 13 subunits in the adult rat brain. *Neuroscience* **101,** 815–850.

Porcello, D. M., Huntsman, M. M., Mihalek, R. M., Homanics, G. E., and Huguenard, J. R. (2003). Intact synaptic GABAergic inhibition and altered neurosteroid modulation of thalamic relay neurons in mice lacking delta subunit. *J. Neurophysiol.* **89,** 1378–1386.

Rittmaster, R. S., Antonian, L., New, M. I., and Stoner, E. (1994). Effect of finasteride on adrenal steroidogenesis in men. *J. Androl.* **15,** 298–301.

Sanna, E., Talani, G., Busonero, F., Pisu, M. G., Purdy, R. H., Serra, M., and Biggio, G. (2004). Brain steroidogenesis mediates ethanol modulation of GABAA receptor activity in rat hippocampus. *J. Neurosci.* **24,** 6521–6530.

Sanna, E., Mostallino, M. C., Murru, L., Carta, M., Talani, G., Zucca, S., Mura, M. L., Maciocco, E., and Biggio, G. (2009). Changes in expression and function of extrasynaptic GABAA receptors in the rat hippocampus during pregnancy and after delivery. *J. Neurosci.* **29,** 1755–1765.

Semyanov, A., Walker, M. C., Kullmann, D. M., and Silver, R. A. (2004). Tonically active GABA A receptors: Modulating gain and maintaining the tone. *Trends Neurosci.* **27,** 262–269.

Shen, H., Gong, Q. H., Yuan, M., and Smith, S. S. (2005). Short-term steroid treatment increases delta GABAA receptor subunit expression in rat CA1 hippocampus: Pharmacological and behavioral effects. *Neuropharmacology* **49,** 573–586.

Skouteris, H., Wertheim, E. H., Rallis, S., Milgrom, J., and Paxton, S. J. (2009). Depression and anxiety through pregnancy and the early postpartum: An examination of prospective relationships. *J. Affect. Disord.* **113,** 303–308.

Smith, S. S., Gong, Q. H., Hsu, F. C., Markowitz, R. S., ffrench-Mullen, J. M., and Li, X. (1998a). GABA(A) receptor alpha4 subunit suppression prevents withdrawal properties of an endogenous steroid. *Nature* **392,** 926–930.

Smith, S. S., Gong, Q. H., Li, X., Moran, M. H., Bitran, D., Frye, C. A., and Hsu, F. C. (1998b). Withdrawal from 3alpha-OH-5alpha-pregnan-20-One using a pseudopregnancy model alters the kinetics of hippocampal GABAA-gated current and increases the GABAA receptor alpha4 subunit in association with increased anxiety. *J. Neurosci.* **18,** 5275–5284.

Smith, S. S., Ruderman, Y., Frye, C., Homanics, G., and Yuan, M. (2006). Steroid withdrawal in the mouse results in anxiogenic effects of 3alpha,5beta-THP: A possible model of premenstrual dysphoric disorder. *Psychopharmacology (Berl)* **186,** 323–333.

Stell, B. M., Brickley, S. G., Tang, C. Y., Farrant, M., and Mody, I. (2003). Neuroactive steroids reduce neuronal excitability by selectively enhancing tonic inhibition mediated by delta subunit-containing GABAA receptors. *Proc. Natl. Acad. Sci. USA* **100,** 14439–14444.

Sundstrom-Poromaa, I., Smith, D. H., Gong, Q. H., Sabado, T. N., Li, X., Light, A., Wiedmann, M., Williams, K., and Smith, S. S. (2002). Hormonally regulated alpha(4)beta(2)delta GABA(A) receptors are a target for alcohol. *Nat. Neurosci.* **5,** 721–722.

Wafford, K. A., Thompson, S. A., Thomas, D., Sikela, J., Wilcox, A. S., and Whiting, P. J. (1996). Functional characterization of human gamma-aminobutyric acidA receptors containing the alpha 4 subunit. *Mol. Pharmacol.* **50,** 670–678.

Wei, W., Zhang, N., Peng, Z., Houser, C. R., and Mody, I. (2003). Perisynaptic localization of delta subunit-containing GABA(A) receptors and their activation by GABA spillover in the mouse dentate gyrus. *J. Neurosci.* **23,** 10650–10661.

Whiting, P. J., Bonnert, T. P., McKernan, R. M., Farrar, S., Le Bourdelles, B., Heavens, R. P., Smith, D. W., Hewson, L., Rigby, M. R., Sirinathsinghji, D. J., Thompson, S. A., and Wafford, K. A. (1999). Molecular and functional diversity of the expanding GABA-A receptor gene family. *Ann. N. Y. Acad. Sci.* **868,** 645–653.

Wohlfarth, K. M., Bianchi, M. T., and Macdonald, R. L. (2002). Enhanced neurosteroid potentiation of ternary GABA(A) receptors containing the delta subunit. *J. Neurosci.* **22,** 1541–1549.

Zuluaga, M. J., Agrati, D., Pereira, M., Uriarte, N., Fernandez-Guasti, A., and Ferreira, A. (2005). Experimental anxiety in the black and white model in cycling, pregnant and lactating rats. *Physiol. Behav.* **84,** 279–286.

EARLY POSTNATAL STRESS AND NEURAL CIRCUIT UNDERLYING EMOTIONAL REGULATION

Machiko Matsumoto,* Mitsuhiro Yoshioka,† and Hiroko Togashi*

*Department of Pharmacology, School of Pharmaceutical Science, Health Sciences University of Hokkaido, Ishikari-Tobetsu, Japan

†Department of Neuropharmacology, Hokkaido University Graduate School of Medicine, Sapporo, Japan

Several lines of evidence have shown that traumatic events during the early postnatal period precipitate long-lasting alterations in the functional properties underlying emotional expression that are attributable to the pathophysiology of stress-related disorders. In this chapter, we summarize our recent work elucidating whether early postnatal stress alters the neural circuits underlying emotional regulation. Rats exposed to footshock stress (FS) during the second (2W) or the third (3W) postnatal week were subjected to unconditioned and conditioned stresses at the postadolescent period (10–12 weeks). No differences in locomotor activity or hippocampal synaptic changes were observed between the FS-groups and non-FS controls during exposure to open field stress. Fear-related freezing behavior during exposure to contextual fear conditioning (CFC) was markedly attenuated in the 2W-FS group, presumably due to disturbance of the retention for fear memory, an effect that was attributable to synaptic changes in the hippocampal CA1 field. The 3W-FS group exhibited attenuation of the decreases in freezing behavior induced by CFC extinction trials. The deficits in extinction was abolished by repeated treatment with the partial *N*-methyl-D-aspartate receptor agonist D-cycloserine, suggesting that aversive stress exposure during the third postnatal week impaired extinction of context-dependent fear memory. Taken together, the altered behavior observed in adulthood is likely the result of

DOI: 10.1016/S0074-7742(09)85007-1

0074-7742/09 $35.00

neurodevelopmental perturbations elicited by early life stress. Thus, a "critical period" exists for neural circuits involved in emotional expression that may contribute to lifelong susceptibility to stress.

I. Introduction

Several studies have provided evidence that stressful life events increase the susceptibility to emotional stress, termed emotional vulnerability, and contribute to the pathophysiology of stress-related disorders such as depression and anxiety (McAllister-Williams *et al.*, 1998; Modell *et al.*, 1998). Early life stress precipitates long-lasting alterations in the functional properties underlying emotional expression and consequently may alter the responsiveness to stress later in life (Heim and Nemeroff, 2001). The early postnatal period is a critical time for the functional development of the brain. In rodents, tremendous synaptogenesis and neuronal differentiation occur at 2–3 weeks after birth. The density of synapses in the cortex increases fivefold between postnatal day 10 (PND 10) and PND 15 and gradually approaches adult levels (Micheva and Beaulieu, 1996). However, the hippocampus and prefrontal cortex (PFC), in contrast to the amygdala, are not yet fully developed until the third postnatal week. Neurotransmitters regulating the emotional system, such as serotonin (5-hydroxytryptamine, 5-HT) and γ-aminobutyric acid (GABA), are known to modulate developmental processes, that is, neuronal differentiation, synaptogenesis, and migration (Sodhi and Sanders-Bush, 2004). Several studies have demonstrated the developmental transition from depolarization to hyperpolarization of 5-HT and GABA receptors in the corticolimbic system during the early postnatal period (Béïque *et al.*, 2004, 2007; Tyzio *et al.*, 2008). The differentiation and formation of the neuronal system, including signal transduction, may be influenced by stressful situations during brain development. Thus, traumatic events during early life may affect the neural systems associated with emotional regulation, depending on exposure time, and consequently may lead to altered sensitivity to stress later in life.

Based on this hypothesis, we examined whether early postnatal stress affects the neural circuits underlying emotional regulation using electrophysiological approaches combined with behavioral analysis. In this chapter, we summarize our recent study on the relationship between the behavioral responses to emotional stress and synaptic transmission, including synaptic plasticity in the hippocampus, a structure that is critically involved in learning and memory processes. Behavioral analysis was performed by evaluating hippocampus-dependent tasks: unconditioned stress using the open field test (i.e., novel environment stress) and conditioned stress based on fear memory (i.e., contextual fear conditioning, CFC). Furthermore, the present chapter discusses the relevance of the hippocampal–medial prefrontal cortex (mPFC) pathway as a key neural circuit for understanding early postnatal stress-induced behavioral alterations in extinction processes.

II. Behavioral Response and Neural Circuits in Early Postnatal Stressed Rats

A. Unconditioned Stress: Novel Environment Stress

Rats that were exposed to footshock stress (FS) stimulation during PND 14 to PND 18 and during PND 21 to PND 25 were denoted 2W-FS and 3W-FS, respectively. Pups were acclimated to the FS box and subjected to five FS (shock intensity, 0.5 mA; intershock interval, 30 s; shock duration, 2 s) for 5 days. After the last FS stimulation, each rat remained in the FS box for 5 min. Non-FS controls remained in the FS box for the same time period (12.5 min) without FS stimuli. During the postadolescent period of 10–12 weeks, the open field test was performed as novel environment stress exposure. As shown in Fig. 1, no significant difference in locomotor activity was observed by estimating total crossings during exposure to open field stress between the FS groups and non-FS controls, although the 3W-FS group exhibited different patterns of behavior.

The firing rate of hippocampal CA1 pyramidal cells is known to depend on spatial location, that is, contextual environment information related to exploration behavior is encoded by cell activity (Hollup *et al.*, 2001; Thompson and Best, 1990). In the present study, the population spike amplitude (PSA), measured with simultaneous determination of locomotor activity, in the CA1 field evoked by Schaffer collaterals stimulation was slightly decreased during exposure to the open field (Fig. 1). However, no significant difference in changes of synaptic transmission in the CA1 field was observed between the FS groups and non-FS controls (Koseki *et al.*, 2007).

Hippocampal synaptic plasticity, long-term potentiation (LTP), is considered to be the electrophysiological basis underlying cellular mechanisms of learning and memory (Bliss and Collingridge, 1993; Martin *et al.*, 2000). In general, severe stresses impair hippocampus-dependent learning and memory (Malenka and Nicoll, 1999; Yang *et al.*, 2003). Numerous studies have shown that hippocampal LTP is suppressed by a variety of stressors (Matsumoto *et al.*, 2004a,b; Shakesby *et al.*, 2002; Xu *et al.*, 1997; Yang *et al.*, 2005). We recently demonstrated using freely moving rats that LTP in the CA1 field was suppressed by elevated platform stress (Hirata *et al.*, 2008). LTP induction, however, was not affected after exposure to open field stress. Serum corticosterone, a neurochemical marker of stress, was markedly increased by exposure to elevated platform stress compared with open field stress. Based on these findings, we suggested that stress-induced LTP

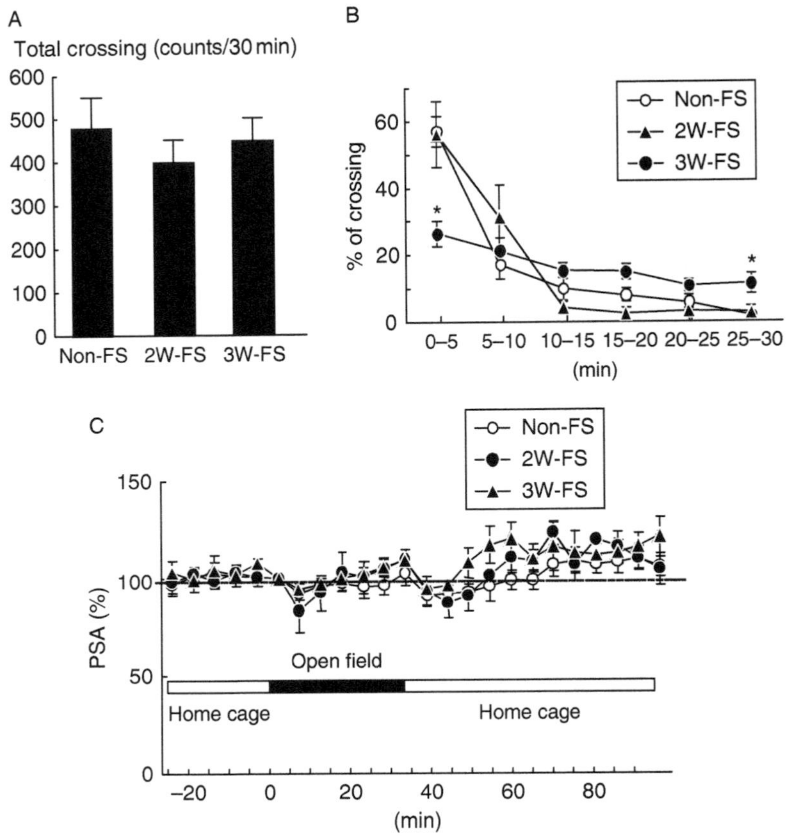

FIG. 1. Behavioral response and hippocampal synaptic transmission during exposure to open field stress in freely moving rats. Behavior analysis and electrophysiological experiments were performed simultaneously during the postadolescent period (10–12 weeks old). (A) Locomotor activity estimated by total crossings for 30 min and (B) time-course of crossings during exposure to open field stress. (C) Time-course of population spike amplitude (PSA) in the hippocampal CA1 field evoked by Schaffer collaterals stimulation. Values are expressed as a percentage of the baseline level before open field stress. Non-FS, pups exposed to the footshock (FS) box without FS; 2W-FS and 3W-FS, pups exposed to FS during the second and third postnatal weeks, respectively. Each value represents the mean ± S.E.M. *$p < 0.05$ versus non-FS controls (modified from Koseki *et al.*, 2007).

suppression in the CA1 field depends on the relative intensity of the stressor (Hirata *et al.*, 2008). The fact that the synaptic response, including synaptic plasticity, was not associated with open field stress may suggest that hippocampal function is not responsible for locomotion. These electrophysiological and behavioral evidence reveal that no link exists between early postnatal stress and intra hippocampal transmission with regard to novel environment stress.

B. Conditioned Stress: Contextual Fear Conditioning

1. *Acquisition and Retention of Contextual Fear Conditioning*

Contextual fear conditioning (CFC) that depends on fear memory has been established as an animal model of anxiety. During the postadolescent period, rats were exposed to CFC as shown in Fig. 2. Fear-related freezing behavior during

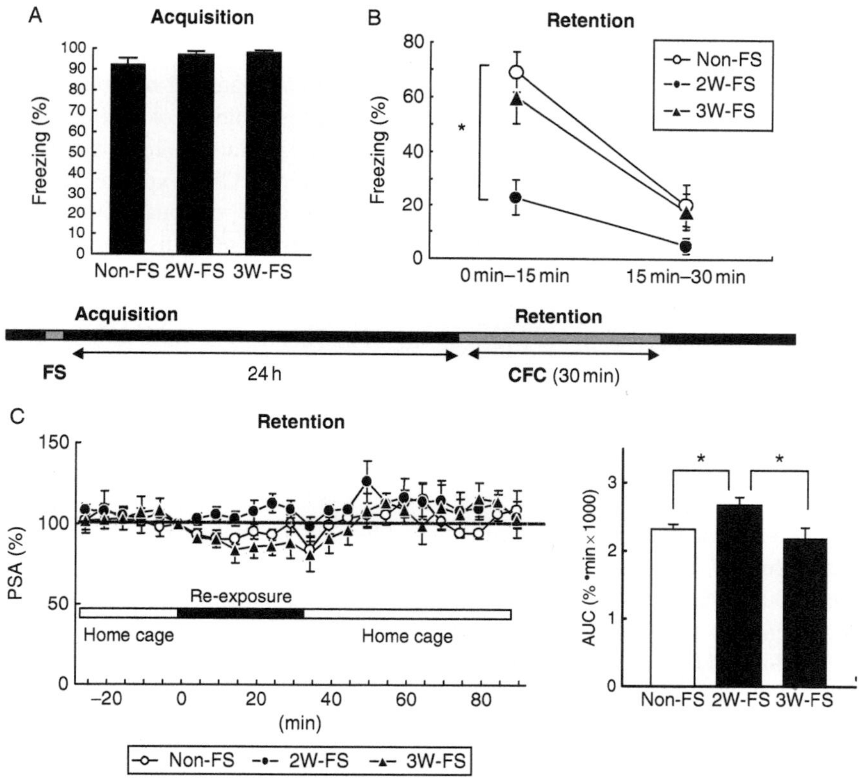

Fig. 2. Behavioral response and hippocampal synaptic transmission during the acquisition and retention of contextual fear conditioning (CFC). Behavior analysis and electrophysiological experiments during the CFC retention session were performed simultaneously during the postadolescent period (10–12 weeks old). Fear-related freezing behavior was determined (A) immediately after footshock (FS) conditioning (i.e., acquisition session) and (B) during exposure to CFC 24 h after FS stimulation (i.e., retention session). (C) Time-course (left panel) and areas-under-the-curve (AUC) (right panel) of the population spike amplitude (PSA) in the hippocampal CA1 field evoked by Schaffer collaterals stimulation. Values are expressed as a percentage of the baseline level of exposure to CFC. AUC (% min × 10^3) of the time-course changes was determined to evaluate the ensemble effect of PSA. Non-FS, pups exposed to the FS box without FS; 2W-FS and 3W-FS, pups exposed to FS during the second and third postnatal weeks, respectively. Each value represents the mean ± S.E.M. *$p < 0.05$ (modified from Koseki *et al.*, 2007).

exposure to CFC was markedly reduced in the 2W-FS group, but not 3W-FS group. Freezing behavior observed immediately after FS conditioning (i.e., acquisition session) in the 2W-FS group was similar to the other groups, indicating that the 2W-FS group acquired the learning process during FS conditioning. Nevertheless, remarkably attenuated freezing behavior was found during the CFC retention session in the 2W-FS group. It suggests that aversive stimulus exposure during the second postnatal week altered the response to emotional stress during the postadolescent period, presumably due to perturbation of the retention and/or consolidation of fear memory (Matsumoto *et al.*, 2005).

The hippocampus is critically involved in associative fear memory evoked by contextual stimuli. The electrophysiological profile of pyramidal neurons in the CA1 field, assessed by input–output curves and the magnitude of LTP, did not exhibit any significant differences between the groups (Matsumoto *et al.*, 2005). Basal synaptic transmission, however, was reduced during CFC exposure in non-FS controls. These results appear to be consistent with a previous report (Tada *et al.*, 2004) showing that the firing rate of CA1 pyramidal neurons was inhibited by CFC. The decreases in synaptic transmission were also observed in the 3W-FS group. In contrast, synaptic transmission did not decrease, but rather increased, during exposure to CFC in the 2W-FS group (Fig. 2). These findings indicate that the synaptic transmission in the CA1 field may have been responsible for the freezing behavior observed during the retention session (Koseki *et al.*, 2007).

LTP induction in the CA1 field was blocked by exposure to CFC under consciousness (Hirata *et al.*, 2008), as well as under anesthesia (Matsumoto *et al.*, 2004a). Interestingly, LTP is suppressed not only by stress paradigms but also by low-frequency stimulation (LFS) prior to LTP-inducing high-frequency stimulation, termed homosynaptic metaplasticity (Abraham and Bear, 1996; Abraham *et al.*, 2001; Zelcer *et al.*, 2006). Although the functional significance of metaplasticity remains to be clarified, we recently reported that the synaptic response induced by CFC was mimicked by LFS, that is, LFS impaired synaptic transmission and subsequent LTP in the CA1 field (Hirata *et al.*, 2009). These findings were supported by the electrophysiological and neurochemical profiles.: Both CFC and LFS increased extracellular levels of hippocampal GABA and serum corticosterone. Thus, metaplasticity in the CA1 field appears to be related to the neural basis of stress experience-dependent fear memory. This historical synaptic event may influence the later efficacy of synaptic transmission and plasticity. Supporting this hypothesis, we recently observed that LFS-induced LTP suppression in the CA1 field was not found in 2W-FS groups, that is, LTP was induced after LFS prior to tetanus (unpublished data). On the other hand, we proposed the possible involvement of the mPFC and/or amygdala in hippocampal function, including metaplasticity, associated with context-specific fear memory based on the electrophysiological and behavioral analysis (Hirata *et al.*, 2009). Moriceau and Sullivan (2006) reported that odor/shock conditioning in preweanling pups is influenced by maternal presence, depending on context. Maternal presence suppressed stress-induced corticosterone release and neuronal activity in the amygdala, which was enhanced by learned odor aversion paired with the initial aversive conditioning (Sullivan *et al.*, 2000). Abnormal fear-related behavior (i.e., low anxiety-like behavior) observed in the 2W-FS group, therefore, may be attributable to neuronal changes in the hippocampus and/or amygdala associated with the retention and/or consolidation of fear memory.

2. *Extinction of Contextual Fear Conditioning*

Fear extinction does not result from loss of the fear memory itself, but rather from a learning process based on the formation of new inhibitory associations between conditioned and unconditioned stimuli (Myers and Davis, 2002; Kim and Jung, 2006). Extinction is influenced by stressors of varying types and intensities. Stressful events are known to affect memory function and traumatic stress is especially believed to result in functional changes in brain regions involved in memory function, such as fear conditioning and extinction. For example, chronic restraint stress impaired the recall of extinction of conditioned fear (Miracle *et al.*, 2006). Shumake *et al.* (2005) reported that rat strains selectively bred for increased susceptibility to learned helplessness showed resistance to extinction of conditioned fear. Early life stress impaired the extinction of tone- and context-dependent fear, an effect that varied by gender and age (Toledo-Rodriguez and Sandi, 2007). In our study, early postnatal stress influenced not only retention, but also extinction of fear memory (Matsumoto *et al.*, 2008). The decrease in freezing behavior was significantly attenuated in the 3W-FS group compared with non-FS controls. The 2W-FS group exhibited a lower level of freezing on day 1, but its decrease did not differ from controls during individual extinction trials (Fig. 3). Thus, exposure to aversive stimuli during the third postnatal week may influence the neural circuits and/or neural mechanisms that mediate extinction of context-dependent fear memory. Accordingly, the third postnatal week may be critical for establishing the neural circuits associated with the extinction processes of contextual fear memory.

Extinction is considered to represent a form of new learning rather than the forgetting or erasure of memory (Myers and Davis, 2002; Quirk and Mueller, 2008). Memory consolidation generally requires gene expression and protein synthesis (Akirav *et al.*, 2006; Kim and Jung, 2006; Quirk *et al.*, 2000). Several studies, however, have provided evidence that hippocampus-dependent extinction is involved in contextual encoding processes (Corcoran and Maren, 2001, 2004; Ji and Maren, 2005) and is mediated by protein synthesis-independent mechanisms (Fischer *et al.*, 2004; Lattal and Abel, 2001). Thus, extinction of context-dependent fear memory may be mediated by different neural mechanisms and/or neural circuits associated with cue-dependent extinction. Recent studies have explored synaptic changes in the hippocampal–mPFC pathway

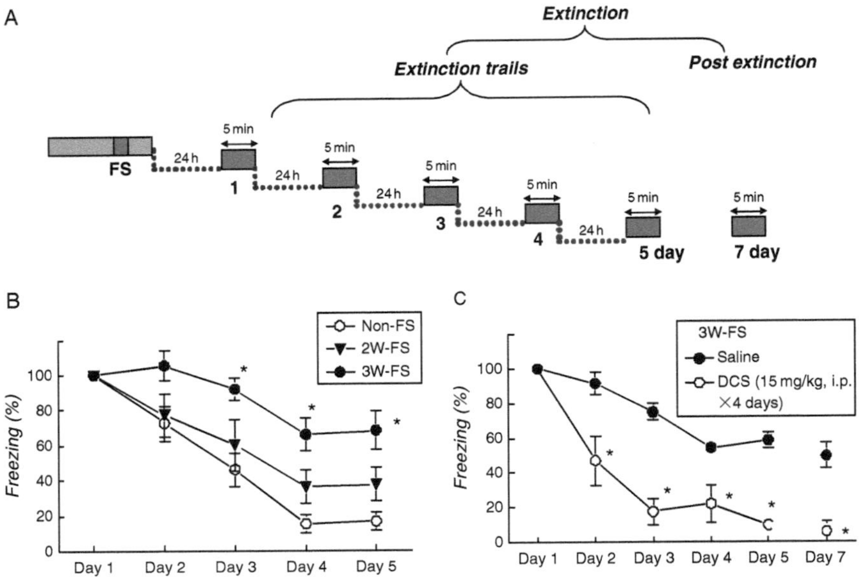

FIG. 3. Behavioral response during the extinction trial of contextual fear conditioning (CFC). (A) Experimental protocol of CFC extinction. Extinction was estimated by measuring freezing behavior consisting of 5 min exposure to the conditioning chamber in the absence of footshock (FS) stimuli. (B) Effects of early postnatal stress on freezing behavior during extinction at the postadolescent period (10–12 weeks old). The expression of freezing behavior was normalized with respect to the initial posttraining session (day 1). Non-FS, pups exposed to the FS box without FS; 2W-FS and 3W-FS, pups exposed to FS during the second and third postnatal weeks, respectively. $*p < 0.05$ versus non-FS controls. (C) Effects of D-cycloserine (DCS) on freezing behavior during extinction in the 3W-FS group. D-Cycloserine (15 mg/kg, i.p., for 4 days) or saline (2 ml/kg, i.p.) was administered 20 min before each extinction trial for 4 consecutive days (from day 2 to day 5). Rats were subjected to recall of extinction memory on the second day (day 7) after the fifth (day 5) extinction trial to determine whether spontaneous recovery of fear or long-term maintenance of extinction occurred. Freezing values were normalized with respect to the initial posttraining freezing session (day 1) without drug treatment. Each value represents the mean ± S.E.M. $*p < 0.05$ versus saline-administered group (modified from Matsumoto *et al.*, 2008). (See Color Insert.)

underlying extinction processes (see review, Corcoran and Quirk, 2007; Quirk and Mueller, 2008). LTP-like responses were exhibited in the hippocampal–mPFC pathway after extinction training. This synaptic response was suppressed by LFS that impaired the recall of tone-dependent fear extinction (Farinelli *et al.*, 2007; Hugues and Garcia, 2007). The hippocampus is a pivotal component responsible for context-dependent fear memory (Corcoran and Maren, 2001; Eldridge *et al.*, 2000), and the hippocampal–mPFC pathway appears to be a key neural circuit in mediating contextual fear extinction. Hypothetically, exposure to aversive stimuli during the third postnatal week may participate in long-lasting neurocircuitry alterations in the hippocampal–mPFC pathway and consequently may elicit impaired extinction of CFC during the postadolescent period (Fig. 4).

Interestingly, treatment with the partial *N*-methyl-D-aspartate (NMDA) receptor agonist D-cycloserine, which is known to facilitate cue-dependent extinction (Ledgerwood *et al.*, 2004; Parnas *et al.*, 2005; Walker *et al.*, 2002), diminished the expression of freezing behavior in the 3W-FS group. These findings suggest that cue- and context-dependent extinction processes are mediated by fundamentally similar molecular mechanisms, including NMDA receptors in which calcium-

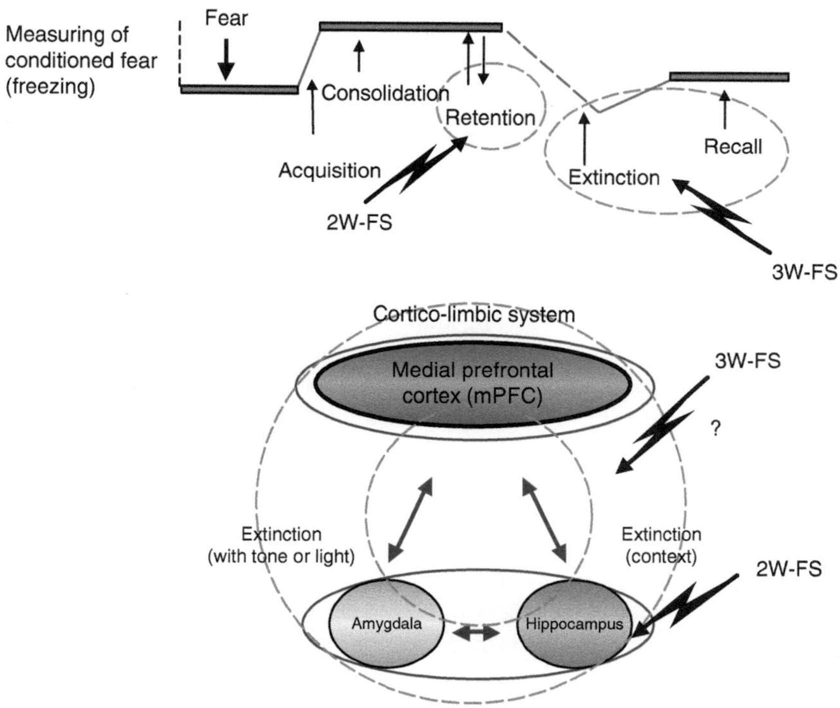

FIG. 4. Hypothetical processes of context-dependent fear memory and neural circuits associated with developmental perturbations caused by early postnatal stress. Fear memory was estimated by the expression of freezing behavior. Early postnatal stress did not affect the acquisition of fear memory, but aversive stress exposure during the second postnatal week (2W-FS) attenuated freezing behavior during the retention of contextual fear conditioning (CFC), presumably due to alterations in the hippocampal synaptic response. Extinction was estimated by the expression of freezing behavior during repeated extinction trials. The decrease in freezing behavior induced by the extinction trial was attenuated by aversive stress exposure during the third postnatal week (3W-FS). Hippocampus-dependent extinction is involved in contextual encoding processes, and the mPFC is an important target of the hippocampus associated with contextual modulation of extinction retrieval. The hippocampal–mPFC pathway, therefore, may be a key neural circuit related to deficits of extinction processes observed in the 3W-FS group. (See Color Insert.)

dependent mechanisms are involved. Recent studies have focused on the clinical relevance of D-cycloserine in facilitating fear extinction in people with anxiety (Guastella *et al.*, 2007; Hofmann *et al.*, 2006; Norberg *et al.*, 2008). Considering the similarities between exposure psychotherapy in humans and fear extinction training in animals, D-cycloserine may have clinical significance in the treatment of anxiety-related disorders, such as posttraumatic stress disorder (PTSD).

III. Conclusions

The behavioral response to CFC in adulthood depends on the early postnatal period when pups are exposed to aversive stress. The 2W-FS group exhibited a remarkable reduction in freezing behavior during CFC retention, an effect that was attributable to changes in synaptic transmission in the hippocampal CA1 field. The electrophysiological and behavioral features of the 2W-FS group further support the hypothesis that spatial learning and CFC tasks are sensitive to hippocampal function. Aversive stress exposure during the third postnatal week (3W-FS) impaired extinction of context-dependent fear memory. Deficits in extinction of conditioned fear are known to cause certain anxiety disorders, such as PTSD (see review, Corcoran and Quirk, 2007; Quirk and Mueller, 2008). Recent clinical and preclinical studies have focused on the functional role of the ventral mPFC, including infralimbic structures, in the neural and molecular mechanisms of extinction underlying the pathophysiology of PTSD and obsessive–compulsive disorder (Krystal, 2007; Norberg *et al.*, 2008; Quirk and Mueller, 2008). The deficits in extinction observed in the 3W-FS group, therefore, might be attributable to dysfunction of the neural circuits including hippocampal–mPFC pathway. Taken together, stressful events experienced during early life may lead to lifelong sensitivity to emotional stress, which can in turn precipitate long-lasting changes in neural circuits. A better understanding of the neural mechanisms underlying emotional regulation induced by early postnatal stress might provide insights into both its physiological consequences and the pathophysiology of neuropsychiatric disorders, such as PTSD.

Acknowledgments

This research was supported by a Grant-in-Aid for Scientific Research from the High Technology Research Program and the Academic Frontier Research Program organized by the Education and Science Department of Japan. The authors thank Dr Hiroyo Koseki, Dr Kasame Higuchi, Dr Riki Hirata, and Mr Kohtaro Konno for their excellent technical supports. We also thank Dr Taku Yamaguchi and Dr Tekeshi Izumi for their helpful advice.

References

Abraham, W. C., and Bear, M. F. (1996). Metaplasticity: The plasticity of synaptic plasticity. *Trends Neurosci.* **19,** 126–130.

Abraham, W. C., Mason-Parker, S. E., Bear, M. F., Webb, S., and Tate, W. P. (2001). Heterosynaptic metaplasticity in the hippocampus *in vivo*: A BCM-like modifiable threshold for LTP. *Proc. Natl. Acad. Sci. USA* **98,** 10924–10929.

Akirav, I., Raizel, H., and Maroun, M. (2006). Enhancement of conditioned fear extinction by infusion of the GABAA agonist muscimol into the rat prefrontal cortex and amygdala. *Eur. J. Neurosci.* **23,** 758–764.

Béïque, J. C., Chapin-Penick, E. M., Mladenovic, L., and Andrade, R. (2004). Serotonergic facilitation of synaptic activity in the developing rat prefrontal cortex. *J. Physiol.* **556,** 739–754.

Béïque, J. C., Imad, M., Mladenovic, L., Gingrich, J. A., and Andrade, R. (2007). Mechanism of the 5-hydroxytryptamine 2A receptor-mediated facilitation of synaptic activity in prefrontal cortex. *Proc. Natl. Acad. Sci. USA* **104,** 9870–9875.

Bliss, T. V., and Collingridge, G. L. (1993). A synaptic model of memory: Long-term potentiation in the hippocampus. *Nature* **361,** 31–39.

Corcoran, K. A., and Maren, S. (2001). Hippocampal inactivation disrupts contextual retrieval of fear memory after extinction. *J. Neurosci.* **21,** 1720–1726.

Corcoran, K. A., and Maren, S. (2004). Factors regulating the effects of hippocampal inactivation on renewal of conditional fear after extinction. *Learn. Mem.* **11,** 598–603.

Corcoran, K. A., and Quirk, G. J. (2007). Recalling safety: Cooperative functions of the ventromedial prefrontal cortex and the hippocampus in extinction. *CNS Spectr.* **12,** 200–206.

Eldridge, L. L., Knowlton, B. J., Furmanski, C. S., Bookheimer, S. Y., and Engel, S. A. (2000). Remembering episodes: A selective role for the hippocampus during retrieval. *Nat. Neurosci.* **3,** 1149–1152.

Farinelli, M., Deschaux, O., Hugues, S., Thevenet, A., and Garcia, R. (2007). Hippocampal train stimulation modulates recall of fear extinction independently of prefrontal cortex synaptic plasticity and lesions. *Learn. Mem.* **13,** 329–334.

Fischer, A., Sanabenesi, F., Schrick, C., Spiess, J., and Radulovic, J. (2004). Distinct roles of hippocampal de novo protein synthesis and actin rearrangement in extinction of contextual fear. *J. Neurosci.* **24,** 1962–1966.

Guastella, A. J., Dadds, M. R., Lovibond, P. F., Mitchell, P., and Richardson, R. (2007). A randomized controlled trial of the effect of D-cycloserine on exposure therapy for spider fear. *J. Psychiatr. Res.* **41,** 466–471.

Heim, C., and Nemeroff, C. B. (2001). The role of childhood trauma in the neurobiology of mood and anxiety disorders: Preclinical and clinical studies. *Biol. Psychiatry* **49,** 1023–1039.

Hirata, R., Togashi, H., Matsumoto, M., Yamaguchi, T., Izumi, T., and Yoshioka, M. (2008). Characterization of stress-induced suppression of long-term potentiation in the hippocampal CA1 field of freely moving rats. *Brain Res.* **1226,** 27–32.

Hirata, R., Matsumoto, M., Judo, C., Yamaguchi, T., Izumi, T., Yoshioka, M., and Togashi, H. (2009). Possible relationship between the stress-induced synaptic response and metaplasticity in the hippocampal CA1 field of freely moving rats. *Synapse* **63,** 549–556.

Hofmann, S. G., Pollack, M. H., and Otto, M. W. (2006). Augmentation treatment of psychotherapy for anxiety disorders with D-cycloserine. *CNS Drug Rev.* **12,** 208–217.

Hollup, S. A., Molden, S., Donnett, J. G., Moser, M. B., and Moser, E. I. (2001). Accumulation of hippocampal place fields at the goal location in an annular watermaze task. *J. Neurosci.* **21,** 1635–1644.

Hugues, S., and Garcia, R. (2007). Reorganization of learning-associated prefrontal synaptic plasticity between the recall of recent and remote fear extinction memory. *Learn. Mem.* **14,** 520–524.

Ji, J., and Maren, S. (2005). Electrolytic lesions of the dorsal hippocampus disrupt renewal of conditional fear after extinction. *Learn. Mem.* **12,** 270–276.

Kim, J. J., and Jung, M. W. (2006). Neural circuits and mechanisms involved in Pavlovian fear conditioning: A critical review. *Neurosci. Biobehav. Rev.* **30,** 188–202.

Koseki, H., Matsumotom, M., Togashi, H., Yamaguchi, T., Izemi, T., and Yoshioka, M. (2007). Effects of aversive stress during brain development on hippocampal synaptic and behavioral responses to emotional stress at postadolescence. *Nihon Shinkei Seishin Yakurigaku Zasshi* **27,** 19–27.

Krystal, J. H. (2007). Neuroplasticity as a target for the pharmacotherapy of psychiatric disorders: New opportunities for synergy with psychotherapy. *Biol. Psychiatry* **62,** 833–834.

Lattal, K. M., and Abel, T. (2001). Different requirements for protein synthesis in acquisition and extinction of spatial preferences and context-evoked fear. *J. Neurosci.* **21,** 5773–5780.

Ledgerwood, L., Richardson, R., and Cranney, J. (2004). D-Cycloserine and the facilitation of extinction of conditioned fear: Consequences for reinstatement. *Behav. Neurosci.* **118,** 505–513.

Malenka, R. C., and Nicoll, R. A. (1999). Long-term potentiation: A decade of progress? *Science* **285,** 1870–1874.

Martin, S. J., Grimwood, P. D., and Morris, R. G. (2000). Synaptic plasticity and memory: An evaluation of the hypothesis. *Annu. Rev. Neurosci.* **23,** 649–711.

Matsumoto, M., Togashi, H., Ohashi, S., Tachibana, K., Yamaguchi, T., and Yoshioka, M. (2004a). Serotonergic modulation of psychological stress-induced alteration in synaptic plasticity in the rat hippocampal CA1 field. *Brain Res.* **1022,** 221–225.

Matsumoto, M., Tachibana, K., Togashi, H., Tahara, K., Kojima, T., Yamaguchi, T., and Yoshioka, M. (2004b). Chronic treatment with milnacipran reverses the impairment of synaptic plasticity induced by conditioned fear stress. *Psychopharmacology (Berl)* **179,** 606–612.

Matsumoto, M., Higuchi, K., Togashi, H., Koseki, H., Yamaguchi, T., Kanno, M., and Yoshioka, M. (2005). Early postnatal stress alters the 5-HTergic modulation to emotional stress at postadolescent periods of rats. *Hippocampus* **15,** 775–781.

Matsumoto, M., Togashi, H., Konno, K., Koseki, H., Hirata, R., Izumi, T., Yamaguchi, T., and Yoshioka, M. (2008). Early postnatal stress alters the extinction of context-dependent conditioned fear in adult rats. *Pharmacol. Biochem. Behav.* **89,** 247–252.

McAllister-Williams, R. H., Ferrier, N. I, and Young, A. H. (1998). Mood and neuropsychological function in depression: The role of corticosteroids and serotonin. *Psychol. Med.* **28,** 573–584.

Micheva, K. D., and Beaulieu, C. (1996). Quantitative aspects of synaptogenesis in the rat barrel field cortex with special reference to GABA circuitry. *J. Comp. Neurol.* **373,** 340–354.

Miracle, A. D., Brace, M. F., Huyck, K. D., Singler, S. A., and Wellman, C. L. (2006). Chronic stress impairs recall of extinction of conditioned fear. *Neurobiol. Learn. Mem.* **85,** 213–218.

Modell, S., Lauer, C. J., Schreiber, W., Huber, J., Krieg, J. C., and Holsboer, F. (1998). Hormonal response pattern in the combined DEX-CRH test is stable over time in subjects at high familial risk for affective disorders. *Neuropsychopharmacol.* **18,** 253–262.

Moriceau, S., and Sullivan, R. M. (2006). Maternal presence serves as a switch between learning fear and attraction in infancy. *Nat. Neurosci.* **9,** 1004–1006.

Myers, K. M., and Davis, M. (2002). Behavioral and neural analysis of extinction. *Neuron* **36,** 567–584.

Norberg, M. M., Krystal, J. H., and Tolin, D. F. (2008). A meta-analysis of D-cycloserine and the facilitation of fear extinction and exposure therapy. *Biol. Psychiatry* **63,** 1118–1126.

Parnas, A. S., Weber, M., and Richardon, R. (2005). Effects of multiple exposures to D-cycloserine on extinction of conditioned fear in rats. *Neurobiol. Learn. Mem.* **83,** 224–231.

Quirk, G. J., and Mueller, D. (2008). Neural mechanisms of extinction learning and retrieval. *Neuropsychopharmacology* **33,** 56–72.

Quirk, G. J., Russo, G. K., Barron, J. L., and Lebron, K. (2000). The role of ventromedial prefrontal cortex in the recovery of extinguished fear. *J. Neurosci.* **20,** 6225–6231.

Shakesby, A. C., Anwyl, R., and Rowan, M. J. (2002). Overcoming the effects of stress on synaptic plasticity in the intact hippocampus: Rapid actions of serotonergic and antidepressant agents. *J. Neurosci.* **22,** 3638–3644.

Shumake, J., Barrett, D., and Gonazalez-Lima, F. (2005). Behavioral characteristics of rats predisposed to learned helplessness: Reduced reward sensitivity, increased novelty seeking, and persistent fear memories. *Behav. Brain Res.* **164,** 222–230.

Sullivan, R. M., Landers, M., Yeaman, B., and Wilson, D. A. (2000). Good memories of bad events in infancy. *Nature* **407,** 38–39.

Sodhi, M. S., and Sanders-Bush, E. (2004). Serotonin and brain development. *Int. Rev. Neurobiol.* **59,** 111–174.

Tada, K., Kasamo, K., Suzuki, T., Matsuzaki, Y., and Kojima, T. (2004). Endogenous 5-HT inhibits firing activity of hippocampal CA1 pyramidal neurons during conditioned fear stress-induced freezing behavior through stimulating 5-HT1A receptors. *Hippocampus* **14,** 143–147.

Thompson, L. T., and Best, P. J. (1990). Long-term stability of the place-field activity of single units recorded from the dorsal hippocampus of freely behaving rats. *Brain Res.* **509,** 299–308.

Toledo-Rodriguez, M., and Sandi, C. (2007). Stress before puberty exerts a sex- and age-related impact on auditory and contextual fear conditioning in the rat. *Neural Plast.* **2007,** 1–12.

Tyzio, R., Minlebaev, M., Rheims, S., Ivanov, A., Jorquera, I., Holmes, G. L., Zilberter, Y., Ben-Ari, Y., and Khazipov, R. (2008). Postnatal changes in somatic γ-aminobutyric acid signalling in the rat hippocampus. *Eur. J. Neurosci.* **27,** 2515–2528.

Walker, D. L., Ressler, K. J., Lu, K. T., and Davis, M. (2002). Facilitation of conditioned fear extinction by systemic administration or intra-amygdala infusions of D-cycloserine as assessed with fear-potentiated startle in rats. *J. Neurosci.* **22,** 2343–2351.

Xu, L., Anwyl, R., and Rowan, M. J. (1997). Behavioural stress facilitates the induction of long-term depression in the hippocampus. *Nature* **387,** 497–500.

Yang, Y., Cao, J., Xiong, W., Zhang, J., Zhou, Q., Wei, H., Liang, C., Deng, J., Li, T., Yang, S., and Xu, L. (2003). Both stress experience and age determine the impairment or enhancement effect of stress on spatial memory retrieval. *J. Endocrinol.* **178,** 45–54.

Yang, C. H., Huang, C. C., and Hsu, K. S. (2005). Behavioral stress modifies hippocampal synaptic plasticity through corticosterone-induced sustained extracellular signal-regulated kinase/mitogen-activated protein kinase activation. *J. Neurosci.* **24,** 11029–11034.

Zelcer, I., Cohen, H., Richter-Levin, G., Lebiosn, T., Grossberger, T., and Barkai, E. (2006). A cellular correlate of learning-induced metaplasticity in the hippocampus. *Cereb. Cortex* **16,** 460–468.

ROLES OF THE HISTAMINERGIC NEUROTRANSMISSION ON METHAMPHETAMINE-INDUCED LOCOMOTOR SENSITIZATION AND REWARD: A STUDY OF RECEPTORS GENE KNOCKOUT MICE

Naoko Takino, Eiko Sakurai, Atsuo Kuramasu, Nobuyuki Okamura, and Kazuhiko Yanai

Department of Pharmacology, Tohoku University Graduate School of Medicine, Sendai 980-8575, Japan

Methamphetamine (METH) is often abused as a psychostimulant, and its administration induces several abnormal behaviors. We propose that neuronal histamine has an inhibitory role on the METH-induced locomotor hyperactivity and development of behavioral sensitization. We examined the roles of the histaminergic neuron system on behavioral sensitization and conditioned place preference (CPP) induced by METH using single and multiple histamine receptors deficient mice. Mice were injected intraperitoneally seven times with METH (1 mg/kg) once in every 3 days. After drug-free intervals of 7 days, METH was rechallenged. The locomotor activities were gradually increased in histamine H1, H3 receptor gene double knockout (H1/H3-DKO), H1, H2, and H3 receptor gene triple knockout (TKO), and their wild-type (WT) mice when METH was repeatedly administrated, suggesting that these mice developed behavioral sensitization. The ratios of the locomotor activity in METH-administrated group to

DOI: 10.1016/S0074-7742(09)85008-3

0074-7742/09 $35.00

saline-treated group were not significantly changed among the different genotypes. The order of ratios were H1/H3-DKO > WT mice > TKO mice. We also examined METH-induced CPP in histamine H1 receptor gene knockout mice (H1-KO), H3 receptor gene knockout mice (H3-KO), and their WT mice. The CPP scores were increased by repeated METH administration. Especially, H1-KO mice showed higher METH-induced CPP scores than those of the WT and H3-KO mice. Our results suggest that the neuronal histamine could inhibit the METH-induced abnormal behaviors through the interactions of H1, H2, and H3 receptors.

I. Introduction

Since the histaminergic neuron system was discovered by Professor Takehiko Watanabe (Watanabe *et al.*, 1984), many studies investigated the roles of histamine in the brain. The neuronal histamine has a variety of physiological functions in the mammalian brain such as sleep–wake cycle, locomotor activity, appetite and drinking, stress response, and learning and memory (Watanabe and Yanai, 2001). In addition, brain histamine can inhibit the methamphetamine (METH)-induced locomotor hyperactivity and development of behavioral sensitization (Ito *et al.*, 1996). Administrations of pyrilamine (an H1 receptor antagonist), zolantidine (an H2 receptor antagonist), or α-fluoromethylhistamine (a specific inhibitor of L-histidine decarboxylase) potentiated METH-induced stereotyped behavior and behavioral sensitization in rats (Ito *et al.*, 1997).

The histamine receptor subtypes (H1, H2, H3, and H4) mediate these neuronal actions. Histamine H1 or H2 receptor gene single knockout mice (H1-KO, H2-KO) were generated by Professor Takeshi Watanabe (Inoue *et al.*, 1996; Kobayashi *et al.*, 2000, respectively). These receptor gene knockout mice exhibited the METH-induced locomotor sensitization, but there was no significant difference in the magnitude of sensitization when compared with their wild-type (WT) mice. However, the H1, H2 receptor gene double knockout (DKO) mice showed significantly higher locomotor activity by repeated administration of METH when compared to that of their WT mice. These data suggest that the neuronal histamine might be able to inhibit the METH-induced behavioral sensitization through both H1 and H2 receptors (Iwabuchi, 2004). Histamine H3 receptor gene knockout mice (H3-KO) were generated by Dr. T. W. Lovenberg, and the mice showed significantly lower locomotor activity than the WT mice when a single dose of METH was administered (Toyota *et al.*, 2002). Although the activity of histaminergic neuron system certainly reduces METH-induced abnormal behavior and sensitization, it is still unclear which histamine receptors are involved in METH-induced abnormal behaviors. In this study, we examined the roles of the histaminergic neurotransmission on METH-induced locomotor sensitization and rewarding using single and multiple histamine receptors deficient mice.

II. Methods and Materials

A. Animals

H1-KO, H3-KO, histamine H1, H3 receptor gene double knockout (H1/H3-DKO), H1, H2, and H3 receptor gene triple knockout (TKO), and their WT mice were used in this study. The mice were bred in our laboratory (Sakurai *et al.*, 2008). All experiments were performed in accordance with institutional guidelines, and experimental protocols were approved by the Animal Care Committee of Tohoku University.

B. Drugs

METH (Dainippon Pharmaceutical Co., Japan) was dissolved in saline and administered intraperitoneally (i.p.) to mice at a dose of 1 mg/kg.

C. Locomotor Activity in Home Cage

Each mouse was placed in an acquainted home cage, and locomotor activity was measured for 3 h after the administration of METH, using an infrared-ray passive sensor system (SUPERMEX®, Muromachi Kikai Co., Tokyo, Japan).

D. Conditioned Place Preference (CPP)

CPP study was conducted in a shuttle box divided into two compartments (Opto-MAX®, Columbus Instruments, Ohio, USA). Mice were placed after injection of METH or saline in one compartment and were placed after injection of saline in the other compartment the next day during the conditioning paradigm.

In sessions for conditioning, mice were injected with METH (1 mg/kg) or saline every other day, and confined for 30 min to a compartment designed to condition the place preference. This procedure was repeatedly performed four times. The CPP scores were calculated using the staying time of mouse in each compartment for 15 min from the counts of before and after the conditioning.

E. Measurement of Histamine and Monoamine

After the behavioral experiments, the brains were removed 1 h after the last injection, and dissected into four parts; the cortex, diencephalon, brain stem, and cerebellum. The levels of dopamine, serotonin, noradrenalin, and histamine were determined using HPLC methods as previously described.

III. Results and Discussion

A. Changes of Locomotor Activity in Histamine Receptor Gene Knockout Mice by METH Administration

Locomotor activities were measured in H1, H3-DKO TKO, and their WT mice after the mice were treated with METH once in every 3 days. METH was rechallenged after the drug-free intervals of 7 days, and the locomotor activities in mice were measured. The locomotor activities were increased in H1/H3-DKO TKO and their WT mice during the repeated administration of METH. This means that these mice developed behavioral sensitization (Fig. 1). The ratios of the locomotor counts by repeated METH administration to those of the saline treatment were calculated in each different genotype. The order of increased

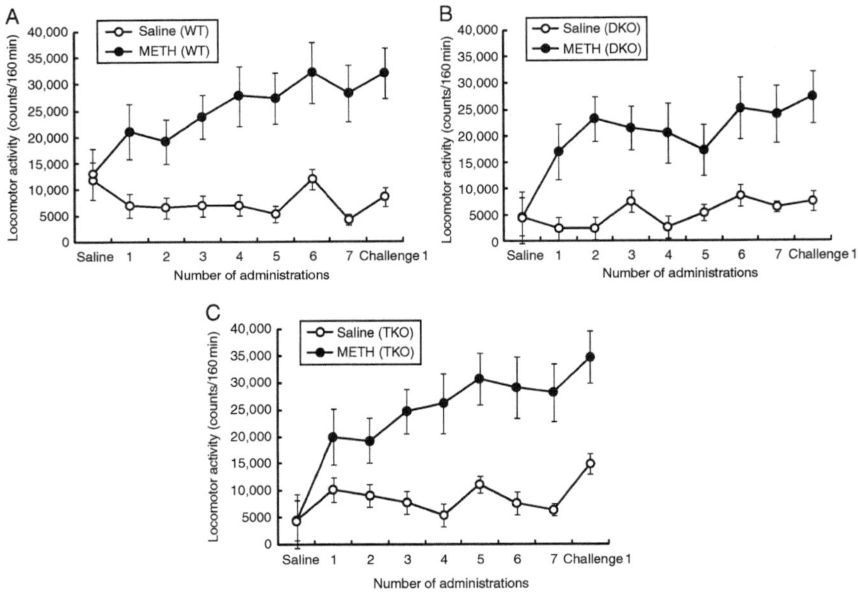

FIG. 1. Effect of repeated administration of METH or saline on the locomotor activity in WT and histamine receptors genes double (H1 and H3) and triple (H1, H2, H3) knockout mice. The locomotor activity of mice was measured for 3 h after administration of METH (1 mg/kg, closed circle) or saline (open circle) in the WT (A), H1/H3 (B), H1/H2/H3 (C) receptor gene knockout mice. Mice were injected seven times once in every 3 days with saline or METH (1mg/kg, i.p.). After drug-free intervals of 7 days, saline or METH were rechallenged. Each value represents the mean ± S.E.M. of six mice.

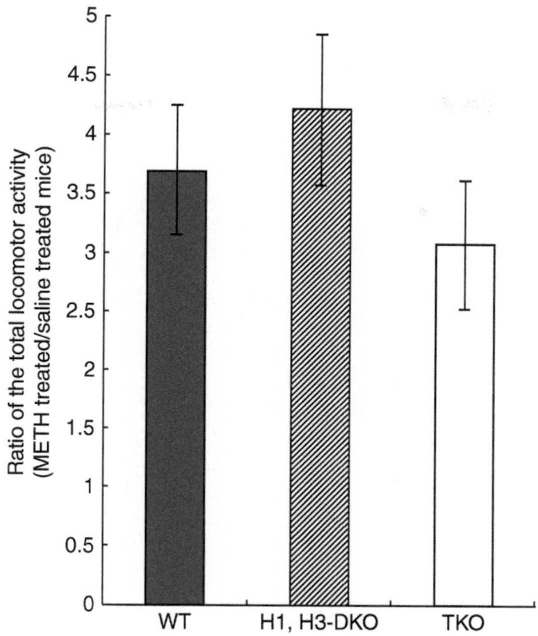

FIG. 2. The increased locomotor activity by repeated treatment with METH when compared to saline in each histamine receptor genotypes. The ratios were calculated from the total locomotor activity count in Fig. 1. The WT (closed column), H1/H3 (diagonal line of column), H1/H2/H3 (open column) receptor gene knockout mice.

ratios by repeated METH treatment were H1/H3-DKO > WT mice > TKO mice, but the difference was statistically insignificant (Fig. 2). These results suggest that the H1, H2, and H3 receptors might not have a decisive role to prevent the METH-induced behavioral sensitization through the interactions of these receptors.

B. METH-Induced CPP in Single Histamine Receptors Genes Knockout Mice

It is well known that METH induces psychotic states such as rewarding behaviors. However, it is still unclear how the histaminergic neuron system acts in rewarding effects. We investigated METH-induced CPP in H1-KO, H3-KO, and WT mice. The CPP scores were significantly increased in METH-treated mice. Especially, H1-KO mice showed higher METH-induced CPP scores than the WT and H3-KO mice (Fig. 3). These results suggest that neuronal histamine might prevent METH-induced rewarding effects through the histamine H1 receptors.

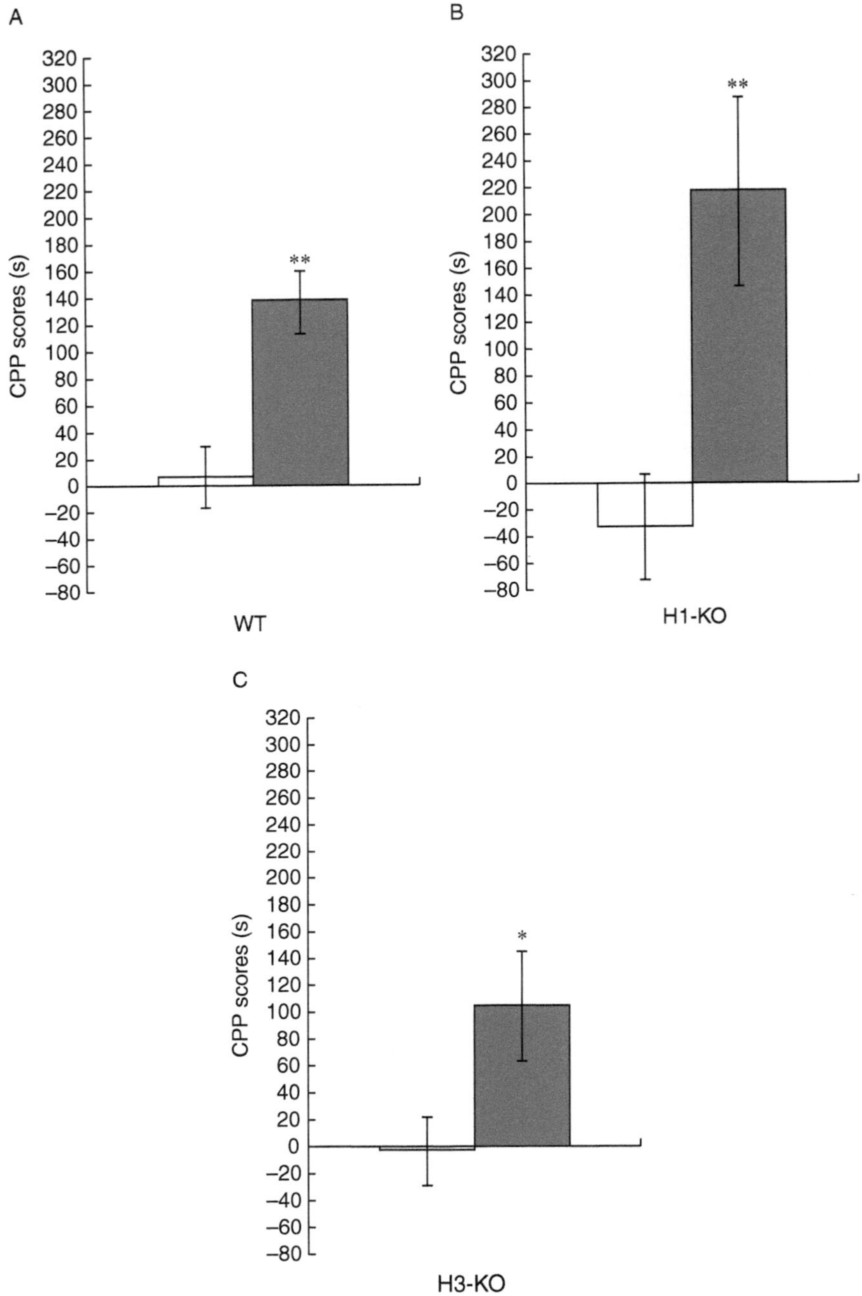

FIG. 3. Comparison of METH-induced CPP in single histamine receptor gene knockout mice. Mice were injected 1 mg/kg of METH or saline every other day, and confined for 30 min to a compartment designed to condition the place preference. The CPP score were calculated using the staying time of mouse in each compartment for 15 min before and after the conditioning. Each value represents the mean ± S.E.M. of 6–18 mice. Statistical analysis was performed by means of one-way ANOVA followed by Tukey's test (*$p < 0.05$, **$p < 0.01$).

C. Effects of Chronic Treatments of METH on the Contents of Histamine and Monoamines in the Brains of WT Mice

Continuous administrations of METH could influence the levels of monoamines and histamine in the brain. In this study, we measured the contents of histamine and monoamines after chronic treatment of METH. The contents of dopamine and noradrenalin in the cortex significantly increased after the repeated administration of METH (Table I). The levels of serotonin and histamine in the cortex were not significantly changed.

IV. Conclusion

From the study of the locomotor measurement, it is suggested that the histamine H1, H2, and H3 receptors might not play a decisive role in suppressing the behavioral sensitization. As for rewarding effects, H1-KO mice showed the enhanced rewarding behaviors, while H3-KO mice exhibited the suppressive CPP responses. These data suggest that H1 receptor might have an important role to prevent METH-induced reward effects.

TABLE I
Effects of Repeated Administration of METH or Saline on Monoamines and Histamine in the Mouse Brain

	Cortex	Diencephalon	Brain stem
Dopamine			
Saline	2.96 (±0.068)	2.21 (±0.102)	2.94 (±0.140)
METH	3.69 (±0.076)**	2.35 (±0.151)	2.64 (±0.024)
Serotonin			
Saline	0.61 (±0.014)	0.98 (±0.077)	1.22 (±0.078)
METH	0.64 (±0.037)	1.02 (±0.032)	1.22 (±0.046)
Noradrenaline			
Saline	0.81 (±0.028)	1.40 (±0.050)	2.61 (±0.052)
METH	0.91 (±0.026)*	1.43 (±0.041)	2.58 (±0.093)
Histamine			
Saline	199.3 (±15.35)	728.3 (±51.72)	172.2 (±29.55)
METH	252.7 (±41.15)	709.7 (±35.03)	125.0 (±23.78)

Mice were injected once in every 3 days during seven times with saline or METH (1 mg/kg, i.p.). After drug-free intervals of 7 days, saline or METH were rechallenged. The brains were removed 1 h after the last injection. Each value represents the mean ± S.E.M. of 6–7 mice. Statistical analysis was performed by means of one-way ANOVA followed by Tukey's test (*$p < 0.05$, **$p < 0.01$).

Acknowledgment

This work is supported by Grants-in-Aid for Scientific Research from the Japan Society of Promotion Science (JSPS).

References

Inoue, I., Yanai, K., Kitamura, D., Taniuchi, I., Kobayashi, T., Niimura, K., Watanabe, T., and Watanabe, T. (1996). Impaired locomotor activity and exploratory behavior in mice lacking histamine H1 receptors. *Proc. Natl. Acad. Sci. USA* **93**(23), 13316–13320.

Ito, C., Onodera, K., Sakurai, E., Sato, M., and Watanabe, T. (1996). Effects of dopamine antagonists on neuronal histamine release in the striatum of rats subjected to acute and chronic treatments with methamphetamine. *J. Pharmacol. Exp. Ther.* **279**(1), 271–276.

Ito, C., Onodera, K., Watanabe, T., and Sato, M. (1997). Effects of histamine agents on methamphetamine-induced stereotyped behavior and behavioral sensitization in rats. *Psychopharmacology (Berl.)* **130**(4), 362–367.

Iwabuchi, K., Kubota, Y., Ito, C., Watanabe, T., Watanabe, T., and Yanai, K. (2004). Methamphetamine and brain histamine: A study using histamine-related gene knockout mice. *Ann. N. Y. Acad. Sci.* **1025,** 129–134.

Kobayashi, T., Tonai, S., Ishihara, Y., Koga, R., Okabe, S., and Watanabe, T. (2000). Abnormal functional and morphological regulation of the gastric mucosa in histamine H2 receptor-deficient mice. *J. Clin. Invest.* **105**(12), 1741–1749.

Sakurai, E., Kuramasu, A., Okamura, N., and Yanai, K. (2008). Effects of histamine receptors gene knockout on behavioral change in mice. *J. Pharmacol. Sci.* **106**(Suppl. I), 90.

Toyota, H., Dugovic, C., Koehl, M., Laposky, A. D., Weber, C., Ngo, K., Wu, Y., Lee, D. H., Yanai, K., Sakurai, E., Watanabe, T., Liu, C., *et al.* (2002). Behavioral characterization of mice lacking histamine H(3) receptors. *Mol. Pharmacol.* **62**(2), 389–397.

Watanabe, T., and Yanai, K. (2001). Studies on functional roles of the histaminergic neuron system by using pharmacological agents, knockout mice and positron emission tomography. *Tohoku J. Exp. Med.* **195**(4), 197–217.

Watanabe, T., Taguchi, Y., Shiosaka, S., Tanaka, J., Kubota, H., Terano, Y., Tohyama, M., and Wada, H. (1984). Distribution of the histaminergic neuron system in the central nervous system of rats; a fluorescent immunohistochemical analysis with histidine decarboxylase as a marker. *Brain Res.* **295**(1), 13–25.

DEVELOPMENTAL EXPOSURE TO CANNABINOIDS CAUSES SUBTLE AND ENDURING NEUROFUNCTIONAL ALTERATIONS

Patrizia Campolongo,* Viviana Trezza,*[,1] Maura Palmery,* Luigia Trabace,† and Vincenzo Cuomo*

*Department of Physiology and Pharmacology "Vittorio Erspamer", Sapienza University of Rome, P.le A. Moro 5, 00185 Rome, Italy
†Department of Biomedical Sciences, University of Foggia, V.le L. Pinto, 71100 Foggia, Italy

Cannabis sativa preparations are among the illicit drugs most commonly used by pregnant women in Western countries. Although they are often considered relatively harmless, increasing evidence suggests that developmental exposure to cannabinoids induces subtle neurofunctional alterations in the offspring. In the present review, we summarize human and animal evidence examining the behavioral and neurobiological effects of exposure to cannabinoids during pregnancy and lactation. These studies show that the endocannabinoid system plays a crucial role in the ontogeny of the central nervous system and its activation, during brain development, can induce subtle and long-lasting neurofunctional alterations.

[1]Current address: Department of Neuroscience and Pharmacology, Rudolf Magnus Institute of Neuroscience, University Medical Center Utrecht, Utrecht, The Netherlands

INTERNATIONAL REVIEW OF NEUROBIOLOGY, VOL. 85
DOI: 10.1016/S0074-7742(09)85009-5

0074-7742/09 $35.00

I. Introduction

Cannabis sativa is one of the first plants used as a medicine, for religious ceremonies and recreationally (Mechoulam, 1986; Pertwee, 2006). Over the last decades its use, as a recreational drug, has been markedly increased, and nowadays *Cannabis sativa* preparations (hashish, marijuana, etc.) are among the illicit drugs most commonly used by pregnant women in Western countries (Bonnin *et al.*, 1994; NIDA, 2005). It has been estimated that, depending on age and ethnicity, 7–17% of women use marijuana during pregnancy (SAMHSA, 2002). This popularity may result from the general opinion that marijuana or other derivatives of the hemp plant are widely tolerated and often considered relatively harmless. However, the main psychoactive ingredient of marijuana, delta-9-tetrahydrocannabinol (THC), in plasma undergoes cross-placental and milk transfer upon marijuana smoking during pregnancy and lactation (Hutchings *et al.*, 1989). Consequently, developmental exposure to cannabis is a major health problem due to its harmful effects, including high rates of fetal distress, growth retardation, and subtle neurofunctional alterations (Day *et al.*, 1994a; Fried, 1995; Hurd *et al.*, 2005).

Despite the significant advances made in recent years in the understanding of the neurobiology of the endocannabinoid system (Howlett *et al.*, 2002; Piomelli, 2003), clinical reports on the effects of marijuana exposure during development are still controversial and far from definitive (Fried and Smith, 2001). Most prospective longitudinal studies on the effects of maternal marijuana consumption in the offspring are, indeed, currently underway, and the identification of potential changes in the adult expression of behavioral functions is far from being achieved (Trezza *et al.*, 2008b). However, such information is essential when considering the current debates as to the potential legalization of marijuana in Western countries and the establishment of national policies related to marijuana's effects on health. In this scenario, animal models may provide a useful tool for examining the potential long-term effects of developmental exposure to cannabinoids. In the area of drug abuse, animal research is indeed of unquestionable value because it allows to examine, while strictly controlling confounding variables, parameters as dosage, timing of exposure, genetic factors, and possible underlying neurological causes for behavioral alterations (Fried, 2002).

In this review, we summarize recent human and animal studies on the ontogeny of the endocannabinoid system, discussing how its stimulation, by exogenous agonists, might result in enduring neurofunctional alterations. Particular emphasis will be given to the long-term cognitive and emotional implications of low–moderate cannabinoid exposure during critical stages of brain development.

II. Ontogeny of the Endocannabinoid System

Cannabis sativa preparations affect biological functions through the binding of the active components to cannabinoid receptors. Two G protein-coupled receptors (GPCRs), cannabinoid receptor type 1 (CB1) and cannabinoid receptor type 2 (CB2), have been cloned so far (Matsuda *et al.*, 1990; Munro *et al.*, 1993). CB1 receptors are highly expressed in several brain regions and in lower amounts in a more widespread fashion, whereas CB2 receptors have a more restricted distribution, being found in a number of immune cells and recently at low numbers in the brainstem (Van Sickle *et al.*, 2005) and other brain regions (Gong *et al.*, 2006). Both CB1 and CB2 receptors primarily signal through inhibitory G proteins (Howlett *et al.*, 2002). Thus, stimulation of cannabinoid receptors leads to the inhibition of adenylyl cyclase, the activation of mitogen-activated protein kinases, the inhibition of certain voltage-gated calcium channels, and the activation of G protein-linked inwardly rectifying potassium channels (Howlett *et al.*, 2002). The activation of these signaling pathways by CB1 receptors expressed on presynaptic terminals suppresses neuronal excitability and inhibits neurotransmission (Mackie, 2008).

Several studies have elucidated that endogenous ligands for cannabinoid receptors, called endocannabinoids, serve as retrograde messengers at central synapses (Hashimotodani *et al.*, 2007). Only two of such compounds have been thoroughly studied so far: 2-arachidonoylglycerol (2-AG) and anandamide (Di Marzo and Fontana, 1995), which meet key criteria for being considered endogenous agonists of brain CB1 receptors. They are produced in an activity-dependent manner by neurons, modulate synaptic transmission via activation of CB1 receptors, and are rapidly deactivated through transport into cells followed by intracellular hydrolysis (Freund *et al.*, 2003; Piomelli, 2003).

The endocannabinoid system plays a major role for embryo survival and brain development (Fernandez-Ruiz *et al.*, 2000; Paria *et al.*, 2001, 2002; Wang and Dey, 2005); it has been detected from the earliest stages of embryogenesis and throughout pre- and postnatal development (Mato *et al.*, 2003). CB1 and CB2 receptor mRNA has been detected as early as the preimplantation period in the embryo (Paria and Dey, 2000) and in the developing brain at pre- and post-natal ages (Belue *et al.*, 1995; Berrendero *et al.*, 1999; Mailleux and Vanderhaeghen, 1992; McLaughlin *et al.*, 1994; Paria and Dey, 2000). CB1 receptor mRNA can be detected around gestational day 11–14 in rats, a prenatal period which matches to the initial expression of several neurotransmitters (Berrendero *et al.*, 1999). At this gestational age, CB1 receptors are already coupled to signal transduction mechanisms that involve GTP-binding proteins (Berrendero *et al.*, 1998). Of relevance is that binding and mRNA expression appear atypically distributed in the fetal brain as compared with the adult brain, particularly with regard to some brain areas, such as the white matter and the subventricular zones of the forebrain (Berrendero *et al.*, 1998, 1999; Romero *et al.*, 1997).

Similar developmental patterns of CB1 receptors were found during human pre- and postnatal development. Thus, functional CB1 receptors were detected at week 14 of gestation in the human embryo (Biegon and Kerman, 2001). In the 20th week of gestation, an intense expression of CB1 mRNA was found in limbic structures, such as the hippocampal CA region and basal nuclear group of the amygdaloid complex as compared to a much wider expression of CB1 mRNA in the adult human brain (Wang *et al.*, 2003). A progressive increase in the concentrations of CB1 receptors was found in the frontal cortex, hippocampus, basal ganglia, and cerebellum between the fetal period and adulthood (Mato *et al.*, 2003). In accordance to the preclinical data, high CB1 receptor concentrations were present on several white matter neuronal tracts of the human fetus, but had disappeared by infancy (Mato *et al.*, 2003).

Several lines of evidence demonstrate that endocannabinoid signaling modulates central nervous system development by tuning the size of neural progenitors pools generating neurons and glia, by defining the sizes of neuronal contingents undergoing radial or tangential migration to populate the developing cerebrum, and by controlling the morphological and functional specification of developing neurons (Harkany *et al.*, 2007).

Atypical distribution patterns of endocannabinoids and CB1 receptors (i.e., a transient presence during development in regions where none are found at adulthood) were detected in white matter regions including the corpus callosum and anterior commissure in the developing brain (Berrendero *et al.*, 1998, 1999; Mato *et al.*, 2003; Romero *et al.*, 1997). The abundance of endocannabinoids and their receptors in the developing nervous system, and their atypical distribution, suggest that the endocannabinoid system is involved in the regulation of structural and functional brain maturation. In this regard, accumulating evidence indicates that endocannabinoid signaling serves key functions during neurodevelopment and is inherently involved in the control of neurogenesis, neural progenitor proliferation, lineage segregation, and the migration and phenotypic specification of immature neurons (Fernandez-Ruiz *et al.*, 2004; Harkany *et al.*, 2007). Recently, CB1 receptors were observed to be enriched in the axonal growth cones of GABAergic interneurons in rodent cortex toward the end of gestation, suggesting that endocannabinoids in the developing brain may also function as axon guidance cues and may be involved in synaptogenesis (Berghuis *et al.*, 2007).

Consistent with a role for anandamide as an endogenous neuroprotectant in the adult brain, it has been hypothesized that the endocannabinoid system promotes neuroprotection in the developing brain (Fride, 2008; Fride and Shohami, 2002). Several studies support this hypothesis: (i) activation of CB1 receptors in 7-day-old rats by WIN55,212-2 prevented neuronal loss (Martinez-

Orgado *et al.*, 2003), (ii) exogenous anandamide reduced ouabain-induced neuronal damage in 1- and 7-day-old rat pups through activation of CB1 receptors (van der Stelt *et al.*, 2001), and (iii) a dramatic increase of anandamide precursors has been observed in the infant rat brain after head trauma (Hansen *et al.*, 2001).

III. Morphological and Neurofunctional Outcomes Induced by Developmental Exposure to Cannabinoids

A. Morphological Changes

Cannabis sativa preparations are among the most commonly used illicit substances, also during pregnancy and lactation. A number of clinical studies have investigated the relationship between heavy prenatal marijuana exposure and outcome at birth. The results, unfortunately, are equivocal. Some studies have reported correlations between marijuana use during pregnancy and smaller size at birth (Fergusson *et al.*, 2002; Linn *et al.*, 1983); several of these studies, however, failed to control adequately for other illicit drug use (Day and Richardson, 1991). Moreover, a relationship between prenatal marijuana use and the premature gestational age of the infant has been reported (Linn *et al.*, 1983), as with growth, however, other studies have not corroborated these findings (Day and Richardson, 1991). In addition, two studies have highlighted an increase in morphologic abnormalities in children exposed *in utero* to cannabis, while others have reported no relationship with either minor or major morphologic abnormalities (Day and Richardson, 1991).

The scenario emerging from clinical studies on the morphological effects induced by *in utero* exposure to cannabis is still a matter of controversy. However, many studies carried out in animal models, which allow to examine, while strictly controlling confounding variables, parameters as dosage, timing of exposure, and genetic factors, demonstrated that maternal exposure to high doses of cannabinoids results in low birth weight and postnatal morbidity and mortality in the offspring (Abel, 1985; Abel *et al.*, 1981; Woodall *et al.*, 1996).

B. Neurofunctional Changes

1. *Clinical Studies*

Clinical reports on the neurofunctional effects induced by marijuana exposure during development are still controversial and far from definitive (Fried and Smith, 2001). Most prospective longitudinal studies are currently underway, and the identification of potential changes in the adult expression of behavioral functions is far from being achieved. To date, there are two longitudinal cohort studies, the Ottawa Prenatal Prospective Study (OPPS) and the Maternal Health Practices and Child Development Study (MHPCD), that have documented the neurobehavioral and developmental effects of prenatal exposure to cannabis past young adulthood (Fried, 2002; Fried and Smith, 2001; Goldschmidt *et al.*, 2004).

The OPPS, initiated in 1978, is examining the effects of cannabis and tobacco use in low-risk, white, predominantly middle-class families from pregnancy to young adulthood. The OPPS study suggests that *in utero* exposure to marijuana affects performance in tasks related to problem-solving situations, which require visual integration and analytical skills as well as sustained attention (Fried, 2002; Smith *et al.*, 2006).

The MHPCD, initiated in 1982, is focusing upon the consequences of prenatal use of marijuana, alcohol, and cocaine. The subjects in this high-risk cohort are of low socioeconomic status and just over half are African American. Alterations in growth, cognitive development, temperament, and emotionality have been reported in offspring up to the age of 10 (Goldschmidt *et al.*, 2000).

Concerning preschool ages, in the MHPCD cohort, third trimester use of one or more joints a day was associated with lower mental scores on the Bayley scale at 9 months of age, but not at 18 months (Richardson *et al.*, 1995). In the OPPS cohort, global measures, particularly between the ages of 1 and 3 years, did not reveal any negative association with prenatal marijuana exposure (Fried and Smith, 2001). This initial, apparent absence of effect during early childhood should not be interpreted as *in utero* marijuana exposure having only transient effects. Indeed, as the children became older, aspects of neuropsychological functioning did discriminate between marijuana-exposed and control children. At 48 months of age, significantly lower scores in verbal and memory domains were associated with maternal marijuana use after adjusting for confounding variables (Fried *et al.*, 2002). This negative relationship is the first reported association beyond the neonatal stage, and may represent a long-term effect of the drug upon complex behavior that, at a younger age, had not developed and/or could not be assessed (Fried and Watkinson, 1990).

At 4 years of age, although global tests of intelligence did not differentiate between marijuana-exposed and control children, decreased verbal ability and poor performances on tasks requiring visual analysis, visual hypothesis testing, and facets attention were found in children prenatally exposed to cannabis (Fried, 1995). These observations are in accordance to findings reported from the MHPCD cohort in which the 3-year-old offspring of daily marijuana users were impaired on short-term memory, verbal, and abstract/visual skills (Day *et al.*, 1994b).

When considering the clinical data on the school-aged exposed offspring, an interesting scenario is emerging. In the OPPS sample, at 5 and 6 years of age, no differences were found on global tests of cognition and language (Fried *et al.*, 1992).

Two reports arising from the OPPS cohort focused on specific cognitive characteristics and strategies on 9–12-year-old subjects. As with the data collected from the OPPS sample at earlier ages, there was no association between the full-scale IQ and *in utero* marijuana exposure (Fried and Smith, 2001). However, at 10 years of age, prenatal marijuana use in the MHCPD cohort was reported to induce poorer abstract/visual reasoning (Fried and Smith, 2001). Of interest is a report based on the 9–12-year-old OPPS cohort which supports the hypothesis that prenatal marijuana may impact upon "top-down" neurocognitive functioning (Fried and Watkinson, 2000), a key aspect of executive functions. Executive functions comprise capacities such as cognitive flexibility, sustained and focused attention, and working memory. In the latter report, visuoperceptual tasks, ranging from those involving basic capabilities to those requiring considerable integration and cognitive manipulation, were utilized. The deficits in executive functions are still present in 18–22-year-old young adults from the OPPS study, which showed an altered neuronal functioning during visuospatial working memory processing (Smith *et al.*, 2006).

It is worth of note that, apart from executive functions, different aspects of attention appeared to be affected in both the OPPS- and MHPCD-exposed children at school age (Fried and Smith, 2001); these findings are accompanied by data demonstrating that children exposed prenatally to marijuana seem to be impulsive, inattentive, and hyperactive (Fried and Smith, 2001; Goldschmidt *et al.*, 2000).

To date, studies of the consequences of prenatal marijuana use have predominantly focused on the cognitive development of the children. Research on other aspects of child neurobehavioral development, such as psychiatric symptomatology, is emerging just recently. However, it is worth noting that prenatal exposure to cannabis in the first and third trimesters predicted significantly levels of self-report anxiety and depressive symptoms in the offspring (Goldschmidt *et al.*, 2004; Gray *et al.*, 2005).

At the moment, the longitudinal OPPS and MHPCD studies are the two major sources of information about the consequences of prenatal exposure to cannabis. An emergent theme arising from these two longitudinal investigations is that developmental exposure to marijuana does not alter the standardized derived IQ scores but may impact upon high-order cognitive processes, leading to attention deficits and impairments in problem-solving tasks that require complex visuoperceptual integration. However, an emergent psychiatric symptomatology in children exposed *in utero* to cannabis has to be taken into account.

2. *Preclinical Studies*

Although significant advances have been made in recent years in the clinical evaluation of detrimental effects induced by developmental marijuana exposure, prospective longitudinal studies are currently underway. So far, they have preferentially investigated the cognitive effects of prenatal exposure to cannabis, with scarce or few attention on other aspects of human behavior relevant, for example, for the development of psychiatric disorders (i.e., affective sphere). Moreover, the clinical studies that are currently underway have so far documented alterations until young adulthood (Trezza *et al.*, 2008b). Therefore, the identification of potential changes in other aspects of neurobehavioral development and in the adult expression of cognitive functions is far from being achieved.

In this scenario, animal models may provide a useful tool for examining the potential long-term effects of *in utero* exposure to cannabinoids (Fried, 2002). Indeed, due to the limited life span of rodents, it is possible to undertake long-term studies, analyzing the effects of a prenatal exposure since early developmental stages till adulthood in a limited period of time (i.e., 6–12 months).

Many studies carried out in laboratory animals have demonstrated that developmental exposure to high doses of cannabinoids may induce morphological and functional abnormalities in the offspring (Dalterio, 1986; Trezza *et al.*, 2008b). However, the neurofunctional consequences of developmental exposure to low/moderate doses of cannabinoids have not completely clarified yet.

Maternal exposure to cannabinoids induces alterations in the correct development of the catecholaminergic and indolaminergic systems (Molina-Holgado *et al.*, 1993, 1996; Navarro *et al.*, 1994, 1995; Rodriguez de Fonseca *et al.*, 1990, 1991; Walters and Carr, 1986). The aminoacidergic neurotransmission appears to be altered by developmental exposure to cannabis as well. In particular, rats perinatally exposed to low-moderate doses of cannabinoids, displayed significant changes of GAD and GABA immunoreactivities in some GABAergic neuronal systems of the adult rat cerebellar cortex (Benagiano *et al.*, 2007). Concerning glutamatergic neurotransmission, exposure to cannabinoids during critical phases of brain development induces several alterations: prenatal cannabinoid exposure downregulates glutamate transporters (GLAST and EAAC1) and the expression of AMPA glutamate receptor subunits GluR1 and GluR2/3 in the rat cerebellum (Suarez *et al.*, 2004a,b). In agreement with this findings, we have recently reported that prenatal exposure to cannabinoids induces an increase in glutamate uptake through overexpression of GLT1 and EAAC1 glutamate transporter subtypes in rat frontal cerebral cortex (Castaldo *et al.*, 2007). Moreover, basal and K^+-evoked extracellular glutamate levels were significantly lower in cortical cell cultures obtained from rat pups prenatal exposed to cannabinoids (Antonelli *et al.*, 2005). In a latter study, we investigated the impact of perinatal cannabinoid exposure on both glutamatergic and noradrenergic neurotransmission, highlighting how perinatal THC may induce long-lasting alterations in the expression of cortical genes related to glutamatergic and noradrenergic systems, associated with a decrease in the cortical extracellular levels of both neurotransmitters (Campolongo *et al.*, 2007).

Finally, it seems worth noting that several reports have described ontogenic effects of maternal exposure to cannabinoids on several neuropeptides and hormones (Garcia-Gil *et al.*, 1997; Kumar *et al.*, 1986, 1990; Rubio *et al.*, 1995; Wenger *et al.*, 1992).

Several animal studies have highlighted how developmental exposure to cannabinoids may differentially affect the ontogeny of motor behavior. Borgen *et al.* (1973) described an increased motor activity in the lactating offspring of mothers receiving THC (gestational days 6–12). Recent findings are in agreement with this previous report, demonstrating how rats, exposed to cannabinoids during pregnancy and/or lactation, were characterized by motor hyperactivity during infantile and juvenile, but not adult, periods (Mereu *et al.*, 2003; Navarro *et al.*, 1995). However, other studies found lower activity in lactating rats perinatally exposed to cannabinoids (Fried, 1976) or no effects at all (Brake *et al.*, 1987; Trezza *et al.*, 2008a; Vardaris *et al.*, 1976). Due to the different protocols used in these studies (i.e., timing of exposure, route of administration, compound used, age of testing, etc.) a comprehensive evaluation of the data is very difficult and the results appear, therefore, to be still controversial; it is relevant, however, that maternal administration of cannabinoids induces alterations in the ontogeny of spontaneous behavior leading preferentially to a hyperactive profile when the offspring is exposed to low/moderate drug doses (Navarro *et al.*, 1995). These preclinical results are in line with human data showing that children prenatally exposed to marijuana are hyperactive and impulsive (Fried and Smith, 2001; Goldschmidt *et al.*, 2004).

Cognitive performances, particularly falling in the domain of executive functions, have been found to be altered in children and adolescents prenatal exposed to cannabis. However, clinical data on the potential changes in the long-term expression of cognitive functions in the exposed offspring are not available yet. Due to these observations, cognitive outcomes in the offspring prenatal exposed to cannabis have been investigated in animal models, since neonatal age to adulthood.

It has been demonstrated that basal and K^+-evoked extracellular glutamate levels were significantly lower in cortical cell cultures obtained from pups exposed to the cannabinoid agonist WIN55,212-2 (0.5 mg/kg from gestational days 5–20) with respect to the control group (Antonelli *et al.*, 2005, 2006). The addition of NMDA to cortical cell cultures from neonates born from vehicle-treated dams increased glutamate levels, and this was absent in cell cultures obtained from WIN-exposed pups. In line with these findings, WIN-exposed rats revealed a poorer performance in the homing behavior (10–12 days of age), which is a simple form of learning during the early phases of postnatal life (Antonelli *et al.*, 2005). These findings provide evidence for a deficit in cortical glutamatergic neurotransmission and behavior in the newborn rat prenatal exposed to cannabinoids. In line with these observations, we found that the same protocol of prenatal exposure to WIN55,212-2, induced a disruption of memory retention in 40- and 80-day-old offspring subjected to a passive avoidance task (Antonelli *et al.*, 2005;

Mereu *et al.*, 2003). The memory impairment caused by gestational exposure to WIN55,212-2 was correlated with alterations of hippocampal long-term potentiation (LTP) and glutamate release (Mereu *et al.*, 2003). In line with LTP results, *in vivo* microdialysis showed a significant decrease in basal and K^+-evoked extracellular glutamate levels in the hippocampus and cortex of rats born from dams treated with WIN55,212-2 (Antonelli *et al.*, 2004; Mereu *et al.*, 2003). The decrease in hippocampal glutamate outflow appears to be the cause of LTP disruption, which in turn might underlie the long-lasting impairment of cognitive functions caused by developmental exposure to cannabinoids. Taken together, these findings show that prenatal exposure to cannabinoids may induce enduring cognitive effects in the offspring, and provide a partial explanation for the cognitive deficits observed in children exposed *in utero* to cannabis (Fried and Smith, 2001). However, concerning the clinical relevance of preclinical studies, it is important to estimate, by extrapolation, whether the dose of the synthetic cannabinoid agonist compares with that of THC absorbed by cannabis users. Previous studies have estimated that 5 mg/kg of THC in rats corresponds to a moderate exposure of the drug in humans, correcting for the differences in route of administration and body weight surface area (Garcia-Gil *et al.*, 1997, 1998). WIN55,212-2 has been found to be 3–10 times more potent than THC, depending on the administration route and the endpoints considered (Compton *et al.*, 1992; French *et al.*, 1997; Hampson *et al.*, 2000). Based on these considerations, the dose of WIN55,212-2 used in the studies described above might correspond to a moderate, or even to a low, exposure to cannabis in humans (Mereu *et al.*, 2003). In line with this observation, we have recently demonstrated that moderate exposure to THC, the natural constituent of *Cannabis sativa* preparations, causes enduring long-term memory impairment in the inhibitory avoidance test, as well as disruptions in short-term olfactory memory in the social discrimination test (Campolongo *et al.*, 2007). Notably, specific and enduring cognitive deficits, likely due to impairment of working memory function, were found in rats directly treated with THC in the early postnatal life (O'Shea and Mallet, 2005; O'Shea *et al.*, 2006).

Taken together, these results suggest that, independently of the timing of exposure or of the cannabinoid compounds used, cannabinoid exposure during critical developmental periods induces subtle and specific enduring cognitive alterations in the offspring.

Research on other aspects of neurobehavioral development, such as emotionality, is emerging just recently, in line with the crucial role played by endocannabinoids in the modulation of emotional states (Millan, 2003; Witkin *et al.*, 2005).

In this scenario, Antonelli *et al.* (2005) found a decrease in the rate of separation-induced ultrasonic emission in the 10-day-old rat pups prenatal exposed to WIN55,212-2, revealing a decreased emotional reactivity in such offspring. In a recent study, we focused on the effects induced by exposure to

THC during both pregnancy and lactation. This study provided new evidence that perinatal exposure to THC, at a dose (5 mg/kg, os, from gestational day 15 to postnatal day 9) that is not associated with overt signs of toxicity, produces subtle but enduring neurobehavioral changes in the offspring emotionality (Trezza *et al.*, 2008a). An increased emotional reactivity has been found, indeed, at neonatal, juvenile, and adult ages in the offspring, with perinatal THC increasing the number of ultrasounds emitted by rat pups removed from the nest, inhibiting social interaction in the adolescent offspring, and inducing an anxiogenic-like profile in the adult offspring tested in the elevated plus-maze test. These observations are in line with human data showing that children prenatally exposed to cannabis self-report anxiety symptoms (Goldschmidt *et al.*, 2000). Moreover, a decreased level of social interaction, indicative of an increased emotionality, has been found in the rat offspring exposed to the cannabinoid receptor agonist CP55,940 from postnatal day 4 to postnatal day 25 (O'Shea *et al.*, 2006). On the other hand, in a recent study, Newsom and Kelly (2008) reported that exposure to THC during early postnatal life (2 mg/kg, s.c., from gestational day 1 to postnatal day 10) increases the level of social interaction, which could be interpreted as a decrease in emotional reactivity.

Due to the controversies existing among the preclinical studies and to the scarce human data, further studies are needed to clarify the emerging scenario on the effects induced by cannabinoid exposure on the offspring emotionality. One possible explanation for the contrasting data may be related to the different timing of exposure. It is well known, indeed, how different time windows of exposure to a psychotropic agent can induce even opposite neurofunctional effects (Costa *et al.*, 2004). Moreover, differences in cannabinoid agonist used, tested doses, and treatment schedules could also account for the apparent discrepancies.

In spite of the controversies, it should be taken into account that, depending on the dose, route of administration, and environmental contexts, acute cannabinoid administration may induce both anxiolytic- and anxiogenic-like effects. However, it is crucial to consider that all studies, focusing on the offspring emotionality, agree on the evidence that cannabinoid exposure during critical developmental periods may induce, even if in an opposite way, enduring changes in the emotional reactivity of the offspring.

IV. Conclusions

The endocannabinoid system plays a relevant role in brain organization during pre- and post-natal life. *Cannabis sativa* preparations are among the illicit drugs most commonly used by pregnant women in Western countries. Developmental exposure to cannabis is a major health problem due to its harmful effects, which are subtle and often long-lasting. The few clinical studies available so far raise marked interpretative issues that must be recognized. However, the available data have led to the observations that the impact of cannabis exposure during the course of pregnancy and upon the neonate appears to be considerably moderated with limited effects upon fetal growth and central nervous system functioning. Conversely, beyond preschool age, several clinical findings suggest an association between prenatal marijuana exposure and specific aspects of cognitive behavior that fall under the umbrella of executive functions. Moreover, an emergent psychiatric symptomatology in children exposed *in utero* to cannabis has to be taken into account. Although there is a degree of consistency in the prenatal literature, the very limited number of findings emphasizes the need for further, well-controlled follow-up studies in this area. Relevant information are available from several preclinical studies, highlighting how even low–moderate doses of cannabinoids, when administered during particular periods of brain development, can have profound consequences for brain maturation, leading to long-lasting alterations of cognitive functions and emotional behaviors. However, although animal models allow to monitor the confounding factors usually present in human studies, they do not take into account environmental and social issues that could influence the neurobiological effects of cannabis exposure during development. Combined preclinical and clinical approaches should be encouraged in the near future to further clarifying the potential relationship between developmental cannabis exposure and long-lasting neurofunctional outcomes.

Acknowledgments

This study was supported by FIRB 2006 and PRIN 2007 from MIUR (Ministero dell'Istruzione, dell'Università e della Ricerca—Italy).

References

Abel, E. L. (1985). Effects of prenatal exposure to cannabinoids. *NIDA Res. Monogr.* **59,** 20–35.

Abel, E. L., Bush, R., Dintcheff, B. A., and Ernst, C. A. (1981). Critical periods for marihuana-induced intrauterine growth retardation in the rat. *Neurobehav. Toxicol. Teratol.* **3,** 351–354.

Antonelli, T., Tanganelli, S., Tomasini, M. C., Finetti, S., Trabace, L., Steardo, L., Sabino, V., Carratu, M. R., Cuomo, V., and Ferraro, L. (2004). Long-term effects on cortical glutamate release induced by prenatal exposure to the cannabinoid receptor agonist (*R*)-(+)-[2,3-dihydro-5-methyl-3-(4-morpholinyl-methyl)pyrrolo[1,2,3-de]-1,4-benzoxazin-6-yl]-1-naphthalenylmethanone: An *in vivo* microdialysis study in the awake rat. *Neuroscience* **124,** 367–375.

Antonelli, T., Tomasini, M. C., Tattoli, M., Cassano, T., Tanganelli, S., Finetti, S., Mazzoni, E., Trabace, L., Steardo, L., Cuomo, V., and Ferraro, L. (2005). Prenatal exposure to the CB1 receptor agonist WIN 55,212-2 causes learning disruption associated with impaired cortical

NMDA receptor function and emotional reactivity changes in rat offspring. *Cereb. Cortex* doi:10.1093/cercor/bhi076.

Antonelli, T., Tomasini, M. C., Tattoli, M., Cassano, T., Finetti, S., Mazzoni, E., Trabace, L., Carratu, M. R., Cuomo, V., Tanganelli, S., and Ferraro, L. (2006). Prenatal exposure to the cannabinoid receptor agonist WIN 55,212-2 and carbon monoxide reduces extracellular glutamate levels in primary rat cerebral cortex cell cultures. *Neurochem. Int.* **49,** 568–576.

Belue, R. C., Howlett, A. C., Westlake, T. M., and Hutchings, D. E. (1995). The ontogeny of cannabinoid receptors in the brain of postnatal and aging rats. *Neurotoxicol. Teratol.* **17,** 25–30.

Benagiano, V., Lorusso, L., Flace, P., Girolamo, F., Rizzi, A., Sabatini, R., Auteri, P., Bosco, L., Cagiano, R., and Ambrosi, G. (2007). Effects of prenatal exposure to the CB-1 receptor agonist WIN 55212-2 or CO on the GABAergic neuronal systems of rat cerebellar cortex. *Neuroscience* **149,** 592–601.

Berghuis, P., Rajnicek, A. M., Morozov, Y. M., Ross, R. A., Mulder, J., Urban, G. M., Monory, K., Marsicano, G., Matteoli, M., Canty, A., Irving, A. J., Katona, I., *et al.* (2007). Hardwiring the brain: Endocannabinoids shape neuronal connectivity. *Science* **316,** 1212–1216.

Berrendero, F., Garcia-Gil, L., Hernandez, M. L., Romero, J., Cebeira, M., de Miguel, R., Ramos, J. A., and Fernandez-Ruiz, J. J. (1998). Localization of mRNA expression and activation of signal transduction mechanisms for cannabinoid receptor in rat brain during fetal development. *Development* **125,** 3179–3188.

Berrendero, F., Sepe, N., Ramos, J. A., Di Marzo, V., and Fernandez-Ruiz, J. J. (1999). Analysis of cannabinoid receptor binding and mRNA expression and endogenous cannabinoid contents in the developing rat brain during late gestation and early postnatal period. *Synapse* **33,** 181–191.

Biegon, A., and Kerman, A. (2001). Autoradiographic study of pre- and postnatal distribution of cannabinoid receptors in human brain. *Neuroimage* **14,** 1463–1468.

Bonnin, A., de Miguel, R., Rodriguez-Manzaneque, J. C., Fernandez-Ruiz, J. J., Santos, A., and Ramos, J. A. (1994). Changes in tyrosine hydroxylase gene expression in mesencephalic catecholaminergic neurons of immature and adult male rats perinatally exposed to cannabinoids. *Brain Res. Dev. Brain Res.* **81,** 147–150.

Borgen, L. A., Lott, G. C., and Davis, W. M. (1973). Cannabis-induced hypothermia: A dose–effect comparison of crude marihuana extract and synthetic 9-tetrahydrocannabinol in male and female rats. *Res. Commun. Chem. Pathol. Pharmacol.* **5,** 621–626.

Brake, S. C., Hutchings, D. E., Morgan, B., Lasalle, E., and Shi, T. (1987). Delta-9-tetrahydrocannabinol during pregnancy in the rat. II. Effects on ontogeny of locomotor activity and nipple attachment in the offspring. *Neurotoxicol. Teratol.* **9,** 45–49.

Campolongo, P., Trezza, V., Cassano, T., Gaetani, S., Morgese, M. G., Ubaldi, M., Soverchia, L., Antonelli, T., Ferraro, L., Massi, M., Ciccocioppo, R., and Cuomo, V. (2007). Perinatal exposure to delta-9-tetrahydrocannabinol causes enduring cognitive deficits associated with alteration of cortical gene expression and neurotransmission in rats. *Addict. Biol.* **12,** 485–495.

Castaldo, P., Magi, S., Gaetani, S., Cassano, T., Ferraro, L., Antonelli, T., Amoroso, S., and Cuomo, V. (2007). Prenatal exposure to the cannabinoid receptor agonist WIN 55,212-2 increases glutamate uptake through overexpression of GLT1 and EAAC1 glutamate transporter subtypes in rat frontal cerebral cortex. *Neuropharmacology* **53,** 369–378.

Compton, D. R., Gold, L. H., Ward, S. J., Balster, R. L., and Martin, B. R. (1992). Aminoalkylindole analogs: Cannabimimetic activity of a class of compounds structurally distinct from delta 9-tetrahydrocannabinol. *J. Pharmacol. Exp. Ther.* **263,** 1118–1126.

Costa, L. G., Steardo, L., and Cuomo, V. (2004). Structural effects and neurofunctional sequelae of developmental exposure to psychotherapeutic drugs: Experimental and clinical aspects. *Pharmacol. Rev.* **56,** 103–147.

Dalterio, S. L. (1986). Cannabinoid exposure: Effects on development. *Neurobehav. Toxicol. Teratol.* **8,** 345–352.

Day, N. L., and Richardson, G. A. (1991). Prenatal marijuana use: Epidemiology, methodologic issues, and infant outcome. *Clin. Perinatol.* **18,** 77–91.

Day, N. L., Richardson, G. A., Geva, D., and Robles, N. (1994a). Alcohol, marijuana, and tobacco: Effects of prenatal exposure on offspring growth and morphology at age six. *Alcohol Clin. Exp. Res.* **18,** 786–794.

Day, N. L., Richardson, G. A., Goldschmidt, L., Robles, N., Taylor, P. M., Stoffer, D. S., Cornelius, M. D., and Geva, D. (1994b). Effect of prenatal marijuana exposure on the cognitive development of offspring at age three. *Neurotoxicol. Teratol.* **16,** 169–175.

Di Marzo, V., and Fontana, A. (1995). Anandamide, an endogenous cannabinomimetic eicosanoid: 'Killing two birds with one stone'. *Prostaglandins Leukot. Essent. Fatty Acids* **53,** 1–11.

Fergusson, D. M., Horwood, L. J., and Northstone, K. (2002). Maternal use of cannabis and pregnancy outcome. *BJOG* **109,** 21–27.

Fernandez-Ruiz, J., Berrendero, F., Hernandez, M. L., and Ramos, J. A. (2000). The endogenous cannabinoid system and brain development. *Trends Neurosci.* **23,** 14–20.

Fernandez-Ruiz, J., Gomez, M., Hernandez, M., de Miguel, R., and Ramos, J. A. (2004). Cannabinoids and gene expression during brain development. *Neurotox. Res.* **6,** 389–401.

French, E. D., Dillon, K., and Wu, X. (1997). Cannabinoids excite dopamine neurons in the ventral tegmentum and substantia nigra. *Neuroreport* **8,** 649–652.

Freund, T. F., Katona, I., and Piomelli, D. (2003). Role of endogenous cannabinoids in synaptic signaling. *Physiol. Rev.* **83,** 1017–1066.

Fride, E. (2008). Multiple roles for the endocannabinoid system during the earliest stages of life: Pre- and postnatal development. *J. Neuroendocrinol.* **20**(Suppl. 1), 75–81.

Fride, E., and Shohami, E. (2002). The endocannabinoid system: Function in survival of the embryo, the newborn and the neuron. *Neuroreport* **13,** 1833–1841.

Fried, P. A. (1976). Short and long-term effects of pre-natal cannabis inhalation upon rat offspring. *Psychopharmacology (Berl.)* **50,** 285–291.

Fried, P. A. (1995). Prenatal exposure to marihuana and tobacco during infancy, early and middle childhood: Effects and an attempt at synthesis. *Arch. Toxicol. Suppl.* **17,** 233–260.

Fried, P. A. (2002). Conceptual issues in behavioral teratology and their application in determining long-term sequelae of prenatal marihuana exposure. *J. Child Psychol. Psychiatry* **43,** 81–102.

Fried, P. A., and Smith, A. M. (2001). A literature review of the consequences of prenatal marihuana exposure. An emerging theme of a deficiency in aspects of executive function. *Neurotoxicol. Teratol.* **23,** 1–11.

Fried, P. A., and Watkinson, B. (1990). 36- and 48-month neurobehavioral follow-up of children prenatally exposed to marijuana, cigarettes, and alcohol. *J. Dev. Behav. Pediatr.* **11,** 49–58.

Fried, P. A., and Watkinson, B. (2000). Visuoperceptual functioning differs in 9- to 12-year olds prenatally exposed to cigarettes and marihuana. *Neurotoxicol. Teratol.* **22,** 11–20.

Fried, P. A., Watkinson, B., and Gray, R. (1992). A follow-up study of attentional behavior in 6-year-old children exposed prenatally to marihuana, cigarettes, and alcohol. *Neurotoxicol. Teratol.* **14,** 299–311.

Garcia-Gil, L., De Miguel, R., Munoz, R. M., Cebeira, M., Villanua, M. A., Ramos, J. A., and Fernandez-Ruiz, J. J. (1997). Perinatal delta(9)-tetrahydrocannabinol exposure alters the responsiveness of hypothalamic dopaminergic neurons to dopamine-acting drugs in adult rats. *Neurotoxicol. Teratol.* **19,** 477–487.

Garcia-Gil, L., Ramos, J. A., Rubino, T., Parolaro, D., and Fernandez-Ruiz, J. J. (1998). Perinatal delta9-tetrahydrocannabinol exposure did not alter dopamine transporter and tyrosine hydroxylase mRNA levels in midbrain dopaminergic neurons of adult male and female rats. *Neurotoxicol. Teratol.* **20,** 549–553.

Goldschmidt, L., Day, N. L., and Richardson, G. A. (2000). Effects of prenatal marijuana exposure on child behavior problems at age 10. *Neurotoxicol. Teratol.* **22,** 325–336.

Goldschmidt, L., Richardson, G. A., Cornelius, M. D., and Day, N. L. (2004). Prenatal marijuana and alcohol exposure and academic achievement at age 10. *Neurotoxicol. Teratol.* **26,** 521–532.

Gong, J. P., Onaivi, E. S., Ishiguro, H., Liu, Q. R., Tagliaferro, P. A., Brusco, A., and Uhl, G. R. (2006). Cannabinoid CB2 receptors: Immunohistochemical localization in rat brain. *Brain Res.* **1071,** 10–23.

Gray, K. A., Day, N. L., Leech, S., and Richardson, G. A. (2005). Prenatal marijuana exposure: Effect on child depressive symptoms at ten years of age. *Neurotoxicol. Teratol.* **27,** 439–448.

Hampson, R. E., Mu, J., and Deadwyler, S. A. (2000). Cannabinoid and kappa opioid receptors reduce potassium K current via activation of G(s) proteins in cultured hippocampal neurons. *J. Neurophysiol.* **84,** 2356–2364.

Hansen, H. H., Schmid, P. C., Bittigau, P., Lastres-Becker, I., Berrendero, F., Manzanares, J., Ikonomidou, C., Schmid, H. H., Fernandez-Ruiz, J. J., and Hansen, H. S. (2001). Anandamide, but not 2-arachidonoylglycerol, accumulates during *in vivo* neurodegeneration. *J. Neurochem.* **78,** 1415–1427.

Harkany, T., Guzman, M., Galve-Roperh, I., Berghuis, P., Devi, L. A., and Mackie, K. (2007). The emerging functions of endocannabinoid signaling during CNS development. *Trends Pharmacol. Sci.* **28,** 83–92.

Hashimotodani, Y., Ohno-Shosaku, T., and Kano, M. (2007). Endocannabinoids and synaptic function in the CNS. *Neuroscientist* **13,** 127–137.

Howlett, A. C., Barth, F., Bonner, T. I., Cabral, G., Casellas, P., Devane, W. A., Felder, C. C., Herkenham, M., Mackie, K., Martin, B. R., Mechoulam, R., and Pertwee, R. G. (2002). International Union of Pharmacology. XXVII. Classification of cannabinoid receptors. *Pharmacol. Rev.* **54,** 161–202.

Hurd, Y. L., Wang, X., Anderson, V., Beck, O., Minkoff, H., and Dow-Edwards, D. (2005). Marijuana impairs growth in mid-gestation fetuses. *Neurotoxicol. Teratol.* **27,** 221–229.

Hutchings, D. E., Martin, B. R., Gamagaris, Z., Miller, N., and Fico, T. (1989). Plasma concentrations of delta-9-tetrahydrocannabinol in dams and fetuses following acute or multiple prenatal dosing in rats. *Life Sci.* **44,** 697–701.

Kumar, A. M., Solomon, J., Patel, V., Kream, R. M., Drieze, J. M., and Millard, W. J. (1986). Early exposure to delta 9-tetrahydrocannabinol influences neuroendocrine and reproductive functions in female rats. *Neuroendocrinology* **44,** 260–264.

Linn, S., Schoenbaum, S. C., Monson, R. R., Rosner, R., Stubblefield, P. C., and Ryan, K. J. (1983). The association of marijuana use with outcome of pregnancy. *Am. J. Public Health* **73,** 1161–1164.

Mackie, K. (2008). Cannabinoid receptors: Where they are and what they do. *J. Neuroendocrinol.* **20**(Suppl. 1), 10–14.

Mailleux, P., and Vanderhaeghen, J. J. (1992). Localization of cannabinoid receptor in the human developing and adult basal ganglia. Higher levels in the striatonigral neurons. *Neurosci. Lett.* **148,** 173–176.

Martinez-Orgado, J., Fernandez-Frutos, B., Gonzalez, R., Romero, E., Uriguen, L., Romero, J., and Viveros, M. P. (2003). Neuroprotection by the cannabinoid agonist WIN-55212 in an *in vivo* newborn rat model of acute severe asphyxia. *Brain Res. Mol. Brain Res.* **114,** 132–139.

Mato, S., Del Olmo, E., and Pazos, A. (2003). Ontogenetic development of cannabinoid receptor expression and signal transduction functionality in the human brain. *Eur. J. Neurosci.* **17,** 1747–1754.

Matsuda, L. A., Lolait, S. J., Brownstein, M. J., Young, A. C., and Bonner, T. I. (1990). Structure of a cannabinoid receptor and functional expression of the cloned cDNA. *Nature* **346,** 561–564.

McLaughlin, C. R., Martin, B. R., Compton, D. R., and Abood, M. E. (1994). Cannabinoid receptors in developing rats: Detection of mRNA and receptor binding. *Drug Alcohol Depend.* **36,** 27–31.

Mechoulam, R. (1986). Interview with Prof. Raphael Mechoulam, codiscoverer of THC. Interview by Stanley Einstein. *Int. J. Addict.* **21,** 579–587.

Mereu, G., Fa, M., Ferraro, L., Cagiano, R., Antonelli, T., Tattoli, M., Ghiglieri, V., Tanganelli, S., Gessa, G. L., and Cuomo, V. (2003). Prenatal exposure to a cannabinoid agonist produces memory deficits linked to dysfunction in hippocampal long-term potentiation and glutamate release. *Proc. Natl. Acad. Sci. USA* **100,** 4915–4920.

Millan, M. J. (2003). The neurobiology and control of anxious states. *Prog. Neurobiol.* **70,** 83–244.

Molina-Holgado, F., Molina-Holgado, E., Leret, M. L., Gonzalez, M. I., and Reader, T. A. (1993). Distribution of indoleamines and [^{3}H]paroxetine binding in rat brain regions following acute or perinatal delta 9-tetrahydrocannabinol treatments. *Neurochem. Res.* **18,** 1183–1191.

Molina-Holgado, F., Amaro, A., Gonzalez, M. I., Alvarez, F. J., and Leret, M. L. (1996). Effect of maternal delta 9-tetrahydrocannabinol on developing serotonergic system. *Eur. J. Pharmacol.* **316,** 39–42.

Munro, S., Thomas, K. L., and Abu-Shaar, M. (1993). Molecular characterization of a peripheral receptor for cannabinoids. *Nature* **365,** 61–65.

Navarro, M., Rodriguez de Fonseca, F., Hernandez, M. L., Ramos, J. A., and Fernandez-Ruiz, J. J. (1994). Motor behavior and nigrostriatal dopaminergic activity in adult rats perinatally exposed to cannabinoids. *Pharmacol. Biochem. Behav.* **47,** 47–58.

Navarro, M., Rubio, P., and de Fonseca, F. R. (1995). Behavioural consequences of maternal exposure to natural cannabinoids in rats. *Psychopharmacology (Berl.)* **122,** 1–14.

Newsom, R. J., and Kelly, S. J. (2008). Perinatal delta-9-tetrahydrocannabinol exposure disrupts social and open field behavior in adult male rats. *Neurotoxicol. Teratol.* **30,** 213–219.

NIDA (2005). NIDA Research Monograph Series: Marijuana Abuse NIH, Washington, DC.

O'Shea, M., and Mallet, P. E. (2005). Impaired learning in adulthood following neonatal delta9-THC exposure. *Behav. Pharmacol.* **16,** 455–461.

O'Shea, M., McGregor, I. S., and Mallet, P. E. (2006). Repeated cannabinoid exposure during perinatal, adolescent or early adult ages produces similar longlasting deficits in object recognition and reduced social interaction in rats. *J. Psychopharmacol.* **20,** 611–621.

Paria, B. C., and Dey, S. K. (2000). Ligand–receptor signaling with endocannabinoids in preimplantation embryo development and implantation. *Chem. Phys. Lipids* **108,** 211–220.

Paria, B. C., Song, H., Wang, X., Schmid, P. C., Krebsbach, R. J., Schmid, H. H., Bonner, T. I., Zimmer, A., and Dey, S. K. (2001). Dysregulated cannabinoid signaling disrupts uterine receptivity for embryo implantation. *J. Biol. Chem.* **276,** 20523–20528.

Paria, B. C., Wang, H., and Dey, S. K. (2002). Endocannabinoid signaling in synchronizing embryo development and uterine receptivity for implantation. *Chem. Phys. Lipids* **121,** 201–210.

Pertwee, R. G. (2006). Cannabinoid pharmacology: The first 66 years. *Br. J. Pharmacol.* **147**(Suppl. 1), S163–S171.

Piomelli, D. (2003). The molecular logic of endocannabinoid signalling. *Nat. Rev. Neurosci.* **4,** 873–884.

Richardson, G. A., Day, N. L., and Goldschmidt, L. (1995). Prenatal alcohol, marijuana, and tobacco use: Infant mental and motor development. *Neurotoxicol. Teratol.* **17,** 479–487.

Rodriguez de Fonseca, F., Cebeira, M., Hernandez, M. L., Ramos, J. A., and Fernandez-Ruiz, J. J. (1990). Changes in brain dopaminergic indices induced by perinatal exposure to cannabinoids in rats. *Brain Res. Dev. Brain Res.* **51,** 237–240.

Rodriguez de Fonseca, F., Cebeira, M., Fernandez-Ruiz, J. J., Navarro, M., and Ramos, J. A. (1991). Effects of pre- and perinatal exposure to hashish extracts on the ontogeny of brain dopaminergic neurons. *Neuroscience* **43,** 713–723.

Romero, J., Garcia-Palomero, E., Berrendero, F., Garcia-Gil, L., Hernandez, M. L., Ramos, J. A., and Fernandez-Ruiz, J. J. (1997). Atypical location of cannabinoid receptors in white matter areas during rat brain development. *Synapse* **26,** 317–323.

Rubio, P., Rodriguez de Fonseca, F., Munoz, R. M., Ariznavarreta, C., Martin-Calderon, J. L., and Navarro, M. (1995). Long-term behavioral effects of perinatal exposure to delta 9-tetrahydrocannabinol in rats: Possible role of pituitary–adrenal axis. *Life Sci.* **56,** 2169–2176.

SAMHSA (2002). Substance Abuse and Mental Health Service Administration. Results from the 2001 National Household Survey on Drug Abuse. NHSDA *In* "NHSDA Series H-17." DHHS Publication No. SMA 02-3758, Rockville, MD.

Smith, A. M., Fried, P. A., Hogan, M. J., and Cameron, I. (2006). Effects of prenatal marijuana on visuospatial working memory: An fMRI study in young adults. *Neurotoxicol. Teratol.* **28,** 286–295.

Suarez, I., Bodega, G., Fernandez-Ruiz, J., Ramos, J. A., Rubio, M., and Fernandez, B. (2004a). Down-regulation of the AMPA glutamate receptor subunits GluR1 and GluR2/3 in the rat cerebellum following pre- and perinatal delta9-tetrahydrocannabinol exposure. *Cerebellum* **3,** 66–74.

Suarez, I., Bodega, G., Rubio, M., Fernandez-Ruiz, J. J., Ramos, J. A., and Fernandez, B. (2004b). Prenatal cannabinoid exposure down-regulates glutamate transporter expressions (GLAST and EAAC1) in the rat cerebellum. *Dev. Neurosci.* **26,** 45–53.

Trezza, V., Campolongo, P., Cassano, T., Macheda, T., Dipasquale, P., Carratu, M. R., Gaetani, S., and Cuomo, V. (2008a). Effects of perinatal exposure to delta-9-tetrahydrocannabinol on the emotional reactivity of the offspring: A longitudinal behavioral study in Wistar rats. *Psychopharmacology (Berl.)* **198,** 529–537.

Trezza, V., Cuomo, V., and Vanderschuren, L. J. (2008b). Cannabis and the developing brain: Insights from behavior. *Eur. J. Pharmacol.* **585,** 441–452.

van der Stelt, M., Veldhuis, W. B., van Haaften, G. W., Fezza, F., Bisogno, T., Bar, P. R., Veldink, G. A., Vliegenthart, J. F., Di Marzo, V., and Nicolay, K. (2001). Exogenous anandamide protects rat brain against acute neuronal injury *in vivo*. *J. Neurosci.* **21,** 8765–8771.

Van Sickle, M. D., Duncan, M., Kingsley, P. J., Mouihate, A., Urbani, P., Mackie, K., Stella, N., Makriyannis, A., Piomelli, D., Davison, J. S., Marnett, L. J., Di Marzo, V., *et al.* (2005). Identification and functional characterization of brainstem cannabinoid CB2 receptors. *Science* **310,** 329–332.

Vardaris, R. M., Weisz, D. J., Fazel, A., and Rawitch, A. B. (1976). Chronic administration of delta-9-tetrahydrocannabinol to pregnant rats: Studies of pup behavior and placental transfer. *Pharmacol. Biochem. Behav.* **4,** 249–254.

Walters, D. E., and Carr, L. A. (1986). Changes in brain catecholamine mechanisms following perinatal exposure to marihuana. *Pharmacol. Biochem. Behav.* **25,** 763–768.

Wang, H., and Dey, S. K. (2005). Lipid signaling in embryo implantation. *Prostaglandins Other Lipid Mediat.* **77,** 84–102.

Wang, X., Dow-Edwards, D., Keller, E., and Hurd, Y. L. (2003). Preferential limbic expression of the cannabinoid receptor mRNA in the human fetal brain. *Neuroscience* **118,** 681–694.

Wenger, T., Croix, D., Tramu, G., and Leonardelli, J. (1992). Marijuana and reproduction. Effects on puberty and gestation in female rats. Experimental results. *Ann. Endocrinol. (Paris)* **53,** 37–43.

Witkin, J. M., Tzavara, E. T., and Nomikos, G. G. (2005). A role for cannabinoid CB1 receptors in mood and anxiety disorders. *Behav. Pharmacol.* **16,** 315–331.

Woodall, S. M., Breier, B. H., Johnston, B. M., and Gluckman, P. D. (1996). A model of intrauterine growth retardation caused by chronic maternal undernutrition in the rat: Effects on the somatotrophic axis and postnatal growth. *J. Endocrinol.* **150,** 231–242.

NEURONAL MECHANISMS FOR PAIN-INDUCED AVERSION: BEHAVIORAL STUDIES USING A CONDITIONED PLACE AVERSION TEST

Masabumi Minami

Department of Pharmacology, Graduate School of Pharmaceutical Sciences, Hokkaido University, Sapporo 060–0812, Japan

Pain consists of sensory discriminative and negative affective components. Although the neural systems responsible for the sensory component of pain have been studied extensively, the neural basis of the affective component is not well understood. Recently, behavioral studies using conditioned place aversion (CPA) tests have successfully elucidated the neural circuits and mechanisms underlying the negative affective component of pain. Excitotoxic lesions of the anterior cingulate cortex (ACC), central amygdaloid nucleus, basolateral amygdaloid nucleus (BLA), or bed nucleus of the stria terminalis (BNST) suppressed intraplantar formalin-induced aversive responses. Glutamatergic transmission within the ACC and BLA via NMDA receptors was shown to play a critical role in the affective component of pain. In the BNST, especially its ventral part, noradrenergic transmission via β-adrenergic receptors was demonstrated as important for pain-induced aversion. Because persistent pain is frequently associated with psychological and emotional dysfunction, studies of the neural circuits and the molecular mechanisms involved in the affective component of pain may have considerable clinical importance in the treatment of chronic pain.

INTERNATIONAL REVIEW OF NEUROBIOLOGY, VOL. 85
DOI: 10.1016/S0074-7742(09)85010-1

I. Introduction

Pain is comprised of sensory discriminative and negative affective components. Although the neural systems responsible for the sensory component of pain have been studied extensively, the neural basis of the affective component is not well understood. Recently, behavioral studies using the conditioned place paradigm have successfully elucidated the neural circuits and mechanisms underlying the negative affective component of pain. This review focuses on the behavioral studies that have used conditioned place aversion (CPA) tests to elucidate the neuronal mechanisms underlying pain-induced aversion.

II. Anterior Cingulate Cortex

Accumulating evidence shows that the anterior cingulate cortex (ACC) is related to emotion and pain. Electrophysiological studies have demonstrated that some ACC neurons respond to peripheral noxious stimuli (Sikes and Vogt, 1992; Yamamura *et al.*, 1996). Positron emission tomography and functional magnetic resonance imaging have shown the activation of ACC after noxious stimulation (Peyron *et al.*, 2000). Johansen *et al.* (2001) used a conditioning place paradigm to investigate the involvement of the ACC in the affective component of pain, and demonstrate that excitotoxic lesions of this brain area suppressed CPA induced by intraplantar injection of formalin (F-CPA) without reducing formalin-induced nociceptive behaviors. This result suggests that the ACC plays a crucial role in the affective component of pain. Gao *et al.* (2004) also reported the inhibition of F-CPA by bilateral excitotoxic lesions of the ACC.

Johansen and Fields (2004) reported the involvement of glutamatergic transmission within the ACC in the affective component of pain. They demonstrated that intra-ACC injection of kynurenic acid, a nonselective glutamate receptor antagonist, before formalin injection suppressed the acquisition of F-CPA. Lei *et al.* (2004) showed that F-CPA was suppressed by intra-ACC injection of either 2-amino-5-phosphonovalerate (AP5), an NMDA receptor antagonist, but not 6,7-dinitroquinoxaline-2, 3-dione (DNQX), an AMPA/KA receptor antagonist. F-CPA was also inhibited by intra-ACC administration of 7-chlorokynurenate, an antagonist of the glycine site of NMDA receptors (Ren *et al.*, 2006). These findings suggest that intra-ACC glutamatergic transmission via NMDA receptors plays a pivotal role in the affective component of pain.

III. Amygdala

The amygdala is a forebrain structure comprised of several distinct nuclei, including the basolateral (BLA) and central (CeA) amygdaloid nuclei, and is thought to be a key neural substrate underlying emotional responses such as anxiety, fear, and depression (Gallagher and Chiba, 1996). Electrophysiological studies have demonstrated that painful stimuli activate the amygdaloid neurons (Bernard and Besson, 1990; Neugebauer and Li, 2002). Furthermore, recent functional neuroimaging studies in humans have shown amygdala responses to painful stimuli (Becerra *et al.*, 2001; Bornhovd *et al.*, 2002). These findings suggest the involvement of this brain area in the affective component of pain.

Nakagawa *et al.* (2003) reported that intraplantar injection of formalin as a chemical somatic noxious stimulus increased c-fos mRNA expression in the BLA, but not in the CeA. On the other hand, intraperitoneal injection of acetic acid as a chemical visceral noxious stimulus induced this mRNA in the CeA, but not in the BLA. These results suggest that distinct amygdaloid nuclei differentially contribute to the affective components of somatic and visceral pain. This notion was supported by a lesion study (Tanimoto *et al.*, 2003). Bilateral excitotoxic lesions of either the BLA or CeA significantly suppressed F-CPA, while lesions of the CeA, but not the BLA, suppressed intraperitoneal acetic acid-induced CPA (A-CPA). Somatic sensory, including noxious, information from the spinal dorsal horn is transmitted through the lateral thalamus to such cortical areas as the primary and secondary somatosensory cortex and the insular cortex and then reaches the amygdala, particularly the BLA (Bushnell *et al.*, 1999; Shi and Cassell, 1998; Shi and Davis, 1999; Smith *et al.*, 2000). On the other hand, the CeA is directly linked to nociceptive relay nuclei in the spinal cord and brain stem through the spino–parabrachio–amygdaloid nociceptive pathway (Bernard and Besson, 1990; Bernard *et al.*, 1992, 1996; Jasmin *et al.*, 1997; Neugebauer and Li, 2002), which is reported to be involved in visceral pain (Bernard *et al.*, 1994). Visceral sensory information also reaches the nucleus of the solitary tract, which contains neurons projecting to the parabrachial nuclei, hypothalamus, bed nucleus of the stria terminalis (BNST), medial preoptic area, and amygdala, particularly the CeA (Van Giersbergen *et al.*, 1992). Taken together, somatic noxious information evoked by intraplantar formalin might be transmitted through the spino–thalamo–cortico–amygdaloid pathway to the BLA and then the CeA to induce aversive responses, while visceral noxious information evoked by intraperitoneal acetic acid might be transmitted through the spino–parabrachio–amygdaloid pathway and/or the nucleus of the solitary tract–amygdaloid pathway directly to the CeA, not via the BLA, to induce aversion (Fig. 1).

Deyama *et al.* (2007b) have discussed the role of glutamatergic transmission within the BLA in F-CPA. Intra-BLA injection of MK-801, an NMDA receptor

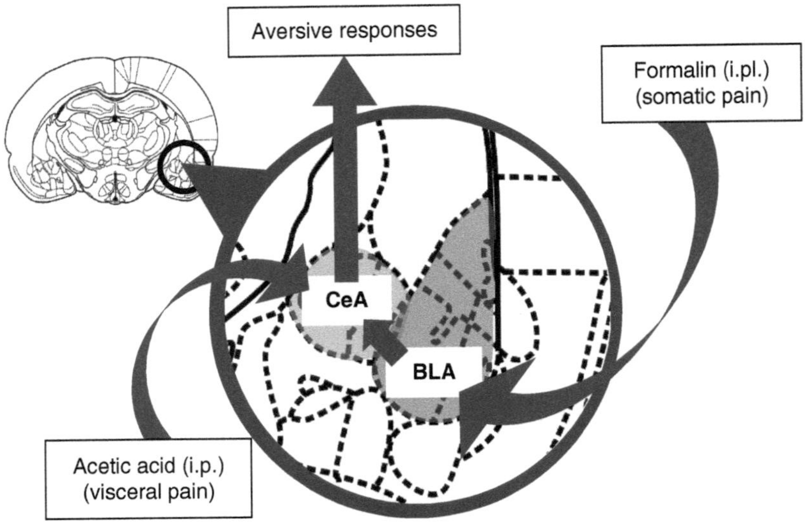

FIG. 1. Roles of the CeA and BLA in somatic and visceral pain-induced aversion. Please see the text for details.

antagonist, but not 6-cyano-7-nitroquinoxaline-2, 3-dione (CNQX), an AMPA/KA receptor antagonist, dose-dependently attenuated F-CPA without affecting nociceptive behaviors. Microdialysis experiments revealed that intraplantar injection of formalin induced an increase in the extracellular glutamate level within the BLA. This increase in glutamate level was suppressed by morphine perfusion via the microdialysis probe. Moreover, intra-BLA injection of morphine significantly attenuated F-CPA without affecting nociceptive behaviors. These findings suggest that glutamatergic transmission via NMDA receptors in the BLA plays a crucial role in pain-induced aversion, and that morphine may also influence the affective component of pain through an inhibitory effect on intra-BLA glutamatergic transmission (Fig. 2), in addition to its well-characterized effects on the sensory component of pain.

IV. Bed Nucleus of the Stria Terminalis

The BNST is a forebrain structure that represents one of the key neural substrates regulating stress responses and negative affective states, such as anxiety, fear, and aversion (Cecchi *et al.*, 2002; Delfs *et al.*, 2000; Fendt *et al.*, 2005; Nakagawa *et al.*, 2005; Sahuque *et al.*, 2006; Sullivan *et al.*, 2004; Walker and Davis, 1997). Anatomical studies have demonstrated that the BNST receives

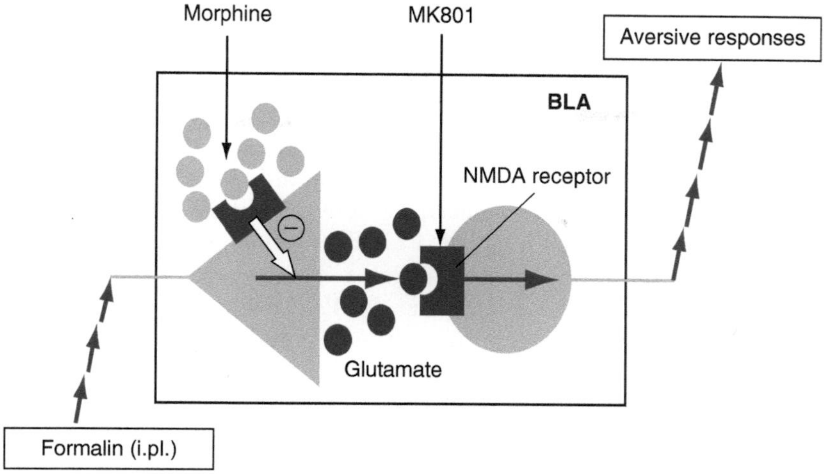

FIG. 2. Roles of intra-BLA glutamatergic transmission via NMDA receptors in pain-induced aversion. Please see the text for details.

direct and indirect nociceptive inputs from the spinal dorsal horn, as well as from such limbic regions as the ventromedial hypothalamus and amygdala (Braz *et al.*, 2005; Gauriau and Bernard, 2002). These findings suggest that this brain region is involved in the negative affective component of pain. Deyama *et al.* (2007a) addressed this issue in a study combining BNST lesions and a CPA test. In this study, they bilaterally destroyed the BNST by local injection of ibotenic acid in male Sprague–Dawley rats. Bilateral BNST lesions significantly suppress both the F-CPA and A-CPA, whereas the lesions did not suppressed formalin- or acetic acid-induced nociceptive behaviors, suggesting the crucial role of the BNST in the negative affective, but not sensory, component of pain.

The BNST, especially its ventral part (vBNST), is densely innervated by noradrenergic fibers arising mainly from the medullary A1/A2 cell groups (Forray *et al.*, 2000; Woulfe *et al.*, 1990). Fendt *et al.* (2005) reported that exposure to trimethylthiazoline, a component of fox odor, increased noradrenaline release in rats, and that intra-vBNST administration of the α_2-adrenoceptor agonist clonidine suppressed both the enhanced release of noradrenaline and the trimethylthiazoline-induced potentiation of freezing behavior. Cecchi *et al.* (2002) showed that noradrenaline release within the BNST of rats was increased by acute restraint stress, and that intra-vBNST injection of a cocktail of β_1 and β_2-adrenoceptor antagonists blocked stress-induced anxiety-like behaviors in the elevated plus maze test. Furthermore, Delfs *et al.* (2000) reported that microinjection of β-antagonists or an α_2-agonist into the BNST markedly attenuated opiate withdrawal-induced CPA in rats. They also demonstrated that lesions of the ventral noradrenergic bundle suppressed opiate withdrawal-induced CPA, suggesting the involvement of noradrenergic fibers arising from the A1/A2 medullary cell group. These findings suggest that noradrenergic transmission within the vBNST is important for mediating negative emotions such as fear, anxiety, and aversion.

The involvements of intra-vBNST noradrenergic transmission via β-adrenoceptors in F-CPA and A-CPA were examined (Deyama *et al.*, 2008, 2009). *In vivo* microdialysis showed that extracellular noradrenaline levels within the vBNST significantly increased after intraplantar formalin and intraperitoneal acetic acid injection. Intra-vBNST injection of timolol, a β-adrenoceptor antagonist, dose-dependently attenuated the F-CPA and A-CPA without reducing nociceptive behaviors. These results suggest that enhanced noradrenergic transmission via β-adrenoceptors within the vBNST plays a pivotal role in the negative affective, but not sensory, component of pain.

Because the CPA test is based on associative learning between a noxious stimulus-induced aversive affect and a neutral environmental context, it is difficult to determine whether the attenuation of CPA is due to the impairment of associative learning or the suppression of the primary aversive affect. To address this issue, we examined whether intra-vBNST injection of isoproterenol, a β-adrenoceptor agonist, produced CPA even in the absence of formalin-induced noxious stimulus. The results showed the aversive effect of intra-vBNST injection of isoproterenol, indicating that the activation of β-adrenoceptors within the vBNST is sufficient to produce the negative affective states. Therefore, the attenuation of F-CPA and A-CPA by the blockade of β-adrenoceptors within the vBNST may be due to the reduction of the primary aversive affect.

It is well know that β-adrenoceptors couple to adenylate cyclases to activate a protein kinase A (PKA), but no direct evidence exists for the involvement of the β-adrenoceptor–PKA signaling pathway in the affective component of pain. Thus, we examined the effect of intra-vBNST administration of a selective PKA inhibitor on isoproterenol- and pain-induced aversion. CPA induced by the intra-vBNST injection of isoproterenol was reversed by the coinjection of Rp-cyclic adenosine monophosphorothioate (Rp-cAMPS), a selective PKA inhibitor. Furthermore, intra-vBNST injection of Rp-cAMPS dose-dependently attenuated the F-CPA. These data suggest that PKA activation within the vBNST via the enhancement of β-adrenergic transmission is important for the negative affective component of pain (Fig. 3).

V. Other Brain Regions

The medial thalamic neurons receive the projection arising from the spinal dorsal horn through the spino–thalamic tract and then project to cortical structures, including the ACC. Wang *et al.* (2007) used a CPA paradigm to examine the

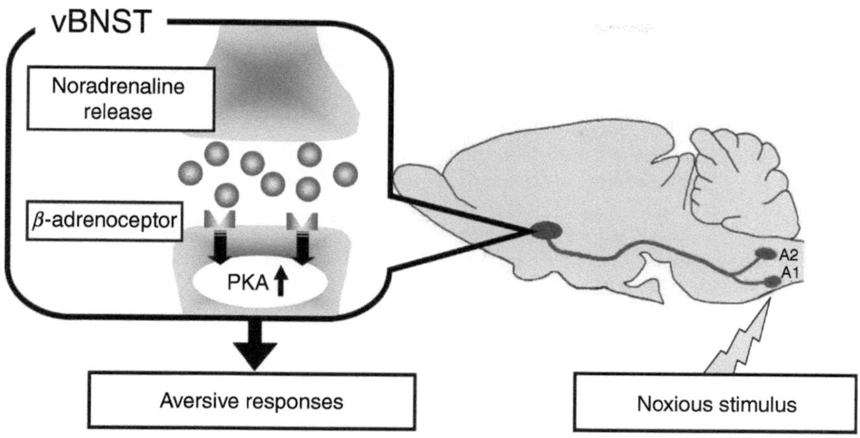

FIG. 3. Activation of the β-adrenoceptor–protein kinase A (PKA) signaling pathway within the vBNST mediates the negative affective component of pain. Please see the text for details.

role of this brain area in the affective component of pain. Bilateral lesion of the medial thalamus slightly attenuated the intraperitoneal acetic acid-induced nociceptive behaviors, but did not affect A-CPA. These results suggest that the medial thalamus is not involved in the affective component of pain.

VI. Conclusion

The conditioned place paradigm has been successfully used to reveal the neural substrate and neurotransmitters involved in the affective component of pain. Because persistent pain is frequently associated with psychological and emotional dysfunction (McWilliams *et al.*, 2003), studies of the neural circuits and the molecular mechanisms involved in the affective component of pain may have considerable clinical importance for the treatment of chronic pain.

References

Becerra, L., Breiter, H. C., Wise, R., Gonzalez, R. G., and Borsook, D. (2001). Reward circuitry activation by noxious thermal stimuli. *Neuron* **32,** 927–946.

Bernard, J. F., and Besson, J. M. (1990). The spino(trigemino)pontoamygdaloid pathway: Electrophysiological evidence for an involvement in pain processes. *J. Neurophysiol.* **63,** 473–490.

Bernard, J. F., Huang, G. F., and Besson, J. M. (1992). Nucleus centralis of the amygdala and the globus pallidus ventralis: Electrophysiological evidence for an involvement in pain processes. *J. Neurophysiol.* **68,** 551–569.

Bernard, J. F., Huang, G. F., and Besson, J. M. (1994). The parabrachial area: Electrophysiological evidence for an involvement in visceral nociceptive processes. *J. Neurophysiol.* **71,** 1646–1660.

Bernard, J. F., Bester, H., and Besson, J. M. (1996). Involvement of the spino-parabrachio-amygdaloid and -hypothalamic pathways in the autonomic and affective emotional aspects of pain. *Prog. Brain Res.* **107,** 243–255.

Bornhovd, K., Quante, M., Glauche, V., Bromm, B., Weiller, C., and Buchel, C. (2002). Painful stimuli evoke different stimulus-response functions in the amygdala, prefrontal, insula and somatosensory cortex: A single-trial fMRI study. *Brain* **125,** 1326–1336.

Braz, J. M., Nassar, M. A., Wood, J. N., and Basbaum, A. I. (2005). Parallel "pain" pathways arise from subpopulations of primary afferent nociceptor. *Neuron* **47,** 787–793.

Bushnell, M. C., Duncan, G. H., Hofbauer, R. K., Ha, B., Chen, J. I., and Carrier, B. (1999). Pain perception: Is there a role for primary somatosensory cortex? *Proc. Natl. Acad. Sci. USA* **96,** 7705–7709.

Cecchi, M., Khoshbouei, H., Javors, M., and Morilak, D. A. (2002). Modulatory effects of norepinephrine in the lateral bed nucleus of the stria terminalis on behavioral and neuroendocrine responses to acute stress. *Neuroscience* **112,** 13–21.

Delfs, J. M., Zhu, Y., Druhan, J. P., and Aston-Jones, G. (2000). Noradrenaline in the ventral forebrain is critical for opiate withdrawal-induced aversion. *Nature* **403,** 430–434.

Deyama, S., Nakagawa, T., Kaneko, S., Uehara, T., and Minami, M. (2007a). Involvement of the bed nucleus of the stria terminalis in the negative affective component of visceral and somatic pain in rats. *Behav. Brain Res.* **176,** 367–371.

Deyama, S., Yamamoto, J., Machida, T., Tanimoto, S., Nakagawa, T., Kaneko, S., Satoh, M., and Minami, M. (2007b). Inhibition of glutamatergic transmission by morphine in the basolateral amygdaloid nucleus reduces pain-induced aversion. *Neurosci. Res.* **59,** 199–204.

Deyama, S., Katayama, T., Ohno, A., Nakagawa, T., Kaneko, S., Yamaguchi, T., Yoshioka, M., and Minami, M. (2008). Activation of the beta-adrenoceptor-protein kinase A signaling pathway within the ventral bed nucleus of the stria terminalis mediates the negative affective component of pain in rats. *J. Neurosci.* **28,** 7728–7736.

Deyama, S., Katayama, T., Kondoh, N., Nakagawa, T., Kaneko, S., Yamaguchi, T., Yoshioka, M., and Minami, M. (2009). Role of enhanced noradrenergic transmission within the ventral bed nucleus of the stria terminalis in visceral pain-induced aversion in rats. *Behav. Brain Res.* **197,** 279–283.

Fendt, M., Siegl, S., and Steiniger-Brach, B. (2005). Noradrenaline transmission within the ventral bed nucleus of the stria terminalis is critical for fear behavior induced by trimethylthiazoline, a component of fox odor. *J. Neurosci.* **25,** 5998–6004.

Forray, M. I., Gysling, K., Andres, M. E., Bustos, G., and Araneda, S. (2000). Medullary noradrenergic neurons projecting to the bed nucleus of the stria terminalis express mRNA for the NMDA-NR1 receptor. *Brain Res. Bull.* **52,** 163–169.

Gallagher, M., and Chiba, A. A. (1996). The amygdala and emotion. *Curr. Opin. Neurobiol.* **6,** 221–227.

Gao, Y. J., Ren, W. H., Zhang, Y. Q., and Zhao, Z. Q. (2004). Contributions of the anterior cingulate cortex and amygdala to pain- and fear-conditioned place avoidance in rats. *Pain* **110,** 343–353.

Gauriau, C., and Bernard, J. F. (2002). Pain pathways and parabrachial circuits in the rat. *Exp. Physiol.* **87,** 251–258.

Jasmin, L., Burkey, A. R., Card, J. P., and Basbaum, A. I. (1997). Transneuronal labeling of a nociceptive pathway, the spino-(trigemino-)parabrachio-amygdaloid, in the rat. *J. Neurosci.* **17,** 3751–3765.

Johansen, J. P., and Fields, H. L. (2004). Glutamatergic activation of anterior cingulate cortex produces an aversive teaching signal. *Nat. Neurosci.* **7,** 398–403.

Johansen, J. P., Fields, H. L., and Manning, B. H. (2001). The affective component of pain in rodents: Direct evidence for a contribution of the anterior cingulate cortex. *Proc. Natl. Acad. Sci. USA* **98,** 8077–8082.

Lei, L. G., Sun, S., Gao, Y. J., Zhao, Z. Q., and Zhang, Y. Q. (2004). NMDA receptors in the anterior cingulate cortex mediate pain-related aversion. *Exp. Neurol.* **189,** 413–421.

McWilliams, L. A., Cox, B. J., and Enns, M. W. (2003). Mood and anxiety disorders associated with chronic pain: An examination in a nationally representative sample. *Pain* **106,** 127–133.

Nakagawa, T., Katsuya, A., Tanimoto, S., Yamamoto, J., Yamauchi, Y., Minami, M., and Satoh, M. (2003). Differential patterns of c-fos mRNA expression in the amygdaloid nuclei induced by chemical somatic and visceral noxious stimuli in rats. *Neurosci. Lett.* **344,** 197–200.

Nakagawa, T., Yamamoto, R., Fujio, M., Suzuki, Y., Minami, M., Satoh, M., and Kaneko, S. (2005). Involvement of the bed nucleus of the stria terminalis activated by the central nucleus of the amygdala in the negative affective component of morphine withdrawal in rats. *Neuroscience* **134,** 9–19.

Neugebauer, V., and Li, W. (2002). Processing of nociceptive mechanical and thermal information in central amygdala neurons with knee-joint input. *J. Neurophysiol.* **87,** 103–112.

Peyron, R., Garcia-Larrea, L., Gregoire, M. C., Convers, P., Richard, A., Lavenne, F., Barral, F. G., Mauguiere, F., Michel, D., and Laurent, B. (2000). Parietal and cingulate processes in central pain. A combined positron emission tomography (PET) and functional magnetic resonance imaging (fMRI) study of an unusual case. *Pain* **84,** 77–87.

Ren, W. H., Guo, J. D., Cao, H., Wang, H., Wang, P. F., Sha, H., Ji, R. R., Zhao, Z. Q., and Zhang, Y. Q. (2006). Is endogenous d-serine in the rostral anterior cingulate cortex necessary for pain-related negative affect? *J. Neurochem.* **96,** 1636–1647.

Sahuque, L. L., Kullberg, E. F., McGeehan, A. J., Kinder, J. R., Hicks, M. P., Blanton, M. G., Janak, P. H., and Olive, M. F. (2006). Anxiogenic and aversive effects of corticotropin-releasing factor (CRF) in the bed nucleus of the stria terminalis in the rat: Role of CRF receptor subtypes. *Psychopharmacology (Berl.)* **186,** 122–132.

Shi, C. J., and Cassell, M. D. (1998). Cascade projections from somatosensory cortex to the rat basolateral amygdala via the parietal insular cortex. *J. Comp. Neurol.* **399,** 469–491.

Shi, C., and Davis, M. (1999). Pain pathways involved in fear conditioning measured with fear-potentiated startle: Lesion studies. *J. Neurosci.* **19,** 420–430.

Sikes, R. W., and Vogt, B. A. (1992). Nociceptive neurons in area 24 of rabbit cingulate cortex. *J. Neurophysiol.* **68,** 1720–1732.

Smith, Y., Pare, J. F., and Pare, D. (2000). Differential innervation of parvalbumin-immunoreactive interneurons of the basolateral amygdaloid complex by cortical and intrinsic inputs. *J. Comp. Neurol.* **416,** 496–508.

Sullivan, G. M., Apergis, J., Bush, D. E., Johnson, L. R., Hou, M., and Ledoux, J. E. (2004). Lesions in the bed nucleus of the stria terminalis disrupt corticosterone and freezing responses elicited by a contextual but not by a specific cue-conditioned fear stimulus. *Neuroscience* **128,** 7–14.

Tanimoto, S., Nakagawa, T., Yamauchi, Y., Minami, M., and Satoh, M. (2003). Differential contributions of the basolateral and central nuclei of the amygdala in the negative affective component of chemical somatic and visceral pains in rats. *Eur. J. Neurosci.* **18,** 2343–2350.

Van Giersbergen, P. L., Palkovits, M., and De Jong, W. (1992). Involvement of neurotransmitters in the nucleus tractus solitarii in cardiovascular regulation. *Physiol. Rev.* **72,** 789–824.

Walker, D. L., and Davis, M. (1997). Double dissociation between the involvement of the bed nucleus of the stria terminalis and the central nucleus of the amygdala in startle increases produced by conditioned versus unconditioned fear. *J. Neurosci.* **17,** 9375–9383.

Wang, H. C., Chai, S. C., Wu, Y. S., and Wang, C. C. (2007). Does the medial thalamus play a role in the negative affective component of visceral pain in rats? *Neurosci. Lett.* **420,** 80–84.
Woulfe, J. M., Flumerfelt, B. A., and Hrycyshyn, A. W. (1990). Efferent connections of the A1 noradrenergic cell group: A DBH immunohistochemical and PHA-L anterograde tracing study. *Exp. Neurol.* **109,** 308–322.
Yamamura, H., Iwata, K., Tsuboi, Y., Toda, K., Kitajima, K., Shimizu, N., Nomura, H., Hibiya, J., Fujita, S., and Sumino, R. (1996). Morphological and electrophysiological properties of ACCx nociceptive neurons in rats. *Brain Res.* **735,** 83–92.

Bv8/PROKINETICINS AND THEIR RECEPTORS: A NEW PRONOCICEPTIVE SYSTEM

Lucia Negri, Roberta Lattanzi, Elisa Giannini, Michela Canestrelli, Annalisa Nicotra, and Pietro Melchiorri

Department of Physiology and Pharmacology, University "Sapienza" of Roma, P.le Aldo Moro 5, 00185 Rome, Italy

Bv8 is a small protein secreted by frog skin. Mammalian homologues of Bv8, the prokineticins PK1 and PK2, and their G-protein coupled receptors prokineticin receptor 1 (PKR1) and prokineticin receptor 2 (PKR2) have been identified and linked to several biological effects as gut motility, neurogenesis, angiogenesis, circadian rhythms, hematopoiesis, and nociception. Emerging evidences indicated that prokineticins are also associated with pathologies of the reproductive and nervous system, myocardial infarction, and tumorigenesis. Bv8 elicits a dose-dependent reduction in nociceptive threshold to thermal, mechanical, and chemical stimuli. The prokineticin receptors are present in a fraction of C- and Aδ-fiber neurons also expressing the vanilloid receptors, TRPV1 and TRPA1. Mice lacking PKR genes exhibit impaired Bv8-induced hyperalgesia, develop deficient responses to noxious heat, capsaicin, and protons and show reduced thermal and mechanical hypersensitivity to paw inflammation, indicating a requirement for PKR signaling in activation and sensitization of primary afferent fibers. Bv8/PK2 is highly expressed by neutrophils and other inflammatory cells and must be considered as new pronociceptive mediators in inflamed tissues. Bv8-like hyperalgesic activity was demonstrated in extracts of rat inflammatory granulocytes. Bv8 stimulates macrophage and T lymphocyte to differentiate towards an inflammatory and

INTERNATIONAL REVIEW OF
NEUROBIOLOGY, VOL. 85
DOI: 10.1016/S0074-7742(09)85011-3

Th1 profile indicating that Bv8/PK2 plays a role in immunoinflammatory responses. Blockade of PKRs may represent a novel therapeutic strategy in acute and inflammatory pain conditions.

I. Introduction

A small protein, named Bv8 to indicate its origin from the skin secretion of *Bombina variegata* and its molecular weight (8 kDa), is the first amphibian member of the Bv8-Prokineticin family. Homologues of Bv8 are present in skin secretions of other amphibians such as *Bombina bombina*, *Bombina orientalis*, *Bombina maxima*, lizards, and fishes of *Takifugu* species. A Bv8 homologue, mamba intestinal toxin (MIT-1), is a component of the venom of the black mamba, *Dendroaspis polylepsis*. Striking characteristics of these proteins are their identical amino terminal sequence, AVITG, and the presence of 10 cysteines with identical spacing that define a five disulphide-bridged motif called a colipase fold (Kaser *et al.*, 2003).

The high degree of identity between amphibian Bv8 peptides, fish peptides, and mamba MIT-1 (58%) suggested that similar peptides could also be present in other species, including mammals. In the mouse, rat, cattle, monkey, and man, cDNA cloning identified orthologues of Bv8 (Table I). The two mammalian proteins similar to Bv8 were named prokineticin 1 (PK1 or EG-VEGF) and prokineticin 2 (PK2 or mBv8). A second form of PK2 has been identified and named PK2b (basic) on account of an insert of 21 basic amino acids in its sequence (Wechselberger *et al.*, 1999). The name prokineticin refers to the ability of these peptides to contract guinea pig ileum, a property shared with amphibian Bv8.

II. Bv8-Related Mammalian Peptides

Expression patterns of rodent and human mRNAs for PK1 and PK2 have been reported in peripheral tissues (dorsal root ganglia (DRG), gastrointestinal tract, endocrine glands, spleen, human, and murine leucocytes) and in the central nervous system.

In neonatal and adult rat and mouse brain, both PK1 and PK2 are clearly expressed in the olfactory bulb. Olfactory bulb neurogenesis may depend on PK2 signaling because PK2-null mice display a marked reduction in the size of the olfactory bulb, a loss of normal olfactory bulb architecture, and an accumulation of neuronal progenitors in the rostral migratory stream. PK1 and PK2 are

TABLE I
Bv8-Related Peptides of Frogs, Fishes, Lizards, and Snakes

Peptides	Primary structure	Species
		Amphibians
	1----------------------------------38	
Bv8	**AVITGA**CDKDVQCGSGTCCAASAWSRNIRFCIPLGNSG	*Bombina variegata*
Bm8a	AVITGV**C**DRDAQ**C**GSGT**CC**AASAFSRNIRF**C**VPLGNNG	*Bombina maxima*
Bm8b	AVITGVRDRDAQ**C**GSGT**CC**AASAFSRNIRF**C**VPLGNNG	*Bombina maxima*
Bm8c	AVITGV**C**DRDAQ**C**GSGT**CC**AASAFSRNVRF**C**VPLGNNG	*Bombina maxima*
Bm8d	AVITGV**C**DRDAQ**C**GSGT**CC**AASAFSRNIRF**C**VPLGNNG	*Bombina maxima*
Bm8e	AVITGV**C**DRDAQ**C**GSGT**CC**AASAFSRNIRF**C**VPLGNNG	*Bombina maxima*
Bm8f	AVITGV**C**DRDAQ**C**GSGT**CC**AASAFSRNIRF**C**VPLGNNG	*Bombina maxima*
Bo8	AVITGA**C**DRDVQ**C**GSGT**CC**AASAWSRNIRF**C**VPLGNSG	*Bombina orientalis*
	39----------------------------------77	
Bv8	ED**C**HPASHKVPYDGKRLSSL**C**P**C**KSGLT**C**SKSGEKFK**C**S	*Bmbina variegata*
Bm8a	EE**C**HPASHKVPYNGKRLSSL**C**P**C**NTGLT**C**SKSGEKFQ**C**S	*Bombina maxima*
Bm8b	EE**C**HPASHKVPYNGKRLSSL**C**P**C**NTGLT**C**SKSGEKYQ**C**S	*Bombina maxima*
Bm8c	EE**C**HPASHKVPYNGKRLSSL**C**P**C**NTGLT**C**SKSGEKFQ**C**S	*Bombina maxima*
Bm8d	EE**C**HPASHKVPYNGKRLSSL**C**P**C**NTGLT**C**SKSGEKSQ**C**S	*Bombina maxima*
Bm8e	EE**C**HPASHKVPYNGKRLSSL**C**P**C**NTGLT**C**PKSGEKFQ**C**S	*Bombina maxima*
Bm8f	EE**C**HPASHKVPSDGKRLSSL**C**P**C**NTGLT**C**SKSGEKYQ**C**S	*Bombina maxima*
Bo8	EE**C**HPASHKVPYDGKRLSSL**C**P**C**KSGLT**C**SKSGAKFK**C**S	*Bombina orientalis*
		Fishes
	1----------------------------------38	
Fugu-1	AVITGACERDVQCGLGLCCAVSLWLRGLRMCAPRGLEG	*Takifugu bimaculatus*
Fugu-2	AVITGACEKDSQCGGGMCCAVSLWIRSLRMCTPMGREG	*Takifugu chinensis*
	39----------------------------------77	
Fugu-1	DECHPFSHKVPYPGKRQHHTCPCLPHLVCTRDRDSKYRC	*Takifugu bimaculatus*
Fugu-2	DDCHPMSHTVPFFGKRLHHTCPCLPNLSCIPMDEGRAKC	*Takifugu Chinensis*
	78------------94	
Fugu-1	TDDFKNVDLYEVGQTLR	*Takifugu bimaculatus*
Fugu-2	LSTYKYPDYYL	*Takifugu Chinensis*
		Snakes
	1----------------------------------38	
MIT-1	AVITGACERDLQCGKGTCCAVSLWIKSVRVCTPVGTSGE	*Dendroaspis polylepsis*
	39----------------------------------77	
MIT-1	D**C**HPASHKIPFSGQRKMHHT**C**P**C**APNLA**C**VQTSPKKFK**C**	*Dendroaspis polylepsis*
	–80	
MIT-1	LSK	*Dendroaspis polylepsis*
		Lizards
	1----------------------------------38	
VAR-1	AVITGACDKDLQCGEGMCCAVSLWIRSIRICTPLGSSGE	*Varanus varius*
VAR-2	AVITGA**C**DKDLQ**C**GEGM**CC**AVSLWIRSIRICTPLGSSGE	*Varanus varius*
	39----------------------------------77	
VAR-1	DCHPLSHKVPFDGQRKHHTCPCLPNLVCGQTSPGKYKCL	*Varanus varius*
VAR-2	DCHPLSHKVPFDGQRKHHTCPCLPNLVCGQTSPGKHKCL	*Varanus varius*
	78---84	
VAR-1	PEFKNVF	*Varanus varius*
VAR-2	PEFKNVF	*Varanus varius*

also clearly expressed in Calleja islands and in suprachiasmatic nucleus (SCN). The PK2 expression pattern in the SCN of mice and rats is rhythmic according to circadian cycle (being lowest in the dark phase) and is severely blunted in mutant mice deficient in clock or cryptochrome genes. PK2, therefore, has been indicated as a candidate SCN clock output signal that regulates circadian locomotor rhythm. PK2-null mice also displayed significantly reduced rhythmicity for a variety of other physiological and behavioral parameters, including sleep–wake cycle, body temperature, feeding circulating glucocorticoid, and glucose levels, as well as the expression of peripheral clock genes (Negri *et al.*, 2004; Zhou, 2006). Thus, PK2, acting as a SCN output factor, is important for the maintenance of robust circadian rhythms.

Prokineticins are also expressed in neurons of medial preoptic area, nucleus of solitary tract, trigeminal and facial nuclei, and DRG (Fig. 1).

PK1 has recently been detected in the mucosa/mesenchyme of mouse embryonic gut. It modulates both proliferation and differentiation of enteric neural crest cells (NCC) during enteric nervous system development, which eventually contribute to the formation of the myenteric and submucosal enteric plexus of the bowel. Hence, PK1 may be useful to patients exhibiting congenital colonic aganglionosis also known as Hirschprung disease (Ngan *et al.*, 2007, 2008).

PK1, also called endocrine gland-derived vascular endothelial growth factor (EG-VEGF), is a highly specific angiogenic mitogen, promotes proliferation and survival of endothelial cells, facilitates implantation, and acts as a novel placental growth factor for trophoblast differentiation (Maldonado-Perez *et al.*, 2007). PK1 is also predominately expressed in the Leydig cells of human testis and believed to promote the interstitial angiogenesis to support testicular endocrine activity of the testis. The expression of PK2, on the other hand, is restricted only in the primary spermatocytes, whereas its biological roles still remained to be defined (LeCouter *et al.*, 2003; Samson *et al.*, 2004; Wechselberger *et al.*, 1999). Elevated EG-VEGF/PK1 expression has also been implicated in colorectal cancer, where it promotes angiogenesis, cell proliferation, and liver metastasis (Goi *et al.*, 2004), and in prostate cancer, localized in the glandular epithelial cells of hyperplastic and malignant prostate tissues (Pasquali *et al.*, 2006). A high expression level of prokineticin receptor 1 (PKR1) is associated with malignant clinical characteristics of human neuroblastoma, a pediatric tumor derived from improperly differentiated NCC (Ngan *et al.*, 2007a).

Another role for PKs was revealed in hematopoiesis and in regulation of the immune response and in cardiomiocytes survival. They promote survival and differentiation of the granulocytic and monocytic lineage as well as stimulate hematopoietic cell mobilization. PKs, via Akt activation, protect cardiomyocytes against oxidative stress and rescue the myocardium against myocardial infarction in mouse model (Urayama *et al.*, 2007).

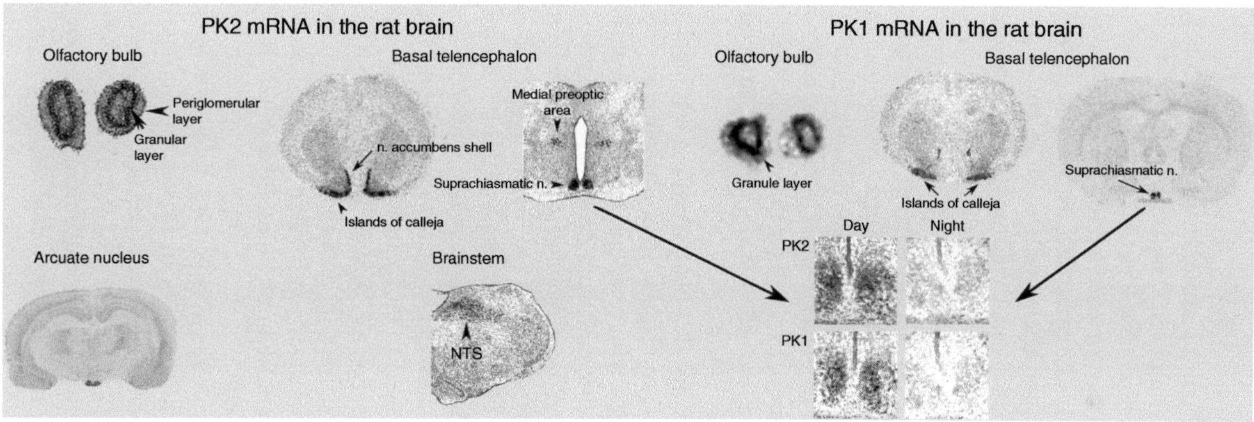

Fig. 1. Localization of prokineticin mRNAs in rat brain. PK1 and PK2 are both expressed in the olfactory bulb, island of Calleja, and SCN where they oscillate following the day–night cycle. PK2 mRNA is present also in the shell of nucleus accumbens, in nucleus arcuatus, and in nucleus of solitary tract. (See Color Insert.)

III. Bv8-Prokineticin Receptors

The two G-protein-coupled receptors for Bv8-PKs, PKR1 and PKR2, have an overall identity in their amino acid sequences of 85%, with most differences at the N-terminal and are about 80% identical to the previously described mouse orphan receptor gpr73. They are encoded by genes located on human chromosome region 2q14 and 20p13, respectively (Kaser *et al.*, 2003). In specific endothelial cells, neurons, and transfected cells, these receptors couple to Gq/o, Gi, and Gs to mediate the intracellular calcium mobilization and phosphoinositol turnover, activation of p44/42 MAPK signaling pathways, serine/threonine kinase Akt, and cyclic AMP accumulation (Chen *et al.*, 2005). Receptor binding studies showed that PKR2 is a MIT-preferring receptor, indeed affinity of MIT for PKR2 (in the range of pM) is about 10 times higher than that of PK2 and 50 times higher than that of PK1. PKR1 is a MIT and PK2-preferring receptor. Affinity of MIT for PKR1 (about 5–10 times lower than for PKR2) is comparable to that of PK2 and 60 times higher than that of PK1. Affinity of Bv8 for the receptors is comparable to that of PK2 and is about 40 times higher than that of PK1.

In rat embryos from day 12, both receptors are highly expressed in the neuroepithelium lining ventricles, the olfactory bulb, the Gasser-ganglion, and DRG. One day after birth, PKR2 receptors are still expressed at high levels in the olfactory bulb, in the neuroepithelium lining the ventricles, in the striatum, hippocampus, thalamic and hypothalamic paraventricular nuclei, SCN, amygdala, and cortex, whereas PKR1 receptors are only found in the cortex (Negri *et al.*, 2007). In adult rats, only PKR2 is abundantly or moderately expressed in several discrete brain regions (Fig. 2). The presence of PKR2 in the nucleus arcuatus explains the anorexogenic effect of Bv8 (Negri *et al.*, 2004). The dipsogenic effect of Bv8 depends on its binding to PKR2 in the subfornical organ (SFO). Central and peripheral administration of Bv8, in rats, induces GRH, oxitocin, and vasopressin release and vasopressin-dependent antidiuresis probably by acting on PKR2 in the paraventricular hypothalamic nucleus (Lattanzi *et al.*, 2001, and work in preparation). Stimulation of PKR2 in the SCN by Bv8 induces sleep in rodents (Negri *et al.*, 2004).

Human Kallmann syndrome (KS) which combine anosmia, related to defective olfactory bulb morphogenesis, with hypogonadism, due to gonadotropin-releasing hormone deficiency, appears to be related to mutations in the genes coding for PK2 and PKR2. In a cohort of 192 patients affected by KS, ten and four different point mutations were identified in the genes encoding PKR2 and PK2, respectively. These findings reveal that insufficient prokineticin-signaling through PKR2 leads to abnormal development of the olfactory system and reproductive axis also in man (Pitteloud *et al.*, 2007).

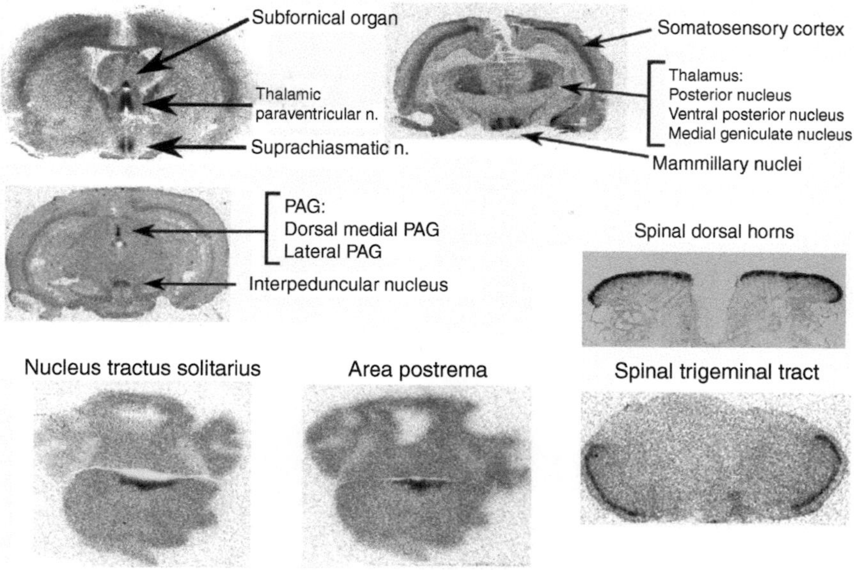

FIG. 2. Distribution of PKR2 in rat barin. Autoradiography of brain slices showing the binding sites of ^{125}I-MIT, a PKR2-selective ligand.

IV. Role of the Bv8-PK2/PKR System in the Neurobiology of Pain

PKR1 and PKR2 mRNAs are expressed in DRG of neonatal and adult rats: PKR1 is mainly expressed in small and medium size neurons and PKR2 in large neurons. PKR proteins are present in DRG, in the outer layers of the dorsal horns of the spinal cord and in peripheral terminals of nociceptor axons (Negri *et al.*, 2006a,b).

A. Pain Threshold

Activation of nociceptor PKRs by Bv8 in rats and mice produces nociceptive sensitization to thermal and mechanical stimuli, without inducing any spontaneous, overt nocifensive behavior, or local inflammation. Very low doses of Bv8 (50 fmol) injected into the paw induce a decrease in the nociceptive threshold that reaches the maximum in 1 h and disappears in 2–3 h. The same dose i.th., or higher doses by systemic routes (s.c. and i.v.), induces hyperalgesia with a characteristic biphasic time-course: the first peak occurs in 1 h and the second peak invariably in 4–5 h. The first phase depends on a direct action on nociceptors, because it resembles that

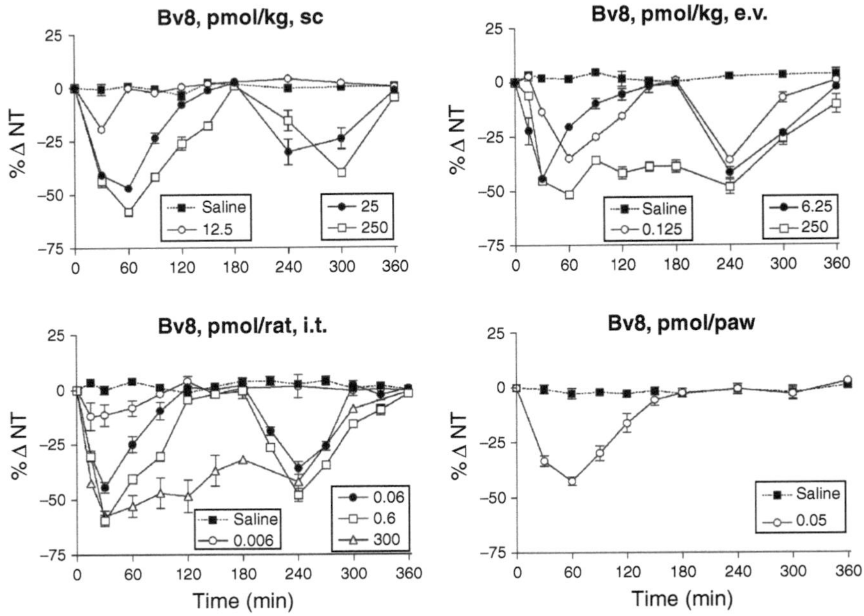

FIG. 3. Very low doses of Bv8 (50 fmol) injected into the paw of the rat induce a decrease in the nociceptive threshold that reaches the maximum in 1 h and disappears in 2–3 h. The same dose i.th., or higher doses by systemic routes (s.c. and i.v.), induces hyperalgesia with a characteristic biphasic time-course: the first peak occurs in 1 h and the second peak invariably in 4–5 h.

produced by intrapaw injection of a few femtomoles of Bv8. The second phase may depend by central and/or peripheral sensitization (Fig. 3).

A physiological role of Bv8-Prokineticins as peripheral and central pain modulators is supported by the observation that mice lacking the PKRs or PK2 are less sensitive to noxious stimuli than wild-type (WT) mice and exhibit impaired development of hyperalgesia after tissue injury (Hu *et al.*, 2006; Negri *et al.*, 2006a).

The molecular mechanism of the Bv8-induced hyperalgesia has been investigated in primary cultures of DRG neurons. Of the neurons responding to Bv8, 90% also responded to capsaicin and more than 70% also respond to mustard oil, showing a very high degree of colocalization of functional PKRs with the vanilloid receptors TRPV1 and TRPA1. Half of the Bv8-responding neurons also express calcitonin-gene-related peptide (CGRP) and substance P (SP) and release these neuropeptides upon exposure to Bv8 (De Felice *et al.*, submitted). Patch clamp experiments showed that a brief exposure to Bv8 tremendously potentiated the inward current activated by capsaicin in DRG neurons, an effect that is lacking in neurons from PKR1-KO mice. Bv8 causes translocation of PKCε to the neuronal membrane of nociceptors and Bv8-activated PKCε sensitizes TRPV1 to activating agents by phosphorylating two serine residues of the channel. Indeed, the Bv8-induced capsaicin potentiation is partially inhibited by the PKC inhibitors staurosporine and RO 31–8220. These data indicate that Bv8-induced hyperalgesia results, at least in part, from increased sensitivity of TRPV1 to heat and acid and from expression and release of excitatory transmitters (CGRP and substance P) at the spinal dorsal horn (Negri *et al.*, 2006a,b; Vellani *et al.*, 2006).

Mice lacking the PKRs are both less sensitive than WT mice to Bv8 induced heat hyperalgesia. But only PKR2-KO are significantly less sensitive to Bv8-induced tactile allodynia and mustard oil-induced thermal hyperalgesia confirm the prevalent distribution of PKR2 in larger neurons and their cooperativity with TRPA1 (Lattanzi *et al.*, submitted). PKR1 and PKR2-null mice both displace comparable impaired responses to capsaicin and acetic acid, compounds that activate TRPV1. Moreover, blocking the TRPV1 pathway with the antagonist capsazepine or deleting the TRPV1 gene attenuates but does not abolish the Bv8-induced hyperalgesia. These results confirm, *in vivo*, the functional cooperativity, between PKRs and TRPV1 and suggest that Bv8-induced pain sensitization must also involve other pain mediators.

About 18% of Bv8 responding neurons are COX1-positive (Vellani, personal communication) whereas no peripheral somatosensory neurons labeled for COX2 (Chopra *et al.*, 2000). Further data strongly suggest an involvement of prostanoid in Bv8-induced hyperalgesia (Negri *et al.*, 2007, and work in preparation). The Bv8-induced hyperalgesia was abolished by pretreatment with the prostaglandin receptor (EP1) antagonist SC51322, with PLA2 and COX1 inhibitors, whereas the COX2 inhibitor, NS392, was ineffective. COX1-null mice were significantly less sensitive than WT mice to Bv8-induced thermal hyperalgesia, whereas COX2-null mice showed the same sensitivity as WT mice. In accordance with the involvement of PKA signaling in prostaglandin-induced sensitization of nociceptors, the PKA inhibitors (H89 and WIPTIDE) impair the hyperalgesic action of Bv8. As already demonstrated for capsaicin, the hyperalgesic response of mice to intraplantar injection of a threshold dose of PGE2 (10 ng) increased markedly when mice were previously treated with 50 fmol Bv8 (Negri *et al.*, 2006b).

B. Inflammation and Inflammatory Pain

PK2 is expressed by inflammatory cells in human inflamed tonsil and appendix (LeCouter *et al.*, 2004) and in rat inflamed paw (Fig. 4).

CFA-induced chronic inflammation strongly increases the expression of PK2 and PK2b in mouse paw skin (Negri *et al.*, 2006a,b) and the time-course of CFA-induced hyperalgesia correlates with the dramatic increase in PK2

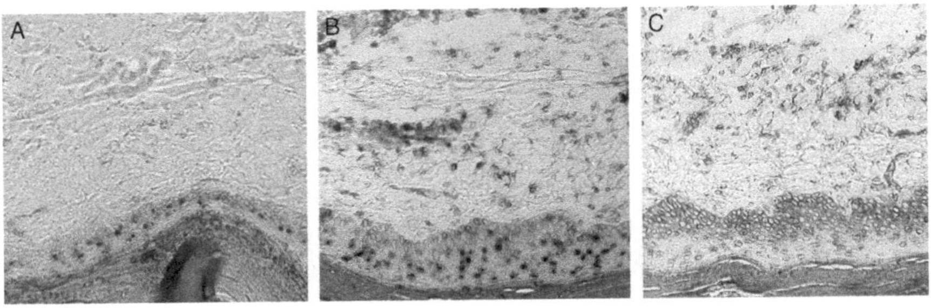

FIG. 4. PK2 transcript (DIG-PK2 antisense riboprobe) associated with infiltrating cells, possibly neutrophils and/or macrophages in rat paw, 12 h after CFA injection (B). Sense probe (C) or saline injection (A) did not generate specific signal. (See Color Insert.)

expression in mouse and rat paw skin. By *in situ* hybridization studies, performed on inflamed paw sections, and RT-PCR experiments, performed on FACS sorted cells, we demonstrated that neutrophils are the major source of PK2 and more notably that inflammatory condition induce an upregulation of PK2 both in neutrophils and in macrophages. Neutrophil extracts, fractionated using ionic exchange chromatography, gel filtration, and reverse phase (RP) chromatography, displayed Bv8-like activity, displaced ^{125}I-MIT binding from PKR1-transfected CHO cell membranes, and produced the Bv8-characteristic hyperalgesia when injected intrathecally in rats (Giannini *et al.*, submitted).

Deletion of the PKR1 gene strongly reduces the inflammation-induced heat hypersensitivity and PK2 upregulation. Conversely, deletion of PKR2 gene strongly reduces the inflammation-induced heat hypersensitivity but inflammation-induced PK2 upregulation is comparable to that of WT mice (Fig. 5).

Murine macrophages and splenocytes express PKR1 in large amounts. We demonstrated that Bv8 (at a concentration as low as 10^{-12} M) induces the macrophage to migrate and to acquire a proinflammatory phenotype. Bv8 (10^{-11} M) also stimulates T lymphocyte differentiation towards a Th1 profile, both directly and indirectly by reducing Th2 cytokines (Franchi *et al.*, 2008; Martucci *et al.*, 2006). Studies conducted using PKR1 knockout mice clearly indicate that all the activities exerted by Bv8 on macrophages and splenocytes are mediated by PKR1 (Martucci *et al.*, 2006), leading to propose that this receptor plays a pivotal role in initiating and maintaining inflammatory responses and pain.

V. Conclusion

Prokineticin receptors are potential targets for drugs which block the nociceptive information before it reaches the brain. Identifying of the structural determinants required for receptor binding and hyperalgesic activity of Bv8-Prokineticins

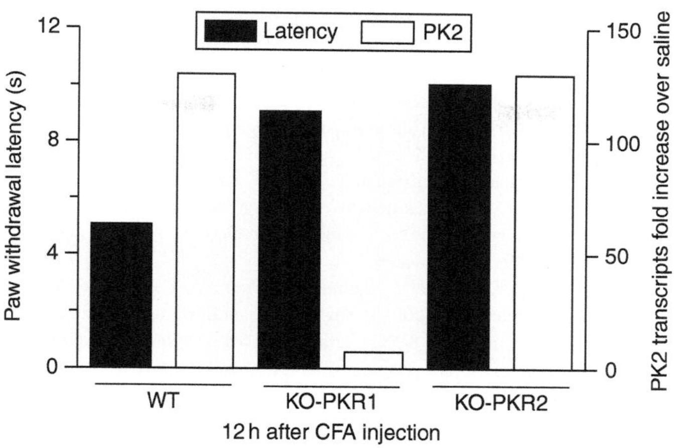

FIG. 5. CFA-induced heat hyperalgesia (evaluated as paw withdrawal latency) is significantly lower in PKR1- and PKR2-KO mice than in WT mice. Deletion of PKR1 but not of PKR2 gene strongly reduces the inflammation-induce PK2 upregulation.

is thus mandatory for the design of PKR antagonists. The highly conserved amino terminal sequence AVITGA and the tryptophan residue in position 24 in all members of the Bv8/PK family are required for biological activity: deletions and substitutions in these conserved residues produces antagonist molecules (Bullock *et al.*, 2004; Negri *et al.*, 2005). However, nonpeptide molecules are usually preferred as drugs. Studies are in progress to evaluate the possible antihyperalgesic/analgesic efficacy of some trazine-dione derivatives.

Acknowledgments

This work was supported by grants from the Italian Ministry of University and Scientific Research (PRIN 2004057339, PRIN 2004037781) and from the University "Sapienza" of Rome.

References

Bullock, C. M., Li, J. D., and Zhou, Q. Y. (2004). Structural determinants required for bioactivities of prokineticins and identification of prokineticin receptor antagonists. *Mol. Pharmacol.* **65,** 582–588.

Chen, J., Kuei, C., Sutton, S., Wilson, S., Yu, J., Kamme, F., Mazur, C., Lovenberg, T., and Liu, C. (2005). Identification and pharmacological characterization of prokineticin 2 beta as a selective ligand for prokineticin receptor 1. *Mol. Pharmacol.* **67,** 2070–2076.

Cheng, M. J., Bullock, C. M., Li, C., Lee, A. G., Bermak, J. C., Belluzzi, J., Weaver, D. R., Leslie, F. M., and Zhou, Q. Y. (2002). Prokineticin 2 transmits the behavioural circadian rhythm of the suprachiasmatic nucleus. *Nature* **417,** 405–410.

Chopra, B., Giblett, S., Little, J. G., Donaldson, L. F., Tate, S., Evans, R. J., Grubb, B. D. (2000). Cyclooxygenase-1 is a marker for a subpopulation of putative nociceptive neurons in rat dorsal root ganglia. *Eur. J. Neurosci.* **12,** 911–920.

Franchi, S., Giannini, E., Lattuada, D., Lattanzi, R., Tian, H., Melchiorri, P., Negri, L., Panerai, A. E., and Sacerdote, P. (2008). The prokineticin receptor agonist Bv8 decreases IL-10 and IL-4 production in mice splenocytes by activating prokineticin receptor-1. *BMC Immunol.* 2008 PMID: 18957080 [PubMed—in process].

Goi, T., Fujioka, M., Satoh, Y., Tabata, S., Koneri, K., Nagano, H., Hirono, Y., Katayama, K., Hirose, K., and Yamaguchi, A. (2004). Angiogenesis and tumor proliferation/metastasis of human colorectal cancer cell line SW620 transfected with endocrine glands-derived-vascular endothelial growth factor, as a new angiogenic factor. *Cancer Res.* **64,** 1906–1910.

Hu, W. P., Zhang, C., Li, J. D., Luo, Z. D., Amadesi, S., Bunnett, N., and Zhou, Q. Y. (2006). Impaired pain sensation in mice lacking prokineticin 2. *Mol. Pain* **2,** 35.

Kaser, A., Winklmayr, M., Lepperdinger, G., and Kreil, G. (2003). The AVIT protein family. Secreted cysteine-rich vertebrate proteins with diverse functions. *EMBO Rep.* **4,** 469–473.

Lattanzi, R., Giannini, E., Melchiorri, P., and Negri, L. (2001). Pharmacology of Bv8: A new peptide from amphibian skin. *Br. J. Pharmacol.* **133,** 45P.

LeCouter, J., Kowalski, J., Foster, J., Hass, P., Zhang, Z., Dillard-Telm, L., Frantz, G., Rangell, L., Deguzman, L., Keller, G., Peale, F., Gurney, A., *et al.* (2001). Identification of an angiogenic mitogen selective for endocrine gland endothelium. *Nature* **412,** 876–884.

LeCouter, J., Lin, R., Tejada, M., Frantz, G., Peale, F., Hillan, K. J., and Ferrara, N. (2003). The endocrine-gland-derived VEGF homologue Bv8 promotes angiogenesis in the testis: Localization of Bv8 receptors to endothelial cells. *Proc. Natl. Acad. Sci. USA* **100,** 2685–2690.

LeCouter, J., Zlot, C., Tejada, M., Peale, F., and Ferrara, N. (2004). Bv8 and endocrine gland-derived vascular endothelial growth factor stimulate hematopoiesis and hematopoietic cell mobilization. *Proc. Natl. Acad. Sci. USA* **101,** 16813–16818.

Maldonado-Perez, D., Evans, J., Denison, F., Millar, R. P., and Jabbour, H. N. (2007). Potential roles of the prokineticins in reproduction. *Trends Endocrinol. Metab.* [Epub ahead of print].

Martucci, C., Franchi, S., Giannini, E., Tian, H., Melchiorri, P., Negri, L., and Sacerdote, P. (2006). Bv8, the amphibian homologue of the mammalian prokineticins, induces a proinflammatory phenotype of mouse macrophages. *Br. J. Pharmacol.* **147,** 225–234.

Negri, L., Lattanzi, R., Giannini, E., De Felice, M., Colucci, A., and Melchiorri, P. (2004). Bv8, the amphibian homologue of the mammalian prokineticins, modulates ingestive behaviour in rats. *Br. J. Pharmacol.* **142,** 141–151.

Negri, L., Lattanzi, R., Giannini, E., Colucci, A., Mignogna, G., Barra, D., Grohovaz, F., Codazzi, F., Kaiser, A., Kreil, G., and Melchiorri, P. (2005). Biological activities of Bv8 analogues. *Br. J. Pharmacol.* **146,** 625–632.

Negri, L., Lattanzi, R., Giannini, E., Colucci, M., Margheriti, F., Melchiorri, P., Vellani, V., Tian, H., De Felice, M., and Porreca, F. (2006a). Impaired nociception and inflammatory pain sensation in mice lacking the prokineticin receptor PKR1: Focus on interaction between PKR1 and the capsaicin receptor TRPV1 in pain behavior. *J. Neurosci.* **26,** 6716–6727.

Negri, L., Lattanzi, R., Giannini, E., and Melchiorri, P. (2006b). Modulators of Pain: Bv8 and Prokineticins. *Curr. Neuropharmacol.* **4,** 207–215.

Negri, L., Lattanzi, R., Giannini, E., and Melchiorri, P. (2007). Bv8/Prokineticin proteins and their receptors. *Life Sci.* **81,** 1103–1116.

Ngan, E. S., Lee, K. Y., Sit, F. Y., Poon, H. C., Chan, J. K., Sham, M. H., Lui, V. C., and Tam, P. K. (2007). Prokineticin-1 modulates proliferation and differentiation of enteric neural crest cells. *Biochim. Biophys. Acta* **1773,** 536–545.

Ngan, E. S., Sit, F. Y., Lee, K. L., Miao, X., Yuan, Z., Wang, W., Nicholls, J. M., Wong, K. K., Garcia-Barcelo, M., Lui, V. C., and Tam, P. K. (2007a). Implications of endocrine gland-derived vascular endothelial growth factor/prokineticin-1 signaling in human neuroblastoma progression. *Clin. Cancer Res.* **13,** 868–875.

Ngan, E. S., Shum, C. K., Poon, H. C., Sham, M. H., Garcia-Barcelo, M. M., Lui, V. C., and Tam, P. K. (2008). Prokineticin-1 (ProK-1) works coordinately with glial cell line-derived neurotrophic factor (GDNF) to mediate proliferation and differentiation of enteric neural crest cell. *Biochim. Biophys. Acta. Mol. Cell Res.* **1783,** 467–478.

Pasquali, D., Rossi, V., Staibano, S., De Rosa, G., Chieffi, P., Prezioso, D., Mirone, V., Mascolo, M., Tramontano, D., Bellastella, A., and Sinisi, A. A. (2006). The endocrine-gland-derived vascular endothelial growth factor (EG-VEGF)/prokineticin 1 and 2 and receptor expression in human prostate: Up-regulation of EG-VEGF/prokineticin 1 with malignancy. *Endocrinology* **147,** 4245–4251.

Pitteloud, N., Zhang, C., Pignatelli, D., Li, J. D., Raivio, T., Cole, L. W., Plummer, L., Jacobson-Dickman, E. E., Mellon, P. L., Zhou, Q. Y., and Crowley, W. F. Jr. (2007). Loss-of-function mutation in the prokineticin 2 gene causes Kallmann syndrome and normosmic idiopathic hypogonadotropic hypogonadism. *Proc. Natl. Acad. Sci. USA* **104,** 17447–17452.

Samson, M., Peale, F. V. Jr, Frantz, G., Rioux-Leclercq, N., Rajpert-De Meyts, E., and Ferrara, N. (2004). Human endocrine gland-derived vascular endothelial growth factor: Expression early in development and in Leydig cell tumors suggests roles in normal and pathological testis angiogenesis. *J. Clin. Endocrinol. Metab.* **89,** 4078–4088.

Urayama, K., Guilini, C., Messaddeq, N., Hu, K., Steenman, M., Kurose, H., Ert, G., and Nebigil, C. G. (2007). The prokineticin receptor-1 (GPR73) promotes cardiomyocyte survival and angiogenesis. *FASEB J.* [Epub ahead of print].

Vellani, V., Colucci, M., Lattanzi, R., Giannini, E., Negri, L., Melchiorri, P., and McNaughton, P. A. (2006). Sensitization of transient receptor potential vanilloid 1 by the prokineticin receptor agonist Bv8. *J. Neurosci.* **26,** 5109–5116.

Wechselberger, C., Puglisi, R., Engel, E., Lepperdinger, G., Boitani, C., and Kreil, G. (1999). The mammalian homologues of frog Bv8 are mainly expressed in spermatocytes. *FEBS Lett.* **462,** 177–181.

Zhou, Q. Y. (2006). The Prokineticins. A novel pair of regulatory peptides. *Mol. Interv.* **6,** 330.

$P2Y_6$-EVOKED MICROGLIAL PHAGOCYTOSIS

Kazuhide Inoue,* Schuichi Koizumi,† Ayako Kataoka,* Hidetoshi Tozaki-Saitoh,* and Makoto Tsuda*

*Department of Molecular and System Pharmacology, Graduate School of Pharmaceutical Sciences, Kyushu University, 3-1-1 Maidashi, Higashi, Fukuoka 812-8582, Japan
†Department of Pharmacology, Interdisciplinary Graduate School of Medicine and Engineering, University of Yamanashi, 1110 Shimokato, Chuo, Yamanashi 409-3893, Japan

While it was reported that microglia is engaged in the clearance of dead cells or dangerous debris, the mechanism of phagocytosis is still unclear. Recently, we found that purinergic system has a very important role for the chemotaxis and phagocytosis of microglia. When neighboring cells are injured, the cells release or leak ATP into extracellular space and microglia rapidly move toward or extend a process to the nucleotides as chemotaxis through $P2Y_{12}$ receptors of microglia. In the meanwhile, microglia expressing metabotropic $P2Y_6$ receptors show phagocytosis by the stimulation of uridine 5′-diphosphate (UDP), an agonist of $P2Y_6$. UDP/UTP is leaked when hippocampal neurons are damaged by kainic acid (KA) *in vivo* and *in vitro*. Systemic administration of KA in rats results in neuronal cell death in the hippocampal CA1 and CA3 regions, where mRNA for $P2Y_6$ receptors increases activated microglia. Thus, the $P2Y_6$ receptor is upregulated when neurons are damaged, and would function as a sensor for phagocytosis by sensing diffusible UDP signals.

I. Introduction

Accumulating findings indicate that nucleotides play an important role in neuron-to-glia communication through P2 purinoceptors. P2 purinoceptors are divided into two families, ionotropic receptors (P2X) and metabotropic receptors (P2Y). P2X receptors (seven types; $P2X_1$–$P2X_7$) contain intrinsic pores that open

INTERNATIONAL REVIEW OF
NEUROBIOLOGY, VOL. 85
DOI: 10.1016/S0074-7742(09)85012-5

0074-7742/09 $35.00

by binding with ATP. P2Y (eight types; $P2Y_{1, 2, 4, 6, 11, 12, 13, \text{ and } 14}$) are activated by nucleotides and couple to intracellular second-messenger systems through heteromeric G-proteins (Inoue, 2006). Microglia express $P2X_4$, $P2X_7$, $P2Y_2$, $P2Y_6$, and $P2Y_{12}$ (Inoue, 2006) and are known as resident macrophages in central nervous system (CNS), accounting for 5–10% of the total population of glia (Kreutzberg, 1996; Stoll and Jander, 1999). When neurons are injured or dead, microglia are activated, resulting in their interaction with immune cells, active migration to the site of injury, release of proinflammatory substances, and the phagocytosis of damaged cells or debris. In the series of microglial functions, growing evidence indicates that extracellular nucleotides have a central role in regulating the action of microglia.

II. Chemotaxis

In the adult normal conditions, microglia are ubiquitously distributed throughout the CNS and represent a morphologically unique type of cell which has a small soma bearing thin and branched processes. Although such microglia are called as "resting microglia," recent studies using a transgenic mouse line that expressed green fluorescent protein in microglia together with two-photon microscopy have revealed that "resting" microglial processes are highly dynamic in the brain (Davalos *et al.*, 2005; Nimmerjahn *et al.*, 2005). The processes of microglia rapidly move toward the site of injury, an effect that is mimicked by local injection of ATP and can be inhibited either by the ATP-hydrolyzing enzyme apyrase or by blockers of P2YRs (Davalos *et al.*, 2005). Microglia in $P2Y_{12}$ receptor knockout mice show significantly diminished directional branch extension toward sites of cortical damage in the living mouse (Haynes *et al.*, 2006). Thus, microglia appear to act as sensors using the nucleotides/$P2Y_{12}$ system.

Microglial chemotaxis by ATP via $P2Y_{12}$ receptors was originally found by Honda *et al.* (2001), and has recently been confirmed *in vivo* using $P2Y_{12}$ receptor knockout animals (Haynes *et al.*, 2006). Haynes *et al.* (2006) show that microglia from $P2Y_{12}$ receptor knockout mice exhibit normal baseline motility but are unable to chemotaxis toward nucleotides *in vitro* or *in vivo*. Moreover, $P2Y_{12}$ expression is robust in the "resting" state, but dramatically reduced after microglial activation. These results imply that $P2Y_{12}$ is a primary site at which nucleotides act to induce microglial chemotaxis at early stages of the response to local CNS injury (Haynes *et al.*, 2006). These data suggest that microglial chemotaxis requires ATP signaling via a $P2Y_{12}$-independent mechanism (Haynes *et al.*, 2006; Ohsawa *et al.*, 2007). Neuronal injury results in the release or leakage of ATP that appears to be a "find-me" signal from damaged neurons to microglia to cause chemotaxis.

III. Phagocytosis

Phagocytosis is a specialized form of endocytosis taking relatively large particles ($>1.0\ \mu m$) into vacuoles and has a central role in tissue remodeling, inflammation, and the defense against infectious agents (Tjelle *et al.*, 2000). Phagocytosis is initiated by the activation of cell-surface phagocytosis receptors, including Fc receptors, complement receptors, integrins, endotoxin receptors (CD18 and CD14), mannose receptors, and scavenger receptors (Lauber *et al.*, 2003), which are activated by corresponding extracellular ligands called as "eat-me" signals. Since recognition is the most important step for phagocytosis, extensive studies on phagocytosis receptors have been reported. With regard to apoptotic cells, it is well known that dying cells express so-called "eat-me" signals such as phosphatidylserine (PS) on their surface membrane (Lauber *et al.*, 2003), by which microglia recognize the apoptotic cells in order to catch and remove them (Lauber *et al.*, 2003). As for amyloid β protein (Aβ), a key molecule that mediates Alzheimer's disease, microglia remove Aβ presumably via Fc receptor-dependent phagocytosis (Bard *et al.*, 2000; Schenk *et al.*, 1999). It is, however, unclear how phagocytotic cells come to the target cells or debris. Our findings (Koizumi *et al.*, 2007) suggest that nucleotides might be the molecules to guide phagocytotic cells to the targets as below.

We found that application of uridine 5′-diphosphate (UDP) caused microglial phagocytosis through $P2Y_6$ in a concentration-dependent manner, and that neuronal injury caused by kainic acid (KA) upregulated $P2Y_6$ receptors in microglia, the KA-evoked neuronal injury resulted in an increase in extracellular UTP, which was immediately metabolized into UDP *in vivo* and *in vitro* (Koizumi *et al.*, 2007). We also found that UDP leaked from injured neurons caused $P2Y_6$ receptor-dependent phagocytosis *in vivo* and *in vitro* (Koizumi *et al.*, 2007). Thus, UDP might be a diffusible molecule that signals the crisis of damaged neurons to microglia to show phagocytosis.

IV. Conclusion

Nucleotides seem to have the ability to act as "eat-us" signals for necrotic cells suffering traumatic or ischemic injury because such necrotic cells cause swelling, followed by shrinkage, leading to the leakage of cytoplasmic molecules including a large amount of ATP and UTP, and extracellular nucleotides are immediately degraded by ecto-nucleotidases, suggesting that leaked nucleotides could be transient and localized signals that alert to the crisis created by the presence of the necrotic cells. These findings suggest that microglia might be attracted by

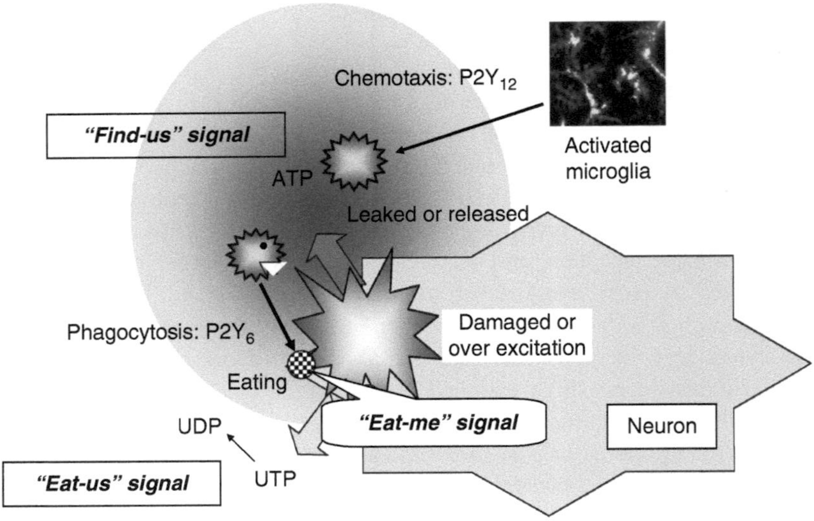

FIG. 1. Illustration of nucleotide-activated microglial chemotaxis and phagocytosis. Damaged neurons released or leaked ATP and UTP. The concentration of ATP is higher than that of UTP. UTP is easily converted to UDP by ecto-nucleotidases. Activated microglia might be attracted by ATP and move to the damaged area, and then subsequently recognize UDP as "eat-us" signal to start for finding "eat-me" signals attached to the targets and engulf them.

ATP/ADP (Davalos *et al.*, 2005; Haynes *et al.*, 2006; Honda *et al.*, 2001; Nimmerjahn *et al.*, 2005) and move to the damaged area, and then subsequently recognize UDP as "eat-us" signal to start for finding "eat-me" signals attached to the targets and engulf them (Fig. 1). It is interesting that ATP/ADP is not able to efficiently activate $P2Y_6$ receptors, nor can UDP act on $P2Y_{12}$ receptors. Thus, adenine and uridine nucleotides would regulate microglial chemotaxis and phagocytosis in a coordinated fashion.

References

Bard, F., Cannon, C., Barbour, R., Burke, R. L., Games, D., Grajeda, H., Guido, T., Hu, K., Huang, J., Johnson-Wood, K., Khan, K., Kholodenko, D., *et al.* (2000). Peripherally administered antibodies against amyloid beta-peptide enter the central nervous system and reduce pathology in a mouse model of Alzheimer disease. *Nat. Med.* **6,** 916–919.

Davalos, D., Grutzendler, J., Yang, G., Kim, J. V., Zuo, Y., Jung, S., Littman, D. R., Dustin, M. L., and Gan, W. B. (2005). ATP mediates rapid microglial response to local brain injury *in vivo*. *Nat. Neurosci.* **8,** 752–758.

Haynes, S. E., Hollopeter, G., Yang, G., Kurpius, D., Dailey, M. E., Gan, W. B., and Julius, D. (2006). The P2Y12 receptor regulates microglial activation by extracellular nucleotides. *Nat. Neurosci.* **9,** 1512–1519.

Honda, S., Sasaki, Y., Ohsawa, K., Imai, Y., Nakamura, Y., Inoue, K., and Kohsaka, S. (2001). Extracellular ATP or ADP induce chemotaxis of cultured microglia through Gi/o-coupled P2Y receptors. *J. Neurosci.* **21,** 1975–1982.

Inoue, K. (2006). The function of microglia through purinergic receptors: Neuropathic pain and cytokine release. *Pharmacol. Ther.* **109,** 210–226.

Koizumi, S., Shigemoto-Mogami, Y., Nasu-Tada, K., Shinozaki, Y., Ohsawa, K., Tsuda, M., Joshi, B. V., Jacobson, K. A., Kohsaka, S., and Inoue, K. (2007). UDP acting at P2Y6 receptors is a mediator of microglial phagocytosis. *Nature* **446,** 1091–1095.

Kreutzberg, G. W. (1996). Microglia: A sensor for pathological events in the CNS. *Trends Neurosci.* **19,** 312–318.

Lauber, K., Bohn, E., Krober, S. M., Xiao, Y. J., Blumenthal, S. G., Lindemann, R. K., Marini, P., Wiedig, C., Zobywalski, A., Baksh, S., Xu, Y., Autenrieth, I. B., *et al.* (2003). Apoptotic cells induce migration of phagocytes via caspase-3-mediated release of a lipid attraction signal. *Cell* **113,** 717–730.

Nimmerjahn, A., Kirchhoff, F., and Helmchen, F. (2005). Resting microglial cells are highly dynamic surveillants of brain parenchyma *in vivo*. *Science* **308,** 1314–1318.

Ohsawa, K., Irino, Y., Nakamura, Y., Akazawa, C., Inoue, K., and Kohsaka, S. (2007). Involvement of P2X(4) and P2Y(12) receptors in ATP-induced microglial chemotaxis. *Glia* **55,** 604–616.

Schenk, D., Barbour, R., Dunn, W., Gordon, G., Grajeda, H., Guido, T., Hu, K., Huang, J., Johnson-Wood, K., Khan, K., Kholodenko, D., Lee, M., *et al.* (1999). Immunization with amyloid-beta attenuates Alzheimer-disease-like pathology in the PDAPP mouse. *Nature* **400,** 173–177.

Stoll, G., and Jander, S. (1999). The role of microglia and macrophages in the pathophysiology of the CNS. *Prog. Neurobiol.* **58,** 233–247.

Tjelle, T. E., Lovdal, T., and Berg, T. (2000). Phagosome dynamics and function. *Bioessays* **22,** 255–263.

PPAR AND PAIN

Takehiko Maeda and Shiroh Kishioka

Department of Pharmacology, Wakayama Medical University, Wakayama 641–0012, Japan

Peroxisome proliferator-activated receptors (PPARs) are ligand-activated transcription factor belonging to a nuclear hormone receptor superfamily, containing three isoforms (α, β/δ, and γ). PPARs play a critical physiological role as a primary lipid sensor and regulator of lipid metabolism. Thus, its ligands are clinically used for treatment of type 2 diabetes and hyperlipidemia. On the other hand, PPAR ligands exert the antineuroinflammatory activity through preventing upregulation of inflammatory mediators in animal models for neurodegenerative disease and autoimmune disease. Neuropathic pain and inflammatory pain, clinically important one, are chronically progressed and underlain by neuroinflammation. In a few years, some studies using experimental models emerge that administration of PPAR ligands reduces inflammatory pain and neuropathic pain. PPAR ligands repress expression of genes for inflammatory mediators involved in both pains, such as proinflammatory cytokines, by a molecular mechanism termed ligand-dependent direct transrepression. Alternative mechanism is independent of transcriptional regulation of target genes, such as inhibition of activity of ion channels involved in the development of inflammatory pain and neuropathic pain, and therefore the analgesic effect occurs with rapid onset. The effects of PPAR ligands on neuroinflammation in animal models suggest their possible use for treating human inflammatory pain and neuropathic pain.

INTERNATIONAL REVIEW OF
NEUROBIOLOGY, VOL. 85
DOI: 10.1016/S0074-7742(09)85013-7

I. Introduction

Pain is an unpleasant sensory and emotional experience associated with actual or potential tissue damage or described in terms of such damage. Neuropathic pain and inflammatory pain are pathophysiological one and of clinical significance. Neuropathic pain is characterized by pain in the absence of a stimulus and by reduced nociceptive thresholds so that normally innocuous stimuli produce pain; a burdensome and potentially debilitating pain state. Neuropathic pain is the chronic pain state caused by significant pathological changes in the nervous system. It can occur secondarily to injury of the central nervous system (CNS) but it occurs most commonly in association with injury to the peripheral nervous system (PNS). On the other hand, peripheral inflammation induces inflammatory pain through nociceptor activation. Inflammatory pain is caused by various inflammatory mediators and, subsequently, spontaneous pain and hyperalgesia occurs. Recently, it is proposed that neuroinflammation underlies both neuropathic pain and inflammatory pain (Sweitzer *et al.*, 2002). Neuroinflammation is an inflammatory molecule's mediated process within the nervous system that can be provoked by direct injury to the nervous system. Numerous studies using animal models have proposed candidates for therapeutic targets to reduce neuropathic pain and inflammatory pain. However, currently, there are often no good pharmacotherapies for both of them, especially neuropathic pain (Scholz and Woolf, 2007).

Peroxisome proliferator-activated receptors (PPARs) are nuclear hormone receptors that are activated by naturally occurring substance (fatty acids and their derivatives) and medicines (fibrates and thiazolidinedione). PPAR consists of three isotypes named α, β/δ, and γ in vertebrates. PPARs primary function as an important fatty acid sensor that regulates lipid, carbohydrates, and amino acid metabolism. PPARs also play an important role in various pathophysiology; immunity, inflammation, apoptosis, and cell differentiation (Evans *et al.*, 2004). Since evidence shows that PPAR is expressed in nervous system such as neurons (Maeda *et al.*, 2008; Moreno *et al.*, 2004) and glia (Cristiano *et al.*, 2001; Cullingford *et al.*, 1998; Heneka *et al.*, 1999), it is possible that PPAR stimulation also potentially inhibits neuroinflammation. Therefore, PPARs may control neuroinflammation to subside inflammatory pain and neuropathic pain. In this context, it has recently been demonstrated that PPARs may also play a role in the control of the nociceptive response and neuropathic pain (Churi *et al.*, 2008; Costa *et al.*, 2008; LoVerme *et al.*, 2006; Maeda *et al.*, 2008; Oliveira *et al.*, 2007; Park *et al.*, 2007; Russo *et al.*, 2007; Suardiaz *et al.*, 2007). In this chapter, we will summarize the current understanding of the role of PPAR agonists in neuroinflammation and discuss their potential for the treatment of neuropathic pain and inflammatory pain.

II. Structure and Function of PPAR

PPARs belong to the nuclear receptor superfamily that are transcription factors activated by ligands; naturally occurring fatty acids or their derivatives (Delerive *et al.*, 2000; Devchand *et al.*, 1996; Forman *et al.*, 1997; Kliewer *et al.*, 1995; Nagy *et al.*, 1998). PPARs control lipid and glucose metabolism, energy homeostasis, adipocyte, and macrophage differentiation through regulation of specific target gene transcription (Issemann and Green, 1990). Three PPAR subtypes have been identified: PPARα, PPARβ/δ, and PPARγ. The lipid-lowering fibrates and the antidiabetic thiazolidinedione are synthetic ligands for PPARα and PPARγ, respectively (Forman *et al.*, 1997; Lehmann *et al.*, 1995).

The PPARs share a high homology and the organization in five different functional domains (Escher and Wahli, 2000). The two major domains are the highly conserved DNA-binding domain (DBD) and ligand-binding domain (LBD). The DBD can be considered the hallmark of the nuclear receptor superfamily. It contains the two zinc finger-like motifs that recognize the DNA target, composed of a pair of six nucleotide sequences. The spacing and the relative orientation of the two sequences determine which receptor of the superfamily will bind to a given response element. Although less conserved than DBD, the LBD of the nuclear receptors conserves a common three-dimensional structure, consisting of an antiparallel α-helical sandwich of 12 helices, organized in three layers with a central hydrophobic pocket. Crystallographic analysis has revealed a particularly large-binding cavity, of which only 30–40% is occupied by the ligand. This allows a relatively free nonspecific interaction between the cavity and the hydrophobic domains of the ligand, thus, explaining the relatively low ligand specificity of PPARs.

All three PPARs bind with retinoid X receptor (RXR) as a heterodimeric partner to specific DNA sequence elements termed PPAR response elements (PPREs) (Tugwood *et al.*, 1992). PPREs consist of a direct repeat of the nuclear receptor hexameric recognition sequence separated by one or two nucleotides (DR-1 and DR-2) (Gervois *et al.*, 1999). The binding of the agonist induces structural changes that allow the receptor to dissociate from corepressors that prevent the interaction of the receptor with DNA and coactivators. PPARs may also control gene expression independently of PPRE binding, through a mechanism not yet fully understood and termed transrepression. In the case of PPARγ, transrepression can be achieved through the physical interaction of the receptor with several transcriptional factors, including nuclear factor-κB (NF-κB), activator protein-1 (AP-1), and signal transducer and activator of transcription-1 (STAT-1) (Kielian and Drew, 2003).

The three PPARs share a high homology, but differ for tissue distribution and ligand specificity. PPARα is expressed mainly in tissues with high catabolic rates of fatty acids, such as the liver, muscle, and heart, whereas PPARδ is present in a wide range of tissues including heart, adipose tissue, brain, intestine, muscle, spleen, lung, and adrenal glands. PPARγ exists in two isoforms (PPARγ1 and γ2) with no functional differences, and is highly expressed in adipose tissue, where it plays a central role in the regulation of adipogenesis (Willson *et al.*, 2001). In addition, PPARγ is found in cells of the immune system, including lymphocytes and macrophages. In peripheral monocytes, PPARγ expression is induced during the process of extravasation from blood vessels into the tissues, and in the course of activation by proinflammatory stimuli. PPARγ has been shown to be important for the differentiation of monocytes into macrophages and to act as a negative regulator of macrophage activation (Daynes and Jones, 2002; Ricote *et al.*, 1998a,b; Tontonoz *et al.*, 1998).

III. PPAR and Inflammation

There has been a great deal of interest in the involvement of PPAR in inflammation and immune system. Of considerable interest, ligands for PPARγ and PPARα have therapeutic activity in several rodent models of inflammatory and autoimmune disease (Straus and Glass, 2007), suggesting that they might have similar activity in human disease as well. Devchand *et al.* (1996) were the first to report that PPARα deficient mice presented an increased inflammatory response induced by leukotriene B4 and arachidonic acid. Fenofibrate inhibits the production of IL-6 and prostaglandins and the expression of cyclooxygenase-2 gene induced by IL-1β in smooth muscle cells (Delerive *et al.*, 2000). Other PPARα agonists also inhibit the expression of genes for IL-6 and IL-8 in human keratinocytes irradiated with ultraviolet. On the other hand, there are evidences from *in vitro* studies demonstrating that PPARγ agonists also present anti-inflammatory activity (Konturek *et al.*, 2003; Maggi *et al.*, 2000; Ricote *et al.*, 1998b; Shiojiri *et al.*, 2002). Ligands for PPARδ have been studied less extensively but have been reported to have therapeutic activity in experimental autoimmune encephalomyelitis (Polak *et al.*, 2005). In addition to studies performed with PPAR ligands in mice with normal PPAR expression, mice with mutations disrupting one of the PPAR genes have provided evidence for the anti-inflammatory effects of PPARs *in vivo*. For example, mice with a mutation knocking out one copy of the PPARγ gene (PPAR$\gamma^{+/-}$ heterozygotes) (Desreumaux *et al.*, 2001), mice with a targeted disruption of the PPARγ gene in macrophages (Shah *et al.*, 2007), and mice that have a targeted disruption of the PPARγ gene in the intestinal epithelium (Adachi *et al.*, 2006), all have increased susceptibility to chemically induced colitis. Similarly, mice that have both copies of the PPARα gene knocked out (PPAR$\alpha^{-/-}$ homozygotes) have a worsened course of disease in several models for inflammatory and autoimmune disease, including chemically induced colitis (Cuzzocrea *et al.*, 2004), allergic asthma (Woerly *et al.*, 2003), carrageenan-induced edema and pleurisy (Cuzzocrea *et al.*, 2006), cerulein-induced pancreatitis (Genovese *et al.*, 2006), and experimental autoimmune encephalomyelitis (Dunn *et al.*, 2007). Remarkably, in all of these studies, a difference between control and PPAR-deficient mice in disease severity was observed in the absence of exogenously administered PPAR ligand, suggesting an anti-inflammatory role for either unbound PPAR or PPAR bound to endogenous ligands.

IV. Neuroinflammation and Pain

Inflammation is a pathophysiological state associated with pain (Scholz and Woolf, 2007). The free nerve endings of peripheral nerve fibers in systemic tissues respond directly to inflammatory mediators to generate electrical activity that is interpreted as painful. Neuroinflammation is a restrictive term referring to inflammatory states within nervous system that can give rise to the serious problem of neuropathic pain, burdensome pain state for which there is often no effective therapy. Neuropathic pain is the chronic pain state caused by significant pathological changes in the nervous system and occurs most commonly in association with injury of the PNS. These injuries can be caused by tumors compressing peripheral nerves, toxins used as chemotherapy, metabolic or viral diseases, severe ischemic insults and trauma and disc herniation. Neuropathic pain is mediated through neuroinflammatory mechanisms affecting nervous tissue that is controlled by inflammatory responses to the initial insult. The local inflammation provokes a reaction in peripheral immune cells and glia in the pain pathway macrophages and Schwann cells facilitate the wallerian degeneration of injured nerve fibers distal to a nerve lesion (Myers *et al.*, 1993; Stoll *et al.*, 2002); an immune response in the dorsal root ganglion is driven by macrophages, lymphocytes, and satellite cells (Hu and McLachlan, 2003); activation of spinal microglia dominates the early glial responses in the CNS to peripheral nerve injury, which is followed by activation and proliferation of astrocytes (Colburn *et al.*, 1999; Jin *et al.*, 2003). Proinflammatory molecule activation in the nervous system may develop neuropathic pain.

V. Role of PPAR in Pain

A. PPARα

The first relevance on pain was activation of PPARα in the spinal cord of rats with peripheral inflammation (Benani *et al.*, 2004). Electrophoretic mobility-shift assay was employed to compare the DNA-binding activity toward a PPRE. The PPARα isoform was observed to be activated in the rat spinal cord after complete Freund's adjuvant injection, which could elicit hyperalgesia. PPARα was provided as a new player in the molecular modeling of pain pathways, although it was discussed that inhibitors of PPARα activation might be relevant antinociceptive drugs.

LoVerme *et al.* (2006) gave the first suggestion that PPARα agonist represents a novel class of analgesics. The PPARα synthetic agonists and naturally occurring agonist palmitoylethanolamide reduced nociceptive behaviors elicited in mice by intraplantar injection of formalin or i.p. injection of magnesium sulfate. These effects were absent in PPARα-null mice. However, the effect occurred within minutes of agonist administration in wild-type mice, suggesting that they were mediated rather through a transcription-independent mechanism. In addition to modulating nociception, PPARα agonist reduced hyperalgesic responses in mice with chronic constriction injury of sciatic nerve, when administered once immediately before behavioral test. These results suggest that PPARα agonists are rapid, broad-spectrum analgesics.

It is interesting that PPARα is a target for fenofibrate, antihyperlipidemic drug. Inflammation could link life style disease to pain, because inflammation can underlie both life style disease and pain. In this regard, the effect of acute or prolonged treatment with fenofibrate on thermal and inflammatory nociception was examined (Oliveira *et al.*, 2007). Prolonged, but not acute, treatment with fenofibrate inhibited the second phase of formaldehyde-induced nociceptive response in mice, whereas it did not inhibit the nociceptive response in the hot-plate model. Mechanical allodynia induced by intraplantar carrageenan in rats was inhibited by prolonged treatment with fenofibrate, but not by its acute treatment. These results give support to the potential use of PPARα agonist in the treatment of different inflammatory diseases.

The analgesic properties of cannabinoid receptor agonists are well characterized. However, numerous side effects limit the therapeutic potential of these agents. Russo *et al.* (2007) reported a synergistic antinociceptive interaction between the endogenous cannabinoid receptor agonist anandamide and the synthetic PPARα agonist (to be indicated by the authors) in a model of acute chemical-induced pain. However, PPARα agonist exerted the antinociception in transcription-independent way; PPARα agonist acted on the large-conductance potassium channel. These findings revealed an interaction between the endocannabinoid and PPARα systems.

X-ray crystal structures of the PPARs LBDs revealed that the receptors contain a much larger ligand-binding pocket than other nuclear receptors (Gampe *et al.*, 2000; Nolte *et al.*, 1998; Uppenberg *et al.*, 1998; Xu *et al.*, 1999, 2001). The size of this pocket may explain the ability of the PPARs to bind a variety of naturally occurring and synthetic lipophilic acids. Especially, PPARα binds to a wide range of lipids (Xu *et al.*, 1999). Oleoylethanolamide (OEA) is a natural fatty acid amide that mainly modulates feeding and energy homeostasis by binding to PPARα (Fu *et al.*, 2003). Additionally, it has been proposed that OEA could act via other receptors, including the vanilloid receptor (Wang *et al.*, 2005). These data suggest that OEA might subserve other physiological roles, including pain perception. Thus, the effects of OEA in two types of nociceptive responses evoked by visceral and inflammatory pain were evaluated in rodents. OEA has analgesic properties reducing the nociceptive responses produced by administration of acetic acid and formalin in two experimental animal models. The actions of OEA were absent in mice null for the PPARα receptor gene. However, the antinociceptive effect of OEA through PPARα was transcription independent; the participation of blocking glutamatergic transmission (to authors: pleas, make it clearer) (Suardiaz *et al.*, 2007). Some endocannabinoids also have an antinociceptive effect. Anandamide and *N*-palmitoylethanolamine, endocannabinoids are ligands for PPARα and the levels of the endocannabinoids are increased in the local inflamed tissue. In fact, GW6471, a PPARα antagonist, blocked the inhibitory effects of anandamide and *N*-palmitoylethanolamine on hyperalgesia (Jhaveri *et al.*, 2008; Sagar *et al.*, 2008). Thus, it is interesting that these endogenous PPARα ligands contribute to their antinociceptive effects in the model of visceral pain and inflammatory hyperalgesia, because these findings provide a framework for understanding its biological functions and endogenous targets in inflammatory pain.

B. PPARγ

Thiazolidinediones (TZDs) are potent synthetic agonists of PPARγ and medicine for type 2 diabetes. TZDs were shown to induce neuroprotection after cerebral ischemia by blocking inflammation (Culman *et al.*, 2007). Spinal cord injury (SCI), another type of neurodegenerative disease, also induces massive inflammation that precipitates secondary neuronal death. Park *et al.* (2007) analyzed the therapeutic efficacy of TZDs, pioglitazone, and rosiglitazone, after SCI in adult rats. Both TZDs decreased various histological alterations after SCI: the lesion size, motor neuron loss, myelin loss, astrogliosis, and microglial activation. TZDs also enhanced the motor function recovery. Chronic thermal hyperalgesia after SCI was decreased significantly in the pioglitazone group compared with the vehicle group. The pioglitazone group showed significantly less induction of inflammatory genes: interleukin (IL)-6, IL-1β, monocyte chemoattractant protein-1, intracellular adhesion molecule-1, and early growth response-1. Pioglitazone also enhanced the post-SCI induction of neuroprotective heat shock proteins and antioxidant enzymes. Those effects were prevented by pretreatment with a PPARγ antagonist. This was the first to report involvement of the PPARγ agonist in neuropathic pain.

Considering that PPAR activation directly regulates gene transcription, the repeated dosing regime might be required for treatment for chronic pain. Churi *et al.* (2008), however, reported that a single administration of TZDs reduced allodynia in peripheral nerve injury model. PPARγ agonists and/or antagonists were intrathecally administered in rats after transection of the tibial and common peroneal branches of the sciatic nerve. Single injection of rosiglitazone dose-dependently decreased mechanical and cold hypersensitivity. These effects peaked at approximately 60 min after injection, a rapid event suggestive of transcription-independent mechanisms of action. The 15-deoxy-δ(12,14)-prostaglandin J_2 (15d-PGJ_2) is an endogenous ligand of PPARγ and is now recognized as a potent anti-inflammatory mediator (Kliewer *et al.*, 1995). However, information regarding the influence of 15d-PGJ_2 on pain is still unknown. Single injection of 15d-PGJ_2 decreased mechanical and cold hypersensitivity (Churi *et al.*, 2008). They concluded that ligand-induced activation of spinal PPARγ rapidly reversed nerve injury-induced mechanical and cold allodynia. The mechanisms on rapid onset of analgesia remain unclear, but are needed to be clarified in the future study. The effect of 15d-PGJ_2 on inflammatory pain has also been reported; 15d-PGJ_2 reduced inflammatory hypernociception elicited by injection carrageenan, formalin, and PGE_2 through PPARγ activation (Napimoga *et al.*, 2008).

More recently, we presented histological distribution of PPARγ in the pain pathway and found the antiallodynia effect of pioglitazone in the partial ligation of sciatic nerve of mice (PSL), a peripheral nerve injury model (Maeda *et al.*, 2008). It is well known that much adipocyte, such as visceral fat and express PPARγ, while we found that PPARγ were expressed in the adipocytes of perineurium of sciatic nerve. PPARγ was also expressed in the neurons of dorsal root ganglion and dorsal horn of spinal cord. These may be a supportive evidence for functional role of PPARγ in the pain as previously reported (Moreno *et al.*, 2004). Repeated administration of pioglitazone after PSL reduced tactile allodynia and thermal hyperalgesia, whereas single administration of pioglitazone had no significant effect. The inhibitory mechanism may be regulation of the altered expression of the well accepted molecules underlying neuropathic pain; we found that pioglitazone reversed PSL-induced increase of IL-6 and TNF-α expressions in the pain pathway, which are proinflammatory cytokines and essential for development of neuropathic pain (Scholz and Woolf, 2007).

VI. Conclusion

The experimental research on the role of PPARs in inflammatory pain and neuropathic pain is quite primitive. In the past few years, the outline of functional involvement of PPARs in both chronic pains is just now emerging; natural and synthetic PPAR ligands have been reported to have significant effect on both chronic pains in animal models. However, there are some problems that are left unresolved in basic research. First, it is unclear how PPARs prevent the development of pain. Although it is well known that PPARs are nuclear receptor and regulate transcription of target genes, the transcription-independent mechanisms are increasingly reported. In some pain research above mentioned, the effect of PPAR ligands occurred with rapid onset (Churi *et al.*, 2008; LoVerme *et al.*, 2006; Suardiaz *et al.*, 2007) and with the regimen of a single administration (Churi *et al.*, 2008; Russo *et al.*, 2007). This may complicate the understanding of mechanism underlying the effects of PPAR ligands. There could be something else that should be considered for elucidation of the mechanism. Some conventional synthetic and natural ligands exert anti-inflammatory effect through both transcription-dependent and independent mechanism upon PPAR activation. These findings might make it difficult to develop rationally PPAR ligands as new analgesics. The extensive basic research on PPAR activation and subsequent intracellular signaling is required for explanation of the mechanism of PPAR analgesics. Second, currently available drugs that target PPARs exhibit several undesirable activities that limit their application to treatment of chronic pain. For example, long-term use of PPARγ agonists promotes adiposity and fluid retention (Kahn *et al.*, 2006), and concern has been raised recently that some agonists might increase the risk of myocardial infarction (Diamond *et al.*, 2007; Nissen and Wolski, 2007). These undesirable activities are linked to positive regulation of target gene transcription, suggesting that the identification of small molecules that selectively retain the ability to stimulate transrepression would lead to development of analgesics with reduced side effects. Recent insight into the molecular mechanisms mediating transrepression suggests that this might be possible. A deep knowledge of the molecular mechanisms of the receptor activation evoked by PPAR ligands is mandatory for the development of novel PPAR ligands with increasing efficacy and safety. The promising clinical use might include prevention of the progress of both chronic pains induced by major surgery and chemotherapy and the treatment of established pain such as diabetic peripheral neuropathy.

References

Adachi, M., Kurotani, R., Morimura, K., Shah, Y., Sanford, M., Madison, B. B., Gumucio, D. L., Marin, H. E., Peters, J. M., Young, H. A., and Gosnzalez, F. J. (2006). Peroxisome proliferator activated receptor gamma in colonic epithelial cells protects against experimental inflammatory bowel disease. *Gut* **55,** 1104–1113.

Benani, A., Heurtaux, T., Netter, P., and Minn, A. (2004). Activation of peroxisome proliferator-activated receptor alpha in rat spinal cord after peripheral noxious stimulation. *Neurosci. Lett.* **369,** 59–63.

Churi, S. B., Abdel-Aleem, O. S., Tumber, K. K., Scuderi-Porter, H., and Taylor, B. K. (2008). Intrathecal rosiglitazone acts at peroxisome proliferator-activated receptor-gamma to rapidly inhibit neuropathic pain in rats. *J. Pain* **9,** 639–649.

Colburn, R. W., Rickman, A. J., and DeLeo, J. A. (1999). The effect of site and type of nerve injury on spinal glial activation and neuropathic pain behavior. *Exp. Neurol.* **157,** 289–304.

Costa, B., Comelli, F., Bettoni, I., Colleoni, M., and Giagoni, G. (2008). The endogenous fatty acid amide, palmitoylethanolamide, has anti-allodynic and anti-hyperalgesic effects in a murine model of neuropathic pain: Involvement of CB(1), TRPV1 and PPARgamma receptors and neurotrophic factors. *Pain* **139,** 541–550.

Cristiano, L., Bernardo, A., and Ceru, M. P. (2001). Peroxisome proliferator-activated receptors (PPARs) and peroxisomes in rat cortical and cerebellar astrocytes. *J. Neurocytol.* **30,** 671–683.

Cullingford, T. E., Bhakoo, K., Peuchen, S., Dolphin, C. T., Patel, R., and Clark, J. B. (1998). Distribution of mRNAs encoding the peroxisome proliferator-activated receptor alpha, beta, and gamma and the retinoid X receptor alpha, beta, and gamma in rat central nervous system. *J. Neurochem.* **70,** 1366–1375.

Culman, J., Zhao, Y., Gohlke, P., and Herdegen, T. (2007). PPAR-gamma: Therapeutic target for ischemic stroke. *Trends Pharmacol. Sci.* **28,** 244–249.

Cuzzocrea, S., Di Paloa, R., Mazzon, E., Genovese, T., Muia, C., Centorrino, T., and Caputi, A. P. (2004). Role of endogenous and exogenous ligands for the peroxisome proliferators activated receptors alpha (PPAR-alpha) in the development of inflammatory bowel disease in mice. *Lab. Invest.* **84,** 1643–1654.

Cuzzocrea, S., Mazzon, E., Di Paola, R., Peli, A., Bonato, A., Britti, D., Genovese, T., Muia, C., Crisafulli, C., and Caputi, A. P. (2006). The role of the peroxisome proliferator-activated receptor-alpha (PPAR-alpha) in the regulation of acute inflammation. *J. Leukoc. Biol.* **79,** 999–1010.

Daynes, R. A., and Jones, D. C. (2002). Emerging roles of PPARs in inflammation and immunity. *Nat. Rev. Immunol.* **2,** 748–759.

Delerive, P., Furman, C., Teissier, E., Fruchart, J., Duriez, P., and Staels, B. (2000). Oxidized phospholipids activate PPARalpha in a phospholipase A2-dependent manner. *FEBS Lett.* **471,** 34–38.

Desreumaux, P., Dubuquoy, L., Nutten, S., Peuchmaur, M., Englaro, W., Schoonjans, K., Derijard, B., Desvergne, B., Wahli, W., Chambon, P., Leibowitz, M. D., Colombel, J. F., *et al.* (2001). Attenuation of colon inflammation through activators of the retinoid X receptor (RXR)/peroxisome proliferator-activated receptor gamma (PPARgamma) heterodimer. A basis for new therapeutic strategies. *J. Exp. Med.* **193,** 827–838.

Devchand, P. R., Keller, H., Peters, J. M., Vazquez, M., Gonzalez, F. J., and Wahli, W. (1996). The PPARalpha-leukotriene B4 pathway to inflammation control. *Nature* **384,** 39–43.

Diamond, G. A., Bax, L., and Kaul, S. (2007). Uncertain effects of rosiglitazone on the risk for myocardial infarction and cardiovascular death. *Ann. Intern. Med.* **147,** 578–581.

Dunn, S. E., Ousman, S. S., Sobel, R. A., Zuniga, L., Baranzini, S. E., Youssef, S., Crowell, A., Loh, J., Oksenberg, J., and Steinman, L. (2007). Peroxisome proliferator-activated receptor (PPAR)alpha expression in T cells mediates gender differences in development of T cell-mediated autoimmunity. *J. Exp. Med.* **204,** 321–330.

Escher, P., and Wahli, W. (2000). Peroxisome proliferator-activated receptors: Insight into multiple cellular functions. *Mutat. Res.* **448,** 121–138.

Evans, R. M., Barish, G. D., and Wang, Y. X. (2004). PPARs and the complex journey to obesity. *Nat. Med.* **10,** 355–361.

Forman, B. M., Chen, J., and Evans, R. M. (1997). Hypolipidemic drugs, polyunsaturated fatty acids, and eicosanoids are ligands for peroxisome proliferator-activated receptors alpha and delta. *Proc. Natl. Acad. Sci. USA* **94,** 4312–4317.

Fu, J., Gaetani, S., Oveisi, F., Lo Verme, J., Serrano, A., Rodriguez De Fonseca, F., Rosengarth, A., Luecke, H., Di Giacomo, B., Tarzia, G., and Piomelli, D. (2003). Oleylethanolamide regulates feeding and body weight through activation of the nuclear receptor PPAR-alpha. *Nature* **425,** 90–93.

Gampe, R. T. Jr., Montana, V. G., Lambert, M. H., Miller, A. B., Bledsoe, R. K., Milburn, M. V., Kliewer, S. A., Willson, T. M., and Xu, H. E. (2000). Asymmetry in the PPARgamma/RXRalpha crystal structure reveals the molecular basis of heterodimerization among nuclear receptors. *Mol. Cell.* **5,** 545–555.

Genovese, T., Mazzon, E., Di Paola, R., Muia, C., Crisafulli, C., Malleo, G., Esposito, E., and Cuzzocrea, S. (2006). Role of peroxisome proliferator-activated receptor-alpha in acute pancreatitis induced by cerulein. *Immunology* **118,** 559–570.

Gervois, P., Torra, I. P., Chinetti, G., Grotzinger, T., Dubois, G., Fruchart, J. C., Fruchart-Najib, J., Leitersdorf, E., and Staels, B. (1999). A truncated human peroxisome proliferator-activated receptor alpha splice variant with dominant negative activity. *Mol. Endocrinol.* **13,** 1535–1549.

Heneka, M. T., Feinstein, D., Galea, E., Gleichmann, M., Wullner, U., and Klockgether, T. (1999). Peroxisome proliferator-activated receptor gamma agonists protect cerebellar granule cells from cytokine-induced apoptotic cell death by inhibition of inducible nitric oxide synthase. *J. Neuroimmunol.* **100,** 156–168.

Hu, P., and McLachlan, E. M. (2003). Distinct functional types of macrophage in dorsal root ganglia and spinal nerves proximal to sciatic and spinal nerve transections in the rat. *Exp. Neurol.* **184,** 590–605.

Issemann, I., and Green, S. (1990). Activation of a member of the steroid hormone receptor superfamily by peroxisome proliferators. *Nature* **347,** 645–650.

Jhaveri, M. D., Richardson, D., Robinson, I., Garle, M. J., Patel, A., Sun, Y., Sagar, D. R., Bennett, A. J., Alexander, S. P., Kendall, D. A., Barrett, D. A., and Chapman, V. (2008). Inhibition of fatty acid amide hydrolase and cyclooxygenase-2 increases levels of endocannabinoid related molecules and produces analgesia via peroxisome proliferator-activated receptor-alpha in a model of inflammatory pain. *Neuropharmacology* **55,** 85–93.

Jin, S. X., Zhuang, Z. Y., Woolf, C. J., and Ji, R. R. (2003). p38 mitogen-activated protein kinase is activated after a spinal nerve ligation in spinal cord microglia and dorsal root ganglion neurons and contributes to the generation of neuropathic pain. *J. Neurosci.* **23,** 4017–4022.

Kahn, S. E., Haffner, S. M., Heise, M. A., Herman, W. H., Holman, R. R., Jones, N. P., Kravitz, B. G., Lachin, J. M., O'Neill, M. C., Zinman, B., and Viberti, G. (2006). Glycemic durability of rosiglitazone, metformin, or glyburide monotherapy. *N. Engl. J. Med.* **355,** 2427–2443.

Kielian, T., and Drew, P. D. (2003). Effects of peroxisome proliferator-activated receptor-gamma agonists on central nervous system inflammation. *J. Neurosci. Res.* **71,** 315–325.

Kliewer, S. A., Lenhard, J. M., Willson, T. M., Patel, I., Morris, D. C., and Lehmann, J. M. (1995). A prostaglandin J2 metabolite binds peroxisome proliferator-activated receptor gamma and promotes adipocyte differentiation. *Cell* **83,** 813–819.

Konturek, P. C., Brzozowski, T., Kania, J., Konturek, S. J., Kwiecien, S., Pajdo, R., and Hahn, E. G. (2003). Pioglitazone, a specific ligand of peroxisome proliferator-activated receptor-gamma, accelerates gastric ulcer healing in rat. *Eur. J. Pharmacol.* **472,** 213–220.

Lehmann, J. M., Moore, L. B., Smith-Oliver, T. A., Wilkison, W. O., Willson, T. M., and Kliewer, S. A. (1995). An antidiabetic thiazolidinedione is a high affinity ligand for peroxisome proliferator-activated receptor gamma (PPAR gamma). *J. Biol. Chem.* **270,** 12953–12956.

LoVerme, J., Russo, R., La Rana, G., Fu, J., Farthing, J., Mattace-Raso, G., Meli, R., Hohmann, A., Calignano, A., and Piomelli, D. (2006). Rapid broad-spectrum analgesia through activation of peroxisome proliferator-activated receptor-alpha. *J. Pharmacol. Exp. Ther.* **319,** 1051–1061.

Maeda, T., Kiguchi, N., Kobayashi, Y., Ozaki, M., and Kishioka, S. (2008). Pioglitazone attenuates tactile allodynia and thermal hyperalgesia in mice subjected to peripheral nerve injury. *J. Pharmacol. Sci.* **108,** 341–347.

Maggi, L. B. Jr., Sadeghi, H., Weigand, C., Scarim, A. L., Heitmeier, M. R., and Corbett, J. A. (2000). Anti-inflammatory actions of 15-deoxy-delta 12,14-prostaglandin J2 and troglitazone: Evidence for heat shock-dependent and -independent inhibition of cytokine-induced inducible nitric oxide synthase expression. *Diabetes* **49,** 346–355.

Moreno, S., Farioli-Vecchioli, S., and Ceru, M. P. (2004). Immunolocalization of peroxisome proliferator-activated receptors and retinoid X receptors in the adult rat CNS. *Neuroscience* **123,** 131–145.

Myers, R. R., Yamamoto, T., Yaksh, T. L., and Powell, H. C. (1993). The role of focal nerve ischemia and Wallerian degeneration in peripheral nerve injury producing hyperesthesia. *Anesthesiology* **78,** 308–316.

Nagy, L., Tontonoz, P., Alvarez, J. G., Chen, H., and Evans, R. M. (1998). Oxidized LDL regulates macrophage gene expression through ligand activation of PPARgamma. *Cell* **93,** 229–240.

Napimoga, M. H., Souza, G. R., Cunha, T. M., Ferrari, L. F., Clemente-Napimoga, J. T., Parada, C. A., Verri, W. A. Jr., Cunha, F. Q., and Ferreira, S. H. (2008). 15d-prostaglandin J2 inhibits inflammatory hypernociception: Involvement of peripheral opioid receptor. *J. Pharmacol. Exp. Ther.* **324,** 313–321.

Nissen, S. E., and Wolski, K. (2007). Effect of rosiglitazone on the risk of myocardial infarction and death from cardiovascular causes. *N. Engl. J. Med.* **356,** 2457–2471.

Nolte, R. T., Wisely, G. B., Westin, S., Cobb, J. E., Lambert, M. H., Kurokawa, R., Rosenfeld, M. G., Willson, T. M., Glass, C. K., and Milburn, M. V. (1998). Ligand binding and co-activator assembly of the peroxisome proliferator-activated receptor-gamma. *Nature* **395,** 137–143.

Oliveira, A. C., Bertollo, C. M., Rocha, L. T., Nascimento, E. B. Jr., Costa, K. A., and Coelho, M. M. (2007). Antinociceptive and antiedematogenic activities of fenofibrate, an agonist of PPAR alpha, and pioglitazone, an agonist of PPAR gamma. *Eur. J. Pharmacol.* **561,** 194–201.

Park, S. W., Yi, J. H., Miranpuri, G., Satriotomo, I., Bowen, K., Resnick, D. K., and Vemuganti, R. (2007). Thiazolidinedione class of peroxisome proliferator-activated receptor gamma agonists prevents neuronal damage, motor dysfunction, myelin loss, neuropathic pain, and inflammation after spinal cord injury in adult rats. *J. Pharmacol. Exp. Ther.* **320,** 1002–1012.

Polak, P. E., Kalinin, S., Dello Russo, C., Gavrilyuk, V., Sharp, A., Peters, J. M., Richardson, J., Willson, T. M., Weinberg, G., and Feinstein, D. L. (2005). Protective effects of a peroxisome proliferator-activated receptor-beta/delta agonist in experimental autoimmune encephalomyelitis. *J. Neuroimmunol.* **168,** 65–75.

Ricote, M., Huang, J., Fajas, L., Li, A., Welch, J., Najib, J., Witztum, J. L., Auwerx, J., Palinski, W., and Glass, C. K. (1998a). Expression of the peroxisome proliferator-activated receptor gamma (PPARgamma) in human atherosclerosis and regulation in macrophages by colony stimulating factors and oxidized low density lipoprotein. *Proc. Natl. Acad. Sci. USA* **95,** 7614–7619.

Ricote, M., Li, A. C., Willson, T. M., Kelly, C. J., and Glass, C. K. (1998b). The peroxisome proliferator-activated receptor-gamma is a negative regulator of macrophage activation. *Nature* **391,** 79–82.

Russo, R., LoVerme, J., La Rana, G., D'Agostino, G., Sasso, O., Calignano, A., and Piomelli, D. (2007). Synergistic antinociception by the cannabinoid receptor agonist anandamide and the PPAR-alpha receptor agonist GW7647. *Eur. J. Pharmacol.* **566,** 117–119.

Sagar, D. R., Kendall, D. A., and Chapman, V. (2008). Inhibition of fatty acid amide hydrolase produces PPAR-alpha-mediated analgesia in a rat model of inflammatory pain. *Br. J. Pharmacol.* **155,** 1297–1306.

Scholz, J., and Woolf, C. J. (2007). The neuropathic pain triad: Neurons, immune cells and glia. *Nat. Neurosci.* **10,** 1361–1368.

Shah, Y. M., Morimura, K., and Gonzalez, F. J. (2007). Expression of peroxisome proliferator-activated receptor-gamma in macrophage suppresses experimentally induced colitis. *Am. J. Physiol. Gastrointest. Liver Physiol.* **292,** G657–G666.

Shiojiri, T., Wada, K., Nakajima, A., Katayama, K., Shibuya, A., Kudo, C., Kadowaki, T., Mayumi, T., Yura, Y., and Kamisaki, Y. (2002). PPAR gamma ligands inhibit nitrotyrosine formation and inflammatory mediator expressions in adjuvant-induced rheumatoid arthritis mice. *Eur. J. Pharmacol.* **448,** 231–238.

Stoll, G., Jander, S., and Myers, R. R. (2002). Degeneration and regeneration of the peripheral nervous system: From Augustus Waller's observations to neuroinflammation. *J. Peripher. Nerv. Syst.* **7,** 13–27.

Straus, D. S., and Glass, C. K. (2007). Anti-inflammatory actions of PPAR ligands: New insights on cellular and molecular mechanisms. *Trends Immunol.* **28,** 551–558.

Suardiaz, M., Estivill-Torrus, G., Goicoechea, C., and Bilbao, A. (2007). Analgesic properties of oleoylethanolamide (OEA) in visceral and inflammatory pain. *Pain* **133,** 99–110.

Sweitzer, S. M., Hickey, W. F., Rutkowski, M. D., Pahl, J. L., and DeLeo, J. A. (2002). Focal peripheral nerve injury induces leukocyte trafficking into the central nervous system: Potential relationship to neuropathic pain. *Pain* **100,** 163–170.

Tontonoz, P., Nagy, L., Alvarez, J. G., Thomazy, V. A., and Evans, R. M. (1998). PPARgamma promotes monocyte/macrophage differentiation and uptake of oxidized LDL. *Cell* **93,** 241–252.

Tugwood, J. D., Issemann, I., Anderson, R. G., Bundell, K. R., McPheat, W. L., and Green, S. (1992). The mouse peroxisome proliferator activated receptor recognizes a response element in the 5′ flanking sequence of the rat acyl CoA oxidase gene. *EMBO J.* **11,** 433–439.

Uppenberg, J., Svensson, C., Jaki, M., Bertilsson, G., Jendeberg, L., and Berkenstam, A. (1998). Crystal structure of the ligand binding domain of the human nuclear receptor PPARgamma. *J. Biol. Chem.* **273,** 31108–31112.

Wang, X., Miyares, R. L., and Ahern, G. P. (2005). Oleoylethanolamide excites vagal sensory neurones, induces visceral pain and reduces short-term food intake in mice via capsaicin receptor TRPV1. *J. Physiol.* **564,** 541–547.

Willson, T. M., Lambert, M. H., and Kliewer, S. A. (2001). Peroxisome proliferator-activated receptor gamma and metabolic disease. *Annu. Rev. Biochem.* **70,** 341–367.

Woerly, G., Honda, K., Loyens, M., Papin, J. P., Auwerx, J., Staels, B., Capron, M., and Dombrowicz, D. (2003). Peroxisome proliferator-activated receptors alpha and gamma down-regulate allergic inflammation and eosinophil activation. *J. Exp. Med.* **198,** 411–421.

Xu, H. E., Lambert, M. H., Montana, V. G., Parks, D. J., Blanchard, S. G., Brown, P. J., Sternbach, D. D., Lehmann, J. M., Wisely, G. B., Willson, T. M., Kliewer, S. A., and Milburn, M. V. (1999). Molecular recognition of fatty acids by peroxisome proliferator-activated receptors. *Mol. Cell.* **3,** 397–403.

Xu, H. E., Lambert, M. H., Montana, V. G., Plunket, K. D., Moore, L. B., Collins, J. L., Oplinger, J. A., Kliewer, S. A., Gampe, R. T. Jr., McKee, D. D., Moore, J. T., and Willson, T. M. (2001). Structural determinants of ligand binding selectivity between the peroxisome proliferator-activated receptors. *Proc. Natl. Acad. Sci. USA* **98,** 13919–13924.

INVOLVEMENT OF INFLAMMATORY MEDIATORS IN NEUROPATHIC PAIN CAUSED BY VINCRISTINE

Norikazu Kiguchi, Takehiko Maeda, Yuka Kobayashi, Fumihiro Saika, and Shiroh Kishioka

Department of Pharmacology, Wakayama Medical University, 811–1 Kimiidera, Wakayama 641–0012, Japan

Elucidation of the mechanism of neuropathic pain caused by vincristine is required because long-term treatment with this anticancer agent often causes neuropathic pain. We refer to the involvement of inflammatory mediators in vincristine-induced neuropathic pain in this review.

Several reports using rodents have shown that long-lasting neuropathic pain (mechanical allodynia) is caused by repeated systemic injection of vincristine. Vincristine damaged Schwann cells and DRG neurons in this model. Vincristine-induced macrophage infiltration in the peripheral nervous system (PNS) and macrophage-derived IL-6 elicited mechanical allodynia. These findings proved that inhibition of IL-6 function prevented neuropathic pain caused by vincristine. In the central nervous system (CNS), activation of microglia and astrocytes in the spinal cord were demonstrated after long-term vincristine treatment. TNF-α was upregulated in activated microglia and astrocytes, and inhibition of TNF-α function attenuated neuropathic pain caused by vincristine. These results suggest that vincristine induces macrophage infiltration to the damaged PNS, and that macrophage-derived inflammatory cytokines such as IL-6 elicits neuroinflammation. Signal transduction of pain from the PNS to the CNS activates microglia and

INTERNATIONAL REVIEW OF NEUROBIOLOGY, VOL. 85
DOI: 10.1016/S0074-7742(09)85014-9

astrocytes, and these activated glial cells release inflammatory cytokines such as TNF-α. In the CNS, these inflammatory cytokines have an important role in the neuropathic pain caused by vincristine.

Immune-modulating agents that prevent activation of immune cells and/or the inhibitory agents of inflammatory cytokines could prevent the neuropathic pain caused by vincristine. These agents could increase the tolerability of vincristine when used for the treatment of leukemia and lymphoma.

I. Introduction

Vincristine, a vinca-alkaloid, prevents proliferation of tumor cells through the inhibition of tubulin polymerization. Vincristine is used as an anticancer agent for leukemia and lymphoma (Himes *et al.*, 1976; Owellen *et al.*, 1976). Clinical use of vincristine is often limited by its adverse effects, which include painful peripheral neuropathy (i.e., neuropathic pain) (Casey *et al.*, 1973; Sandler *et al.*, 1969). Elucidation of the detailed mechanism of neuropathic pain caused by vincristine is needed to improve quality-of-life for patients, and to make vincristine more tolerable for cancer treatment.

It is well known that neuropathic pain is elicited by nerve injury and manifests as chronic severe pain. The typical symptoms of neuropathic pain are hyperalgesia (excessive pain caused by usual noxious stimuli), allodynia (burning pain caused by innocuous stimuli such as touching), and spontaneous pain (Marchand *et al.*, 2005). In general, nociceptive pain has an important role in avoiding tissue damage (Scholz and Woolf, 2002), but severe pain should be treated by analgesics. Neuropathic pain is resistant to standard analgesics such as opioids and nonsteroidal anti-inflammatory drugs (NSAIDs) (Gilron *et al.*, 2005; Ueda and Rashid, 2003), so an effective therapeutic strategy must therefore be established. Animal models of neuropathic pain have been established to study the molecular mechanisms of neuropathic pain (Bennett *et al.*, 2003). Inflammatory mediators (e.g., cytokines) in the peripheral nervous system (PNS) and central nervous system (CNS) have been used as therapeutic targets of neuropathic pain (Moalem and Tracey, 2006). It was reported that immune cells such as macrophages and lymphocytes infiltrate into the injured PNS and release inflammatory cytokines. In the CNS, glial cells such as microglia and astrocytes are activated and elicit neuroinflammation, leading to neuropathic pain (Scholz and Woolf, 2007). Participation of the inflammatory cytokines interleukin (IL)-1β, IL-6, and tumor necrosis factor-α (TNF-α) in the development and maintenance of neuropathic pain has been studied (Okamoto *et al.*, 2001). Inhibition of these mediators could attenuate neuropathic pain, and local injections of recombinant cytokines elicited neuropathic-like pain (Lindenlaub *et al.*, 2000; Ramer *et al.*, 1998; Wei *et al.*, 2007).

Experimental models of neuropathic pain caused by repeated systemic treatment of vincristine were established to clarify therapeutic targets and potential therapeutic agents (Authier *et al.*, 1999; Tanner *et al.*, 1998). Studies characterized neuropathic pain caused by vincristine, and indicated that the involvement of inflammatory cytokines was similar to neuropathic pain induced by nerve injury (Kiguchi *et al.*, 2008a,b).

We focus on the aspects of neuropathic pain caused by vincristine. We refer to the roles of inflammatory cytokines, which may be critical in the development and maintenance of neuropathic pain.

II. Characterization of Neuropathic Pain Caused by Vincristine

Vincristine causes chronic pain that is correlated with the dose and the period of vincristine treatment (Aley *et al.*, 1996). In an experimental model using rodents, neuropathic pain was also caused by repeated systemic injection of vincristine, but not by a single injection (Kiguchi *et al.*, 2008b). In general, nerve injury-induced neuropathic pain shows mechanical allodynia and thermal hyperalgesia (Marchand *et al.*, 2005). Neuropathic pain caused by vincristine is different from nerve injury-induced neuropathic pain; mechanical allodynia developed after long-term treatment with vincristine (Luo *et al.*, 2002; Nozaki-Taguchi *et al.*, 2001), whereas the cause of thermal hyperalgesia was not ascertained. There are contradictory reports whether hyperalgesia develops (Beyreuther *et al.*, 2007; Rahn *et al.*, 2007; Weng *et al.*, 2003). These results may be based on the differences of injection routes and schedules of vincristine treatment. The neuropathic pain caused by vincristine persists for several weeks after cessation of long-term vincristine treatment (Xiao *et al.*, 2007), suggesting neuronal plasticity caused by vincristine.

III. Effects of Vincristine on the PNS

A. General Concepts

The detailed molecular mechanism of neuropathic pain caused by repeated systemic injection of vincristine is poorly understood. In general, systemically injected vincristine poorly penetrates into the CNS because of the blood–brain barrier (BBB) (Calabresi and Parks, 1985). It is thought that vincristine may act on peripheral tissue and/or the PNS. Several reports have indicated that vincristine affects the neuronal and glial cells of the PNS. For instance, *in vitro*, vincristine induces a reduction of laminin in Schwann cells and neurite retraction in the dorsal root ganglion (DRG), which are located in primary afferent neurons (Konings *et al.*, 1994a,b). *In vivo*, vincristine prevents development of the normal conformation of myelin through damage to Schwann cells (Djaldetti *et al.*, 1996). In our study, it was clarified that activating transcription factor-3, a marker of injured neurons, was expressed in the DRG after repeated systemic injection of vincristine (unpublished data). These results suggest that vincristine elicits disorders in the PNS. After peripheral nerve injury, injured Schwann cells produce and release the inflammatory cytokines IL-1β, IL-6, and TNF-α (Shamash *et al.*, 2002; Thacker *et al.*, 2007) and chemokines such as monocyte chemoattractant protein-1 (MCP-1) (Tofaris *et al.*, 2002). It is therefore predicted that Schwann cells and DRG neurons may also produce these cytokines and chemokines after long-term vincristine treatment.

B. Infiltration and Activation of Macrophages

When the PNS is damaged, immune cells such as macrophages and lymphocytes infiltrate into the injured region and release inflammatory mediators (Hu and McLachlan, 2002; Scholz and Woolf, 2007). It was recently demonstrated that macrophages infiltrated into the sciatic nerve and DRG in mice after vincristine treatment (Uceyler *et al.*, 2006). Nerve injury-induced neuropathic pain was prevented by injection of the macrophage-suppressing agent clodronate (Liu *et al.*, 2000). Detailed evidence regarding this effect caused by vincristine is lacking, but macrophages may have an essential role in vincristine-induced neuropathic pain. Vincristine elicits macrophage infiltration in the PNS (Kiguchi *et al.*, 2008b; Uceyler *et al.*, 2006). Whether macrophage infiltration is due to mediators derived from the injured PNS caused by vincristine treatment is unknown. It was reported that the released MCP-1-induced macrophage chemotaxis in the injured region when neuropathic pain developed after nerve injury (White *et al.*, 2005, 2007). MCP-1 is expressed in macrophages and also in injured Schwann cells and DRG neurons; MCP-1 is the key regulator eliciting chemotaxis and the infiltration of macrophages (Abbadie, 2005; Jung *et al.*, 2008; Tofaris *et al.*, 2002). In neuropathic pain caused by vincristine, MCP-1 may participate in the chemotaxis and infiltration of macrophages in the PNS.

Lymphocytes and leukocytes also infiltrate into the injured region after nerve injury (Morin *et al.*, 2007; Thacker *et al.*, 2007). It was demonstrated that T-cells infiltrated into the injured nerve at the same time or before macrophage infiltration, and promoted macrophage chemotaxis (Kleinschnitz *et al.*, 2006). There are no reports showing lymphocyte infiltration in the PNS after long-term vincristine treatment, but lymphocytes may play a part in neuropathic pain caused by vincristine.

C. Role of Interleukin-6

It is considered that cytokines and chemokines have an important role in the activation of immune cells and neuropathic pain (White *et al.*, 2005, 2007). Macrophage-derived IL-6 elicits mechanical allodynia because infiltrating macrophages show high expression of this cytokine. This was confirmed when perineural injection (surrounding of the sciatic nerve) of the neutralizing antibody for IL-6 prevented mechanical allodynia caused by vincristine, and mechanical allodynia was not observed in IL-6 knockout mice (Kiguchi *et al.*, 2008b). These results were consistent with those of nerve injury-induced neuropathic pain. Blockade of inflammatory cytokines derived from various immune cells can prevent neuropathic pain after nerve injury (Lindenlaub *et al.*, 2000; Ma and Quirion, 2006). There are several reports showing the relationship between IL-6 and nerve injury-induced neuropathic pain (Arruda *et al.*, 2000; DeLeo *et al.*, 1996; Murphy *et al.*, 1999).

There are no reports concerning the mediators in neuropathic pain caused by vincristine. The detailed mechanisms of the development of neuropathic pain mediated by IL-6 after vincristine-induced nerve damage are incompletely understood (including intracellular signaling pathways). We hypothesize that janus kinase (Jak), a signal transducer and activator of the transcription-3 (STAT3) pathway, and a second messenger responding to IL-6 signaling (Qiu *et al.*, 2005), may participate in neuropathic pain caused by vincristine. It was recently reported that increased levels of IL-6 in the spinal cord after peripheral nerve injury activated the Jak-STAT3 pathway, and that the activation may participate in neuropathic pain (Dominguez *et al.*, 2008).

IV. Effects of Vincristine on the CNS

A. General Concepts

It is well known that signal transduction of pain initiates in primary afferent neurons and transmits to the dorsal horn of the spinal cord (Scholz and Woolf, 2002). Signal transduction in the spinal cord is important for nociception, but dynamic change (i.e., central sensitization) and neuronal plasticity associated with glial activation also have important roles in neuropathic pain (Inoue *et al.*, 2007; Tsuda *et al.*, 2005). Central sensitization was demonstrated in neuropathic pain caused by vincristine, and some candidate molecules for therapeutic targets were predicted (Kiguchi *et al.*, 2008a; Sweitzer *et al.*, 2006). Central sensitization after long-term vincristine treatment may be due to an indirect action of vincristine because it poorly penetrates the BBB (Calabresi and Parks, 1985).

B. Activation of Microglia and Astrocytes

Microglia is one of the glial cells in the CNS. It acts as a scavenger cell for the maintenance of physiological homeostasis, like macrophages in the PNS (Koizumi *et al.*, 2007). It was recently demonstrated that ATP produced in the spinal cord after peripheral nerve injury acts on the $P2X_4$ receptor (ligand-gated cation channel) and microglia are activated (Tsuda *et al.*, 2003). The mechanisms of microglia activation were subsequently studied in-depth, and some mediators for microglia activation determined (e.g., lysophosphatidic acid, fibronectin) (Fujita *et al.*, 2008; Tsuda *et al.*, 2008). In nerve injury-induced neuropathic pain, astrocyte activation is observed after microglia activation. Astrocyte activation participates in neuronal plasticity in the spinal cord, and produces various cytokines (Xu *et al.*, 2006; Zhuang *et al.*, 2006). Astrocytes are therefore important cells that contribute to the development and maintenance of neuropathic pain after nerve injury.

Activation of microglia and astrocytes in the spinal cord was also observed after long-term vincristine treatment (Kiguchi *et al.*, 2008a; Sweitzer *et al.*, 2006). It was reported that propentofylline prevented the activation of microglia and astrocytes associated with attenuation of neuropathic pain caused by vincristine (Sweitzer *et al.*, 2006). These reports suggest that activation of microglia and astrocytes is an important factor for neuropathic pain caused by vincristine.

C. Role of TNF-α

There are several reports concerning the participation of microglia and astrocyte-derived neurotrophic factors and inflammatory cytokines in neuropathic pain (Coull *et al.*, 2005; Inoue *et al.*, 2007; Scholz and Woolf, 2007). TNF-α was upregulated in the PNS and CNS after nerve injury, and contributed to neuropathic pain (Scholz and Woolf, 2007). The p38 MAP kinase was activated by TNF-α after nerve injury, and elicited neuropathic pain (Jin *et al.*, 2003; Schafers *et al.*, 2003). These findings were confirmed by studies using neutralizing antibody for TNF-α and p38 MAP kinase inhibitor (Schafers *et al.*, 2003). A local injection of recombinant TNF-α elicited neuropathic-like pain (Wei *et al.*, 2007). These reports indicate that TNF-α signaling has a critical role in the development and maintenance of neuropathic pain after nerve injury. These results were consistent with those of vincristine-induced neuropathic pain.

In our previous report, TNF-α was temporarily upregulated in activated microglia and astrocytes of the spinal cord after long-term vincristine treatment. This upregulation was accompanied by the development of neuropathic pain. Intrathecal injection of neutralizing antibody for TNF-α attenuated mechanical allodynia, indicating that TNF-α in the spinal cord contributed to neuropathic pain caused by vincristine (Kiguchi *et al.*, 2008a).

V. Other Anticancer Agents that Elicit Neuropathic Pain

Taxanes and platinum compounds also cause neuropathic pain as an adverse effect in clinical and experimental studies (Authier *et al.*, 2003; Polomano *et al.*, 2001). There are several reports noting the high incidence of neuropathic pain with the taxane paclitaxel (Chentanez *et al.*, 2003; Lynch *et al.*, 2004). Neuropathic pain caused by paclitaxel consists of not only mechanical allodynia, but also thermal hyperalgesia; the neuropathic pain developed after short-term treatment with paclitaxel in comparison with vincristine treatment (Flatters and Bennett, 2004; Matsumoto *et al.*, 2006). Some of inflammatory mediators related to the neuroinflammation observed in the PNS and CNS after paclitaxel treatment are identical to those observed during vincristine treatment (Cho *et al.*, 1983; Peters *et al.*, 2007). This suggests a common mechanism (at least in part) by which vincristine and paclitaxel cause neuropathic pain, even though their mode of action is different (vincristine inhibits tubulin polymerization and paclitaxel inhibits tubulin depolymerization) (Kumar, 1981; Owellen *et al.*, 1976). Modulation of the immune system in the PNS and CNS may participate in neuropathic pain caused by vincristine and paclitaxel in a similar way.

VI. Conclusion

Systemic treatment with vincristine damages Schwann cells and DRG neurons. The inflammatory mediators released by these cells may recruit macrophages in the PNS. Macrophages also release the inflammatory cytokine IL-6, which elicits neuroinflammation, leading to neuropathic pain. Secondary signal transduction of pain to the CNS activates microglia and astrocytes, and these activated glial cells contribute to the development and maintenance of neuropathic pain in the CNS through production of inflammatory cytokines (e.g., TNF-α). These key mediators have been identified but various problems remain (Fig. 1).

Immune-modulating agents that prevent activation of immune cells (e.g., macrophages) and/or the inhibitory agents of inflammatory cytokines (e.g., neutralizing antibody) could prevent the neuropathic pain caused by vincristine. These agents could increase the tolerability of vincristine when used for the treatment of leukemia and lymphoma.

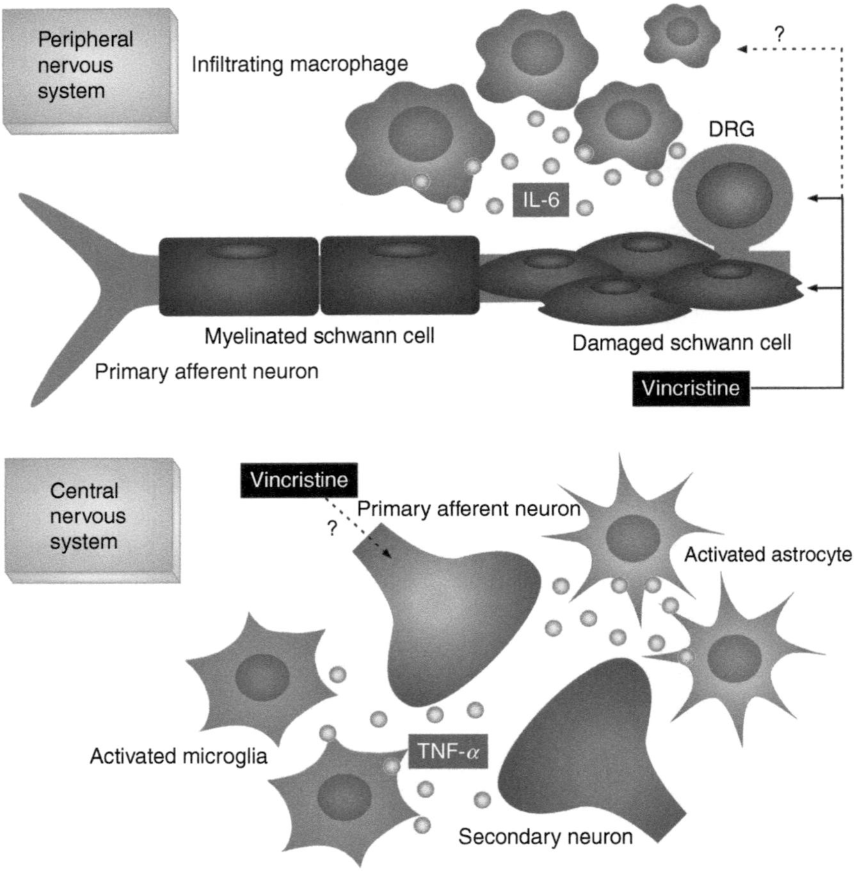

FIG. 1. Peripheral and central mechanism of neuropathic pain caused by vincristine. The upper diagram shows the effect of vincristine on the peripheral nervous system (comprising Schwann cells and the dorsal root ganglion (DRG)) and the involvement of interleukin (IL)-6 derived from infiltrating macrophages in neuropathic pain caused by vincristine. The lower diagram shows the effect of vincristine on the central nervous system, and the involvement of tumor necrosis factor-α (TNF-α) derived from activated microglia and astrocytes in neuropathic pain caused by vincristine.

References

Abbadie, C. (2005). Chemokines, chemokine receptors and pain. *Trends Immunol.* **26,** 529–534.

Aley, K. O., Reichling, D. B., and Levine, J. D. (1996). Vincristine hyperalgesia in the rat: A model of painful vincristine neuropathy in humans. *Neuroscience* **73,** 259–265.

Arruda, J. L., Sweitzer, S., Rutkowski, M. D., and DeLeo, J. A. (2000). Intrathecal anti-IL-6 antibody and IgG attenuates peripheral nerve injury-induced mechanical allodynia in the rat: Possible immune modulation in neuropathic pain. *Brain Res.* **879,** 216–225.

Authier, N., Coudore, F., Eschalier, A., and Fialip, J. (1999). Pain related behaviour during vincristine-induced neuropathy in rats. *Neuroreport* **10,** 965–968.

Authier, N., Gillet, J. P., Fialip, J., Eschalier, A., and Coudore, F. (2003). An animal model of nociceptive peripheral neuropathy following repeated cisplatin injections. *Exp. Neurol.* **182,** 12–20.

Bennett, G. J., Chung, J. M., Honore, M., and Seltzer, Z. (2003). Models of neuropathic pain in the rat. *Curr. Protoc. Neurosci.* Chapter **9,** Unit 9 14.

Beyreuther, B. K., Callizot, N., Brot, M. D., Feldman, R., Bain, S. C., and Stohr, T. (2007). Antinociceptive efficacy of lacosamide in rat models for tumor- and chemotherapy-induced cancer pain. *Eur. J. Pharmacol.* **565,** 98–104.

Calabresi, P., and Parks, R. E. Jr. (1985). Chemotherapy of neoplastic diseases. *In* "Goodman and Gilman's The Pharmacological Basis of Therapeutics" (A. G. Gilman, L. S. Goodman, T. W. Rall, and F. Murad, Eds.), 7th Ed. pp. 1240–1306. Macmillan Publishing Company, New York.

Casey, E. B., Jellife, A. M., Le Quesne, P. M., and Millett, Y. L. (1973). Vincristine neuropathy. Clinical and electrophysiological observations. *Brain* **96,** 69–86.

Chentanez, V., Sanguanrungsirigul, S., and Panyasawad, N. (2003). Effects of ganglioside on paclitaxel (Taxol) induced neuropathy in rats. *J. Med. Assoc. Thai.* **86,** 449–456.

Cho, E. S., Lowndes, H. E., and Goldstein, B. D. (1983). Neurotoxicology of vincristine in the cat. Morphological study. *Arch. Toxicol.* **52,** 83–90.

Coull, J. A., Beggs, S., Boudreau, D., Boivin, D., Tsuda, M., Inoue, K., Gravel, C., Salter, M. W., and De Koninck, Y. (2005). BDNF from microglia causes the shift in neuronal anion gradient underlying neuropathic pain. *Nature* **438,** 1017–1021.

DeLeo, J. A., Colburn, R. W., Nichols, M., and Malhotra, A. (1996). Interleukin-6-mediated hyperalgesia/allodynia and increased spinal IL-6 expression in a rat mononeuropathy model. *J. Interferon Cytokine Res.* **16,** 695–700.

Djaldetti, R., Hart, J., Alexandrova, S., Cohen, S., Beilin, B. Z., Djaldetti, M., and Bessler, H. (1996). Vincristine-induced alterations in Schwann cells of mouse peripheral nerve. *Am. J. Hematol.* **52,** 254–257.

Dominguez, E., Rivat, C., Pommier, B., Mauborgne, A., and Pohl, M. (2008). JAK/STAT3 pathway is activated in spinal cord microglia after peripheral nerve injury and contributes to neuropathic pain development in rat. *J. Neurochem.* **107,** 50–60.

Flatters, S. J., and Bennett, G. J. (2004). Ethosuximide reverses paclitaxel- and vincristine-induced painful peripheral neuropathy. *Pain* **109,** 150–161.

Fujita, R., Ma, Y., and Ueda, H. (2008). Lysophosphatidic acid-induced membrane ruffling and brain-derived neurotrophic factor gene expression are mediated by ATP release in primary microglia. *J. Neurochem.* **107,** 152–160.

Gilron, I., Bailey, J. M., Tu, D., Holden, R. R., Weaver, D. F., and Houlden, R. L. (2005). Morphine, gabapentin, or their combination for neuropathic pain. *N. Engl. J. Med.* **352,** 1324–1334.

Himes, R. H., Kersey, R. N., Heller-Bettinger, I., and Samson, F. E. (1976). Action of the vinca alkaloids vincristine, vinblastine, and desacetyl vinblastine amide on microtubules *in vitro*. *Cancer Res.* **36,** 3798–3802.

Hu, P., and McLachlan, E. M. (2002). Macrophage and lymphocyte invasion of dorsal root ganglia after peripheral nerve lesions in the rat. *Neuroscience* **112,** 23–38.

Inoue, K., Koizumi, S., and Tsuda, M. (2007). The role of nucleotides in the neuron--glia communication responsible for the brain functions. *J. Neurochem.* **102,** 1447–1458.

Jin, S. X., Zhuang, Z. Y., Woolf, C. J., and Ji, R. R. (2003). p38 mitogen-activated protein kinase is activated after a spinal nerve ligation in spinal cord microglia and dorsal root ganglion neurons and contributes to the generation of neuropathic pain. *J. Neurosci.* **23,** 4017–4022.

Jung, H., Toth, P. T., White, F. A., and Miller, R. J. (2008). Monocyte chemoattractant protein-1 functions as a neuromodulator in dorsal root ganglia neurons. *J. Neurochem.* **104,** 254–263.

Kiguchi, N., Maeda, T., Kobayashi, Y., and Kishioka, S. (2008a). Up-regulation of tumor necrosis factor-alpha in spinal cord contributes to vincristine-induced mechanical allodynia in mice. *Neurosci. Lett.* **445,** 140–143.

Kiguchi, N., Maeda, T., Kobayashi, Y., Kondo, T., Ozaki, M., and Kishioka, S. (2008b). The critical role of invading peripheral macrophage-derived interleukin-6 in vincristine-induced mechanical allodynia in mice. *Eur. J. Pharmacol.* **592,** 87–92.

Kleinschnitz, C., Hofstetter, H. H., Meuth, S. G., Braeuninger, S., Sommer, C., and Stoll, G. (2006). T cell infiltration after chronic constriction injury of mouse sciatic nerve is associated with interleukin-17 expression. *Exp. Neurol.* **200,** 480–485.

Koizumi, S., Shigemoto-Mogami, Y., Nasu-Tada, K., Shinozaki, Y., Ohsawa, K., Tsuda, M., Joshi, B. V., Jacobson, K. A., Kohsaka, S., and Inoue, K. (2007). UDP acting at P2Y6 receptors is a mediator of microglial phagocytosis. *Nature* **446,** 1091–1095.

Konings, P. N., Makkink, W. K., van Delft, A. M., and Ruigt, G. S. (1994a). Reversal by NGF of cytostatic drug-induced reduction of neurite outgrowth in rat dorsal root ganglia in vitro. *Brain Res.* **640,** 195–204.

Konings, P. N., Philipsen, R. L., Veeneman, G. H., and Ruigt, G. S. (1994b). Alpha-sialyl cholesterol increases laminin in Schwann cell cultures and attenuates cytostatic drug-induced reduction of laminin. *Brain Res.* **654,** 118–128.

Kumar, N. (1981). Taxol-induced polymerization of purified tubulin. Mechanism of action. *J. Biol. Chem.* **256,** 10435–10441.

Lindenlaub, T., Teuteberg, P., Hartung, T., and Sommer, C. (2000). Effects of neutralizing antibodies to TNF-alpha on pain-related behavior and nerve regeneration in mice with chronic constriction injury. *Brain Res.* **866,** 15–22.

Liu, T., van Rooijen, N., and Tracey, D. J. (2000). Depletion of macrophages reduces axonal degeneration and hyperalgesia following nerve injury. *Pain* **86,** 25–32.

Luo, Z. D., Calcutt, N. A., Higuera, E. S., Valder, C. R., Song, Y. H., Svensson, C. I., and Myers, R. R. (2002). Injury type-specific calcium channel alpha 2 delta-1 subunit up-regulation in rat neuropathic pain models correlates with antiallodynic effects of gabapentin. *J. Pharmacol. Exp. Ther.* **303,** 1199–1205.

Lynch, J. J. 3rd, Wade, C. L., Zhong, C. M., Mikusa, J. P., and Honore, P. (2004). Attenuation of mechanical allodynia by clinically utilized drugs in a rat chemotherapy-induced neuropathic pain model. *Pain* **110,** 56–63.

Ma, W., and Quirion, R. (2006). Increased calcitonin gene-related peptide in neuroma and invading macrophages is involved in the up-regulation of interleukin-6 and thermal hyperalgesia in a rat model of mononeuropathy. *J. Neurochem.* **98,** 180–192.

Marchand, F., Perretti, M., and McMahon, S. B. (2005). Role of the immune system in chronic pain. *Nat. Rev. Neurosci.* **6,** 521–532.

Matsumoto, M., Inoue, M., Hald, A., Xie, W., and Ueda, H. (2006). Inhibition of paclitaxel-induced A-fiber hypersensitization by gabapentin. *J. Pharmacol. Exp. Ther.* **318,** 735–740.

Moalem, G., and Tracey, D. J. (2006). Immune and inflammatory mechanisms in neuropathic pain. *Brain Res. Rev.* **51,** 240–264.

Morin, N., Owolabi, S. A., Harty, M. W., Papa, E. F., Tracy, T. F. Jr., Shaw, S. K., Kim, M., and Saab, C. Y. (2007). Neutrophils invade lumbar dorsal root ganglia after chronic constriction injury of the sciatic nerve. *J. Neuroimmunol.* **184,** 164–171.

Murphy, P. G., Ramer, M. S., Borthwick, L., Gauldie, J., Richardson, P. M., and Bisby, M. A. (1999). Endogenous interleukin-6 contributes to hypersensitivity to cutaneous stimuli and changes in neuropeptides associated with chronic nerve constriction in mice. *Eur. J. Neurosci.* **11,** 2243–2253.

Nozaki-Taguchi, N., Chaplan, S. R., Higuera, E. S., Ajakwe, R. C., and Yaksh, T. L. (2001). Vincristine-induced allodynia in the rat. *Pain* **93,** 69–76.

Okamoto, K., Martin, D. P., Schmelzer, J. D., Mitsui, Y., and Low, P. A. (2001). Pro- and anti-inflammatory cytokine gene expression in rat sciatic nerve chronic constriction injury model of neuropathic pain. *Exp. Neurol.* **169,** 386–391.

Owellen, R. J., Hartke, C. A., Dickerson, R. M., and Hains, F. O. (1976). Inhibition of tubulin-microtubule polymerization by drugs of the Vinca alkaloid class. *Cancer Res.* **36,** 1499–1502.

Peters, C. M., Jimenez-Andrade, J. M., Jonas, B. M., Sevcik, M. A., Koewler, N. J., Ghilardi, J. R., Wong, G. Y., and Mantyh, P. W. (2007). Intravenous paclitaxel administration in the rat induces a peripheral sensory neuropathy characterized by macrophage infiltration and injury to sensory neurons and their supporting cells. *Exp. Neurol.* **203,** 42–54.

Polomano, R. C., Mannes, A. J., Clark, U. S., and Bennett, G. J. (2001). A painful peripheral neuropathy in the rat produced by the chemotherapeutic drug, paclitaxel. *Pain* **94,** 293–304.

Qiu, J., Cafferty, W. B., McMahon, S. B., and Thompson, S. W. (2005). Conditioning injury-induced spinal axon regeneration requires signal transducer and activator of transcription 3 activation. *J. Neurosci.* **25,** 1645–1653.

Rahn, E. J., Makriyannis, A., and Hohmann, A. G. (2007). Activation of cannabinoid CB1 and CB2 receptors suppresses neuropathic nociception evoked by the chemotherapeutic agent vincristine in rats. *Br. J. Pharmacol.* **152,** 765–777.

Ramer, M. S., Murphy, P. G., Richardson, P. M., and Bisby, M. A. (1998). Spinal nerve lesion-induced mechanoallodynia and adrenergic sprouting in sensory ganglia are attenuated in interleukin-6 knockout mice. *Pain* **78,** 115–121.

Sandler, S. G., Tobin, W., and Henderson, E. S. (1969). Vincristine-induced neuropathy. A clinical study of fifty leukemic patients. *Neurology* **19,** 367–374.

Schafers, M., Svensson, C. I., Sommer, C., and Sorkin, L. S. (2003). Tumor necrosis factor-alpha induces mechanical allodynia after spinal nerve ligation by activation of p38 MAPK in primary sensory neurons. *J. Neurosci.* **23,** 2517–2521.

Scholz, J., and Woolf, C. J. (2002). Can. we conquer pain? *Nat. Neurosci.* **5**(Suppl), 1062–1067.

Scholz, J., and Woolf, C. J. (2007). The neuropathic pain triad: Neurons, immune cells and glia. *Nat. Neurosci.* **10,** 1361–1368.

Shamash, S., Reichert, F., and Rotshenker, S. (2002). The cytokine network of Wallerian degeneration: Tumor necrosis factor-alpha, interleukin-1alpha, and interleukin-1beta. *J. Neurosci.* **22,** 3052–3060.

Sweitzer, S. M., Pahl, J. L., and DeLeo, J. A. (2006). Propentofylline attenuates vincristine-induced peripheral neuropathy in the rat. *Neurosci. Lett.* **400,** 258–261.

Tanner, K. D., Reichling, D. B., and Levine, J. D. (1998). Nociceptor hyper-responsiveness during vincristine-induced painful peripheral neuropathy in the rat. *J. Neurosci.* **18,** 6480–6491.

Thacker, M. A., Clark, A. K., Marchand, F., and McMahon, S. B. (2007). Pathophysiology of peripheral neuropathic pain: Immune cells and molecules. *Anesth. Analg.* **105,** 838–847.

Tofaris, G. K., Patterson, P. H., Jessen, K. R., and Mirsky, R. (2002). Denervated Schwann cells attract macrophages by secretion of leukemia inhibitory factor (LIF) and monocyte chemoattractant protein-1 in a process regulated by interleukin-6 and LIF. *J. Neurosci.* **22,** 6696–6703.

Tsuda, M., Inoue, K., and Salter, M. W. (2005). Neuropathic pain and spinal microglia: A big problem from molecules in "small" glia. *Trends Neurosci.* **28,** 101–107.

Tsuda, M., Shigemoto-Mogami, Y., Koizumi, S., Mizokoshi, A., Kohsaka, S., Salter, M. W., and Inoue, K. (2003). P2X4 receptors induced in spinal microglia gate tactile allodynia after nerve injury. *Nature* **424,** 778–783.

Tsuda, M., Toyomitsu, E., Komatsu, T., Masuda, T., Kunifusa, E., Nasu-Tada, K., Koizumi, S., Yamamoto, K., Ando, J., and Inoue, K. (2008). Fibronectin/integrin system is involved in P2X(4) receptor upregulation in the spinal cord and neuropathic pain after nerve injury. *Glia* **56,** 579–585.

Uceyler, N., Kobsar, I., Biko, L., Ulzheimer, J., Levinson, S. R., Martini, R., and Sommer, C. (2006). Heterozygous P0 deficiency protects mice from vincristine-induced polyneuropathy. *J. Neurosci. Res.* **84,** 37–46.

Ueda, H., and Rashid, M. H. (2003). Molecular mechanism of neuropathic pain. *Drug News. Perspect.* **16,** 605–613.

Wei, X. H., Zang, Y., Wu, C. Y., Xu, J. T., Xin, W. J., and Liu, X. G. (2007). Peri-sciatic administration of recombinant rat TNF-alpha induces mechanical allodynia via upregulation of TNF-alpha in dorsal root ganglia and in spinal dorsal horn: The role of NF-kappa B pathway. *Exp. Neurol.* **205,** 471–484.

Weng, H. R., Cordella, J. V., and Dougherty, P. M. (2003). Changes in sensory processing in the spinal dorsal horn accompany vincristine-induced hyperalgesia and allodynia. *Pain* **103,** 131–138.

White, F. A., Jung, H., and Miller, R. J. (2007). Chemokines and the pathophysiology of neuropathic pain. *Proc. Natl. Acad. Sci. USA* **104,** 20151–20158.

White, F. A., Sun, J., Waters, S. M., Ma, C., Ren, D., Ripsch, M., Steflik, J., Cortright, D. N., Lamotte, R. H., and Miller, R. J. (2005). Excitatory monocyte chemoattractant protein-1 signaling is up-regulated in sensory neurons after chronic compression of the dorsal root ganglion. *Proc. Natl. Acad. Sci. USA* **102,** 14092–14097.

Xiao, W., Boroujerdi, A., Bennett, G. J., and Luo, Z. D. (2007). Chemotherapy-evoked painful peripheral neuropathy: Analgesic effects of gabapentin and effects on expression of the alpha-2-delta type-1 calcium channel subunit. *Neuroscience* **144,** 714–720.

Xu, J. T., Xin, W. J., Zang, Y., Wu, C. Y., and Liu, X. G. (2006). The role of tumor necrosis factor-alpha in the neuropathic pain induced by Lumbar 5 ventral root transection in rat. *Pain* **123,** 306–321.

Zhuang, Z. Y., Wen, Y. R., Zhang, D. R., Borsello, T., Bonny, C., Strichartz, G. R., Decosterd, I., and Ji, R. R. (2006). A peptide c-Jun N-terminal kinase (JNK) inhibitor blocks mechanical allodynia after spinal nerve ligation: Respective roles of JNK activation in primary sensory neurons and spinal astrocytes for neuropathic pain development and maintenance. *J. Neurosci.* **26,** 3551–3560.

NOCICEPTIVE BEHAVIOR INDUCED BY THE ENDOGENOUS OPIOID PEPTIDES DYNORPHINS IN UNINJURED MICE: EVIDENCE WITH INTRATHECAL *N*-ETHYLMALEIMIDE INHIBITING DYNORPHIN DEGRADATION

Koichi Tan-No,* Hiroaki Takahashi,* Osamu Nakagawasai,* Fukie Niijima,* Shinobu Sakurada,[†] Georgy Bakalkin,[‡] Lars Terenius,[§] and Takeshi Tadano*

*Department of Pharmacology, Tohoku Pharmaceutical University, 4-4-1 Komatsushima, Aoba-ku, Sendai 981-8558, Japan

[†]Department of Physiology and Anatomy, Tohoku Pharmaceutical University, 4-4-1 Komatsushima, Aoba-ku, Sendai 981-8558, Japan

[‡]Division of Biological Research on Drug Dependence, Department of Pharmaceutical Biosciences, Uppsala University, Uppsala S-751 24, Sweden

[§]Department of Clinical Neuroscience, Section of Alcohol and Drug Dependence Research, Karolinska Institute, Stockholm S-171 76, Sweden

Dynorphins, the endogenous opioid peptides derived from prodynorphin may participate not only in the inhibition, but also in facilitation of spinal nociceptive transmission. However, the mechanism of pronociceptive dynorphin actions, and the comparative potential of prodynorphin processing products to induce these actions were not fully elucidated. In our studies, we examined pronociceptive effects of prodynorphin fragments dynorphins A and B and big dynorphin consisting of dynorphins A and B, and focused on the mechanisms underlying these effects. Our principal finding was that big dynorphin was the most potent pronociceptive dynorphin; when administered intrathecally into mice at extremely low doses (1–10 fmol), big dynorphin produced nociceptive behavior through the activation of the NMDA receptor ion-channel complex by acting on the polyamine recognition site. We next examined whether the endogenous dynorphins participate in the spinal nociceptive transmission using *N*-ethylmaleimide (NEM)

INTERNATIONAL REVIEW OF NEUROBIOLOGY, VOL. 85
DOI: 10.1016/S0074-7742(09)85015-0

that blocks dynorphin degradation by inhibiting cysteine proteases. Similar to big dynorphin and dynorphin A, NEM produced nociceptive behavior mediated through inhibition of the degradation of endogenous dynorphins, presumably big dynorphin that in turn activates the NMDA receptor ion-channel complex by acting on the polyamine recognition site. Our findings support the notion that endogenous dynorphins are critical neurochemical mediators of spinal nociceptive transmission in uninjured animals. This chapter will review above-described phenomena and their mechanism.

I. Introduction

Dynorphin A, dynorphin B, and α-neoendorphin opioid peptides derived from prodynorphin, are present in the dorsal spinal cord and the periaqueductal gray where modulation of nociceptive information occurs (Tan-No *et al.*, 1997). It is known that intrathecally (i.t.) (Tan-No *et al.*, 1996, 2005a) and intracerebroventricularly (Tan-No *et al.*, 2001a) administered dynorphin A or dynorphin B produce antinociceptive effects in mice through κ-opioid receptors. Aside from dynorphin A, dynorphin B, and α-neoendorphin, the dynorphin family includes big dynorphin consisting of dynorphins A and B. Big dynorphin was identified in the porcine and rat pituitary and rat brain (Day and Akil, 1989; Fischli *et al.*, 1982; Kangawa *et al.*, 1981; Seizinger *et al.*, 1984; Xie and Goldstein, 1987) and human brain, and cerebrospinal fluid (Merg *et al.*, 2006). Although the affinity of big dynorphin to the human κ-opioid receptor is similar to that of dynorphin A, big dynorphin activates G proteins through human κ-opioid receptor with much greater potency, efficacy, and selectivity than dynorphins A and B (Merg *et al.*, 2006). Prodynorphin and its possible processing products are shown in Fig. 1.

Dynorphins are characterized by a dual action on the spinal nociceptive transmission. In addition to the antinociceptive effects, dynorphins may elicit hyperalgesia and allodynia secondary to the nerve and tissue injury. Thus, chronic pain states induced in the rat are associated with increased levels of prodynorphin mRNA and dynorphin A in the dorsal horn (Draisci *et al.*, 1991; Dubner and Ruda, 1992; Iadarola *et al.*, 1988; Kajander *et al.*, 1990; Malan *et al.*, 2000; Riley *et al.*, 1996; Ruda *et al.*, 1988; Wagner and Deleo, 1996), and with the enhanced release of dynorphins (Pohl *et al.*, 1997; Riley *et al.*, 1996). Intrathecal injection of dynorphin A produces behavioral signs similar to that of nerve injury-induced pain in rats (Laughlin *et al.*, 1997, 2001; Vanderah *et al.*, 1996). Pronociceptive effects of dynorphin A are not blocked by naloxone but inhibited by antagonists of the *N*-methyl-D-aspartate (NMDA) receptors implying that NMDA receptors but not opioid receptors are involved (Laughlin *et al.*, 1997, 2001; Vanderah *et al.*, 1996;

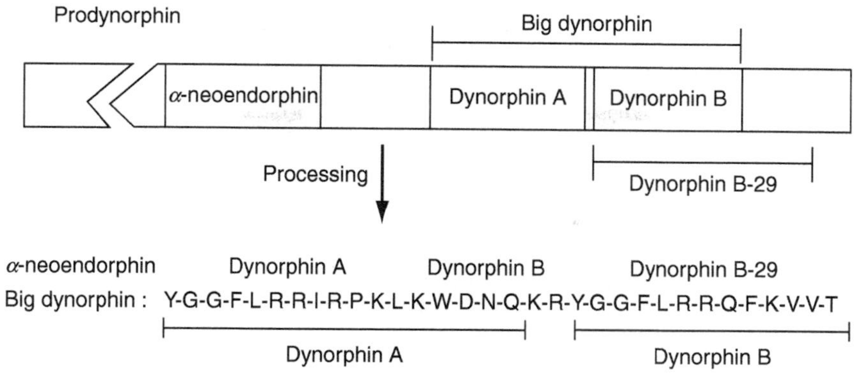

FIG. 1. Prodynorphin and its processing products.

Wang *et al.*, 2001). Intrathecal administration of dynorphin A-antiserum reverses neuropathic pain in nerve-injured rats and mice (Malan *et al.*, 2000; Wang *et al.*, 2001), suggesting that upregulation of the levels of this peptide is critical for development or maintenance of allodynia and hyperalgesia in tissue/nerve injury models of neuropathic pain. Indeed, prodynorphin-deficient mice develop only transient allodynia and hyperalgesia after nerve ligation, suggesting that prodynorphin-derived peptides are pronociceptive and required for maintenance of persistent neuropathic pain (Wang *et al.*, 2001; Xu *et al.*, 2004). These findings lead us to suggest that dynorphins may participate not only in the inhibition, but also in facilitation of nociceptive transmission. To test this hypothesis, we examined pronociceptive actions of dynorphins and spinal mechanisms underlying these actions. In this review, we describe the effects of i.t.-administered prodynorphin-derived peptides dynorphin A, dynorphin B, and big dynorphin in mice. In addition, our findings showing that, the endogenous dynorphins play a pronociceptive role in the spinal cord are presented in detail.

II. Interaction Between Dynorphins and the NMDA Receptor Ion-Channel Complex

The NMDA receptor ion-channel complex has been demonstrated to play an important role in spinal nociceptive transmission (Dickenson *et al.*, 1997; Zhang *et al.*, 2002). In addition to possessing the NMDA receptor, to which glutamate binds, the NMDA receptor ion-channel complex contains several allosteric sites such as polyamine and glycine recognition sites that modulate the receptor function (for a review, see Williams *et al.*, 1991). Our behavioral studies indicate that both spinal polyamine and glycine recognition sites of the NMDA receptor ion-channel complex play the crucial role in nociceptive transmission (Tan-No *et al.*, 2000, 2007).

TABLE I
PRINCIPAL NONOPIOID EFFECTS OF PRODYNORPHIN-DERIVED PEPTIDES

Phenomena	Peptides	References
1. Hindlimb paralysis	Dynorphin A	Bakshi and Faden (1990)
2. Loss of neuronal cell bodies	Dynorphin A (1–13)	Skilling *et al.* (1992)
3. Induction of apoptosis	Big dynorphin	Tan-No *et al.* (2001b)
4. Allodynia	Dynorphin A	Vanderah *et al.* (1996)
5. Nociceptive behavior	Big dynorphin Dynorphin A	Tan-No *et al.* (2002)
6. Memory facilitation, locomotor activation, and reduction of anxiety-like behavior	Big dynorphin	Kuzmin *et al.* (2006)

Many dynorphin actions are not mediated through the κ-opioid receptors (for review, see Shukla and Lemaire, 1994). The principal nonopioid effects of prodynorphin-derived peptides are summarized in Table I. The involvement of NMDA receptors has been suggested to explain some of the nonopioid effects since NMDA receptor antagonists protect against dynorphin-induced hindlimb paralysis (Bakshi and Faden, 1990; Long *et al.*, 1994; Skilling *et al.*, 1992), loss of neuronal cell bodies (Skilling *et al.*, 1992), allodynia (Laughlin *et al.*, 1997; Vanderah *et al.*, 1996), and memory facilitation, locomotor activation, and reduction of anxiety-like behavior (Kuzmin *et al.*, 2006). Moreover, dynorphin A has dual effects on NMDA receptor-mediated currents in the CA3 pyramidal cells of the guinea pig hippocampus, increasing currents at low concentrations, and decreasing currents at high concentrations (Caudle *et al.*, 1994; Caudle and Dubner, 1998). The inhibition of NMDA receptor-mediated currents by dynorphin A at a high concentration (5 μM) is mediated through κ2-opioid receptors (Caudle *et al.*, 1994), whereas excitation by a low concentration (100 nM) is inhibited by ifenprodil (Caudle and Dubner, 1998). It has been indicated in earlier studies that ifenprodil blocks the polyamine recognition site on the NMDA receptor ion-channel complex and, therefore the latter response may be interpreted as mediated through this site.

III. Nociceptive Behavior Induced by i.t.-Administered Prodynorphin-Derived Peptides and Polycationic Compounds

Intrathecal administration of big dynorphin (1–10 fmol) produced a characteristic behavioral response, the biting and/or licking of the hindpaw and the tail along with a slight hindlimb scratching directed toward the flank (Tan-No *et al.*, 2002).

Neither dynorphin A nor dynorphin B produced any significant response at a dose of 3 fmol. However, dynorphin A (30 pmol) produced the behavioral response whereas dynorphin B was practically inactive even at 1000 pmol. The behavioral response induced by 3 fmol big dynorphin peaked at 5–15 min after the injection. Pretreatment with morphine (0.125–2 mg/kg, i.p.) inhibited big dynorphin-induced behavior in a dose-dependent manner, suggesting that the behavioral response is related to nociception. Big dynorphin-induced nociceptive behavior was not mediated via opioid receptors since naloxone was inactive. D-APV (1–4 nmol), a competitive NMDA receptor antagonist, and MK-801 (0.25–4 nmol), an NMDA ion-channel blocker, dose-dependently inhibited big dynorphin-induced nociceptive behavior, suggesting that peptide effects are mediated through the NMDA receptor ion-channel complex. Neither 7-chlorokynurenic acid (4 nmol), an antagonist of glycine recognition site, nor CNQX (0.5 nmol), an antagonist of non-NMDA glutamate receptors, inhibited big dynorphin-induced nociceptive behavior, and, therefore, the glycine recognition site on the NMDA receptor ion-channel complex and non-NMDA glutamate receptors appear not to be involved.

Polycationic compounds including poly-L-arginine and poly-L-lysine have high affinity for the polyamine recognition site on the NMDA receptor ion-channel complex (Hashimoto *et al.*, 1994). Big dynorphin, dynorphin A, and dynorphin B contain 10, five, and three basic amino acids, respectively. Positive charge of these peptides correlates well with the rank order of behavioral potency, big dynorphin >> dynorphin A > dynorphin B, suggesting that their effects are mediated via the polyamine recognition site. The endogenous polycations spermine and spermidine, at concentrations in the low micromolar range, enhance the binding of the NMDA channel ligands [^{3}H]-MK-801 and [^{3}H]-TCP to rat brain membranes (Ransom and Stec, 1988; Sacaan and Johnson, 1990; Williams *et al.*, 1989). Spermine at low concentrations (1–10 μM) also enhances NMDA-elicited currents in the membranes of cultured cortical neurons (Rock and Macdonald, 1992). These findings indicate that the polyamine recognition site facilitates activity of the NMDA receptor ion-channel complex. We have also found that i.t.-administered spermine (0.1–10000 fmol) (Tan-No *et al.*, 2000) and poly-L-lysine (12 and 36 pg) (Tan-No *et al.*, 2004) produce nociceptive behavior, which is similar to that observed with big dynorphin. Nociceptive behavior induced by either of three polycationic compounds, poly-L-lysine, spermine, or big dynorphin was efficiently inhibited by D-APV, MK-801, and ifenprodil (Tan-No *et al.*, 2000, 2002, 2004). Ifenprodil selectively blocks the NMDA receptors in a noncompetitive, voltage-independent, and activity-dependent manner via binding to the NR1 subunit and/or NR2B subunit, which is allosterically linked with the polyamine recognition site on the NR1 subunit (for a review, see Chizh *et al.*, 2001a). Although the NR2B-containing NMDA receptors are present in the spinal cord, several *in vitro* and *in vivo* studies of adult rodent spinal cord preparations have failed to find substantial involvement of the NR2B subunit in nociceptive transmission (for a review, see Chizh *et al.*, 2001a). Antinociceptive effects of ifenprodil may be predominantly mediated through supraspinal but not spinal mechanisms (Chizh *et al.*, 2001b). Therefore, these findings lead us to suggest that the nociceptive behavior produced by these polycationic compounds may be mediated through the polyamine recognition site on the NR1 subunit of the NMDA receptor ion-channel complex but not via the NR2B-containing NMDA receptors.

It has been reported that dynorphin A at extremely low concentrations (0.01–0.1 nM) strongly facilitates substance P (SP) release evoked by capsaicin from rat trigeminal primary afferent C-fibers (Arcaya *et al.*, 1999). The release was inhibited by MK-801, but not by opioid antagonists nor-binaltorphimine and naloxone, suggesting that dynorphin A increases SP release from C-fibers by the activation of presynaptic NMDA receptors, and, as a result, produces antianalgesia and allodynia. These observations prompted us to test whether big dynorphin-induced nociceptive behavior is mediated through the tachykinin system. [D-Phe^7, D-His^9]-SP (6–11) (2 nmol), a specific antagonist for SP (NK1) receptors, and MEN-10,376 (2 nmol), a tachykinin NK2 receptor antagonist, failed to inhibit big dynorphin-induced nociceptive behavior, suggesting that big dynorphin-induced nociceptive behavior is not mediated through the tachykinin system in the spinal cord (Tan-No *et al.*, 2002).

In summary, evidence is presented that i.t.-administered big dynorphin at extremely low doses produces nociceptive behavior. This effect seems to be mediated through the activation of the NMDA receptor ion-channel complex by acting on the polyamine recognition site because the peptide has a strong positive charge.

IV. Degradation of Dynorphins by Cysteine Proteases

Similar to other neuropeptides, prodynorphin-derived peptides are converted to shorter, bioactive fragments and/or degraded with loss of activity by several proteases. Among the proteases degrading dynorphin-related peptides, dynorphin-converting enzymes that convert dynorphins to shorter opioid peptides have been purified from bovine and human spinal cord (Silberring and Nyberg, 1989; Silberring *et al.*, 1992, 1993). These proteases belong to the cysteine proteases family and cleave dynorphin A and dynorphin B between the Arg^6-Arg^7 and, to a lesser degree, the Leu^5-Arg^6 bonds, thus generating [Leu^5]enkephalin-Arg^6 as a major product and small amounts of [Leu^5]enkephalin, both of which are primarily active at δ-opioid receptors. We have previously shown that *p*-hydroxymercuribenzoate (PHMB), a general cysteine protease inhibitor, and

Boc-Tyr-Gly-NHO-Bz, a representative of a novel class of cysteine protease inhibitors, when coadministered with dynorphin A or dynorphin B, significantly prolong antinociception induced by i.t. injection of these dynorphins in the mouse formalin and capsaicin tests (Tan-No *et al.*, 1996, 2005a). This observation indicates that cysteine proteases may be important for terminating dynorphin A- and dynorphin B-induced antinociception.

Chronic use of morphine results in the development of antinociceptive tolerance. The development of morphine tolerance is suppressed by dynorphins (Hooke *et al.*, 1995; Schmauss and Herz, 1987) and U-50,488H, a selective κ-opioid receptor agonist (Tsuji *et al.*, 2000; Yamamoto *et al.*, 1988). These results indicate that peptides acting at κ-opioid receptors may play an inhibitory role in the development of antinociceptive tolerance to morphine. Moreover, chronic administration of morphine increases the conversion of dynorphin A to [Leu^5] enkephalin-Arg^6 in rat cerebrospinal fluid (Persson *et al.*, 1989) and dynorphin-converting enzyme activity in primary culture of rat cerebral cortex (Vlaskovska *et al.*, 1997). We have recently reported that i.t. administration of cysteine protease inhibitors suppresses the development of tolerance to morphine antinociception presumably through the inhibition of dynorphin degradation by decreasing the elevated dynorphin-converting enzyme activity (Tan-No *et al.*, 2008). Therefore, cysteine proteases seem to play a crucial role for modulating the dynorphin system-mediated functions.

V. *N*-Ethylmaleimide-Induced Nociceptive Behavior Mediated Through Inhibition of Dynorphin Degradation

As described above, i.t.-administered big dynorphin produces nociceptive behavior through the activation of the NMDA receptor ion-channel complex by acting on the polyamine recognition site (Tan-No *et al.*, 2002). However, the effects of exogenously administered dynorphins do not necessarily reflect the physiological and/or pathological roles of the spinal dynorphin system in nociceptive transmission. Therefore, we examined whether the endogenous dynorphin system is also pronociceptive using *N*-ethylmaleimide (NEM), an inhibitor of cysteine proteases involved in dynorphin degradation.

Intrathecal administration of NEM (15 nmol) into mice produced the behavioral response consisting of biting and/or licking of the hindpaw and the tail along with a slight hindlimb scratching directed toward the flank which is similar to that observed with big dynorphin (Tan-No *et al.*, 2005b). The behavioral response peaked at 10–25 min and almost disappeared at 30 min after the injection. Pretreatment with morphine (0.125–2 mg/kg, i.p.) inhibited NEM-induced behavior in a dose-dependent manner, suggesting that the behavioral response is related to nociception. On the other hand, the characteristic NEM-evoked response was not observed in prodynorphin knockout mice. Pretreatment with dynorphin A- or dynorphin B-antiserum inhibited NEM- and big dynorphin-induced nociceptive behavior in the concentration-dependent manner. Chemical measurements demonstrated that dynorphin B-antiserum efficiently binds dynorphin B and big dynorphin, that has the dynorphin B sequence at its C-terminus (Day and Akil, 1989; Dores and Akil, 1985, 1987), and that big dynorphin is also a target for cysteine proteases (Marinova *et al.*, 2004). Collectively, these observations suggest that NEM-evoked nociceptive behavior is induced by the protection of endogenous dynorphin A and big dynorphin against degradation by cysteine proteases.

Similar to effects of big dynorphin, NEM-induced nociceptive behavior was dose-dependently inhibited by ifenprodil (0.25–4 nmol) and MK-801 (0.04–4 nmol). Arcaine (62.5–125 pmol) and agmatine (3–30 pmol), more selective antagonists of the polyamine recognition site than ifenprodil, also caused a dose-dependent inhibition of NEM-induced nociceptive behavior. However, Ro25-6981 (4 nmol), a potent and selective antagonist of the NMDA receptor subtype containing the NR2B subunit, failed to inhibit NEM-induced nociceptive behavior. Therefore, these results indicate that NEM-induced nociceptive behavior is mediated through the polyamine recognition site of the NMDA receptor ion-channel complex but not the NMDA receptor subtype containing NR2B subunit in the mouse spinal cord.

A dual inhibitory and facilitative role of the dynorphin system in the nociceptive transmission has been proposed (for review, see Caudle and Mannes, 2000; Costigan and Woolf, 2002; Xu *et al.*, 2004). Several lines of evidence demonstrate that dynorphins inhibit nociceptive transmission in the spinal cord via interaction with the κ-opioid receptor. Thus, activation of the κ-opioid receptor by dynorphins inhibits voltage activated Ca^{2+} channels in mouse dorsal root ganglion cells (Werz and Macdonald, 1985), inhibits synaptic transmission of nociceptive neurons in the spinal dorsal horn (Randic *et al.*, 1995), induces hyperpolarization of neurons (Ogura and Kita, 2000), suppresses calcium currents and calcium-dependent secretion (Rusin *et al.*, 1997; Wiley *et al.*, 1997), and inhibits SP release in the spinal cord (Zachariou and Goldstein, 1997). In neuropathic pain models, pharmacological antagonists of the κ-opioid receptor enhance allodynia and hyperalgesia (Obara *et al.*, 2003; Xu *et al.*, 2004) and κ-opioid receptor-deficient mice developed increased tactile allodynia and thermal hyperalgesia (Xu *et al.*, 2004). Consistently, prevention of degradation of endogenous dynorphins by PHMB produced κ-opioid receptor-mediated antinociceptive effects in the mouse capsaicin test, a model of acute chemogenic nociception (Tan-No *et al.*, 1998).

Another set of evidence supports the notion on the pronociceptive functions of dynorphin A in the spinal cord. Behavioral hyperalgesia as a result of inflammation or tissue injury is accompanied by elevations in spinal dynorphin content

(Draisci *et al.*, 1991; Dubner and Ruda, 1992; Iadarola *et al.*, 1988; Kajander *et al.*, 1990; Malan *et al.*, 2000; Riley *et al.*, 1996; Ruda *et al.*, 1988; Wagner and Deleo, 1996). This increase is considered as causative of hyperalgesia and allodynia. Dynorphin A injected i.t. can elicit hyperalgesia and mechanical allodynia (Laughlin *et al.*, 1997, 2001; Vanderah *et al.*, 1996). Prodynorphin knockout mice do not maintain the chronic pain state after spinal nerve ligation compared with wild-type animals (Wang *et al.*, 2001; Xu *et al.*, 2004), whereas either MK-801 or dynorphin A-antiserum rescued the enhanced pain behavior of wild-type animals to that of prodynorphin knockout mice (Wang *et al.*, 2001).

The switch between anti- or pronociceptive effects of dynorphin A may depend on (1) peptide concentrations at the sites of action, and (2) kinetics of peptide interactions with opioid and NMDA receptors (for review, see Caudle and Mannes, 2000; Laughlin *et al.*, 2001). Dynorphins at basal, physiological concentrations may be antinociceptive through the opioid receptors, whereas elevated pathophysiological levels (or less processed prodynorphin-derived peptides) may be pronociceptive acting at the NMDA receptors (Hauser *et al.*, 1999). However, the concentration issue is not apparently relevant for the mouse spinal cord because low, fmol doses of dynorphin A or big dynorphin induce nociceptive behavior after i.t. injection into normal uninjured animals (Tan-No *et al.*, 2002, 2005b). A behavioral generalization of the idea on the time-dependent differences in dynorphin A effects on opioid and NMDA receptors (for review, see Caudle and Mannes, 2000) infers that immediately after injury dynorphins produce analgesia via the κ-opioid receptor so that the animal/person may remove themselves from the tissue damaging situation. Once this has been done, dynorphin A could induce allodynia and hyperalgesia to protect the injured area from further damage by immobilizing the injured limb/part of the body and this action is mediated through the NMDA receptors. Differential effects of cysteine protease inhibitors mediated through the dynorphin system support the notion that endogenous dynorphins play antinociceptive and pronociceptive roles in an acute pain state (Tan-No *et al.*, 1998) and uninjured animals (Tan-No *et al.*, 2005b), respectively.

A consensus of different studies appears to be that the spinal dynorphin system plays an inhibitory role in nociceptive transmission mediated through the κ-opioid receptor in an acute pain state, and a facilitative role mediated through an NMDA receptor mechanism in a chronic pain state when the κ-opioid receptor became tolerant due to sustained activation by endogenous dynorphins (Xu *et al.*, 2004). Our studies suggest that the prodynorphin system also has a pronociceptive function in normal uninjured animals.

In summary, NEM-induced nociceptive behavior may be mediated through inhibition of the degradation of endogenous dynorphins, presumably big dynorphin that in turn activates the NMDA receptor ion-channel complex by acting on the polyamine recognition site (Fig. 2).

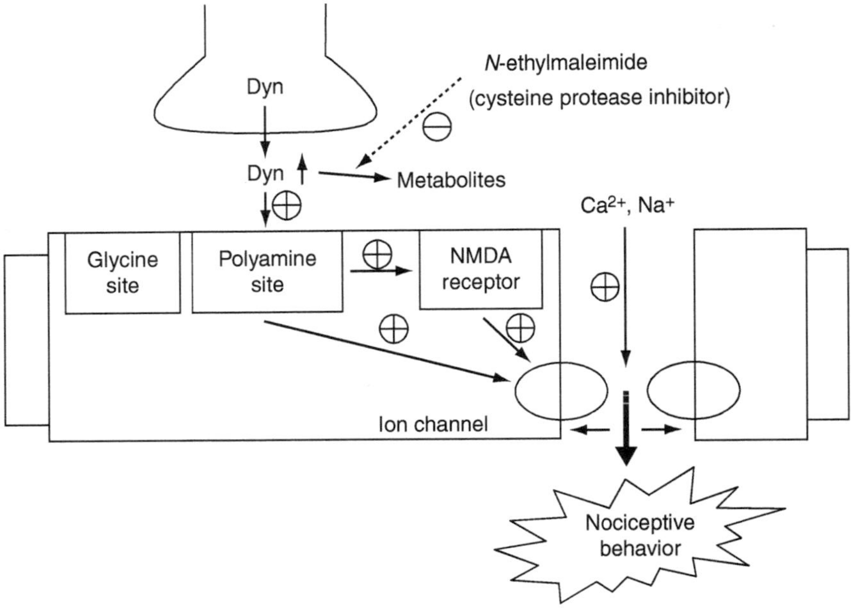

FIG. 2. The possible mechanism of NEM-induced nociceptive behavior. NEM inhibits the degradation of endogenous dynorphins, presumably big dynorphin, and accumulated dynorphins activate the NMDA receptor ion-channel complex by acting on the polyamine recognition site and then induce nociceptive behavior. Dyn: dynorphin.

VI. Conclusion

In the present study, we found that i.t.-administered big dynorphin at extremely low doses produced nociceptive behavior which may be mediated through the activation of the NMDA receptor ion-channel complex by acting on the polyamine recognition site because the peptide has a strong positive charge. Moreover, we demonstrated that endogenous prodynorphin-derived peptides seemed to play a crucial role in the spinal nociceptive transmission because NEM produced nociceptive behavior through the inhibition of the degradation of endogenous dynorphins.

Acknowledgments

We would like to express deep gratitude to our collaborators for their kind help, support, and valuable suggestion in this study: Dr. Akihisa Esashi, Dr. Aki Taira-Ishii, Mr. Kiyoshi Ohshima, and Mr. Masakazu Shimoda, Tohoku Pharmaceutical University. This study was supported in part by a Grant-in-Aid for Scientific Research (C) (Nos. 07672374, 14572062, and 18613016) from the Japan

Society for the Promotion of Science to K.T., and a Grant-in-Aid for High Technology Research Program from the Ministry of Education, Culture, Sports, Science, and Technology, Japan, and by AFA Försäkring, the Swedish Science Research Council and Torsten och Ragnar Söderbergs stiftelser to G.B.

References

Arcaya, J. L., Cano, G., Gomez, G., Maixner, W., and Suarez-Roca, H. (1999). Dynorphin A increases substance P release from trigeminal primary afferent C-fibers. *Eur. J. Pharmacol.* **366,** 27–34.

Bakshi, R., and Faden, A. I. (1990). Competitive and non-competitive NMDA antagonists limit dynorphin A-induced rat hindlimb paralysis. *Brain Res.* **507,** 1–5.

Caudle, R. M., and Dubner, R. (1998). Ifenprodil blocks the excitatory effects of the opioid peptide dynorphin 1–17 on NMDA receptor-mediated currents in the CA3 region of the guinea pig hippocampus. *Neuropeptides* **32,** 87–95.

Caudle, R. M., and Mannes, A. J. (2000). Dynorphin: Friend or foe? *Pain* **87,** 235–239.

Caudle, R. M., Chavkin, C., and Dubner, R. (1994). κ2 Opioid receptors inhibit NMDA receptor-mediated synaptic currents in guinea pig CA3 pyramidal cells. *J. Neurosci.* **14,** 5580–5589.

Chizh, B. A., Headley, P. M., and Tzschentke, T. M. (2001a). NMDA receptor antagonists as analgesics: Focus on the NR2B subtype. *Trends Pharmacol. Sci.* **22,** 636–642.

Chizh, B. A., Reiβmuller, E., Schlutz, H., Scheede, M., Haase, G., and Englberger, W. (2001b). Supraspinal vs spinal sites of the antinociceptive action of the subtype-selective NMDA antagonist ifenprodil. *Neuropharmacology* **40,** 212–220.

Costigan, M., and Woolf, C. J. (2002). No DREAM, no pain. Closing the spinal gate. *Cell* **108,** 297–300.

Day, R., and Akil, H. (1989). The posttranslational processing of prodynorphin in the rat anterior pituitary. *Endocrinology* **124,** 2392–2405.

Dickenson, A. H., Chapman, V., and Green, G. M. (1997). The pharmacology of excitatory and inhibitory amino acid-mediated events in the transmission and modulation of pain in the spinal cord. *Gen. Pharmacol.* **28,** 633–638.

Dores, R. M., and Akil, H. (1985). Steady state levels of pro-dynorphin-related end products in the striatum and substantia nigra of the adult rhesus monkey. *Peptides* **6**(Suppl. 2), 143–148.

Dores, R. M., and Akil, H. (1987). Species-specific processing of prodynorphin in the posterior pituitary of mammals. *Endocrinology* **120,** 230–238.

Draisci, G., Kajander, K. C., Dubner, R., Bennett, G. J., and Iadarola, M. J. (1991). Up-regulation of opioid gene expression in spinal cord evoked by experimental nerve injuries and inflammation. *Brain Res.* **560,** 186–192.

Dubner, R., and Ruda, M. A. (1992). Activity-dependent neuronal plasticity following tissue injury and inflammation. *Trends Neurosci.* **15,** 96–103.

Fischli, W., Goldstein, A., Hunkapiller, M. W., and Hood, L. E. (1982). Isolation and amino acid sequence analysis of a 4,000-dalton dynorphin from porcine pituitary. *Proc. Natl. Acad. Sci. USA* **79,** 5435–5437.

Hashimoto, K., Mantione, C. R., Spada, M. R., Neumeyer, J. L., and London, E. D. (1994). Further characterization of [^{3}H] ifenprodil binding in rat brain. *Eur. J. Pharmacol.* **266,** 67–77.

Hauser, K. F., Foldes, J. K., and Turbek, C. S. (1999). Dynorphin A (1–13) neurotoxicity *in vitro*: Opioid and non-opioid mechanisms in mouse spinal cord neurons. *Exp. Neurol.* **160,** 361–375.

Hooke, L. P., He, L., and Lee, N. M. (1995). Dynorphin A modulates acute and chronic opioid effects. *J. Pharmacol. Exp. Ther.* **273,** 292–297.

Iadarola, M. J., Brady, L. S., Draisci, G., and Dubner, R. (1988). Enhancement of dynorphin gene expression in spinal cord following experimental inflammation: Stimulus specificity, behavioral parameters and opioid receptor binding. *Pain* **35,** 313–326.

Kajander, K. C., Sahara, Y., Iadarola, M. J., and Bennett, G. J. (1990). Dynorphin increases in the dorsal spinal cord in rats with a painful peripheral neuropathy. *Peptides* **11,** 719–728.

Kangawa, K., Minamino, N., Chino, N., Sakakibara, S., and Matsuo, H. (1981). The complete amino acid sequence of α-neo-endorphin. *Biochem. Biophys. Res. Commun.* **99,** 871–878.

Kuzmin, A., Madjid, N., Terenius, L., Ogren, S. O., and Bakalkin, G. (2006). Big dynorphin, a prodynorphin-derived peptide produces NMDA receptor-mediated effects on memory, anxiolytic-like and locomotor behavior in mice. *Neuropsychopharmacology* **31,** 1928–1937.

Laughlin, T. M., Vanderah, T. W., Lashbrook, J., Nichols, M. L., Ossipov, M., Porreca, F., and Wilcox, G. L. (1997). Spinally administered dynorphin A produces long-lasting allodynia: Involvement of NMDA but not opioid receptors. *Pain* **72,** 253–260.

Laughlin, T. M., Larson, A. A., and Wilcox, G. L. (2001). Mechanisms of induction of persistent nociception by dynorphin. *J. Pharmacol. Exp. Ther.* **299,** 6–11.

Long, J. B., Rigamonti, D. D., Oleshansky, M. A., Wingfield, C. P., and Martinez-Arizala, A. (1994). Dynorphin A-induced rat spinal cord injury: Evidence for excitatory amino acid involvement in a pharmacological model of ischemic spinal cord injury. *J. Pharmacol. Exp. Ther.* **269,** 358–366.

Malan, T. P., Ossipov, M. H., Gardell, L. R., Ibrahim, M., Bian, D., Lai, J., and Porreca, F. (2000). Extraterritorial neuropathic pain correlates with multisegmental elevation of spinal dynorphin in nerve-injured rats. *Pain* **86,** 185–194.

Marinova, Z., Yakovleva, T., Meltzig, M., Hallberg, M., Nylander, I., Ray, K., Rodgers, D. W., Hauser, K. F., Ekström, T. J., and Bakalkin, G. (2004). A novel soluble protein factor with non-opioid dynorphin A-binding activity. *Biochem. Biophys. Res. Commun.* **321,** 202–209.

Merg, F., Filliol, D., Usynin, I., Bazov, I., Bark, N., Hurd, Y. L., Yakovleva, T., Kieffer, B. L., and Bakalkin, G. (2006). Big dynorphin as a putative endogenous ligand for the κ-opioid receptor. *J. Neurochem.* **97,** 292–301.

Obara, I., Mika, J., Schafer, M. K., and Przewlocka, B. (2003). Antagonists of the kappa-opioid receptor enhance allodynia in rats and mice after sciatic nerve ligation. *Br. J. Pharmacol.* **140,** 538–546.

Ogura, M., and Kita, H. (2000). Dynorphin exerts both postsynaptic and presynaptic effects in the globus pallidus of the rat. *J. Neurophysiol.* **83,** 3366–3376.

Persson, S., Post, C., Alari, L., Nyberg, F., and Terenius, L. (1989). Increased neuropeptide-converting enzyme activities in cerebrospinal fluid of opiate-tolerant rats. *Neurosci. Lett.* **107,** 318–322.

Pohl, M., Ballet, S., Collin, E., Mauborgne, A., Bourgoin, S., Benoliel, J. J., Hamon, M., and Cesselin, F. (1997). Enkephalinergic and dynorphinergic neurons in the spinal cord and dorsal root ganglia of the polyarthritic rat—*in vivo* release and cDNA hybridization studies. *Brain Res.* **749,** 18–28.

Randic, M., Cheng, G., and Kojic, L. (1995). Kappa-opioid receptor agonists modulate excitatory transmission in substantia gelatinosa neurons of the rat spinal cord. *J. Neurosci.* **15,** 6809–6826.

Ransom, R. W., and Stec, N. L. (1988). Cooperative modulation of [^{3}H] MK-801 binding to the *N*-methyl-D-aspartate receptor-ion channel complex by L-glutamate, glycine, and polyamines. *J. Neurochem.* **51,** 830–836.

Riley, R. C., Zhao, Z. Q., and Duggan, A. W. (1996). Spinal release of immunoreactive dynorphin A (1–8) with the development of peripheral inflammation in the rat. *Brain Res.* **710,** 131–142.

Rock, D. M., and Macdonald, R. L. (1992). The polyamine spermine has multiple actions on *N*-methyl-D-aspartate receptor single-channel currents in cultured cortical neurons. *Mol. Pharmacol.* **41,** 83–88.

Ruda, M. A., Iadarola, M. J., Cohen, L. V., and Young, W. S. III (1988). *In situ* hybridization histochemistry and immunocytochemistry reveal an increase in spinal dynorphin biosynthesis in a rat model of peripheral inflammation and hyperalgesia. *Proc. Natl. Acad. Sci. USA* **85,** 622–626.

Rusin, K. I., Giovannucci, D. R., Stuenkel, E. L., and Moises, H. C. (1997). Kappa-opioid receptor activation modulates Ca^{2+} currents and secretion in isolated neuroendocrine nerve terminals. *J. Neurosci.* **17,** 6565–6574.

Sacaan, A. I., and Johnson, K. M. (1990). Characterization of the stimulatory and inhibitory effects of polyamines on [^{3}H] *N*-(1-[thienyl] cyclohexyl) piperidine binding to the *N*-methyl-D-aspartate receptor ionophore complex. *Mol. Pharmacol.* **37,** 572–577.

Schmauss, C., and Herz, A. (1987). Intrathecally administered dynorphin-(1–17) modulates morphine-induced antinociception differently in morphine-naïve and morphine-tolerant rats. *Eur. J. Pharmacol.* **135,** 429–431.

Seizinger, B. R., Grimm, C., Hollt, V., and Herz, A. (1984). Evidence for a selective processing of proenkephalin B into different opioid peptide forms in particular regions of rat brain and pituitary. *J. Neurochem.* **42,** 447–457.

Shukla, V. K., and Lemaire, S. (1994). Nonopioid effects of dynorphins: Possible role of the NMDA receptor. *Trends Pharmacol. Sci.* **15,** 420–424.

Skilling, S. R., Sun, X., Kurtz, H. J., and Larson, A. A. (1992). Selective potentiation of NMDA-induced activity and release of excitatory amino acids by dynorphin: Possible roles in paralysis and neurotoxicity. *Brain Res.* **575,** 272–278.

Silberring, J., and Nyberg, F. (1989). A novel bovine spinal cord endoprotease with high specificity for dynorphin B. *J. Biol. Chem.* **264,** 11082–11086.

Silberring, J., Castello, M. E., and Nyberg, F. (1992). Characterization of dynorphin A-converting enzyme in human spinal cord. An endoprotease related to a distinct conversion pathway for the opioid heptadecapeptide? *J. Biol. Chem.* **267,** 21324–21328.

Silberring, J., Demuth, H. U., Brostedt, P., and Nyberg, F. (1993). Inhibition of dynorphin converting enzymes from human spinal cord by *N*-peptidyl-*O*-acyl hydroxylamines. *J. Biochem.* **114,** 648–651.

Tan-No, K., Taira, A., Sakurada, T., Inoue, M., Sakurada, S., Tadano, T., Sato, T., Sakurada, C., Nylander, I., Silberring, J., Terenius, L., and Kisara, K. (1996). Inhibition of dynorphin-converting enzymes prolongs the antinociceptive effect of intrathecally administered dynorphin in the mouse formalin test. *Eur. J. Pharmacol.* **314,** 61–67.

Tan-No, K., Terenius, L., Silberring, J., and Nylander, I. (1997). Levels of dynorphin peptides in the central nervous system and pituitary gland of the spontaneously hypertensive rat. *Neurochem. Int.* **31,** 27–32.

Tan-No, K., Taira, A., Inoue, M., Ohshima, K., Sakurada, T., Sakurada, C., Nylander, I., Demuth, H. U., Silberring, J., Terenius, L., Tadano, T., and Kisara, K. (1998). Intrathecal administration of *p*-hydroxymercuribenzoate or phosphoramidon/bestatin-combined induces antinociceptive effects through different opioid mechanisms. *Neuropeptides* **32,** 411–415.

Tan-No, K., Taira, A., Wako, K., Niijima, F., Nakagawasai, O., Tadano, T., Sakurada, C., Sakurada, T., and Kisara, K. (2000). Intrathecally administered spermine produces the scratching, biting and licking behavior in mice. *Pain* **86,** 55–61.

Tan-No, K., Ohshima, K., Taira, A., Inoue, M., Niijima, F., Nakagawasai, O., Tadano, T., Nylander, I., Silberring, J., Terenius, L., and Kisara, K. (2001a). Antinociceptive effect produced by intracerebroventricularly administered dynorphin A is potentiated by *p*-hydroxymercuribenzoate or phosphoramidon in the mouse formalin test. *Brain Res.* **891,** 274–280.

Tan-No, K., Cebers, G., Yakovleva, T., Goh, B. H., Gileva, I., Reznikov, K., Aguilar-Santelises, M., Hauser, K. F., Terenius, L., and Bakalkin, G. (2001b). Cytotoxic effects of dynorphins through nonopioid intracellular mechanisms. *Exp. Cell Res.* **269,** 54–63.

Tan-No, K., Esashi, A., Nakagawasai, O., Niijima, F., Tadano, T., Sakurada, C., Sakurada, T., Bakalkin, G., Terenius, L., and Kisara, K. (2002). Intrathecally administered big dynorphin, a prodynorphin-derived peptide, produces nociceptive behavior through an *N*-methyl-D-aspartate receptor mechanism. *Brain Res.* **952,** 7–14.

Tan-No, K., Esashi, A., Nakagawasai, O., Niijima, F., Sakurada, C., Sakurada, T., Bakalkin, G., Terenius, L., and Tadano, T. (2004). Nociceptive behavior induced by poly-L-lysine and other basic compounds involves the spinal NMDA receptors. *Brain Res.* **1008,** 49–53.

Tan-No, K., Taira, A., Nakagawasai, O., Niijima, F., Demuth, H. U., Silberring, J., Terenius, L., and Tadano, T. (2005a). Differential effects of *N*-peptidyl-*O*-acyl hydroxylamines on dynorphin-induced antinociception in the mouse capsaicin test. *Neuropeptides* **39,** 569–573.

Tan-No, K., Takahashi, H., Nakagawasai, O., Niijima, F., Sato, T., Satoh, S., Sakurada, S., Marinova, Z., Yakovleva, T., Bakalkin, G., Terenius, L., and Tadano, T. (2005b). Pronociceptive role of dynorphins in uninjured animals: *N*-Ethylmaleimide-induced nociceptive behavior mediated through inhibition of dynorphin degradation. *Pain* **113,** 301–309.

Tan-No, K., Esashi, A., Nakagawasai, O., Niijima, F., Furuta, S., Sato, T., Satoh, S., Yasuhara, H., and Tadano, T. (2007). Intrathecally administered D-cycloserine produces nociceptive behavior through the activation of *N*-methyl-D-aspartate receptor ion-channel complex acting on the glycine recognition site. *J. Pharmacol. Sci.* **104,** 39–45.

Tan-No, K., Shimoda, M., Sugawara, M., Nakagawasai, O., Niijima, F., Watanabe, H., Furuta, S., Sato, T., Satoh, S., Arai, Y., Kotlinska, J., Silberring, J., *et al.* (2008). Cysteine protease inhibitors suppress the development of tolerance to morphine antinociception. *Neuropeptides* **42,** 239–244.

Tsuji, M., Yamazaki, M., Takeda, H., Matsumiya, T., Nagase, H., Tseng, L. F., Narita, M., and Suzuki, T. (2000). The novel κ-opioid receptor agonist, TRK-820 has no affect on the development of antinociceptive tolerance to morphine in mice. *Eur. J. Pharmacol.* **394,** 91–95.

Vanderah, T. W., Laughlin, T., Lashbrook, J. M., Nichols, M. L., Wilcox, G. L., Ossipov, M. H., Malan, T. P., and Porreca, F. (1996). Single intrathecal injections of dynorphin A or des-Tyr-dynorphins produce long-lasting allodynia in rats: Blockade by MK-801 but not naloxone. *Pain* **68,** 275–281.

Vlaskovska, M., Nylander, I., Schramm, M., Hahne, S., Kasakov, L., Silberring, J., and Terenius, L. (1997). Opiate modulation of dynorphin conversion in primary cultures of rat cerebral cortex. *Brain Res.* **760,** 85–93.

Wagner, R., and Deleo, J. A. (1996). Pre-emptive dynorphin and *N*-methyl-D-aspartate glutamate receptor antagonism alters spinal immunocytochemistry but not allodynia following complete peripheral nerve injury. *Neuroscience* **72,** 527–534.

Wang, Z., Gardell, L. R., Ossipov, M. H., Vanderah, T. W., Brennan, M. B., Hochgeschwender, U., Hruby, V. J., Malan, T. P., Lai, J., and Porreca, F. (2001). Pronociceptive actions of dynorphin maintain chronic neuropathic pain. *J. Neurosci.* **21,** 1779–1786.

Werz, M. A., and Macdonald, R. L. (1985). Dynorphin and neoendorphin peptides decrease dorsal root ganglion neuron calcium-dependent action potential duration. *J. Pharmacol. Exp. Ther.* **234,** 49–56.

Wiley, J. W., Moises, H. C., Gross, R. A., and MacDonald, R. L. (1997). Dynorphin A-mediated reduction in multiple calcium currents involves a G (o) alpha-subtype G protein in rat primary afferent neurons. *J. Neurophysiol.* **77,** 1338–1348.

Williams, K., Romano, C., and Molinoff, P. B. (1989). Effects of polyamines on the binding of [^{3}H] MK-801 to the *N*-methyl-D-aspartate receptor: Pharmacological evidence for the existence of a polyamine recognition site. *Mol. Pharmacol.* **36,** 575–581.

Williams, K., Romano, C., Dichter, M. A., and Molinoff, P. B. (1991). Modulation of the NMDA receptor by polyamines. *Life Sci.* **48,** 469–498.

Xie, G. X., and Goldstein, A. (1987). Characterization of big dynorphins from rat brain and spinal cord. *J. Neurosci.* **7,** 2049–2055.

Xu, M., Petraschka, M., McLaughlin, J. P., Westenbroek, R. E., Caron, M. G., Lefkowitz, R. J., Czyzyk, T. A., Pintar, J. E., Terman, G. W., and Chavkin, C. (2004). Neuropathic pain activates the endogenous kappa opioid system in mouse spinal cord and induces opioid receptor tolerance. *J. Neurosci.* **24,** 4576–4584.

Yamamoto, T., Ohno, M., and Ueki, S. (1988). A selective κ-opioid agonist, U-50,488, blocks the development of tolerance to morphine analgesia in rat. *Eur. J. Pharmacol.* **156,** 173–176.

Zachariou, V., and Goldstein, B. D. (1997). Dynorphin-(1–8) inhibits the release of substance P-like immunoreactivity in the spinal cord of rats following a noxious mechanical stimulus. *Eur. J. Pharmacol.* **323,** 159–165.

Zhang, Y. Q., Ji, G. C., Wu, G. C., and Zhao, Z. Q. (2002). Excitatory amino acid receptor antagonists and electroacupuncture synergetically inhibit carrageenan-induced behavioral hyperalgesia and spinal fos expression in rats. *Pain* **99,** 525–535.

MECHANISM OF ALLODYNIA EVOKED BY INTRATHECAL MORPHINE-3-GLUCURONIDE IN MICE

Takaaki Komatsu,*[,1] Shinobu Sakurada,[†,1] Sou Katsuyama,* Kengo Sanai,[†] and Tsukasa Sakurada*

*First Department of Pharmacology, Daiichi College of Pharmaceutical Sciences, 22-1 Tamagawa-cho, Minami-ku, Fukuoka 815-8511, Japan
[†]Department of Physiology and Anatomy, Tohoku Pharmaceutical University, 4-4-1 Komatsushima, Aoba-ku, Sendai 981-8558, Japan
[1]These authors contributed equally to this work

Morphine-3-glucuronide (M3G), a main metabolite of morphine, has been proposed as a responsible factor when patients present with the neuroexcitatory side effects (allodynia, hyperalgesia, and myoclonus) observed following systemic administration of large doses of morphine. Indeed, both high-dose morphine (60 nmol/5 μl) and M3G (3 nmol/5 μl) elicit allodynia when administered intrathecally (i.t.) into mice. The allodynic behaviors are not opioid receptor mediated. This chapter reviews the potential mechanism of spinally mediated allodynia evoked by i.t. injection of M3G in mice. We discuss a possible presynaptic release of nociceptive neurotransmitters/neuromodulators such as substance P, glutamate, and dynorphin in the primary afferent fibers following i.t. M3G. It is possible to speculate that i.t. M3G injection could activate indirectly both NK_1 receptor and glutamate receptors that lead to the release of nitric oxide (NO) in the dorsal spinal cord. The NO plays an important role in M3G-induced allodynia. The phosphorylation of extracellular signal-regulated protein kinase (ERK) in the dorsal spinal cord evoked via NO/cGMP/PKG pathway contributes to i.t. M3G-induced allodynia. Furthermore, the increased release of NO observed after i.t. injection of M3G activates astrocytes and induces the release of the proinflammatory cytokine, interleukin-1β. Taken together, these findings suggest that M3G may induce allodynia via activation of NO–ERK pathway, while

INTERNATIONAL REVIEW OF
NEUROBIOLOGY, VOL. 85
DOI: 10.1016/S0074-7742(09)85016-2

0074-7742/09 $35.00

maintenance of the allodynic response may be triggered by NO-activated astrocytes in the dorsal spinal cord. The demonstration of the cellular mechanisms of neuronal–glial interaction underlying M3G-induced allodynia provides a fruitful strategy for improved pain management with high doses of morphine.

I. Introduction

Morphine is endowed with potent analgesic properties and has been widely used for the management of various kinds of pain, ranging from postoperative pain to chronic pain, including cancer pain. These clinical uses of morphine are often required to provide pain treatment for extended periods. However, the use of opioid analgesics for the treatment of chronic pain states is often offset by the development of tolerance. Thus, high doses of the opioid are required to elicit the same level of pain relief in chronic pain state. However, a large number of clinical studies have reported that high doses of morphine can unexpectedly elicit hyperalgesia (enhanced responses to noxious stimulation), allodynia (pain elicited by normal, innocuous, stimuli), and myoclonus (Ali, 1986; Arner *et al.*, 1988; De Conno *et al.*, 1991; Glavina and Robertshaw, 1988; Krames *et al.*, 1985; Parkinson *et al.*, 1990; Penn and Paice, 1987; Potter *et al.*, 1989; Sjogren *et al.*, 1993; Wert and MacDonald, 1982). Behavioral studies have also demonstrated that morphine at doses far higher than those required for antinociception, injected into the spinal subarachnoid space of rats, produce a spontaneous vocalization/sequeaking and agitation as well as hyperalgesia and allodynia as opposed to antinociception at small doses (Alvarez-Vega *et al.*, 1998; Tang and Schoenfeld, 1978; Woolf, 1981; Yaksh *et al.*, 1986). Furthermore, i.t. administration of high-dose morphine into mice was found to induce scratching, biting, and licking resembling that of substance P or *N*-methyl-D-asparatate (NMDA) injected i.t. (Komatsu *et al.*, 2007a; Sakurada *et al.*, 1996a, 2002). It has to be noted that spontaneous behavioral activation induced by high-dose i.t. morphine is irreversible by pretreatment the opioid receptor antagonist naloxone, suggesting a nonopioid mechanism (Sakurada *et al.*, 2002; Watanabe *et al.*, 2003a; Yaksh and Harty, 1988).

Morphine is known to be metabolized by glucuronidation to two biologically active metabolites, morphine-3-glucuronide (M3G) and morphine-6-glucuronide (M6G) (Boerner *et al.*, 1975). M6G has high affinity for the μ-opioid receptor (Loser *et al.*, 1996; Pasternak *et al.*, 1987; Paul *et al.*, 1989) and appears to be a more potent opioid agonist than morphine (Frances *et al.*, 1992; Osborne *et al.*, 2000; Pasternak *et al.*, 1987; Paul *et al.*, 1989). In contrast, M3G does not bind to μ-, δ-, or κ-opioid receptors (Loser *et al.*, 1996; Pasternak *et al.*, 1987) and appears to be devoid of analgesic activity (Pasternak *et al.*, 1987; Yaksh and Harty, 1988).

M3G also does not interact with NMDA, $GABA_A$, or glycine receptors (Bartlett *et al.*, 1994). However, in spite of these apparent lacks of activity, i.t. and intracerebroventricular (i.c.v.) administrations of M3G have been reported to evoke a range of excitatory behaviors including hyperalgesia, allodynia, myoclonus, and seizures in rats (Smith, 2000). These findings are consistent with previous data that one of the main morphine metabolites, for example M3G, may be responsible for the development of hyperalgesia, allodynia, and myoclonus during clinical morphine therapy (De Conno *et al.*, 1991; Sjogren *et al.*, 1993). Therefore, exploring possible mechanisms of M3G-induced nociception may be clinically useful to improve pain management with morphine and opioid analgesics.

II. Mechanism of M3G-Induced Allodynia: Spinal Release of Substance P and Glutamate

Much attention has to be paid that some effects of synthetic or endogenous compounds may be attributed to the action of their metabolites rather than the parent compound. M3G concentrations in the cerebrospinal fluid may account for more than half of morphine administered systemically to rats and humans (Smith, 2000). Additionally, human brain homogenates have been shown to metabolize morphine at nanomolar concentrations to M3G and M6G (Yamada *et al.*, 2003), suggesting the idea that M3G in the central nervous system may be formed there directly from morphine, which penetrates the blood–brain barrier at a greater rate than M3G. It is therefore plausible that allodynia and hyperalgesia evoked by spinal morphine at a high concentration may result from an increasing accumulation of M3G. This view is supported by our recent behavioral study that both high-dose morphine (60 nmol/5 μl) and M3G (3 nmol/5 μl) could elicit allodynia when administered i.t. into mice. M3G-induced allodynia reached the maximum effect at 15 min and declined progressively until cessation at 90 min postinjection (Komatsu *et al.*, 2007b). M3G-induced allodynia was inhibited dose-dependently by i.t. coadministration of the tachykinin neurokinin-1 (NK_1) receptor antagonists, sendide and RP-67580, and the NMDA ion-channel blocker, MK-801. In addition, i.t. injection of high-dose morphine in rats has also been shown to cause significant release of glutamate in the dorsal spinal cord extracellular fluid (Watanabe *et al.*, 2003a). In view of these findings, it is possible to speculate that i.t. injection of M3G could release primary afferent neuromodulators/neurotransmitters such as substance P and glutamate, and activate both NK_1 and glutamate receptors.

Elevation in spinal dynorphin content has also been seen in condition of opioid-induced pain state (Vanderah *et al.*, 2000). Although dynorphine was originally identified as an endogenous κ-opioid agonist and may act as an endogenous antinociceptive peptide under certain conditions, considerable evidence indicates that enhanced expression of spinal dynorphin is pronociceptive. High doses of dynorphin produce hyperalgesia and allodynia, in contrast to analgesia produced by low doses of dynorphin (Laughlin *et al.*, 1997; Vanderah *et al.*, 1996). Pain-related behavior associated with nerve injury is also blocked by antiserum to dynorphin (Bian *et al.*, 1999; Malan *et al.*, 2000; Wagner and Deleo, 1996; Wang *et al.*, 2001). The hyperalgesic effect is considered to be independent of the activation of opioid receptors. There is evidence that increased spinal dynorphin promotes the further release of excitatory transmitters from primary afferent neurons, thus provoking a positive feedback loop that amplifies further sensory input. Microdialysis studies have demonstrated localized dose-dependent release of glutamate elicited by exogenous dynorphin (Faden, 1992; Koetzner *et al.*, 2004; Skilling *et al.*, 1992). Dynorphin also enhances the release of substance P from trigeminal nucleus caudalis slices, the effect of which was blocked by MK801 but not by an opioid antagonist (Arcaya *et al.*, 1999). Thus, it cannot be denied that i.t. high-dose morphine may facilitate dynorphin release from primary afferent neurons, which may also be related to induction of nociceptive response to i.t. high-dose morphine.

III. Mechanism of M3G-Induced Allodynia: Spinal Activation of NO/cGMP/PKG Pathway

Substance P and glutamate released from presynaptic sites in response to i.t. M3G could activate NK_1 and NMDA receptors, which could trigger a feed forward mechanism of stimulation of neuronal nitric oxide synthase (nNOS) activity via mechanism largely dependent on Ca^{2+}. Increases in intracellular Ca^{2+} either by extracellular Ca^{2+} influx through NMDA receptor or Ca^{2+} channels as well as release from intracellular Ca^{2+} stores via production of inositol-1,4,5-triphosphate after activation of G-protein-coupled NK_1 receptors will result in activation of nNOS in the dorsal horn (Berridge,1993; Meller and Gebhart, 1993). Indeed, using *in vivo* microdialysis from rat dorsal spinal cord it has been shown that i.t. injection of high-dose morphine evokes a marked increase of NO metabolites (nitrite/nitrate) in the extracellular fluid (Watanabe *et al.*, 2003a). There is much evidence to suggest that NO is involved in the transmission of nociceptive information in the spinal cord (Meller and Gebhart, 1993). Stimulation of primary afferent C-fibers with capsaicin or formalin could induce nociceptive responses with an increase of NO production in the spinal cord, which is blocked by L-NAME (Sakurada *et al.*, 1996b, 2001; Watanabe *et al.*, 2003b; Wu *et al.*, 1998). Development of allodynia and hyperalgesia in models of chronic pain is also highly sensitive to NOS inhibitors (Morita *et al.*, 2008; Siegan *et al.*, 1996; Yonehara *et al.*, 2003), suggesting a pivotal role for NO in the transitions from acute to chronic pain. Our previous studies showed that the nonselective NOS inhibitors, L-NAME, and TRIM or the selective nNOS inhibitor 7-nitroindazole, dose-dependently blocked high-dose i.t. morphine-induced nociceptive behavior (Komatsu *et al.*, 2007a). The major action of NO is to activate the soluble form of the enzyme guanylate cyclase (sGC), which converts guanosine-5′-triphosphate (GTP) to cyclic guanosine-3′,5′-monophosphate (cGMP) (Qian *et al.*, 1996). The function of NO is also expressed through the production of cGMP and activation of cGMP-dependent protein kinase (PKG) (Schmidtko *et al.*, 2003). The content cGMP is higher in the spinal cord in chronic inflammatory and nerve-injury models (Garry *et al.*, 1994; Siegan *et al.*, 1996). Spinally delivered 8-bromo-cGMP, a membrane-permeable cGMP analog, facilitates nociceptive behavior (Morita *et al.*, 2008; Tegeder *et al.*, 2002). PKG deficiency or spinally delivered PKG inhibitors reduce formalin-induced nociceptive behavior in rats or mice (Schmidtko *et al.*, 2003; Tao *et al.*, 2000; Tegeder *et al.*, 2004). It is therefore likely that both cGMP and PKG act together to spinal nociceptive processing. The M3G-induced allodynia in mice is inhibited dose-dependently by (1) the nonselective NOS inhibitor L-NAME, (2) the nNOS inhibitor N^{ω}-propyl-L-arginine, (3) the sGC inhibitor ODQ, and (4) the PKG inhibitor KT-5823 (Komatsu *et al.*, 2007b). This body of evidence suggests that activation of NK_1 and NMDA receptors and subsequent production of NO, cGMP, and PKG are involved in M3G-induced allodynia.

IV. Mechanism of M3G-Induced Allodynia: Spinal ERK Activation

The mitogen-activated protein kinase (MAPK) is a family of evolutionary conserved molecules that plays a critical role in intracellular signal transduction and consists of ERK (extracellular signal-regulated protein kinase; p44/42 MAPK), p38, and JNK (c-Jun N-terminal kinase). Although ERK was originally implicated only in regulating the mitosis, proliferation, differentiation, and survival of cells during development, they are now widely recognized also to play an important role in neuronal plasticity (Impey *et al.*, 1999; Ji and Woolf, 2001). Recent evidence suggests a role of ERK phosphorylation in the nociceptive processing in the spinal cord (Ji *et al.*, 1999; Karim *et al.*, 2001; Kawasaki *et al.*, 2004). Multiple neurotransmitter/neuromodulator receptors such as NMDA and non-NMDA receptors, metabotropic glutamate receptors, and NK_1 receptors are coupled to ERK activation in the spinal cord (Ji *et al.*, 1999; Karim *et al.*, 2001; Kawasaki *et al.*, 2004; Lever *et al.*, 2003). Activation of the spinal ERK signaling pathway also contributes to naloxone-precipitated withdrawal in morphine-dependent rats (Cao *et al.*, 2005, 2006). M3G-induced ERK phosphorylation reached the maximum effect at 15 min and activated persistently until at 60 min post-i.t. injection which correlates with M3G-induced allodynia (Komatsu *et al.*,

2007b). Behavioral experiments showed that U0126, an ERK inhibitor, dose-dependently attenuated M3G-induced allodynia. The M3G-induced ERK activation and allodynia were completely blocked by L-NAME, N^{ω}-propyl-L-arginine, NO synthase inhibitors, ODQ, a sGC inhibitor, and KT-5823, a PKG inhibitor. These results suggest that M3G-induced allodynia may be triggered by the NO/cGMP/PKG/ERK pathway after increased levels of substance P and glutamate. M3G-induced ERK activation via the NO–cGMP–PKG mechanism translocates onto the membrane and phosphorylates ion channels or receptors increasing membrane excitability and inducing spinal neuronal sensitization. This may be mediated through phosphorylation of ion channels or receptors such as A-type potassium channel Kv4.2 (Hu *et al.*, 2003; Morozov *et al.*, 2003), leading to increased neuronal excitability or to the trafficking of AMPA receptors from the cytoplasm to the subsynaptic membrane (Ji *et al.*, 2003; Zhu *et al.*, 2002). Furthermore, ERK produces not only short-term functional changes by nontranscriptional processing, but also long-term adaptive changes by increasing gene transcription. For example, the activation of ERK in dorsal horn neurons contributes to persistent inflammatory pain, via transcriptional regulation of prodynorphin and NK_1 receptors (Ji *et al.*, 2002). ERK activation is associated with transcription factor, cAMP response element binding protein (CREB) in cultured hippocampal neurons and brain slices (Impey *et al.*, 1998; Obrietan *et al.*, 1999). Phosphorylation of CREB on serine 133 activates cAMP response element-mediated gene expression (Impey *et al.*, 1999; Obrietan *et al.*, 1999). Many inflammation-induced genes, including the immediate early gene *c-fos*, prodynorphin, BDNF, calcitonin gene-related peptide (CGRP), and the alpha subunit of CaMKII and NK_1 receptor, which may contribute to the sensitization of spinal cord neurons, are activated by CREB (Lonze and Ginty, 2002; McClung and Nestler, 2003). Several lines of evidence, *in vitro* and *in vivo*, have indicated that ERK activation and nuclear translocation induced CREB phosphorylation, and CREB-dependent gene expression is required for long-term changes in synaptic plasticity induced by chronic administration of morphine at supraspinal level (Ma *et al.*, 2001). Activation of ERK may contribute a substantial role in the establishment and maintenance of M3G-induced allodynia not only by posttranslational modifications of target protein but also by increasing gene transcription.

V. Mechanism of M3G-Induced Allodynia; Spinal Astrocyte Activation

Glial cells play an important role in the control of pain; in fact, it is known that neuronal plasticity is triggered by many inflammatory mediators and these are mainly produced by glial cells in the central nervous system. Indeed, in the past several years, more and more attention has been paid to neuron–glia interaction as a driving force for the development and maintenance of abnormal pain (Ji and Strichsrtz, 2004; Marchand *et al.*, 2005; Scholz and Woolf, 2007; Tsuda *et al.*, 2005; Watkins *et al.*, 2001). It was also found that spinal injection of a glial inhibitor fluorocitrate reduces hyperalgesia (Meller *et al.*, 1994). Glia activation has been studied in different animal models of neuropathic pain, including those yielded by peripheral nerve (Colburn *et al.*, 1997) or the spinal cord (Hains and Waxman, 2006) injury, inflammatory pain caused by formalin injection into a indlimb (Fu *et al.*, 1999), and cancer pain after inoculation of tumor cells (Honore *et al.*, 2000). Both astorocytes and micloglia are activated in these pain models and interact with neurons in the complex pain pathology. Glial cells express receptors for many neurotransmitters and neuromodulators, synthesize and release numerous transmitters, and produce transporters that either uptake or release transmitters from the extracellular and synaptic space, respectively (Araque *et al.*, 1999; Haydon, 2001). Therefore, glial cells are positioned to regulate neuronal function. It has been recognized that spinal cord glial cells, astrocyte and microglia, are activated by procedures that enhance pain (Raghavendra *et al.*, 2003, 2004; Tsuda *et al.*, 2005; Watkins and Maier, 2003; Watkins *et al.*, 2001). Astrocyte and microglia are also activated by pain-related substances such as NO or substance P, CGRP, ATP, and glutamate from the primary afferent terminal (Watkins *et al.*, 2001). Furthermore, chronic morphine was reported to (1) activate astrocytes as well as microglia (Cui *et al.*, 2006; Raghavendra *et al.*, 2002; Tai *et al.*, 2006) and (2) upregulate proinflammatory cytokines such as IL-1, IL-6, and TNF in spinal cord (Jonston *et al.*, 2004; Raghavendra *et al.*, 2002; Tai *et al.*, 2006). Proinflammatory cytokines such as IL-1β facilitate transmission and processing of noxious inputs at the spinal level (Watkins *et al.*, 2003; Raghavendra *et al.*, 2004; Zhang *et al.*, 2008). IL-1β is reported to be up-regulated in the spinal cord during inflammatory (Raghavendra *et al.*, 2004; Samad *et al.*, 2001) and neuropathic pain (Raghavendra *et al.*, 2003; Winkelstein *et al.*, 2001). Administration (i.t.) of IL-1β induces mechanical and thermal hyperalgesia (Sung *et al.*, 2004). The IL-1 receptor antagonist IL-1ra produces antiallodynic effects in rat models of inflammatory and neuropathic pain (Milligan *et al.*, 2001; Sweitzer *et al.*, 2001). These views are supported by our recent study that there was an upregulation of GFAP, a marker of astrocyte, and IL-1β, a prototype proinflammatory cytokine, in the dorsal spinal cord 30 min after i.t. injection of M3G (Komatsu *et al.*, 2008). The allodynia evoked by M3G was also dose-dependently blocked by fluorocitrate or IL-1ra.

NO has been implicated in mediating the release of neurotransmitter/neuromodulators such as substance P or glutamate (Meller and Gebhart, 1993). In addition, glial activation and cytokine production are found to be induced by NO (Guo *et al.*, 2007; Holguin *et al.*, 2004). The role of NO in activated astrocyte-induced glial cytokines after i.t. injection of M3G was examined by using the NO synthase inhibitor L-NAME. In mice treated with M3G, L-NAME was also very effective in blocking GFAP and IL-1β induction (Komatsu *et al.*, 2008). It is therefore evident that spinal IL-1β produced by activation of NO-astrocyte in the dorsal spinal cord is involved in the maintenance of M3G-induced allodynia.

In conclusion, this review has explored several different possible mechanisms of allodynia to i.t. M3G in rodents. The i.t. administration of M3G results in an increased release of primary afferent neuromodulator/neurotransmitter such as substance P and glutamate, which could lead to the release of NO in the dorsal spinal cord. M3G-induced allodynia may be induced by an activation of the NO–ERK pathway in a cGMP/PKG-dependent manner, while the maintenance may be triggered by spinal IL-1β produced by NO-activated astrocytes in the dorsal spinal cord. It can be anticipated that further understanding of the mechanisms underlying opioid-induced allodynia would advance and improve clinical management of opioid therapy in the management of chronic pain syndromes.

References

Ali, N. M. K. (1986). Hyperalgesic response in a patient receiving high concentrations of spinal morphine. *Anesthesiology* **65,** 449.

Alvarez-Vega, M., Baamonde, A., Gutierrez, M., Hidalgo, A., and Menendes, L. (1998). Comparison of the effects of calmidazolium, morphine and bupivacaine on *N*-methyl-D-aspartate- and septide-induced nociceptive behaviour. *Naunyn-Schmiedeberg's Arch. Pharmacol.* **358,** 628–634.

Araque, A., Parpura, V., Sanzgiri, R. P., and Haydon, P. G. (1999). Triplicate synapses: Glia, the unacknowledged partner. *Trends Neurosci.* **22,** 208–215.

Arcaya, J. L., Cano, G., Gómez, G., Maixner, W., and Suárez-Roca, H. (1999). Dynorphine A increases substance P release from trigeminal primary afferent C-fibers. *Eur. J. Pharmacol.* **366,** 27–34.

Arner, S., Rawal, N., and Gustafsson, L. L. (1988). Clinical experience of long-term treatment with epidural opioids—A nationwide survey. *Acta Anaesthesiol. Scand.* **32,** 253–259.

Bartlett, S. E., Dodd, P. R., and Smith, M. T. (1994). Pharmacology of morphine and morphine-3-glucronide at opioid, excitatory amino acis, GABA and glycine binding site. *Pharmacol. Toxicol.* **75,** 73–81.

Berridge, M. J. (1993). Inositol trisphosphate and calcium signaling. *Nature* **361,** 315–325.

Bian, D. S., Ossipov, M. H., Ibrahim, M., Raffa, R. B., Tallarida, R. J., Malan, T. P., Jr., Lai, J., and Porreca, F. (1999). Loss of anti nociceptive spinal/supraspinal morphine synergy in nerve-injured rats; restoration by MK-801 or dynorphine antiserum. *Brain Res.* **831,** 55–63.

Boerner, U., Abbott, S., and Roe, R. L. (1975). The metabolism of morphine and heroin in man. *Drug Metab. Rev.* **4,** 39–73.

Cao, J. L., He, J. H., Ding, H. L., and Zeng, Y. M. (2005). Activation of the ERK signaling pathway contributes naloxone-precipitated withdrawal in morphine-dependent rats. *Pain* **228,** 336–349.

Cao, J. L., Liu, H. L., Wang, J. K., and Zeng, Y. M. (2006). Cross talk between nitric oxide and ERK1/2 signaling pathway in the spinal cord mediates naloxone-precipitated withdrawal in morphine-dependent rats. *Neuropharmacology* **51,** 315–326.

Colburn, R. W., DeLeo, J. A., Rickman, A. J., Yeager, M. P., Kwon, P., and Hickey, W. F. (1997). Dissociation of microglial activation and neuropathic behaviors following peripheral nerve injury in the rat. *J. Neuroimmunol.* **79,** 163–175.

Cui, Y., Chen, Y., Zhi, J. L., Guo, R. X., Feng, J. Q., and Chen, P. X. (2006). Activation of p38 mitogen-activated protein kinase in spinal microglia mediated morphine antinociceptive tolerance. *Brain Res.* **1069,** 235–243.

DeConno, F., Caraceni, A., Martini, C., Spoldi, E., Salvetti, M., and Ventafridda, V. (1991). Hyperalgesia and myoclonus with intrathecal infusion of high-dose morphine. *Pain* **47,** 337–339.

Faden, A. I. (1992). Dynorphine increases extracellular levels of excitatory amino acids in the brain through a non-opioid mechanism. *J. Neurosci.* **12,** 425–429.

Frances, B., Gout, R., Monsarrat, B., Cros, J., and Zajac, J. M. (1992). Further evidence that morphine-6β-glucronide is a more potent opioid antagonist than morphine. *J. Pharmacol. Exp. Ther.* **262,** 25–31.

Fu, K. Y., Light, A. R., Matusshima, G. K., and Maixner, W. (1999). Micloglial reactions after subcutaneous formalin injection into the rat hind paw. *Neuroscience* **825,** 59–67.

Garry, M. G., Richardson, J. D., and Hargreaves, K. M. (1994). Carrageenan induced inflammation alters the content of i-cGMP and i-cAMP in the dorsal horn of the spinal cord. *Brain Res.* **646,** 135–139.

Glavina, M. J., and Robertshaw, R. (1988). Myoclonic spasms following intrathecal morphine. *Anaesthesia* **43,** 389–390.

Guo, W., Wang, H., Watanabe, M., Shimizu, K., Zou, S., LaGraize, S. C., Wei, F., Dubner, R., and Ren, K. (2007). Glial-cytokine-neuronal interactions underlying the mechanisms of persistent pain. *J. Neurosci.* **27,** 6006–6018.

Hains, B. C., and Waxman, S. G. (2006). Activated microglia contribute to the maintenance of chronic pain after spinal cord injury. *J. Neurosci.* **26,** 4308–4317.

Haydon, P. G. (2001). Glia: Listening and talking to synapse. *Nat. Rev. Neurosci.* **2,** 185–193.

Holguin, A., O'sConnor, K. A., Biedenkapp, J., Campisi, J., Wieseler-Frank, J., Milligan, E. D., Hansen, M. K., Spataro, L., Maksimova, E., Bravmann, C., Martin, D., Fleshner, M., *et al.* (2004). HIV-1 gp120 stimulates proinflammatory cytokine-mediated pain facilitation via activation of nitric oxide synthase-I (nNOS). *Pain* **110,** 517–530.

Honore, P., Rogers, S. D., Schwei, M. J., Salak-Johnson, J. L., Luger, N. M., Sabino, M. C., Clohisy, D. R., and Mantyh, P. W. (2000). Murine models of inflammatory, neuropathic and cancer pain each generates a unique set of neurochemical changes in the spinal cord and sensory neurons. *Neuroscience* **98,** 585–598.

Hu, H., Glauner, K. S., and Gereau, R. W. (2003). ERK integrates PKA and PKC signaling in superficial dorsal horn neurons. I: Modulation of A-type K currents. *J. Neurophysiol.* **90,** 1671–1679.

Impey, S., Obrietan, K., Wong, S. T., Poser, S., Yano, S., Wayman, G., Deloulme, J. C., Chan, G., and Storm, D. R. (1998). Cross talk between ERK and PKA is required for Ca^{2+} stimulation of CREB-dependent transcription and ERK nuclear translocation. *Neuron* **21,** 869–883.

Impey, S., Obrietan, K., and Storm, D. R. (1999). Making new concentration: Role of ERK/MAP kinase signaling in neuronal plasticity. *Neuron* **23,** 11–14.

Ji, R. R., and Strichartz, G. (2004). Cell signaling and the genesis of neuropathic pain. *Sci. STKE* **252,** reE14.

Ji, R. R., and Woolf, C. J. (2001). Neuronal plasticity and signal transduction in nociceptive neurons: Implications for the initiation and maintenance of pathological pain. *Neurobiol. Dis.* **8,** 1–10.

Ji, R. R., Baba, H., Brenner, G. J., and Woolf, C. J. (1999). Nociceptive specific activation of ERK in spinal neurons contributes to pain hypersensitivity. *Nat. Neurosci.* **2,** 1114–1119.

Ji, R. R., Befort, K., Brenner, G. J., and Woolf, C. J. (2002). ERK MAP kinase activation in superficial spinal cord neurons induced prodynorphine and NK-1 upregulation and contributes to persistent inflammatory pain hypersensitivity. *J. Neurosci.* **22,** 478–485.

Ji, R. R., Kohno, T., Moore, K. A., and Woolf, C. J. (2003). Central sensitization and longterm potentiation—Do pain and memory share similar mechanism? *Trends Neurosci.* **26,** 696–705.

Jonston, I. N., Milligan, E. D., Wieseler-Frank, J., Frank, M. G., Zapata, V., Campisi, J., Langer, S., Martin, D., Green, P., Fleeshner, M., Leinwand, L., Maier, S. F., *et al.* (2004). A role for proinflammatory cytokines and fractalkine in analgesia, tolerance, and subsequent pain facilitation induced by chronic intrathecal morphine. *J. Neurosci.* **24,** 7353–7365.

Karim, F., Wang, C. C., and Gereau, R. W. (2001). Metabotropic glutamate receptor subtypes 1 and 5 are activators of extracellular signal-regulated kinase signaling required for inflammatory pain in mice. *J. Neurosci.* **21,** 3771–3779.

Kawasaki, Y., Kohno, T., Zhuang, Z. Y., Brenner, G. J., Wang, H., Van Der Meer, C., Befort, K., Woolf, C. J., and Ji, R. R. (2004). Ionotropic and metabotropic receptors, protein kinase A, protein kinase C, and Src contribute to C-fiber-induced ERK activation and cAMP response element-binding protein phosphorylation in dorsal horn neurons, leading to central sensitization. *J. Neurosci.* **24,** 8310–8321.

Koetzner, L., Hua, X. Y., Lai, J., Porreca, F., and Yaksh, T. (2004). Nonopioid actions of intrathecal dynorphine evoke spinal excitatory amino acids and prodstaglandin E2 release mediated by cyclooxygenase-1 and -2. *J. Neurosci.* **24,** 1451–1458.

Komatsu, T., Sakurada, C., Sasaki, M., Sanai, K., Tsuzuki, M., Bagetta, G., Sakurada, S., and Sakurada, T. (2007a). Extracellular signal-regulated kinase (ERK) and nitric oxide synthase mediate intrathecal morphine-induced nociceptive behavior. *Neuropharmacology* **52,** 1237–1243.

Komatsu, T., Akiyoshi, M., Sasaki, M., Sakurada, C., Sakurada, S., and Sakurada, T. (2007b). Mechanism of allodynia in mice evoked by intrathecal morphine-3-glucronide. *J. Pharmacol. Sci.* **103**(Suppl. I), 125.

Komatsu, T., Shiohira, H., Sanai, K., Sakurada, C., Sakurada, S., and Sakurada, T. (2008). Involvemet of spinal cord astroglia in allodynia induced by morphine-3-glucronide. *J. Pharmacol. Sci.* **106** (Suppl. I), 225.

Krames, E. S., Gershow, J., Glassberg, A., Kenefick, T., Lyons, A., Taylor, P., and Wilkie, D. (1985). Continuous infusion of spinally administered narcotics for pain due to malignant disorder. *Cancer* **56,** 696–702.

Laughlin, T. M., Vanderah, T. W., Lashbrook, J. M., Nichols, M. L., Ossipov, M. H., Porreca, F., and Wilcox, G. L. (1997). Spinally administered dynorphin A produces long-lasting allodynia: Involvement of NMDA but not opioid receptors. *Pain* **72,** 253–260.

Lever, I. J., Pezet, S., McMahon, S. B., and Malcangio, M. (2003). The signaling components of sensory fiber transmission involved in the activation of ERK MAP kinase in the mouse dorsal horn. *Mol. Cell. Neurosci.* **24,** 259–270.

Lonze, B. E., and Ginty, D. D. (2002). Function and regulation of CREB family transcription factors in the nervous system. *Neuron* **35,** 605–623.

Loser, S. V., Meyer, J., Freudenthaler, S., Sattler, M., Desel, C., Meineke, I., and Gundert Remy, U. (1996). Morphine-6-*O*-β-D-glucronide but not morphine-3-*O*-β-D-glucronide binds to μ-, δ-, and κ-supecific opioid binding sites in cerebral membranes. *Naunyn-Schmiedeberg's Arch. Pharmacol.* **354,** 192–197.

Ma, W., Zheng, W. H., Powell, K., Jhamandas, K., and Quirion, R. (2001). Chronic morphine exposure increases the phosphorylation of MAP kinases and the transcription factor CREB in dorsal root ganglion neurons; an *in vitro* and *in vivo* study. *Eur. J. Neurosci.* **14,** 1091–1104.

Malan, T. P., Ossipov, M. H., Gardell, L. R., Ibrahim, M., Bian, D., Lai, J., and Porreca, F. (2000). Extraterritorial neuropathic pain correlates with multisegmental elevation of spinal dynorphin in nerve-injured rats. *Pain* **86,** 185–194.

Marchand, F., Perretti, M., and McMahon, S. B. (2005). Role of the immune system in chronic pain. *Nat. Rev. Neurosci.* **6,** 521–532.

McClung, C. A., and Nestler, E. J. (2003). Regulation of gene expression and cocaine reward by CREB and Delta FosB. *Nat. Neurosci.* **6,** 1208–1215.

Meller, S. T., and Gebhart, G. F. (1993). Nitric oxide (NO) and nociceptive processing in the spinal cord. *Pain* **52,** 127–136.

Meller, S. T., Dykstra, C., Grzybycki, D., Murphy, S., and Gebhart, G. F. (1994). The possible role of glia in nociceptive processing and hyperalgesia in the spinal cord of the rat. *Neuropharmacology* **33,** 1471–1478.

Milligan, E. D., O's Connor, K. A., Nguyen, K. T., Armstrong, C. B., Twining, C., Gaykema, R. P., Holguin, A., Martin, D., Maier, S. F., and Watkins, L. R. (2001). Intrathecal HIV-1 envelope glycoprotein gp120 induces enhanced pain states mediated by spinal cord proinflammatory cytokines. *J. Neurosci.* **21,** 2808–2819.

Morita, K., Kitayama, T., Morioka, N., and Dohi, T. (2008). Glycinergic mediation of tactile allodynia induced by platelet-activating factor (PAF) through glutamate-NO-cyclic GMP signalling in spinal cord in mice. *Pain* **138,** 525–536.

Morozov, A., Muzzio, I. A., Bourtchouladze, R., Van-strien, N., Lapidus, K., Yin, D., Winder, D. G., Adams, J. P., Sweatt, J. D., and Kandel, E. R. (2003). Rap1 couples camp signaling to a distinct pool of p42/44MAPK regulating excitability, synaptic plasticity, learning, and memory. *Neuron* **39,** 309–325.

Obrietan, K., Impey, S., Smith, D., Athos, J., and Storm, D. R. (1999). Circadian regulation of cAMP response element-mediated gene expression in the suparachiasmatic nuclei. *J. Biol. Chem.* **274,** 17748–17756.

Osborne, P. B., Chieng, B., and Christie, M. J. (2000). Morphine-6β-glucronide has a higher efficacy than morphine as mu-opioid receptor agonist in the rat locus coeruleus. *Br. J. Pharmacol.* **131,** 1422–1428.

Parkinson, S. K., Bailey, S. L., Little, W. L., and Mueller, J. B. (1990). Myoclonic seizure activity with chronic high-dose spinal opioid administration. *Anesthesiology* **72,** 743–745.

Pasternak, G. W., Bodnar, R. J., Clark, J. A., and Intrurrisi, C. E. (1987). Morphine-6-glucuronide, a potent mu agonist. *Life Sci.* **41,** 2845–2849.

Paul, D., Standifer, K. M., Intrurrisi, C. E., and Pasternak, G. W. (1989). Pharmacological characterization of morphine-6-glucuronide, a very potent morphine metabolite. *J. Pharmacol. Exp. Ther.* **251,** 477–483.

Penn, R., and Paice, J. (1987). Chronic intrathecal morphine for intractable pain. *J. Neurosurg.* **67,** 182–186.

Potter, J. M., Reid, D. B., Shaw, R. J., Hackett, P., and Hickmann, P. E. (1989). Myoclonus associated with treatment with high doses of morphine: The role of supplemental drugs. *Br. Med. J.* **299,** 150–153.

Qian, Y., Chao, D. S., Santillano, D. R., Cornwell, T. L., Narin, A. C., Greengard, P., Lincoln, T. M., and Bredt, D. S. (1996). cGMP-dependent protein kinase in dorsal root ganglion: Relationship with nitric oxide synthase and nociceptive neurons. *J. Neurosci.* **16,** 3130–3138.

Raghavendra, V., Tanga, F. Y., and DeLeo, J. A. (2002). The role of spinal neuroimmune activation in morphine tolerance/hyperalgesia in neuropathic and sham-operated rats. *J. Neurosci.* **22,** 9980–9989.

Raghavendra, V., Tanga, F. Y., and DeLeo, J. A. (2003). Inhibition of micloglia activation attenuates the development but not existing hypersensitivity in a rat model of neuropathy. *J. Pharmacol. Exp. Ther.* **306,** 624–630.

Raghavendra, V., Tanga, F. Y., and DeLeo, J. A. (2004). Complete Freunds adjuvant-induced peripheral inflammation evokes glial activation and proinflammatory cytokine expression in the CNS. *Eur. J. Neurosci.* **20,** 467–473.

Sakurada, T., Wako, K., Sakurada, C., Manome, Y., Tan-No, K., Sakurada, S., and Kisara, K. (1996a). Spinally-mediated behavioral responses evoked by intrathecal high-dose morphine: Possible involvement of substance P in the mouse spinal cord. *Brain Res.* **724,** 213–221.

Sakurada, T., Sugiyama, A., Sakurada, C., Tan-No, K., Sakurada, S., Kisara, K., Hara, A., and Abiko, Y. (1996b). Involvement of nitric oxide in spinally mediated capsaicin- and glutamate-induced behavioural responses in the mouse. *Neurchem. Int.* **29,** 271–278.

Sakurada, C., Sugiyama, A., Nakayama, M., Yonezawa, A., Sakurada, S., Tan-No, K., Kisara, K., and Sakurada, S. (2001). Antinociceptive effect of spinally injected L-NAME on the acute nociceptive response induced by low concentrations of formalin. *Neurochem. Int.* **38,** 417–423.

Sakurada, T., Watanabe, C., Okuda, K., Sugiyama, A., Moriyama, T., Sakurada, C., Tan-No, K., and Sakurada, S. (2002). Intrathecal high-dose morphine induces spinally-mediated behavioral responses through NMDA receptors. *Mol. Brain Res.* **98,** 111–118.

Samad, T. A., Moore, K. A., Sapirstein, A., Billet, S., Allchorne, A., Poole, S., Bonventre, J. V., and Woolf, C. J. (2001). Interleukin-1 beta-mediated induction of cox-2 in the CNS contributes to inflammatory pain hypersensitivity. *Nature* **410,** 471–475.

Schmidtko, A., Ruth, P., Geisslinger, G., and Tegeder, I. (2003). Inhibition of cyclic guanosine 5′-monophosphate-dependent protein kinase I (PKG-I) in lumbar spinal cord reduces formalin-induced hyperalgesia and PKG upregulation. *Nitric Oxide* **8,** 89–94.

Scholz, J., and Woolf, C. J. (2007). The neuropathic pain triad: Neurons, immune cells and glia. *Nat. Neurosci.* **10,** 1361–1368.

Siegan, J. B., Hama, A. T., and Sagen, J. (1996). Alterations in rat spinal cord cGMP by peripheral nerve injury and adrenal medullary transplantation. *Neurosci. Lett.* **215,** 49–52.

Sjogren, P., Jonsson, T., Jensen, N. H., Drenck, N. E., and Jensen, T. S. (1993). Hyperalgesia and myoclonus in terminal cancer patients treated with continuous intravenous morphine. *Pain* **55,** 93–97.

Skilling, S. R., Sun, X., Kurtz, H. J., and Larson, A. A. (1992). Selective potentiation of NMDA-induced activity and release of excitatory amino acids by dynorphine: Possible roles in paralysis and neurotoxicity. *Brain Res.* **575,** 272–278.

Smith, M. T. (2000). Neuroexcitatory effects of morphine and hydromorphine: Evidence implicating the 3-glucuronide metabolites. *Clin. Exp. Pharmacol. Physiol.* **27,** 524–528.

Sung, C. S., Wen, Z. H., Chang, W. K., Ho, S. T., Tsai, S. K., Chang, Y. C., and Wong, C. S. (2004). Intrathecal interleukin-1beta administration induces themal hyperalgesia by activating inducible nitric oxide synthase expression in the rat spinal cord. *Brain Res.* **1051,** 145–153.

Sweitzer, S., Martin, D., and DeLeo, J. A. (2001). Intrathecal interleukin-1 receptor antagonist in combunation with soluble tumor necrosis factor receptor exhibits an anti-allodynic action in a rat model of neuropathic pain. *Neuroscience* **103,** 529–539.

Tai, Y. H., Wang, Y. H., Wang, J. J., Tao, P. L., Tung, C. S., and Wong, C. S. (2006). Amitriptyline suppresses neuroinflammation and up regulates glutamate transporters in morphine-tolerant rats. *Pain* **124,** 77–86.

Tang, A. H., and Schoenfeld, H. J. (1978). Comparison of subcutaneous and spinal subarachnoid injections of morphine and naloxone in analgesic tests in rat. *Eur. J. Pharmacol.* **52,** 215–223.

Tao, Y. X., Hassan, A., Haddad, E., and Johns, R. A. (2000). Expression and action of cyclic GMP-dependent protein kinase Ialpha in inflammatory hyperalgesia in rat spinal cord. *Neuroscience* **95,** 525–533.

Tegeder, I., Schmidtko, A., Niederberger, E., Ruth, P., and Geisslinger, G. (2002). Dual effects of spinally delivered 8-bromo-cyclic guanosine mono-phosphate (8-bromo-cGMP) in formalin-induced nociception in rats. *Neurosci. Lett.* **332,** 146–150.

Tegeder, I., Del Turco, D., Schmidtko, A., Sausbier, M., Feil, R., Hofmann, F., Deller, T., Ruth, P., and Geisslinger, G. (2004). Reduced inflammatory hyperalgesia with preservation of acute thermal nociception in mice lacking cGMP-dependent protein kinase I. *Proc. Natl. Acad. Sci. USA* **101,** 3253–3257.

Tsuda, M., Inoue, K., and Salter, M. W. (2005). Neuropathic pain and spinal microglia: A big problem from molecules in "small" glia. *Trends Neurosci.* **28,** 101–107.

Vanderah, T. W., Laughlin, T. M., Lashbrook, J. M., Nichols, M. L., Wilcox, G. L., Ossipov, M. H., Malan, T. P., and Porreca, F. (1996). Single intrathecal injections of dynorphin A or des-Tyr-dynorphins produce long-lasting allodynia on rats: Blockade by MK-801 but not naloxone. *Pain* **68,** 275–281.

Vanderah, T. W., Gardell, L. R., Birgess, S. E., Ibrahim, M., Dogrul, A., Zhong, C. M., Zhang, E. T., Malan, T. P., Jr., Ossipov, M. H., Lai, J., and Porreca, F. (2000). Dynorphine promotes abnormal pain and spinal opioid antinociceptive tolerance. *J. Neurosci.* **20,** 7074–7079.

Wagner, R., and Deleo, J. A. (1996). Pre-emptive dynorphine and *N*-methyl-D-asparatate glutamate receptor antagonism alters spinal immunocytochemistry but not allodynia following complete peripheral nerve injury. *Neurocsience* **72,** 527–534.

Wang, Z., Gardell, L. R., Ossipov, M. H., Vanderah, T. W., Brennan, M. B., Hochgeschwender, U., Hruby, V. J., Malan, T. P., Lai, J., and Porreca, F. (2001). Pronociceptive actions of dynorphin maintain chronic neuropathic pain. *J. Neurosci.* **21,** 1779–1786.

Watanabe, C., Sakurada, T., Okuda, K., Sakurada, C., Ando, R., and Sakurada, S. (2003a). The role of spinal nitric oxide and glutamate in nociceptive behaviour evoked by high-dose intrathecal morphine in rats. *Pain* **106,** 269–283.

Watanabe, C., Okuda, K., Sakurada, C., Ando, R., Sakurada, T., and Sakurada, S. (2003b). Evidence that nitric oxide-glutamate cascade modulates spinal antinociceptive effect of morphine: A behavioural and microdialysis study in rats. *Brain Res.* **990,** 77–86.

Watkins, L. R., and Maier, S. F. (2003). Glia: Novel drug discovery target for clinical pain. *Nat. Rev. Drug Discov.* **2,** 973–985.

Watkins, L. R., Milligan, E. D., and Maier, S. F. (2001). Glial activation: A driving force for pathological pain. *Trends Neurosci.* **24,** 450–455.

Watkins, L. R., Milligan, E. D., and Maier, S. F. (2003). Glial proinflammatory cytokines mediate exaggerated pain states: Implications for clinical pain. *Adv. Exp. Med. Biol.* **521,** 1–21.

Wert, M. A., and MacDonald, R. L. (1982). Opiate alkaloids antagonize postsynaptic glycine and GABA responses: Correlation with convulsant action. *Brain Res.* **236,** 107–119.

Winkelstein, B. A., Rutkowski, M. D., Sweitzer, S. M., Pahl, J. L., and Deleo, J. A. (2001). Nerve injury proximal or distal to the DRG induces similar spinal glial activation and selective cytokine expression but differential behavioral responses to pharmacologic treatment. *J. Comp Neurol.* **439,** 127–139.

Woolf, F. J. (1981). Intrathecal high dose morphine produces hyperalgesia in the rat. *Brain Res.* **209,** 491–495.

Wu, J., Lin, Q., McAdoo, D. J., and Willis, W. D. (1998). Nitric oxide contributes to central sensitization following intradermal injection of capsaicin. *NeuroReport* **9,** 589–592.

Yaksh, T. L., and Harty, G. J. (1988). Parmacology of the allodynia in rats evoked by high dose intrathecal morphine. *J. Pharmacol. Exp. Ther.* **244,** 501–507.

Yaksh, T. L., Harty, G. J., Burton, M., and Onofrio, B. M. (1986). High doses of spinal morphine produce a nonopiate receptor-mediated hyperesthesia: Clinical and theoretical implications. *Anesthesiology* **64,** 590–597.

Yamada, H., Ishii, K., Ishii, Y., Ieiri, I., Nishio, S., Morioka, T., and Oguri, K. (2003). Formation of highly analgesic morphine-6-gluclonide following physiologic concentration of morphine in human brain. *J. Toxicol. Sci.* **28,** 395–401.

Yonehara, N., Kudo, C., and Kamisaki, Y. (2003). Involvement of NMDA-nitric oxide pathways in the development of tactile hypersensitivity evoked by the loose-ligation of inferior alveolar nerves in rats. *Brain Res.* **963,** 232–243.

Zhang, R. X., Li, S., Liu, B., Wang, L., Ren, K., Zhang, H., Berman, B. M., and Lao, L. (2008). IL-1ra alleviates inflammatory hyperalgesia through preventing phosphorylation of NMDA receptor NR-1 subunit in rats. *Pain* **135,** 232–239.

Zhu, J. J., Qin, Y., Zhao, M., Van Aelst, L., and Malinow, R. (2002). Ras and Rap control AMPA receptor trafficking during synaptic plasticity. *Cell* **110,** 443–455.

(−)-LINALOOL ATTENUATES ALLODYNIA IN NEUROPATHIC PAIN INDUCED BY SPINAL NERVE LIGATION IN C57/BL6 MICE

Laura Berliocchi,* Rossella Russo,[†] Alessandra Levato,* Vincenza Fratto,* Giacinto Bagetta,[†,‡] Shinobu Sakurada,[§] Tsukasa Sakurada,[¶] Nicola Biagio Mercuri,[||] and Maria Tiziana Corasaniti*

*Department of Pharmacobiological Sciences, University of Catanzaro "Magna Graecia", Catanzaro 88100, Italy

[†]Department of Pharmacobiology, University of Calabria, Arcavacata di Rende, Cosenza 87036, Italy

[‡]University Centre for Adaptive Disorders and Headache (UCADH), Section of Neuropharmacology of Normal and Pathological Neuronal Plasticity, University of Calabria, Arcavacata di Rende, Cosenza 87036, Italy

[§]Department of Anatomy and Physiology, Tohoku Pharmaceutical University, Sendai, Japan

[¶]First Department of Pharmacology, Daiichi College of Pharmaceutical Sciences, Fukuoka, Japan

[||]Department of Neuroscience, University "Tor Vergata", and Laboratory of Experimental Neurology, CERC-IRCCS Santa Lucia, Rome, Italy

(−)-Linalool is a natural compound with anti-inflammatory and antinociceptive properties. The antinociceptive action of linalool has been reported in several models of inflammatory pain. However, its effects in neuropathic pain have not been investigated. Here, we used the spinal nerve ligation (SNL) model of

INTERNATIONAL REVIEW OF NEUROBIOLOGY, VOL. 85
DOI: 10.1016/S0074-7742(09)85017-4

neuropathic pain and studied the effects of acute and chronic administration of an established antinociceptive dose of linalool on mechanical and thermal sensitivity induced by the nerve injury in mice. Linalool did not affect pain behavior triggered by mechanical or thermal stimuli when administered as a single dose before SNL. However, mechanical allodynia was reduced transiently in neuropathic animals when linalool was administered for 7 consecutive days, while no changes were seen in the sensitivity to noxious radiant heat. We investigated the possible involvement of the PI3K/Akt pathway in linalool antinociceptive effect by western blot analysis. Linalool did not induce significant changes in Akt expression and phopshorylation though a trend toward an increased ratio of phosphorylated versus total Akt was observed in SNL animals treated with linalool, in comparison to SNL alone or sham. We then wondered whether linalool could modulate inflammatory processes and investigated spinal glia activation and IL-1β contents following linalool treatment in SNL animals. The data suggest that mechanisms other than an action on inflammatory processes may mediate linalool ability to reduce mechanical allodynia in this model of neuropathic pain.

I. Introduction

(−)-Linalool is the natural occurring enantiomer monoterpene compound commonly found as a major volatile component in the essential oils of several aromatic plants such as bergamot, jasmine, and lavender. This compound showed anti-inflammatory and antinociceptive effects in several models of inflammatory pain (Peana *et al.*, 2002, 2003). Linalool administration in rats inhibited carrageenan-induced edema (Peana *et al.*, 2002) and decreased thermal hyperalgesia induced by carrageenan, L-glutamate and prostaglandin E_2 (Peana *et al.*, 2004b). Also, linalool has been shown to attenuate different pain responses elicited either by a chemical noxious stimulus, such as the acetic acid, or by a thermal noxious stimulus, as well as in tissue injury produced by formalin injection (Peana *et al.*, 2003, 2004a). Although the antinociceptive effects of linalool were initially ascribed to its anti-inflammatory properties, recent evidence suggests that more complex mechanisms may be involved. For instance, muscarinic, opioid, and dopamine receptors antagonists have been shown to reduce the antinociceptive action of linalool, thus suggesting an action of the monoterpene on different neurotransmission systems (Peana *et al.*, 2003, 2004a). Moreover, linalool has been shown to act, possibly indirectly, on K^+ channels (Peana *et al.*, 2004a) and to block NMDA receptors activity (Brum *et al.*, 2001; Elisabetsky *et al.*, 1999; Silva Brum *et al.*, 2001). Also adenosine (A) receptors have been suggested to play a role in the antinociceptive mechanisms of linalool, since A receptor antagonists reduced linalool antinociceptive effects in the hot plate test (Peana *et al.*, 2006).

While the antinociceptive effect exerted by linalool has been extensively investigated in inflammatory pain, no studies exist on the effects of linalool in models of neuropathic pain. To this aim, we used the spinal nerve ligation (SNL) model of neuropathic pain (Kim and Chung, 1992) and studied the effects of acute and chronic administration of (−)-linalool on mechanical and thermal hypersensitivity induced by nerve injury in mice.

Linalool did not show significant effect on pain behavior triggered by mechanical or thermal stimulation when administered as a single dose before SNL. However, mechanical allodynia was reduced transiently in neuropathic animals when linalool was administered for 7 consecutive days. The latter treatment showed no significant effect on hypersensitivity to noxious radiant heat. Linalool did not induce significant changes in Akt expression and phopshorylation though an increased ratio of phosphorylated versus total Akt was observed in SNL animals treated with linalool in comparison to SNL alone or sham. Finally, analysis of spinal IL-1β contents following SNL and in the presence of linalool revealed a modulatory action on the cytokine levels though this was not statistically significant.

II. Materials and Methods

A. Animals

Male C57/BL6 mice (20–22 g) (Charles River, Italy) were housed in a temperature (22 °C)- and humidity (65%)-controlled room and were maintained under a 12 h light/dark cycle (lights on from 7:00 a.m. to 7:00 p.m.), with *ad libitum* access to food and water. The experimental protocol was in accordance to the guidelines of the Ministry of Health for animal care (D.M. 116/1992).

B. Drugs and Experimental Design

(−)-Linalool (Sigma-Aldrich, Milan, Italy) was dissolved in a mixture of polyethylene glycol 200 (Sigma-Aldrich, Milan, Italy) and NaCl (0.9%) saline solution (1:1) and an antinociceptive dose (100 mg/kg; volume of injection 116 μl/kg), established in inflammatory models of pain (see Peana *et al.*, 2002, 2003), was administered by subcutaneous injection (s.c.) 1 h before SNL. An identical volume of vehicle (116 μl/kg, s.c.) was administered to control animals.

Animals (SNL alone, SNL + linalool, SNL + vehicle) were divided in three experimental groups according to the duration of the treatment: group 1 received one single administration of either linalool or vehicle 1 h before ligation, groups 2 and 3 received a daily administration from day 1 to day 7 and day 14, respectively ($n = 10$ for each group treated with linalool, $n = 5$ for each group treated with vehicle, $n = 10$ for each SNL group, $n = 3$ for each sham group).

C. Surgical Procedures for L5 Spinal Nerve Ligation

C57/BL6 mice (20–22 g) were anesthetized using 2% isoflurane. A midline incision was made in the skin of the back at the L2–S2 levels and the left paraspinal muscles separated from the spinal processes, facet joints and transverse processes at the L4–S1 levels. The left L5 spinal nerve was isolated and tightly ligated with 6–0 silk thread (Kim and Chung, 1992). Finally, the wound was sutured. The surgical procedure for the sham-operated group was identical to the SNL group, except that the L5 nerve was not ligated. After surgery, foot posture and general mice behavior were monitored throughout the postoperative period. Behavioral testing was carried out over a 4 weeks period.

D. Behavioral Analysis

Mice were acclimatized in the test chambers for at least 1 h until the exploratory activity had ceased. Behavioral tests were carried out before surgery (baseline) and 1, 3, 7, 10, 14, 17, 21, 24, and 28 days after surgery. Mechanical and thermal sensitivity were evaluated by Von Frey (Chaplan *et al.*, 1994) and the Hargreaves test (Hargreaves *et al.*, 1988), respectively. A series of calibrated Von Frey filaments (0.04–12.75 g) were applied perpendicular to the plantar surface of the left paw. The threshold was determined by using the up–down method as previously described (Chaplan *et al.*, 1994). For the Hargreaves test, the latency of withdrawal to heat was recorded 4–6 times on each paw, with at least 1 min between recordings on the same paw. A cutoff of 20 s was not exceeded to avoid tissue injury. For each group data were expressed as mean of the 50% pain threshold (Von Frey test) or withdrawal latency (Hargreaves test) ±SEM.

E. Western Blotting

Mice ($n = 3$ for each group) were killed 7 days after surgery and the spinal lumbar segment (L4–L5) rapidly removed. Dorsal ipsi- and contra-lateral sides of each spinal cord were dissected and pools of three animals homogenized in a glass homogenizer using 90 μl of ice-cold lysis buffer containing 50 mM Tris–HCl (pH 7.5), 150 mM NaCl, 2 mM EDTA, 2 mM EGTA, 1% Triton, 1 nM okadaic acid, a cocktail of protease inhibitors (code P8340, Sigma, Milan, Italy), and a cocktail of phosphatase inhibitors (code 524625, Calbiochem). Samples were then centrifuged at 10,000*g* for 15 min at 4 °C. Protein concentration was determined in the supernatants by the DC protein assay (Bio-Rad Laboratories, Milan, Italy).

Proteins were resolved by sodium dodecyl sulfate-polyacrylamide gel electrophoresis (SDS-PAGE; 8%) and electrotransferred onto nitrocellulose membranes (Optitran BA-S 83, Schleicher & Schuell Bioscence, Dassel, Germany). Primary antibodies were incubated overnight at 4 °C followed by a horseradish peroxidase-conjugated secondary antibody for 1 h at room temperature. Immunoreactivity was visualized by enhanced chemiluminescent detection (Amersham Biosciences, GE Healthcare, Milan, Italy) and exposure to X-ray films (Hyperfilm ECL, Amersham Biosciences). Autoradiographic films were scanned and densitometric analysis was carried out using Quantiscan software (Biosoft, Cambridge, UK). The following primary antibodies were used: a rabbit polyclonal antibody for p-Akt (Ser473) (1:1000; Cell Signaling Technology, Beverly, MA, USA), a rabbit polyclonal antibody for Akt (1:4000; Cell Signaling Technology), a rabbit anti-GFAP polyclonal antibody (1:1000; Chemicon International), a mouse monoclonal anti-actin antibody (1:1000; clone AC-40 Sigma), and a mouse monoclonal anti-GAPDH antibody (1:4000; clone 6C5 Ambion). Horseradish peroxidase-conjugated goat anti-rabbit or anti-mouse IgG (Pierce Biotechnology, Rockford, IL, USA) were used as secondary antibodies. Densitometry data are expressed as the mean ± SEM of three independent experiments.

F. IL-1β Assay

Lumbar spinal cord sections were homogenized individually in ice-cold phosphate buffer saline (PBS) containing a cocktail of protease inhibitors (Sigma, code P8340) and centrifuged at 14,000*g* for 10 min at 4 °C. The supernatant was then collected and protein concentration determined using the Bio-Rad DC protein assay (Bio-Rad Laboratories, Milan, Italy).

Immunoreactive IL-1β levels were analyzed by a mouse specific, sandwich ELISA. The immunoaffinity-purified polyclonal sheep anti-mouse IL-1β coating antibody (S5/150799/JW, 1 μg/ml), the biotinylated, immunoaffinity-purified, polyclonal sheep anti-mouse IL-1β detecting antibody (329/010800/SP, 1:500 dilution), as well as the recombinant mouse IL-1β standard were kindly provided by Dr Stephen Poole (National Institute of Biological Standards and Controls (NIBSC), Hertfordshire, UK). Poly-horseradish peroxidase-conjugated streptavidin (Sanquin, Amsterdam, The Netherlands) was used at 1:5000 dilution and the color was developed by using the chromogen *o*-phenylenediamine (Sigma).

Optical densities (OD) were read at 492 nm using an automated plate reader (Multiscan MS; Labsystems, Helsinki, Finland) and cytokine levels calculated by interpolation from the standard curve (0.0–64 pg/ml). Data were corrected for protein concentration and the results expressed as pg of IL-1β per mg of protein.

G. Statistical Analysis

All data were presented as mean ± SEM. The data were evaluated statistically for differences by one-way analysis of variance (ANOVA) followed by post hoc Tukey test for multiple comparisons. $p < 0.05$ was regarded as significant. Calculations were done using GraphPad Prism.

III. Results

A. Linalool Transiently Reduced SNL-Induced Mechanical Allodynia

Ligation of the L5 spinal nerve produced an early onset and long lasting mechanical allodynia (Fig. 1) as previously described (Kim and Chung, 1992; Liu *et al.*, 2000). Tactile allodynia determined by Von Frey filaments developed 1 day after SNL and was maintained for up to 28 days (Fig. 1). Systemic administration of a single dose of linalool (100 mg/kg s.c. given 1 h before surgery) produced no significant changes on mechanical allodynia (Fig. 1A). However, repeated administration of linalool (100 mg/kg given s.c. 1 h before surgery and then 1 h before behavioral testing) for 7 consecutive days attenuated hypersensitivity to tactile stimulation (Fig. 1B). Attenuation of mechanical allodynia was observed starting from day 3 and reached statistical significance at day 7 (Fig. 1B). The reduction was transient as hypersensitivity to mechanical stimulation then reverted to values similar to SNL animals (Fig. 1B). A prolonged treatment with linalool for up to 14 consecutive days showed similar results and did not ameliorate further hypersensitivity induced by SNL (data not shown). The anti-allodynic effect was specific for linalool as it was not shown by its vehicle when given alone (Fig. 1B).

Sensitivity to noxious radiant heat, measured as withdrawal latency to heat using the Hargreaves test, developed 1 day after SNL and was maintained for up to 28 days (Fig. 2). A single dose of linalool (100 mg/kg s.c. given 1 h before surgery) produced no major behavioral changes over a period of 28 days (Fig. 2A). Similarly, the paw withdrawal latencies to radiant heat applied to the plantar surface of the hind paw did not change over the same period following repeated linalool administration (100 mg/kg given s.c. 1 h before surgery and then daily 1 h before behavioral testing for 7 consecutive days; Fig. 2B). Also a prolonged

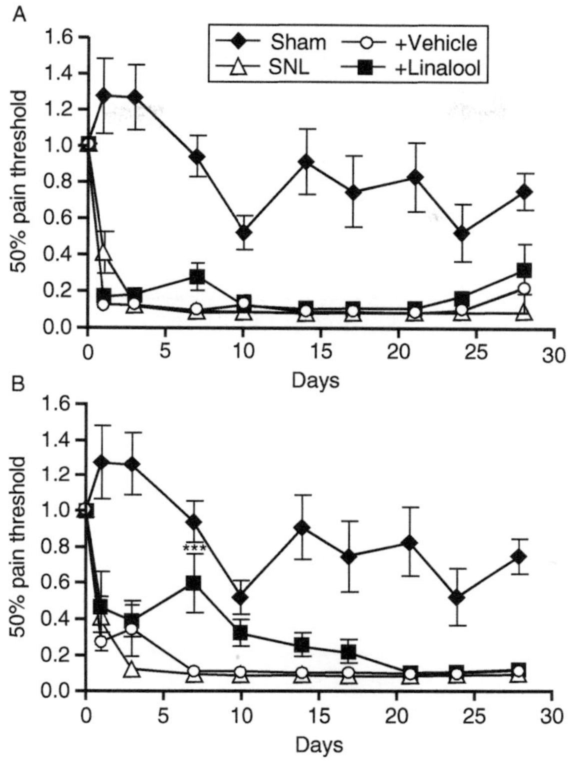

FIG. 1. (−)-Linalool attenuates mechanical allodynia induced by spinal nerve ligation in mice. (A and B) Mechanical allodynia developed and maintained over time following spinal nerve ligation (SNL). (A) A single dose of linalool (100 mg/kg s.c.) did not cause any significant changes compared to SNL and vehicle-treated animals. (B) Linalool administered daily for 7 days attenuated mechanical allodynia compared to SNL animals and SNL animals treated with the vehicle (***$p < 0.001$ vs vehicle; ANOVA+Tukey test). Data are expressed as mean ± SEM of the value corresponding to 50% of pain threshold and are normalized to the basal value of each animal. Differences are evaluated using one-way analysis of variance (ANOVA), followed by post hoc Tukey multiple comparison tests. $p < 0.05$ was regarded as significant.

treatment with linalool for up to 14 consecutive days showed similar results and did not affect sensitivity induced by SNL (data not shown).

B. Linalool Does Not Affect Akt Expression in the Spinal Cord Following SNL

Recently, the phosphatidylinositol 3-kinase (PI3K)-Akt signaling pathway has been implicated in hypersensitivity induced by SNL (Xu *et al.*, 2007). We investigated the possible involvement of this pathway in the antinociceptive effects of

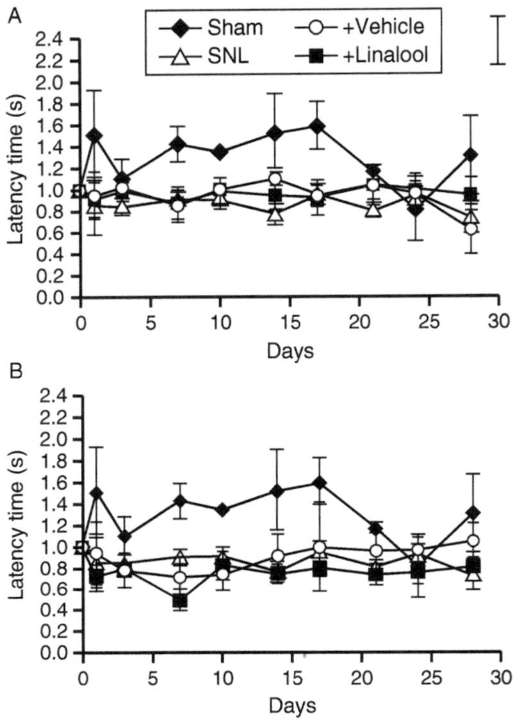

FIG. 2. (−)-Linalool does not affect thermal sensitivity induced by spinal nerve ligation in mice. (A), (B) Thermal sensitivity developed and maintained over time following spinal nerve ligation (SNL). SNL animals treated either with (A) a single dose of linalool (100 mg/kg s.c.) or (B) a daily dose for 7 days did not show significant difference in withdrawal latency to heat compared to SNL animals and SNL animals treated with the vehicle. Data are expressed as mean ± SEM of the value corresponding to 50% of pain threshold and are normalized to the basal value of each animal. Differences are evaluated using one-way ANOVA, followed by post hoc Tukey multiple comparison tests. $p < 0.05$ was regarded as significant.

linalool by western blot analysis of total and phosphorylated Akt (Ser473). Akt expression was studied in the spinal L4–L5 segment of animals undergone SNL either alone and in combination with vehicle or linalool treatment for 7 consecutive days (100 mg/kg given s.c. 1 h before surgery and then daily; Fig. 3). In contrast to previous work (Xu *et al.*, 2007), no significant changes in Akt and p-Akt expression were detected at day 7 following SNL (Fig. 3A). Similarly, levels of Akt and p-Akt were not appreciably altered by treatment with linalool or its vehicle in SNL animals (Fig. 3A). However, densitometric analysis showed a trend toward an increased ratio of phosphorylated versus total Akt in SNL animals treated with linalool, in comparison to SNL alone or sham (Fig. 3B). This seems to be due to a

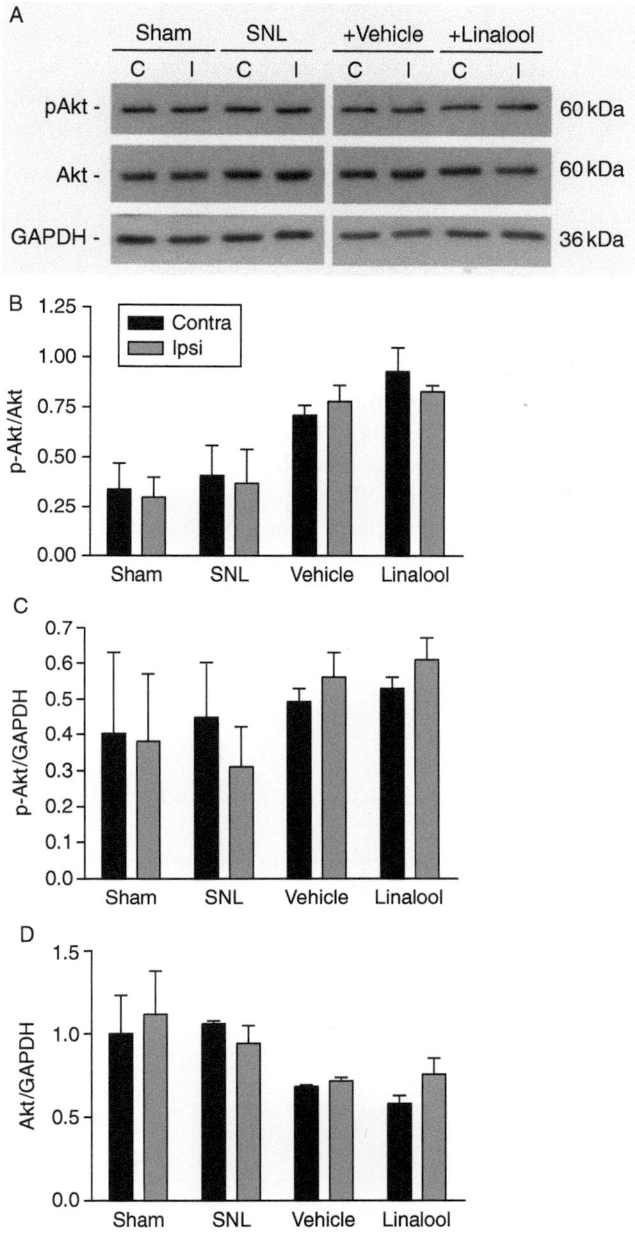

FIG. 3. (−)-Linalool does not affect the expression of p-Akt and Akt in the spinal cord of neuropathic mice. (A) Western blot analysis of lumbar spinal cord (L4–L5 segment) 7 days after SNL alone or in combination with linalool (100 mg/kg s.c.) or vehicle treatment. Homogenates were from the ipsi(I)- and contra(C)-lateral side of the cord (pool of three animals for each experimental group). (B–D) Densitometric analysis of immunoblots probed with anti-Akt, anti-phospho-Akt (Ser473), and anti-GAPDH antibodies from three different experiments (mean ± SEM).

slight increase in p-Akt (Fig. 3C) and to a more pronounced decrease in total levels of Akt in SNL animals treated with linalool, in comparison to sham and SNL (Fig. 3D). However, this trend was present also in vehicle-treated animals. GAPDH was used as loading control.

C. Effect of (−)-Linalool on Spinal GFAP Expression

Nerve injury and inflammation have been shown to induce glial cells activation in the spinal dorsal horn (Hashizume *et al.*, 2000; Watkins *et al.*, 1997). This event seems to participate in the process of central sensitization and contribute to tactile allodynia (Meller *et al.*, 1994). To investigate whether the reported anti-inflammatory proprieties of linalool could be responsible for its ability to reduce tactile hypersensitivity following SNL, expression of glial fibrillary acidic protein (GFAP), an indicator of reactive gliosis after CNS injury (Kajander *et al.*, 1990; Ma and Quirion, 2002) was analyzed by western blot in the spinal dorsal horn (Fig. 4). GFAP expression was markedly reduced by linalool treatment in SNL animals. However, this effect was even more dramatic in the presence of the vehicle alone (Fig. 4) thus suggesting that the vehicle itself may have strong anti-inflammatory proprieties that can mask those of linalool.

D. Modulation of Spinal IL-1β Content by Linalool Following SNL

Glial cells are a source of multiple cytokines, including interleukin-1β (IL-1β), IL-6, and tumor necrosis factor-α (Hanisch, 2002). These cytokines can contribute to different features of pathological pain, although their role within the spinal cord has not been completely understood (DeLeo and Yezierski, 2001; Watkins

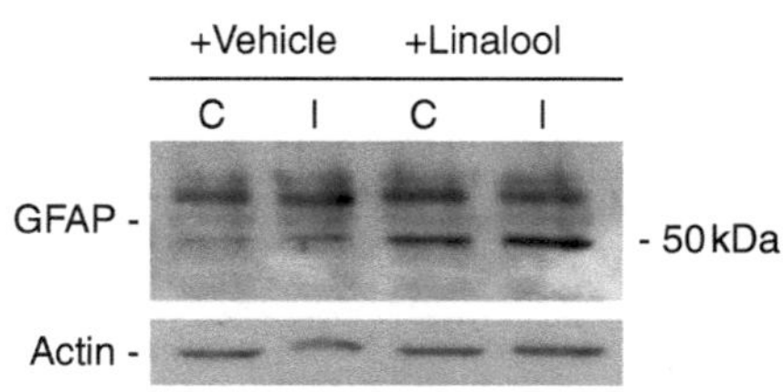

Fig. 4. Effect of (−)-linalool on GFAP expression in the spinal cord of neuropathic mice. GFAP expression was analyzed by western blot in the lumbar spinal cord (L4–L5 segment). Homogenates were from the ipsi(I)- and contra(C)-lateral side of the cord from SNL animals treated either with linalool (100 mg/kg s.c.) or vehicle for 7 days (pool of three animals for each experimental group). Actin was used as loading control.

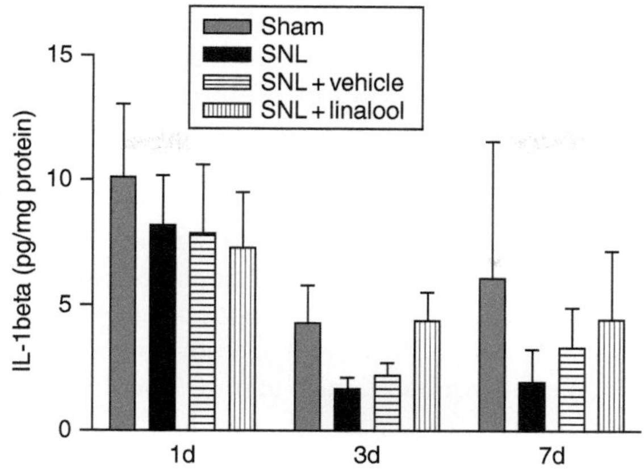

FIG. 5. Modulation of IL-1β spinal content by spinal nerve ligation and (–)-linalool. IL-1β content was determined by ELISA in lumbar spinal cord homogenates. Animals were sacrificed at different time points after SNL alone or in combination with linalool (100 mg/kg s.c.) and vehicle treatment (mean $\pm$ SEM from $n = 2-3$ for each experimental group).

et al., 2001a). Here, we investigated IL-1β levels over time in the spinal cord of animals undergone SNL, in the absence and presence of linalool. No statistically significant differences were seen between experimental groups (Fig. 5). However, IL-1β content decreased in animals undergone SNL when compared to sham animals at days 3 and 7. A trend toward an increase in IL-1β levels was observed at the same time points in SNL animals treated with linalool, in comparison to animals that only had nerve injury. However, at day 7 a similar trend was also present in SNL animals that received the vehicle (Fig. 5).

IV. Discussion

Systemic administration of linalool showed antinociceptive effects in several models of inflammatory pain (Peana *et al.*, 2003, 2004b). Here, we show for the first time that linalool is able to reduce mechanical allodynia in mice following SNL, a widely studied and recognized model of neuropathic pain.

While a single administration had no effect on pain behavior measured by sensitivity to mechanical stimulation (Fig. 1A), repeated treatment with linalool, 1 h before L5 ligation and then daily for 7 days, attenuated tactile allodynia reaching statistical significance at day 7 (Fig. 1B). A prolonged treatment with linalool for up to 14 consecutive days showed similar results and did not ameliorate further hypersensitivity induced by SNL (data not shown). This suggests that linalool may reduce mechanical hypersensitivity only in the early phase of development of central sensitization, but it is not able to attenuate the pain-like behavior during the following period of maintenance. Linalool was able to ameliorate nerve injury-induced tactile hypersensitivity, although transiently, but did not affect SNL-induced thermal sensitivity (Fig. 2A and B).

Allodynia, which is the prominent feature in L5 spinal nerve ligated mice, is thought to be the behavioral expression of pain generated by low threshold A-fibers (Ziegler *et al.*, 1999), but also of the spontaneous activity of uninjured C-fiber nociceptors after injury to the neighboring nerves (Wu *et al.*, 2001). Continuous activation of peripheral afferent fibers by noxious stimulation induces the release of excitatory neurotransmitters into the dorsal horn, and these may elicit pain hypersensitivity or central sensitization. For instance, activation of the NMDA receptor in the spinal dorsal horn has been shown to be essential for the development of central sensitization and, thereby, mechanical allodynia (Bursztajn *et al.*, 2004; Gao *et al.*, 2005; Ultenius *et al.*, 2006). Several signaling pathways and second messengers mediate central sensitization triggered by the activation of postsynaptic excitatory receptors in the dorsal horn. Among these are PKA, PKC, PKG, CAMKII, ERK, and PI3K/Akt (Sun *et al.*, 2006; Xu *et al.*, 2007). In this study, we investigated the involvement of the PI3K/Akt pathway in the anti-allodynic effect of linalool. Linalool did not induce significant changes in Akt expression and phopshorylation (Fig. 3A). However, a slight decrease in total levels of Akt was observed in SNL animals treated with vehicle and linalool in comparison to sham and SNL (Fig. 3D). This was paralleled by a slight increase in p-Akt (Fig. 3C), resulting in a trend toward an increased ratio of phosphorylated versus total Akt in SNL animals treated with linalool in comparison to SNL alone or sham (Fig. 3B). Our results seem to be in contrast with recent work showing Akt activation in ipsilateral L5 and L4 DRG neurons and in L5 spinal dorsal horn following SNL in rat (Xu *et al.*, 2007). However, Akt activation measured by immunohistochemistry appeared to be limited to a very small area of the dorsal horn, reached a peak 3 days after SNL and, although still present, was much reduced at day 7. Thus, the apparent discrepancy may be ascribed to the different techniques' sensitivity as well as to the different animal species used.

Although pain hypersensitivity was originally thought to result exclusively from altered neuronal activity in the primary sensory and spinal cord neurons, there is evidence indicating that glial cells may also play a role in the pathogenesis of pain (DeLeo and Yezierski, 2001; Meller *et al.*, 1994; Watkins *et al.*, 2001b). Several cytokines released from glial cells produce pain hypersensitivity and astrocytes and microglia are activated in the spinal cord after inflammation, nerve injury and in cancer models (Mantyh *et al.*, 2002; Sweitzer *et al.*, 1999; Winkelstein *et al.*, 2001). However, the exact role of glial cells and cytokines within the spinal cord has not been completely understood (DeLeo and Yezierski, 2001; Watkins *et al.*, 2001a). Recently, activation of the NF-κB signal pathway has been suggested to mediate the actions of cytokines such as IL-1, IL-6 and TNF-α following nerve injury (Ledeboer *et al.*, 2005; Sakaue *et al.*, 2001; Tegeder *et al.*, 2004) and PI3K and PKB/Akt seem to be crucial mediators in this pathway (Reddy *et al.*, 1997; Sizemore *et al.*, 1999; Xu *et al.*, 2007). Our data show a trend toward a decrease in spinal IL-1β levels following SNL (Fig. 5) and linalool seems to restore them to the levels of sham animals (Fig. 5). However, at the latest time point a similar trend was also observed in the presence of the vehicle alone (Fig. 5). Analysis of glial activation in the spinal cord of SNL animals treated either with linalool or with its vehicle alone highlighted the possibility that the vehicle itself may have anti-inflammatory proprieties that could mask the effects of linalool. However, the vehicle did not show any antinociceptive effect in the behavioral tests, thus strengthening the observation that the anti-allodynic effects reported in this model of neuropathic pain are proper of linalool. This also suggests that mechanisms other rather than an action on inflammatory processes may mediate linalool ability to reduce mechanical allodynia. Accordingly, further studies are necessary to investigate possible alternative pathways involved in antinociceptive mechanisms of linalool in neuropathic pain.

Acknowledgments

Financial support from the Ministry of Health (RF 2005 to MNB and CMT) and from the University of Calabria (ex quota 60%) is gratefully acknowledged. Mr Guido Fico is gratefully acknowledged for skilful technical assistance.

References

Brum, L. F., Elisabetsky, E., and Souza, D. (2001). Effects of linalool on [(3)H]MK801 and [(3)H] muscimol binding in mouse cortical membranes. *Phytother. Res.* **15,** 422–425.

Bursztajn, S., Rutkowski, M. D., and Deleo, J. A. (2004). The role of the N-methyl-D-aspartate receptor NR1 subunit in peripheral nerve injury-induced mechanical allodynia, glial activation and chemokine expression in the mouse. *Neuroscience* **125,** 269–275.

Chaplan, S. R., Bach, F. W., Pogrel, J. W., Chung, J. M., and Yaksh, T. L. (1994). Quantitative assessment of tactile allodynia in the rat paw. *J. Neurosci. Methods* **53,** 55–63.

DeLeo, J. A., and Yezierski, R. P. (2001). The role of neuroinflammation and neuroimmune activation in persistent pain. *Pain* **90,** 1–6.

Elisabetsky, E., Brum, L. F., and Souza, D. O. (1999). Anticonvulsant properties of linalool in glutamate-related seizure models. *Phytomedicine* **6,** 107–113.

Gao, X., Kim, H. K., Chung, J. M., and Chung, K. (2005). Enhancement of NMDA receptor phosphorylation of the spinal dorsal horn and nucleus gracilis neurons in neuropathic rats. *Pain* **116,** 62–72.

Hanisch, U. K. (2002). Microglia as a source and target of cytokines. *Glia* **40,** 140–155.

Hargreaves, K., Dubner, R., Brown, F., Flores, C., and Joris, J. (1988). A new and sensitive method for measuring thermal nociception in cutaneous hyperalgesia. *Pain* **32,** 77–88.

Hashizume, H., DeLeo, J. A., Colburn, R. W., and Weinstein, J. N. (2000). Spinal glial activation and cytokine expression after lumbar root injury in the rat. *Spine* **25,** 1206–1217.

Kajander, K. C., Sahara, Y., Iadarola, M. J., and Bennett, G. J. (1990). Dynorphin increases in the dorsal spinal cord in rats with a painful peripheral neuropathy. *Peptides* **11,** 719–728.

Kim, S. H., and Chung, J. M. (1992). An experimental model for peripheral neuropathy produced by segmental spinal nerve ligation in the rat. *Pain* **50,** 355–363.

Ledeboer, A., Gamanos, M., Lai, W., Martin, D., Maier, S. F., Watkins, L. R., and Quan, N. (2005). Involvement of spinal cord nuclear factor kappaB activation in rat models of proinflammatory cytokine-mediated pain facilitation. *Eur. J. Neurosci.* **22,** 1977–1986.

Liu, C. N., Wall, P. D., Ben-Dor, E., Michaelis, M., Amir, R., and Devor, M. (2000). Tactile allodynia in the absence of C-fiber activation: Altered firing properties of DRG neurons following spinal nerve injury. *Pain* **85,** 503–521.

Ma, W., and Quirion, R. (2002). Partial sciatic nerve ligation induces increase in the phosphorylation of extracellular signal-regulated kinase (ERK) and c-Jun N-terminal kinase (JNK) in astrocytes in the lumbar spinal dorsal horn and the gracile nucleus. *Pain* **99,** 175–184.

Mantyh, P. W., Clohisy, D. R., Koltzenburg, M., and Hunt, S. P. (2002). Molecular mechanisms of cancer pain. *Nat. Rev. Cancer* **2,** 201–209.

Meller, S. T., Dykstra, C., Grzybycki, D., Murphy, S., and Gebhart, G. F. (1994). The possible role of glia in nociceptive processing and hyperalgesia in the spinal cord of the rat. *Neuropharmacology* **33,** 1471–1478.

Peana, A. T., D'Aquila, P. S., Panin, F., Serra, G., Pippia, P., and Moretti, M. D. (2002). Anti-inflammatory activity of linalool and linalyl acetate constituents of essential oils. *Phytomedicine* **9,** 721–726.

Peana, A. T., D'Aquila, P. S., Chessa, M. L., Moretti, M. D., Serra, G., and Pippia, P. (2003). (−)-Linalool produces antinociception in two experimental models of pain. *Eur. J. Pharmacol.* **460,** 37–41.

Peana, A. T., De Montis, M. G., Nieddu, E., Spano, M. T., D'Aquila, P. S., and Pippia, P. (2004a). Profile of spinal and supra-spinal antinociception of (−)-linalool. *Eur. J. Pharmacol.* **485,** 165–174.

Peana, A. T., De Montis, M. G., Sechi, S., Sircana, G., D'Aquila, P. S., and Pippia, P. (2004b). Effects of (−)-linalool in the acute hyperalgesia induced by carrageenan, L-glutamate and prostaglandin E2. *Eur. J. Pharmacol.* **497,** 279–284.

Peana, A. T., Rubattu, P., Piga, G. G., Fumagalli, S., Boatto, G., Pippia, P., and De Montis, M. G. (2006). Involvement of adenosine A1 and A2A receptors in (−)-linalool-induced antinociception. *Life Sci.* **78,** 2471–2474.

Reddy, S. A., Huang, J. H., and Liao, W. S. (1997). Phosphatidylinositol 3-kinase in interleukin 1 signaling. Physical interaction with the interleukin 1 receptor and requirement in NFkappaB and AP-1 activation. *J. Biol. Chem.* **272,** 29167–29173.

Sakaue, G., Shimaoka, M., Fukuoka, T., Hiroi, T., Inoue, T., Hashimoto, N., Sakaguchi, T., Sawa, Y., Morishita, R., Kiyono, H., Noguchi, K., and Mashimo, T. (2001). NF-kappa B decoy suppresses cytokine expression and thermal hyperalgesia in a rat neuropathic pain model. *Neuroreport* **12,** 2079–2084.

Silva Brum, L. F., Emanuelli, T., Souza, D. O., and Elisabetsky, E. (2001). Effects of linalool on glutamate release and uptake in mouse cortical synaptosomes. *Neurochem. Res.* **26,** 191–194.

Sizemore, N., Leung, S., and Stark, G. R. (1999). Activation of phosphatidylinositol 3-kinase in response to interleukin-1 leads to phosphorylation and activation of the NF-kappaB p65/RelA subunit. *Mol. Cell Biol.* **19,** 4798–4805.

Sun, R. Q., Tu, Y. J., Yan, J. Y., and Willis, W. D. (2006). Activation of protein kinase B/Akt signaling pathway contributes to mechanical hypersensitivity induced by capsaicin. *Pain* **120,** 86–96.

Sweitzer, S. M., Colburn, R. W., Rutkowski, M., and DeLeo, J. A. (1999). Acute peripheral inflammation induces moderate glial activation and spinal IL-1beta expression that correlates with pain behavior in the rat. *Brain Res.* **829,** 209–221.

Tegeder, I., Niederberger, E., Schmidt, R., Kunz, S., Guhring, H., Ritzeler, O., Michaelis, M., and Geisslinger, G. (2004). Specific Inhibition of IkappaB kinase reduces hyperalgesia in inflammatory and neuropathic pain models in rats. *J. Neurosci.* **24,** 1637–1645.

Ultenius, C., Linderoth, B., Meyerson, B. A., and Wallin, J. (2006). Spinal NMDA receptor phosphorylation correlates with the presence of neuropathic signs following peripheral nerve injury in the rat. *Neurosci. Lett.* **399,** 85–90.

Watkins, L. R., Martin, D., Ulrich, P., Tracey, K. J., and Maier, S. F. (1997). Evidence for the involvement of spinal cord glia in subcutaneous formalin induced hyperalgesia in the rat. *Pain* **71,** 225–235.

Watkins, L. R., Milligan, E. D., and Maier, S. F. (2001a). Glial activation: A driving force for pathological pain. *Trends Neurosci.* **24,** 450–455.

Watkins, L. R., Milligan, E. D., and Maier, S. F. (2001b). Spinal cord glia: New players in pain. *Pain* **93,** 201–205.

Winkelstein, B. A., Rutkowski, M. D., Sweitzer, S. M., Pahl, J. L., and DeLeo, J. A. (2001). Nerve injury proximal or distal to the DRG induces similar spinal glial activation and selective cytokine expression but differential behavioral responses to pharmacologic treatment. *J. Comp. Neurol.* **439,** 127–139.

Wu, G., Ringkamp, M., Hartke, T. V., Murinson, B. B., Campbell, J. N., Griffin, J. W., and Meyer, R. A. (2001). Early onset of spontaneous activity in uninjured C-fiber nociceptors after injury to neighboring nerve fibers. *J. Neurosci.* **21,** RC140.

Xu, J. T., Tu, H. Y., Xin, W. J., Liu, X. G., Zhang, G. H., and Zhai, C. H. (2007). Activation of phosphatidylinositol 3-kinase and protein kinase B/Akt in dorsal root ganglia and spinal cord contributes to the neuropathic pain induced by spinal nerve ligation in rats. *Exp. Neurol.* **206,** 269–279.

Ziegler, E. A., Magerl, W., Meyer, R. A., and Treede, R. D. (1999). Secondary hyperalgesia to punctate mechanical stimuli. Central sensitization to A-fibre nociceptor input. *Brain* **122**(Pt 12), 2245–2257.

INTRAPLANTAR INJECTION OF BERGAMOT ESSENTIAL OIL INTO THE MOUSE HINDPAW: EFFECTS ON CAPSAICIN-INDUCED NOCICEPTIVE BEHAVIORS

Tsukasa Sakurada,* Hikari Kuwahata,† Soh Katsuyama,* Takaaki Komatsu,* Luigi Antonio Morrone,‡ Maria Tiziana Corasaniti,§ Giacinto Bagetta,‡ and Shinobu Sakurada†

*First Department of Pharmacology, Daiichi College of Pharmaceutical Sciences, Fukuoka, Japan

†Department of Physiology and Anatomy, Tohoku Pharmaceutical University, Sendai, Japan

‡Department of Pharmacobiology and University Center for Adaptive Disorders and Headache (UCADH), Section of Neuropharmacology of Normal and Pathological Neuronal Plasticity, University of Calabria, 87036 Arcavacata di Rende, Italy

§Department of Pharmacobiological Sciences, University Magna Graecia of Catanzaro, Catanzaro, Italy

Despite the increasing use of aromatherapy oils, there have not been many studies exploring the biological activities of bergamot (*Citrus bergamia*, Risso) essential oil (BEO). Recently, we have investigated the effects of BEO injected into the plantar surface of the hindpaw in the capsaicin test in mice. The intraplantar injection of capsaicin produced an intense and short-lived licking/biting response toward the injected hindpaw. The capsaicin-induced nociceptive response was reduced significantly by intraplantar injection of BEO. The essential oils of Clary Sage (*Salvia sclarea*), Thyme ct. linalool (linalool chemotype of *Thymus vulgaris*), Lavender Reydovan (*Lavandula hybrida reydovan*), and True Lavender (*Lavandula angustifolia*), had similar antinociceptive effects on the capsaicin-induced nociceptive response, while Orange Sweet (*Citrus sinensis*) essential oil was without effect. In contrast to a small number of pharmacological studies of BEO, there is ample evidence regarding isolated components of BEO which are also found in other essential oils. The most abundant compounds found in the volatile fraction are the monoterpene hydrocarbons, such as limonene, γ-terpinene, β-pinene, and oxygenated derivatives, linalool and linalyl acetate. Of these monoterpenes, the

INTERNATIONAL REVIEW OF NEUROBIOLOGY, VOL. 85
DOI: 10.1016/S0074-7742(09)85018-6

0074-7742/09 $35.00

pharmacological activities of linalool have been examined. Following intraperitoneal (i.p.) administration in mice, linalool produces antinociceptive and antihyperalgesic effects in different animal models in addition to anti-inflammatory properties. Linalool also possesses anticonvulsant activity in experimental models of epilepsy. We address the importance of linalool or linalyl acetate in BEO—or the other essential oil-induced antinociception.

I. Introduction

Bergamot essential oil (BEO) as well as other essential oils (Halcon, 2002) is increasingly used in aromatherapy to minimize the effects of stress-induced anxiety and to facilitate sleep induction (Komori *et al.*, 1995; Lehrner *et al.*, 2000; Wiebe, 2000). Aromatherapy oils applied by inhalation do not appear to reduce anxiety (Graham *et al.*, 2003), whereas aromatherapy massage has been shown to relieve symptoms of anxiety in patients with cancer (Comer *et al.*, 1995; Kite *et al.*, 1988; Wilkinson, 1995; Wilkinson *et al.*, 1999). Pharmacological studies have demonstrated that BEO elevates the release of amino acid neurotransmitters, glutamate, GABA, aspartate, glycine, and taurine, in rat hippocampus as measured by *in vivo* microdialysis and by *in vitro* superfusion of isolated nerve terminals (Morrone *et al.*, 2007). Recently, BEO has been shown to reduce neuronal damage caused *in vitro* by excitotoxic stimuli and that neuroprotection is associated with prevention of injury-induced decrease of phospho-Akt and phospho-GSK-3β levels (Corasaniti *et al.*, 2007). The nonvolatile fraction inhibits oxidative stress which occurs in injured arteries and modulates both LOX-1 expression and neointima formation (Mollace *et al.*, 2008). Bergamottin, a furocoumarin isolated from the nonvolatile fraction of BEO, is found to have antianginal and antiarrhythmic effects (Occhiuto and Circosta, 1996, 1997).

However, little information is available in the literature concerning the biological effects of BEO. By contrast, there are many pharmacological reports regarding the effects of isolated components such as psoralens that, among other uses, are employed in the therapy of psoriasis and vitiligo (Gupta and Anderson, 1987). Electrophysiological experiments carried out using the patch-clamp technique demonstrate that psoralens, bergapten and 8-methoxypsorale, block voltage-gated K^+ channels in Ranvier nodes, Schwann cells maintained *in vitro* (During *et al.*, 2000). Inhibition of K^+ currents by alkoxypsoralens has been shown in cultured immune cells together with inhibition of immune cell proliferation response, interferon-γ gene expression and cytokine production, supporting a role in the modulation of the immune response which follows, for instance,

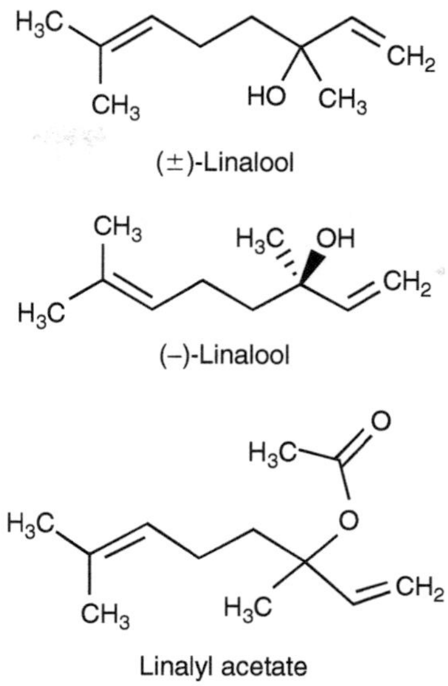

FIG. 1. Structures of (±) linalool, (−) linalool, and linalyl acetate.

encephalitogenic stimuli (Strauss *et al.*, 2000). Thus, natural products regarded as potential source of biologically active compounds have been attractive for developing new pharmacological agents.

The present review will focus on the pharmacological activities of BEO. Particular attention has been devoted to antinociceptive characteristics of BEO to discuss the possible role of linalool or linalyl acetate (Fig. 1), which is present in the essential oil of bergamot as well as other aromatic plants.

II. General Characteristics of Bergamot Essential Oil

Bergamot is a citrus classified as *Citrus bergamia*, Risso which belongs to the Rutaceae, genus Citrus. BEO is obtained by cold pressing of the epicarp and, partly, of the mesocarp of the fresh fruit of bergamot. The citrus is cultivated for its essential oil, a product in great demand by perfumery and cosmetic industries because of a pleasant and refreshing scent. BEO has also been employed by pharmaceutical, food, and confectionery industries. BEO comprises a volatile (93–96% of total) and a nonvolatile (4–7% of total) fraction; the former fraction contains monoterpene and sesquiterpene hydrocarbons such as limonene, α- and β-pinene, β-myrcene, γ-terpinene, terpinolene, sabinene, β-bisabolene, and oxygenated derivatives such as linalool, neral, geranial, linalyl acetate, neryl acetate, and geranyl acetate. The latter fraction contains waxes, polymethoxylated flavones, coumarins, and psoralens, that is, bergapten (5-methoxypsoralen) and bergamottin (5-geranyloxypsoralen) (Dugo *et al.*, 2000; Mondello *et al.*, 1993). The hydrocarbon fraction does not have a fundamental role in determining the olfactory character of BEO, but oxygenated compounds, that is, linalool and linalyl acetate, mark the flavor notes of BEO. The nonvolatile residue is a natural odor fixative which influences the olfactory properties of the oil. However, it contains approximately 0.35% bergapten which is responsible for the phototoxicity of BEO (Ashwood-Smith *et al.*, 1980; Zaynoun *et al.*, 1977). BEO has been widely used as an ingredient in cosmetics, whereas it was restricted or banned in most countries because of certain adverse effects due to photosensitive and melanogenic properties of bergapten. Then, a bergapten-free extract of the essence together with a natural essence deprived of the hydrocarbon fraction and of bergapten are prepared by extractive industries for perfumery and cosmetic uses.

III. Antinociception Induced by the Essential Oil of Bergamot

The paw-licking/biting behavior is evoked in mice by local application of capsaicin (8-methyl-*N*-vanillyl-6-nonenamide), a pungent principle in peppers of the *Capsicum* family, to peripheral tissues (Sakurada *et al.*, 1992). The capsaicin test in mice is a reliable model of peripheral nociception, which produces nociceptive behavior similar to that elicited by intraplantar injection of formalin (Sakurada *et al.*, 1992). The noxious stimulus is a subcutaneous (s.c.) injection of capsaicin under the plantar surface of a hindpaw. The spontaneous nociceptive response is the amount of time the animals spent licking/biting the paw. The s.c. injection of diluted capsaicin into the hindpaw produced an immediate short-lasting nociceptive response (approximately 5 min) that could be dose-dependently reduced by either systemically or locally applied morphine (Pelissier *et al.*, 2002; Sakurada *et al.*, 1994). Capsaicin is known to excite almost specifically the C-fiber population of nociceptive afferents, since capsaicin acts on transient receptor potential vanilloid type-1 (TRPV-1) receptors located mainly in C-fiber-type nociceptors (Di Marzo *et al.*, 2002; Szallasi, 2006). Several experimental results showed that injection of capsaicin could produce a release of endogenous tachykinins, substance P or neurokinin A, and glutamate from primary sensory neurons in the dorsal spinal cord (Holzer, 1991; Kawamata *et al.*, 2001; Sorkin and McAdoo, 1993). Behavioral studies have indicated that intrathecal (i.t.) administration of antiserum against substance P reduces capsaicin-induced nociceptive responses (Sakurada *et al.*, 1992). The competitive NMDA receptor antagonist CPP and the noncompetitive NMDA receptor antagonist MK801 reduce capsaicin-induced nociceptive behavior (Sakurada *et al.*, 1998). Therefore, it seems reasonable that the capsaicin test is useful for evaluating antinociceptive effects of intrathecally injected neurokinin-1 (NK_1) and glutamate receptor antagonists in the spinal cord.

In the capsaicin (50 μg) test, intraplantar injection of BEO (5, 10 and 20 μg) produced a significant antinociceptive effect in normal mice, being ineffective only in a lower dose of 2.5 μg (Fig. 2). Antinociceptive effects of BEO resulting from interactions with peripheral cutaneous nociceptors are particularly important since unwanted side effects associated with the central nervous system (CNS) mechanisms would be avoided. However, the type of afferent fibers affected by BEO is still unknown. The other essential oils of Clary Sage (*Salvia sclarea*), Thyme ct. linalool (*Thymus vulgaris*), Lavender Reydovan (*Lavandula hybrida reydovan*), and True Lavender (*Lavandula angustifolia*), have also been examined in the capsaicin test. The intraplantar injection of essential oils extracted from these plants showed a significant reduction of capsaicin-induced nociceptive behavior (Sakurada *et al.*, 2008). BEO injected into the hindpaw contralateral to intraplantar injection of capsaicin did not attenuate capsaicin-induced nociceptive response (unpublished data), suggesting that the antinociceptive effect of BEO is locally mediated and not systemic. Thus, the local tissues located on the terminals of primary afferent nociceptive neurons innervating the hindpaw may be essential target for BEO in alleviating capsaicin-induced nociception. Table I shows the concentration of linalool and linalyl acetate contained by the essential oils of various aromatic plants. It is noteworthy that above-mentioned essential oils as well as BEO contain

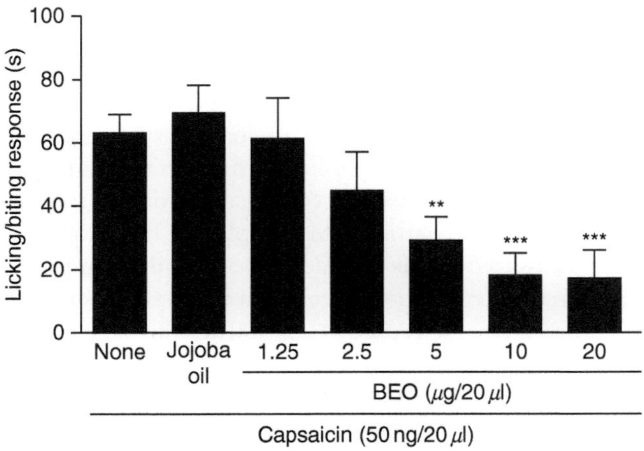

FIG. 2. Antinociceptive effects of intraplantar BEO in the capsaicin test. BEO was injected into the plantar surface of the hindpaw 10 min prior to injection of capsaicin (50 ng/paw). Jojoba oil was used as a control, which was without affecting the capsaicin-induced nociceptive response. Data are means ± S.E.M. ($n = 10$ mice per group). ***$p < 0.001$, **$p < 0.01$ when compared with Jojoba oil-treated controls by Dunnett's test.

TABLE I

Percentage of the Monoterpene Compounds, Linalool and Linalyl Acetate, in the Volatile Fraction Extracted from Aromatic Plant Species

Aromatic plant species	Linalool	Linalyl acetate
Bergamot (*Citrus bergamia*, Risso)	9.8	30.0
Lavender Reydovan (*Lavandula hybrida reydovan*)	15.2	58.7
True Lavender (*Lavandula angustifolia*)	36.6	43.5
Rose Wood (*Aniba rosaeodora*)	83.1	0
Thyme ct. linalool (*Thymus vulgaris*)	81.2	4.6
Clary Sage (*Salvia sclarea*)	21.0	62.5
Orange Sweet (*Citrus sinensis*)	0.3	0

linalool and/or linalyl acetate in high concentrations (Table I). These monoterpenes, particularly linalool, are known to possess several biological activities. Both linalool and linalyl acetate were much more potent than BEO or other essential oils in inhibiting the response to intraplantar capsaicin (Sakurada *et al.*, 2008). It seems possible to speculate that antinociceptive efficacy of BEO and the other essential oils may be dependent on the amount of linalool and/or linalyl acetate. This speculation is supported by the lack of significant antinociceptive activity in the capsaicin test yielded by injection into the hindpaw of the Orange Sweet (*Citrus sinensis*) essential oil (Sakurada *et al.*, 2008) known to contain extremely small amount of linalool and linalyl acetate (Table I). These results provide evidence suggesting that intraplantar application of the essential oils containing linalool and or linalyl acetate may be a promising therapeutic approach in the treatment of clinical pain. Recently, we have reported that combinations of morphine (i.p. and i.t.) and intraplantar BEO or linalool at a concentration that do not produce significant antinociception resulted in a synergistic antinociceptive effect of morphine (Sakurada *et al.*, 2008). Our findings that intraplantar BEO or linalool achieved potentiating effects in combination with morphine suggest that local opioid receptors may contribute to intraplantar BEO- or linalool-induced antinociception as assayed by the capsaicin test. This thought is in line with the result that the effect of s.c. linalool is decreased by pretreatment with the opioid receptor antagonist naloxone (Peana *et al.*, 2004a). Similar potentiating results have been observed in the clinical cases that i.t.-injected mixture of morphine with tetracaine or bupivacaine produced excellent postoperative pain relief of significantly longer duration than the local anesthetics alone (Akerman *et al.*, 1988; Cunningham *et al.*, 1983; Kalso, 1983). Taken together, our results raise the possibility that concomitant use of peripheral BEO or linalool, and opioids may offer advantages for the treatment of painful conditions. Antinociceptive effects of BEO or the other essential oils resulting from interactions with primary afferent neurons are important since unwanted side effects in the CNS would be avoided. However, the peripheral mechanism by which these essential oils produce antinociception is unclear. Further investigation as targets for peripherally acting analgesic compounds is needed.

IV. Linalool-Induced Antinociceptive Activity

The monoterpene alcohol, linalool, and its corresponding ester linalyl acetate are present as major volatile compounds in the essential oils of various aromatic plants. It is inferred that the pharmacological activities of these plants may be, in part at least, derived from the presence of linalool or linalyl acetate. A variety of behavioral and pharmacological actions of linalool have been reported in animal experiments. Linalool displays anticonvulsant, anxiolytic, anti-inflammatory, and antinociceptive effects as well as hypnotic and hypothermic properties (Elisabersky *et al.*, 1995). Of many pharmacological effects elicited by linalool, its antinociceptive action has been investigated extensively (Table II). It is of importance to assess whether linalool given through several administration routes could produce a pharmacological activities. The s.c. administration of linalool in doses ranging from 50 to 100 mg/kg produced a significant attenuation of the acetic acid-induced

TABLE II
RESULT OF REPORTED STUDIES INVESTIGATING THE EFFECT OF NOCICEPTIVE SENSITIVITY OF LINALOOL ADMINISTRATION IN ANIMALS

Effect	Injection routes	Species	Strain	Assay	Doses	References
Antinociception	s.c.	Mouse	CD1	AAW	25–75 mg/kg	Peana *et al.* (2003)
	s.c.	Mouse	CD1	HP	50–100 mg/kg	
Antinociception	s.c.	Mouse	CD1	FT	50–150 mg/kg	Peana *et al.* (2004a)
Antihyperalgesia	s.c.	Rat	Wistar	CTH	50–150 mg/kg	Peana *et al.* (2004b)
	s.c.	Rat	Wistar	GTH	100–200 mg/kg	
	s.c.	Mouse	Wistar	PTH	100–200 mg/kg	
Antinociception	i.p.	Rat	Swiss	GSN	10–200 mg/kg	Batista *et al.* (2008)
	p.o.	Mouse	Swiss	GSN	5–100 mg/kg	
	i.pl.	Mouse	Swiss	GSN	10–300 ng/paw	
	i.t.	Mouse	Swiss	GSN	0.1–3 μg/site	
Antinociception	i.pl.	Mouse	ddY	CSN	2.5–20 μg/paw	Sakurada *et al.* (2008)

Abbreviations of injection routes: s.c., subcutaneous; i.p., intraperitoneal; p.o., per os (oral); i.pl., intraplantar; i.t., intrathecal.

Assay abbreviations: AAW, acetic acid-induced writhing; HP, hot-plate test; FT, formalin test; CTH, carrageenan-induced thermal hyperalgesia; GTH, L-glutamate-induced thermal hyperalgesia; PTH, prostaglandin E_2-induced thermal hyperalgesia; GSN, L-glutamate-induced spontaneous nociception; CSN, capsaicin-induced spontaneous nociception.

writhing, which was not dose dependent (Peana *et al.*, 2003). In addition, they reported that s.c.-administered linalool is antinociceptive in the mouse hot-plate assay. This result is consistent with the recent report that i.p.-administered linalool was effective in inhibiting nociceptive behavioral response evoked by intraplantar injection of glutamate (Batista *et al.*, 2008).

Injection of formalin into the paw in animals elicits an initial licking/biting of the injected paw area (early phase or first phase), presumably in response to direct activation of nociceptors by formalin (Dubuisson and Dennis, 1977; Sakurada *et al.*, 1995). A delay licking/biting response (late phase or second phase) occurs 10–30 min after formalin injection, which is thought to be an inflammatory response. Formalin-induced paw-licking/biting in the late phase is continuous, in contrast to short-lived nociception of the tail-flick and the hot-plate assays, and involves neural pathways different from those underlying tail-flick. Linalool, injected s.c. at doses of 50 and 100 mg/kg, is reported to cause a significant reduction of formalin evoked early-phase response, but not of late-phase nociceptive response (Peana *et al.*, 2004a). The results of Peana *et al.* (2004a) showed that linalool-induced antinociception was reversed significantly by pretreatment with atropine, naloxone, and sulpiride, suggesting that the antinociceptive activity of s.c.-injected linalool may be mediated through the activation of muscarinic M_2, opioid, or dopamine D_2 receptors. Regarding the involvement of peripheral opioid receptors, we have recently reported using the capsaicin test that antinociception induced by intraplantar injection of linalool was reversed significantly by intraplantar or i.p. pretreatment with the opioid receptor antagonist naloxone hydrochloride (Sakurada *et al.*, 2008). Furthermore, pretreatment with naloxone methiodide, an opioid receptor antagonist acting at the peripheral level, antagonized linalool-induced antinociception. Therefore, our findings suggest that a local effect of linalool on cutaneous nociceptors is mediated indirectly through opioid receptors, presumably the release of β-endorphin, which produces analgesia by signaling through μ-opioid receptors on sensory endings. Besides the involvement of opioid receptors, it has been suggested that the inhibition of nitric oxide (NO) synthesis or release *in vitro*, and the activation of adenosine A_1 and A_{2A} receptors *in vivo* may attenuate effectively linalool-induced antinociception (Peana *et al.*, 2006a,b).

Systemic administration of linalool in rats is also active in a model of thermal hyperalgesia induced by intraplantar injection of carrageenan, L-glutamate, and prostaglandin E_2 (Peana *et al.*, 2004b). (−) Linalool as well as its racemate form (±) linalool are found to possess anti-inflammatory activity as assayed by carrageenan-induced oedema (Peana *et al.*, 2002). Linalyl acetate, present in linalool-containing essential oils, was less potent than (−) linalool and (±) linalool in inducing anti-inflammatory activity. This result is in line with our recent data that antinociceptive effect of intraplantar linalool was more potent than linalyl acetate in inhibiting the response to capsaicin (Sakurada *et al.*, 2008). Thus, it appears that peripherally administered linalool is pharmacologically effective in reducing nociception and inflammation. Recent behavioral studies have shown that linalool, administered systemically (i.p. or s.c.) or orally, exhibits antinociceptive activities as measured by intraplantar glutamate-induced paw-licking, but intraplantar injection of linalool (10–300 ng/paw) partially inhibited glutamate-induced nociception (Batista *et al.*, 2008). The result of intraplantar linalool is inconsistent with our data that linalool injected into the paw elicited a dose-dependent inhibition of the licking/biting response induced by intraplantar capsaicin (Sakurada *et al.*, 2008). Thus, linalool produces more potent antinociceptive activity as assessed by intraplantar capsaicin than by intraplantar glutamate. It seems likely that linalool has a differential action on capsaicin and glutamate injected into the hindpaw. This notion is supported by the data that i.p. injection of linalool was effective in attenuating the licking/biting response induced by i.t. injection of glutamate, AMPA, substance P, NMDA, and kainate, but not by *trans*-ACPD. In addition, Batista *et al.* (2008) also reported that i.t. administration of linalool into the spinal subarachnoid space produces a significant reduction of the nociceptive response to intraplantar injection of glutamate.

It has been demonstrated that neuropathic pain is opioid resistant and, indeed, neither systemic nor i.t. administration of opioids reduced effectively neuropathic pain in rats (Bian *et al.*, 1995; Lee *et al.*, 1995; Mao *et al.*, 1995; Ossipov *et al.*, 1995). It seems worth to note that s.c. administration of (−) linalool attenuates mechanical allodynia in spinal nerve ligation model of neuropathic pain in mice (Levato *et al.*, 2006).

The peripherally acting antiallodynic/antihyperalgesic linalool may represent a promising therapeutic approach for alleviating neuropathic pain.

Acknowledgments

This work was supported by The Science Research Promotion Fund from The Promotion and Mutual Aid of Corporation for Private Schools of Japan, a Grant-in-Aid for Science Research (C) (KAKENHI 16590058 and 17590065) from Japan Society for the Promotion of Science, and a Grant-in-Aid for High Technology Research Program from the Ministry of Education, Cultures, Sports, Science, and Technology of Japan. We thank Makiko Akiyoshi and Maki Shigeno for expert technical assistance.

References

Akerman, B., Arwestrom, E., and Post, C. (1988). Local anesthetics potentiate spinal morphine antinociception. *Anesth. Analg.* **67,** 943–948.

Ashwood-Smith, M. J., Poulton, G. A., Barker, M., and Mildenberger, M. (1980). Methoxypsoralen, an ingredient in several suntan preparations, has lethal, mutagenic and clastogenic properties. *Nature* **285,** 407–409.

Batista, P. A., Paula Werner, M. F., Oliveira, E. C., Burgos, L., Pereira, P., Sila Brum, L. F., and Santos, A. R. S. (2008). Evidence for the involvement of ionotropic glutamatergic receptors on the antinociceptive effect of (−)-linalool in mice. *Neurosci. Lett.* **440,** 299–303.

Bian, D., Nichols, M. L., Ossipov, M. H., Lai, J., and Porreca, F. (1995). Characterization of the antiallodynic efficacy of morphine in a model of neuropathic pain in rats. *Neuroreport* **6,** 1981–1994.

Comer, J., Cawley, N., and Hildebrand, S. (1995). An evaluation of the use of massage and essential oils on the wellbeing of cancer patients. *Int. J. Palliat. Nurs.* **1,** 67–73.

Corasaniti, M. T., Maiuolo, J., Maida, S., Fratto, V., Navarra, M., Russo, R., Amantea, D., Morrone, L. A., and Bagetta, G. (2007). Cell signaling pathways in the mechanisms of neuroprotection afforded by bergamot essential oil against NMDA-induced cell death *in vitro. Br. J. Pharmacol.* **151,** 518–529.

Cunningham, A. J., McKenna, J. A., and Skene, D. S. (1983). Single injection spinal anaesthesia with amethocaine and morphine for transurethral prostatectomy. *Br. J. Anaesth.* **55,** 423–426.

Di Marzo, V., Blumberg, P. M., and Szallasi, A. (2002). Endovanilloid signaling in pain. *Curr. Opin. Neurobiol.* **12,** 372–379.

Dubuisson, D., and Dennis, S. G. (1977). The formalin test: A quantitative study of the analgesic effects of morphine, meperidine, and brain stem stimulation in rats and cats. *Pain* **4,** 161–174.

Dugo, P., Mondello, L., Dugo, L., Gugo, L., Stancanelli, R., and Dugo, G. (2000). LC-MS for the identification of oxygen heterocyclic compounds in citrus essential oils. *J. Pharm. Biomed. Anal.* **24,** 147–154.

During, T., Gerst, F., Hansel, W., Wulff, H., and Koppenhofer, E. (2000). Effect of three alkoxypsoralens on voltage gated ion channels in Ranvier nodes. *Gen. Physiol. Biophys.* **19,** 345–364.

Elisabersky, E., Souza, G. P. C., Santos, M. A. C., Siqueira, I. R., and Amador, T. A. (1995). Sedative properties of linalool. *Fitoterapia* **115,** 407–414.

Graham, P. H., Browne, L., Cox, H., and Graham, J. (2003). Inhalation aromatherapy during radiotherapy: Results of a placebo controlled double-blind randomized trial. *J. Clin. Oncol.* **21,** 2372–2376.

Gupta, A. K., and Anderson, T. F. (1987). Psoralen photochemotherapy. *J. Am. Acad. Dermatol.* **17,** 703–734.

Halcon, L. L. (2002). Aromatherapy: Therapeutic applications of plant essential oils. *Minnesota Med.* **85,** 42–46.

Holzer, P. (1991). Capsaicin: Cellular targets, mechanisms of action, and selectivity for thin sensory neurons. *Pharmacol. Rev.* **43,** 143–201.

Kalso, E. (1983). Effects of intrathecal morphine, injected with bupivacaine, on pain after orthopaedic surgery. *Br. J. Anaesth.* **55,** 415–422.

Kawamata, T., Omoto, K., Toriyabe, M., Kawamata, M., and Namiki, A. (2001). Involvement of capsaicin-sensitive fibers in spinal NMDA-induced glutamate release. *Neuroreport* **12,** 3447–3450.

Kite, S. M., Maher, E. J., Anderson, K., Young, T., Young, J., Wood, J., Howell, N., and Bradburn, J. (1988). Development of an aromatherapy service at a Cancer Centre. *Palliat. Med.* **12,** 171–180.

Komori, T., Fujiwara, R., Tanida, M., and Nomura, J. (1995). Potential antidepressant effects of lemon odor in rats. *Eur. Neuropsychopharmacol.* **5,** 477–480.

Lee, Y. W., Chaplan, S. R., and Yaksh, T. L. (1995). Systemic and supraspinal, but not spinal, opiates suppress allodynia in a rat neuropathic pain model. *Neurosci. Lett.* **199,** 111–114.

Lehrner, J., Eckersberger, C., Walla, P., Potsch, G., and Deecke, L. (2000). Ambient odor of orange in a dental office reduces anxiety and improves mood in female patients. *Physiol. Behav.* **71,** 83–86.

Levato, A., Rombola, L., Morrone, L. A., Corasaniti, M. T., Bagetta, G., Wood, J. N., and Berliocchi, L. (2006). (−)-Linalool attenuates allodynia in the spinal nerve ligation model of neuropathic pain in C57BL6 mice. *In* "IX Workshop on Apoptosis in Biology and Medicine: Neuroinflammation in Neuronal Death and Repair", Abstract P17.

Mao, J., Price, D. D., and Mayer, D. J. (1995). Experimental mononeuropathy reduces the antinociceptive effects of morphine: Implications for common intracellular mechanisms involved in morphine tolerance and neuropathic pain. *Pain* **61,** 353–364.

Mollace, V., Ragusa, S., Sacco, I., Muscoli, C., Sculco, F., Visalli, V., Palma, E., Muscoli, S., Mondello, L., Dugo, P., Rotiroti, D., and Romeo, F. (2008). The protective effect of bergamot oil extract on lecitine-like OxyLDL receptor-1 expression in balloon injury-related neointima formation. *J. Cardiovasc. Pharmacol. Ther.* **13,** 120–129.

Mondello, L., Stagno D'Alcontres, I., Del Duce, R., and Crispo, F. (1993). On the genuineness of citrus essential oils. Part XL. The composition of the coumarins and psoralens of calabrian bergamot essential oil (*Citrus bergamia* Risso). *Flavour Fragr. J.* **8,** 17–24.

Morrone, L. A., Rombola, L., Corasaniti, M. T., Zappettini, S., Paudice, P., Bonanno, G., and Bagetta, G. (2007). The essential oil of bergamot enhances the levels of amino acid neurotransmitters in the hippocampus of rat: Implication of monoterpene hydrocarbons. *Pharmacol. Res.* **55,** 255–262.

Occhiuto, F., and Circosta, C. (1996). Antianginal and antiarrhythmic effects of bergamottine, a furocoumarin isolated from bergamot oil. *Phytother. Res.* **10,** 491–496.

Occhiuto, F., and Circosta, C. (1997). Investigations to characterize the antiarrhythmic action of bergamottine, a furocoumarin isolated from bergamot oil. *Phytother. Res.* **11,** 450–453.

Ossipov, M. H., Lopez, Y., Nichols, M. L., Bian, D., and Porreca, F. (1995). The loss of antinociceptive efficacy of spinal morphine in rats with nerve ligation injury is prevented by reducing spinal afferent drive. *Neurosci. Lett.* **199,** 87–90.

Peana, A. T., D'Aquila, P. S., Panin, F., Serra, G., Pippia, P., and Moretti, M. D. L. (2002). Anti-inflammatory activity of linalool and linalyl acetate constituents of essential oils. *Phytomedicine* **9,** 721–726.

Peana, A. T., D'Aquila, P. S., Chessa, M. L., Moretti, M. D. L., Serra, G., and Pippia, P. (2003). (−)-Linalool produces antinociception in two experimental models of pain. *Eur. J. Pharmacol.* **460,** 37–41.

Peana, A. T., De Montis, M. G., Nieddu, E., Spano, T. M., Sechi, S., D'Aquila, P. S., and Pippia, P. (2004a). Profile of spinal and supra-spinal antinociception of (−)-linalool. *Eur. J. Pharmacol.* **485,** 165–174.

Peana, A. T., De Montis, M. G., Sechi, S., Sircana, G., D'Aquila, P. S., and Pippia, P. (2004b). Effects of (−)-linalool in the acute hyperalgesia induced by carrageenan, L-glutamate and prostaglandin E_2. *Eur. J. Pharmacol.* **497,** 279–284.

Peana, A. T., Marzocco, S., Popolo, A., and Pinto, A. (2006a). (−)-Linalool inhibits *in vitro* NO formation: Probable involvement in the antinociceptive activity of this monoterpene compound. *Life Sci.* **78,** 719–723.

Peana, A. T., Rubattu, P., Piga, G. G., Fumagalli, S., Boatto, G., Pippia, P., and De Montis, M. G. (2006b). Involvement of adenosine A1 and A2 receptors in (−)-linalool-induced antinociception. *Life Sci.* **78,** 2471–2474.

Pelissier, T., Pajot, J., and Dallel, R. (2002). The orofacial capsaicin test in rats: Effects of different capsaicin concentrations and morphine. *Pain* **96,** 81–87.

Sakurada, T., Katsumata, K., Tan-No, K., Sakurada, S., and Kisara, K. (1992). The capsaicin test in mice for evaluating tachykinin antagonists in the spinal cord. *Neuropharmacology* **31,** 1279–1285.

Sakurada, T., Yogo, H., Katsumata, K., Tan-No, K., Sakurada, S., Kisara, K., and Ohba, M. (1994). Differential antinociceptive effects of sendide, and morphine in the capsaicin test. *Brain Res.* **649,** 319–322.

Sakurada, T., Katsumata, K., Yogo, H., Tan-No, K., Sakurada, S., Ohba, M., and Kisara, K. (1995). The neurokinin-1 receptor antagonist, sendide, exhibits antinociceptive activity in the formalin test. *Pain* **60,** 175–180.

Sakurada, T., Wako, K., Sugiyama, A., Sakurada, C., Tan-No, K., and Kisara, K. (1998). Involvement of spinal NMDA receptors in capsaicin-induced nociception. *Pharmacol. Biochem. Behav.* **59,** 339–345.

Sakurada, T., Sakurada, S., Morrone, L. A., and Bagetta, G. (2008). Intraplantar injection of bergamot essential oil into the mouse hindpaw: Effects on capsaicin-induced nociceptive behavior. *In* "XI Workshop on Apoptosis in Biology and Medicine", Abstract book.

Sorkin, L. S., and McAdoo, D. J. (1993). Amino acid and serotonin are released into the lumbar spinal cord of the anesthetized cat following intradermal capsaicin injections. *Brain Res.* **607,** 89–98.

Strauss, U., Wissel, K., Jung, S., Wulff, H., Hansel, W., Zhu, J., Rolfs, A., and Mix, E. (2000). K^+ channel-blocking alkoxypsoralens inhibit the immune response of encephalitogenic T line cells and lymphocytes from Lewis rats challenged for experimental autoimmune encephalomyelitis. *Immunopharmacology* **48,** 51–63.

Szallasi, A. (2006). Small molecule vanilloid TRPV1 receptor antagonists approaching drug status: Can they live up to the expectations? *Naunyn Schmiedebergs Arch. Pharmacol.* **373,** 273–286.

Wiebe, E. (2000). A randomized trial of aromatherapy to reduce anxiety before abortion. *Eff. Clin. Pract.* **3,** 166–169.

Wilkinson, S. (1995). Aromatherapy and massage in palliative care: Does it improve patients' quality of life. *Int. J. Palliat. Nurs.* **1,** 15–20.

Wilkinson, S., Aldridge, J., Salmon, I., Cain, E., and Wilson, B. (1999). An evaluation of aromatherapy massage in palliative care. *Palliat. Med.* **13,** 409–417.

Zaynoun, S. T., Johnson, B. E., and Frain-Bell, W. (1977). A study of oil of bergamot and its importance as a phototoxic agent. I. Characterization and quantification of the photoactive component. *Br. J. Dermatol.* **96,** 475–482.

NEW THERAPY FOR NEUROPATHIC PAIN

Hirokazu Mizoguchi, Chizuko Watanabe, Akihiko Yonezawa, and Shinobu Sakurada

Department of Physiology and Anatomy, Tohoku Pharmaceutical University, Sendai 981-8558, Japan

Neuropathic pain is one of the worst painful symptoms in clinic. It contains nerve-injured neuropathy, diabetic neuropathy, chronic inflammatory pain, cancer pain, and postherpes pain, and is characterized by a tactile allodynia and hyperalgesia. Neuropathic pain, especially the nerve-injured neuropathy, the diabetic neuropathy, and the cancer pain, is opioid resistant pain. Since the downregulation of μ-opioid receptors is observed in dorsal spinal cord, morphine and fentanyl could not provide marked antihyperalgesic/antiallodynic effects in the course neuropathic pain states. The downregulation of μ-opioid receptors is suggested to be mediated through the activation of NMDA receptors. Moreover, at the neuropathic pain states, the increased expression of voltage-dependent Na^+ channels and Ca^{2+} channels are observed. Based on the above information concerned with the pathophysiology of neural changes in neuropathic pain states, new drug treatments for neuropathic pain, using ketamine, methadone, and gabapentin, have been developed. These drugs show remarkable effectiveness against hyperalgesia and allodynia during neuropathic pain states. Oxycodone is a μ-opioid receptor agonist, which has different pharmacological profiles with morphine. The remarkable effectiveness of oxycodone for neuropathic pain provides the possibility that μ-opioid receptor agonists, which have different pharmacological profile with morphine, can be used for the management of neuropathic pain.

INTERNATIONAL REVIEW OF NEUROBIOLOGY, VOL. 85
DOI: 10.1016/S0074-7742(09)85019-8

I. Neuropathic Pain

Neuropathic pain is an abnormal pain that results from traumatic, inflammatory, ischemic, metabolic and neoplastic insults to the peripheral or central nervous system, usually related to direct nerve injury, stroke, chronic inflammation of tissue, cancer, diabetes, or other nerve diseases. It is characterized by continuous or intermittent spontaneous pain and abnormal sensitivity of the painful site to a variety of noxious (hyperalgesia) or innocuous (allodynia) stimuli (Yarnitsky and Eisenberg, 1998). Evoked pain that results from light touch by garments, running water or even cold air can be extremely bothersome for many of these patients. Basically, the symptom of neuropathic pain is dependent on each patient pathophysiological background. Since many patients have multiple pathophysiological backgrounds, the diverse pathophysiological background of each patient makes a symptom variable and complicated.

There are several experimental animal models for neuropathic pain, such as diabetic neuropathy, inflammatory neuropathy, nerve-injured neuropathy, neuropathic cancer pain, neuropathic herpes pain, and so on. As experimental animal models for diabetic neuropathy, streptozotocin-induced diabetic rats (Courteix *et al.*, 1998) and mice (Rashid and Ueda, 2002) are used as well as spontaneous diabetic BB/Wor rat (Zhang *et al.*, 2002) and NOD mice (Gabra and Sirois, 2005). Complete Freund's adjuvant-induced chronic inflammation in rat (Okuse *et al.*, 1997) and mouse (Honore *et al.*, 2000) are used for the experimental animal models of inflammatory neuropathy. There are three major experimental animal models for the nerve-injured neuropathy, Seltzer model (Shir and Seltzer, 1990), Bennett model (Bennett and Xie, 1988), and Chung model (Kim and Chung, 1992). In Seltzer model (also known as the partial ligation model) the half of sciatic nerve is tightly ligated, while the whole sciatic nerve is loosely and constrictively ligated in Bennett model (also known as the chronic constriction injury model). In contrast, the segmental spinal nerve (especially L5 or L6) is ligated in the Chung model. Recent growing attention to the experimental animal models for neuropathic cancer pain and neuropathic herpes pain may reflect the importance of pain management for cancer pain and herpes pain in clinic. In the experimental animal models for neuropathic cancer pain, inoculating Meth A sarcoma cells to the immediate proximity of the sciatic nerve leads the progressive compression of the sciatic nerve by growth of the tumor mass resulting in a gradual development of neuropathy (Shimoyama *et al.*, 2002). On the other hand, infection with herpes simplex virus type-1 of the hind paw reveals the neuropathy in infected paw along with herpes zoster in the experimental animal models for herpes pain (Takasaki *et al.*, 2000b). These experimental animal models for neuropathic pain commonly show the hyperalgesia and allodynia, as well as spontaneous pain.

II. Drug Therapy of Neuropathic Pain—Effectiveness of Narcotic Analgesics

The pharmacotherapy of neuropathic pain in humans is frequently unsatisfactory. Currently, the antidepressants (Max, 1990), anticonvulsants (Covington, 1998), and Na^+ channel blockers (Gracely *et al.*, 1992) are used for management of neuropathic pain. However, all have limited success, and effective analgesia is achieved in fewer than half of all patients (Sindrup and Jensen, 1999). Therefore, at present, these drugs are generally used to increase the effectiveness of opioids, narcotic analgesics. The use of opioids is general and major drug treatment for the management of neuropathic pain. The systematic reviews of randomized controlled trials have shown the efficacy of opioids in reducing spontaneous neuropathic pain (Eisenberg *et al.*, 2005; Kalso *et al.*, 2004). However, in clinic, it is generally accepted that the neuropathic pain states are resistant to opioid treatment. The neuropathic pain in patients is not suppressed by ordinary doses of opioids, which suppress nociceptive pain (Arnér and Meyerson, 1988). Higher doses of opioids are required for control of neuropathic pain, but in some case only partial relief of pain is available. This discrepancy may be caused by the difference in the type of pain measured, and in the doses of opioids used. The evoked pain in mechanical allodynia and cold/heat allodynia is prevalent in neuropathic pain. Therefore, the clinical outcome of opioids in patients suffering neuropathic pain may dominantly reflect the effectiveness of opioids against evoked pain in a neuropathic pain state. Recent systematic review of randomized controlled trials on the efficacy of opioid agonists for the treatment of evoked neuropathic pain provided strong evidence to support this hypothesis (Eisenberg *et al.*, 2006). In that review, the effectiveness of opioids in the neuropathic pain is variable according to the type of evoked pain. Opioid treatments are effective against the cold allodynia/hyperalgesia in neuropathic pain. However, the opioid treatments do not work at all against the heat allodynia/hyperalgesia. The effectiveness of opioid treatments in the tactile allodynia/hyperalgesia is much more complicated. The opioid treatment is effective against dynamic allodynia/hyperalgesia, but not static allodynia/hyperalgesia at all. Moreover, in most of the randomized controlled trials, the effectiveness of opioid treatment against neuropathic pain was compared with placebo treatment, but not with the effectiveness against acute nociceptive pain. It means that the positive results in these trials may contain the case, which the effectiveness of opioids is reduced in neuropathic pain states, but opioids are still effective at the higher doses. In fact, a shift of the dose–response curve of opioids to the right beyond the normal therapeutic range, is observed in the neuropathic pain states (Portenoy *et al.*, 1990).

The responses of animal chronic neuropathic pain models to opioid treatment have often not mirrored this opioid resistance observed in clinic and thus the results have been controversial. Some researchers have found that opioids may produce antihyperalgesic/antiallodynic effects in neuropathic pain models but at higher than normal doses (Ossipov *et al.*, 1995). This phenomenon is observed especially in the animal model of the nerve-injured neuropathy (Narita *et al.*, 2008). In animal models of neuropathic cancer pain (Luger *et al.*, 2002) and diabetic neuropathy (Zurek *et al.*, 2001), the effectiveness of opioids is more dramatically decreased in comparison to normal animals and, in some cases, the effectiveness is almost undetectable. Interestingly, the effectiveness of opioids in these neuropathic pain models is variable according to the route of injection of opioids (Lee *et al.*, 1995; Zurek *et al.*, 2001). The antihyperalgesic/antiallodynic effect of opioids is remarkably reduced after subcutaneous injection and is almost abolished for intrathecal injection, while it is completely retained after intracerebroventricular injection. Unlike above neuropathic pain models, the antihyperalgesic/ antiallodynic effects of opioids in the animal models of chronic inflammatory neuropathy or postherpetic neuropathy are completely retained (Luger *et al.*, 2002; Narita *et al.*, 2008; Takasaki *et al.*, 2000a). Especially in the model for chronic inflammatory neuropathy, effectiveness of opioids is increased as compared with normal animals (Hylden *et al.*, 1991; Narita *et al.*, 2008).

III. Alternative Neural Changes in Morphine-Resistant Neuropathic Pain States

The mechanism underlying neuropathic pain is considered to be complicated. Since the effectiveness of opioids is dramatically reduced in intrathecal injection as compared with subcutaneous injection at the neuropathic pain states (Lee *et al.*, 1995; Zurek *et al.*, 2001), one possible mechanism that has been suggested for the occurrence of opioid resistance in the neuropathic pain states is that opioid mechanisms at spinal cord may be disturbed (Bian *et al.*, 1995; DeGroot *et al.*, 1997). Therefore, many studies have focused on the long-term changes in the neural functions of the dorsal spinal cord and these include changes occurring at the receptor level, protein kinases, peptides, and so on, in the course of neuropathic pain states. Fundamental functional changes in dorsal spinal cord during neuropathic pain states involve downregulation or desensitization of μ-opioid receptors. Downregulation and desensitization of μ-opioid receptors in dorsal spinal cord, observed in the nerve-injured neuropathy (DeGroot *et al.*, 1997; Porreca *et al.*, 1998) and diabetic neuropathy (Chen and Pan, 2003; Chen *et al.*, 2002), may be related to increased production of protein kinase C (Mao *et al.*, 1995b; Narita *et al.*, 2004; Roberts and McLean, 1997; Yajima *et al.*, 2003). There are two possible mechanisms for increase of protein kinase C in dorsal spinal cord at neuropathic pain states. One is activation of *N*-methyl-D-aspartate (NMDA) receptors in postsynaptic cells (Mao *et al.*, 1995a,b). The second mechanism is an autophosphorylation of TrkB receptor by brain-derived neurotrophic factor

(BDNF) (Narita *et al.*, 2000; Yajima *et al.*, 2005). In fact, the development of the hyperalgesia and allodynia in neuropathic pain states is suppressed by administration of NMDA receptor antagonists, TrkB/Fc chimera protein (sequesters endogenous BDNF), or protein kinase C inhibitors (Mao *et al.*, 1992, 1995a; Yajima *et al.*, 2005).

An endogenous opioid peptide dynorphin A also shows remarkable changes in dorsal spinal cord at the neuropathic pain states; in fact, immunoreactivity for dynorphin A and mRNA level of prodynorphin are dramatically increased in ipsilateral dorsal spinal cord (Kajander *et al.*, 1990; Wang *et al.*, 2001). Dynorphin A, originally identified as an endogenous κ-opioid peptide, also acts as agonist for NMDA receptor to modulate the transmission of pain. Treatment with antiserum against dynorphin A suppresses hyperalgesia/allodynia after sciatic nerve ligation and in prodynorphin knockout mice the persistent chronic hyperalgesia/allodynia is not observed (Wang *et al.*, 2001). The increase in dynorphin A is also suggested to be involved in opioid resistance in the course of neuropathic pain (Vanderah *et al.*, 2000).

Changes in ion channel expressions also may contribute to the development or maintenance of neuropathic pain. Following peripheral nerve injury, the expression of several subtypes of voltage-dependent Na^+ channels is increased in the primary afferent neuron and secondary dorsal horn neuron (Hains *et al.*, 2004; Yang *et al.*, 2004). Pain-related behaviors after peripheral nerve injury are attenuated by the suppression of Na^+ channel expression by antisense oligodeoxynucleotides (Hains *et al.*, 2004). On the other hand, increased expression of α_2-δ_1 subunit of voltage-dependent Ca^{2+} channels also reported after the peripheral nerve injury (Narita *et al.*, 2007; Yang *et al.*, 2004). Gabapentin, a Ca^{2+} channel blocker binding to the α_2-δ_1 subunit, can inhibit the hyperalgesia/allodynia after the sciatic nerve ligation (Narita *et al.*, 2007). The α_2-δ_1 subunit is a one of the subunit construct N-type voltage-dependent Ca^{2+} channel. Selective antagonists for N-type Ca^{2+} channel also suppress the hyperalgesia/allodynia at the neuropathic pain states (Scott *et al.*, 2002).

IV. New Drug Therapy for Neuropathic Pain

With the growing evidence regarding the alternative neural changes related to neuropathic pain, several drugs effective to neuropathic pain have been identified basing on the information regarding related neural changes. The identified drugs contain ketamine, methadone, gabapentin, and so on. Ketamine and methadone are noncompetitive antagonists for NMDA receptors (Gorman *et al.*, 1997; Wong *et al.*, 1988). The activation of the NMDA-receptor/channel complex at the spinal cord level and in the brain is related to the activation of the excitatory glutamatergic nociceptive pathway. As stated above, activation of NMDA receptors is an important mechanism for the development of hyperalgesia/allodynia and opioid resistance occurring during neuropathic pain. In clinic, likewise under experimental conditions, ketamine shows remarkable antihyperalgesic/antiallodynic effects with some side effect (psychotomimetic effect), which can be treated with benzodiazepines (Okon, 2007). Results from animal experiments suggest the advantage of combination use of NMDA receptor antagonists with morphine. The use-dependent NMDA receptor antagonist, MK-801, potentiates morphine antinociception (Grass *et al.*, 1996) and inhibits the development of tolerance and dependence of morphine (Kim *et al.*, 1996; Trujillo and Akil, 1991). Methadone is proposed as NMDA receptor antagonist for management of neuropathic pain. Methadone is a μ-opioid receptor agonist with an antagonist profile toward NMDA receptor. With its potent antinociceptive effect and low dependence liability, methadone is used in opioid agonist therapy for heroin addictive patients in clinic (Alford *et al.*, 2006; Rhodin *et al.*, 2006). Recent clinical trials show that methadone is very effective in suppressing the hyperalgesia and allodynia in neuropathic pain and to improve the patient's quality of life (Moulin *et al.*, 2005).

Gabapentin was originally developed as an antiepileptic drug. However, it has been extensively used to treat painful neuropathies, including the nerve-injury neuropathy, diabetic neuropathy, postherpetic neuralgia, and neuropathic cancer pain (Dworkin *et al.*, 2007). The mechanism of action of gabapentin is mediated by binding to the α_2-δ_1 subunit of the presynaptic voltage-dependent Ca^{2+} channels, which are upregulated in the dorsal root ganglia and spinal cord under neuropathic pain states (Narita *et al.*, 2007; Yang *et al.*, 2004). Gabapentin may produce antihyperalgesic and antiallodynic effects by inhibiting Ca^{2+} influx via these channels and, subsequently, inhibiting the release of excitatory neurotransmitters (e.g., glutamate, substance P) from the primary afferent nerve fibers sensitized during neuropathic pain states (Taylor, 2004). Because of the lack of hepatic metabolism and low protein binding, gabapentin has no clinically relevant drug interactions (Rose and Kam, 2002). Gabapentin may also prevent opioid tolerance (Gilron *et al.*, 2003) and may have anxiolytic properties (Pollack *et al.*, 1998). The major side effects of gabapentin are somnolence and dizziness, which are reduced by gradual dosage titration, and peripheral edema (Dworkin *et al.*, 2003). However, its generally excellent tolerability, safety, and lack of drug interactions distinguish gabapentin from most other oral medications used for the treatment of chronic neuropathic pain.

As stated above, opioids generally do not show remarkable antihyperalgesic and antiallodynic effects against most of the neuropathic pain. However, among many opioids used in clinic, oxycodone shows excellent antihyperalgesic and antiallodynic effects against the neuropathic pain (Watson and Babul, 1998;

Watson *et al.*, 2003). Oxycodone is a semisynthetic opioid analgesic derived from the natural alkaloid, thebain, and has been in clinical use since 1917. Oxycodone has a high oral bioavailability and seems to provide as much as potent analgesic effect as morphine. Like other opioids, the analgesic effect of oxycodone is mediated through the activation of μ-opioid receptor, although its affinity and intrinsic activity to μ-opioid receptor is remarkably lower than morphine (Lemberg *et al.*, 2006; Narita *et al.*, 2008). At present, the mechanisms underlying the differential antinociceptive profile shown by oxycodone as compared to morphine is not known. It has been recently reported that in addition to μ-opioid receptors, oxycodone shows moderate affinity to morphine-insensitive component (probably putative κ_{2b}-opioid receptor) (Nielsen *et al.*, 2007). Analgesic effect of oxycodone may be mediated through the activation of both μ- and additional opioid receptor components. This hypothesis is supported by the evidence that oxycodone shows asymmetric antinociceptive cross-tolerance with morphine. The latter shows antinociceptive cross-tolerance in oxycodone-tolerant animals, while oxycodone does not show antinociceptive cross-tolerance in morphine-tolerant animals (Nielsen *et al.*, 2000).

More recently, we have reported the analgesic profile of amidino-TAPA, another μ-opioid receptor agonist, extremely effective against neuropathic pain. Unlike other μ-opioid receptor agonists, antiallodynic effect of amidino-TAPA is not altered at all in the mouse model of the nerve-injured neuropathy (unpublished observation). Amidino-TAPA is a tetrapeptide derivative of dermorphin that we recently developed (Ogawa *et al.*, 2002), showing extremely high affinity and relatively high selectivity to μ-opioid receptors (Mizoguchi *et al.*, 2007). Unlike other peptidic analgesics, amidino-TAPA shows a potent and long-lasting antinociceptive effect even following peripheral injections. The potency of amidino-TAPA for antinociception is 10 times, 1.2 times, and 650 times higher than that of morphine in the subcutaneous, oral, and intrathecal routes of injection, respectively (Mizoguchi *et al.*, 2007; Ogawa *et al.*, 2002). We recently reported that the antinociceptive effect of amidino-TAPA injected intrathecally is mediated through mechanisms different from traditional μ-opioid receptor agonist DAMGO (Mizoguchi *et al.*, 2007). Unlike DAMGO and morphine, the antinociception induced by amidino-TAPA is mediated through the release of endogenous κ-opioid peptides dynorphin A, dynorphin B, and α-neo-endorphin, which are revealed by the activation of μ-opioid receptor. Similar antinociceptive profile was reported in endogenous μ-opioid peptide endomorphin-2, which releases dynorphin A via μ-opioid receptor activation (Mizoguchi *et al.*, 2006; Sakurada *et al.*, 2001). Endomorphin-2 shows potent antiallodynic effect against neuropathic pain (Przewlocka *et al.*, 1999). Additional antinociceptive profiles of amidino-TAPA or endomorphin-2 may be related to their potent antiallodynic effect against neuropathic pain.

V. Conclusion

Recent findings in the neural changes related to neuropathic pain states provide new targets for novel drug therapy of neuropathic pain. Among others, methadone, ketamine, and gabapentin are effective drugs that have already been evaluated in clinic, as well as in animal experiments. On the other hand, oxycodone, a μ-opioid receptor agonist with different pharmacological profile as compared to morphine, shows remarkable effectiveness against neuropathic pain. For the management of neuropathic pain, oxycodone provides an alternative therapeutic approach that possesses a differential pharmacological profile as compared with morphine.

Acknowledgments

This work was supported by The Science Research Promotion Fund from The Promotion and Mutual Aid Corporation for Private Schools of Japan, a Grant-in-Aid for Scientific Research (C) (KAKENHI 17590065, 18613015 and 19603011) from the Japan Society for the Promotion of Science, and a Grant-in-Aid for High Technology Research Program from the Ministry of Education, Culture, Sports, Science and Technology Japan.

References

Alford, D. P., Compton, P., and Samet, J. H. (2006). Acute pain management for patients receiving maintenance methadone or buprenorphine therapy. *Ann. Intern. Med.* **144,** 127–134.

Arnér, S., and Meyerson, B. A. (1988). Lack of analgesic effect of opioids on neuropathic and idiopathic forms of pain. *Pain* **33,** 11–23.

Bennett, G. J., and Xie, Y. K. (1988). A peripheral mononeuropathy in rat that produces disorders of pain sensation like those seen in man. *Pain* **33,** 87–107.

Bian, D., Nichols, M. L., Ossipov, M. H., Lai, J., and Porreca, F. (1995). Characterization of the antiallodynic efficacy of morphine in a model of neuropathic pain in rats. *Neuroreport* **6,** 1981–1984.

Chen, S. R., and Pan, H. L. (2003). Antinociceptive effect of morphine, but not mu opioid receptor number, is attenuated in the spinal cord of diabetic rats. *Anesthesiology* **99,** 1409–1414.

Chen, S. R., Sweigart, K. L., Lakoski, J. M., and Pan, H. L. (2002). Functional mu opioid receptors are reduced in the spinal cord dorsal horn of diabetic rats. *Anesthesiology* **97,** 1602–1608.

Courteix, C., Bourget, P., Caussade, F., Bardin, M., Coudore, F., Fialip, J., and Eschalier, A. (1998). Is the reduced efficacy of morphine in diabetic rats caused by alterations of opiate receptors or of morphine pharmacokinetics? *J. Pharmacol. Exp. Ther.* **285,** 63–70.

Covington, E. C. (1998). Anticonvulsants for neuropathic pain and detoxification. *Cleve. Clin. J. Med.* **65**(Suppl. 1), S121–S129, S145–S147.

DeGroot, J. F., Coggeshall, R. E., and Carlton, S. M. (1997). The reorganization of mu opioid receptors in the rat dorsal horn following peripheral axotomy. *Neurosci. Lett.* **233,** 113–116.

Dworkin, R. H., Backonja, M., Rowbotham, M. C., Allen, R. R., Argoff, C. R., Bennett, G. J., Bushnell, M. C., Farrar, J. T., Galer, B. S., Haythornthwaite, J. A., Hewitt, D. J., Loeser, J. D., *et al.* (2003). Advances in neuropathic pain: Diagnosis, mechanisms, and treatment recommendations. *Arch. Neurol.* **60,** 1524–1534.

Dworkin, R. H., O'Connor, A. B., Backonja, M., Farrar, J. T., Finnerup, N. B., Jensen, T. S., Kalso, E. A., Loeser, J. D., Miaskowski, C., Nurmikko, T. J., Portenoy, R. K., Rice, A. S. C., *et al.* (2007). Pharmacologic management of neuropathic pain: Evidence-based recommendations. *Pain* **132,** 237–251.

Eisenberg, E., McNicol, E. D., and Carr, D. B. (2005). Efficacy and safety of opioid agonists in the treatment of neuropathic pain of nonmalignant origin: Systematic review and meta-analysis of randomized controlled trials. *JAMA* **293,** 3043–3052.

Eisenberg, E., McNicol, E. D., and Carr, D. B. (2006). Efficacy of mu-opioid agonists in the treatment of evoked neuropathic pain: Systematic review of randomized controlled trials. *Eur. J. Pain* **10,** 667–676.

Gabra, B. H., and Sirois, P. (2005). Hyperalgesia in non-obese diabetic (NOD) mice: A role for the inducible bradykinin B1 receptor. *Eur. J. Pharmacol.* **514,** 61–67.

Gilron, I., Biederman, J., Jhamandas, K., and Hong, M. (2003). Gabapentin blocks and reverses antinociceptive morphine tolerance in the rat pawpressure and tail-flick tests. *Anesthesiology* **98,** 1288–1292.

Gorman, A. L., Elliott, K. J., and Inturrisi, C. E. (1997). The D- and L-isomers of methadone bind to the non-competitive site on the *N*-methyl D-aspartate (NMDA) receptor in rat forebrain and spinal cord. *Neurosci. Lett.* **223,** 5–8.

Gracely, R. H., Lynch, S. A., and Bennett, G. J. (1992). Painful neuropathy: Altered central processing maintained dynamically by peripheral input. *Pain* **51,** 175–194.

Grass, S., Hoffmann, O., Xu, X. J., and Wiesenfeld-Hallin, Z. (1996). *N*-methyl-D-aspartate receptor antagonists potentiate morphine's antinociceptive effect in the rat. *Acta Physiol. Scand.* **158,** 269–273.

Hains, B. C., Saab, C. Y., Klein, J. P., Craner, M. J., and Waxman, S. G. (2004). Altered sodium channel expression in second-order spinal sensory neurons contributes to pain after peripheral nerve injury. *J. Neurosci.* **24,** 4832–4839.

Honore, P., Rogers, S. D., Schwei, M. J., Salak-Johnson, J. L., Luger, N. M., Sabino, M. C., Clohisy, D. R., and Mantyh, P. W. (2000). Murine models of inflammatory, neuropathic and cancer pain each generates a unique set of neurochemical changes in the spinal cord and sensory neurons. *Neuroscience* **98,** 585–598.

Hylden, J. L., Thomas, D. A., Iadarola, M. J., Nahin, R. L., and Dubner, R. (1991). Spinal opioid analgesic effects are enhanced in a model of unilateral inflammation/hyperalgesia: Possible involvement of noradrenergic mechanisms. *Eur. J. Pharmacol.* **194,** 135–143.

Kajander, K. C., Sahara, Y., Iadarola, M. J., and Bennett, G. J. (1990). Dynorphin increases in the dorsal spinal cord in rats with a painful peripheral neuropathy. *Peptides* **11,** 719–728.

Kalso, E., Edwards, J. E., Moore, R. A., and McQuay, H. J. (2004). Opioids in chronic non-cancer pain: Systematic review of efficacy and safety. *Pain* **112,** 372–380.

Kim, S. H., and Chung, J. M. (1992). An experimental model for peripheral neuropathy produced by segmental spinal nerve ligation in the rat. *Pain* **50,** 355–363.

Kim, H. S., Jang, C. G., and Park, W. K. (1996). Inhibition by MK-801 of morphineinduced conditioned place preference and postsynaptic dopamine receptor supersensitivity in mice. *Pharmacol. Biochem. Behav.* **55,** 11–17.

Lee, Y. W., Chaplan, S. R., and Yaksh, T. L. (1995). Systemic and supraspinal, but not spinal, opiates suppress allodynia in a rat neuropathic pain model. *Neurosci. Lett.* **199,** 111–114.

Lemberg, K. K., Kontinen, V. K., Siiskonen, A. O., Viljakka, K. M., Yli-Kauhaluoma, J. T., Korpi, E. R., and Kalso, E. A. (2006). Antinociception by spinal and systemic oxycodone: Why does the route make a difference? *In vitro* and *in vivo* studies in rats. *Anesthesiology* **105,** 801–812.

Luger, N. M., Sabino, M. A., Schwei, M. J., Mach, D. B., Pomonis, J. D., Keyser, C. P., Rathbun, M., Clohisy, D. R., Honore, P., Yaksh, T. L., and Mantyh, P. W. (2002). Efficacy of systemic morphine suggests a fundamental difference in the mechanisms that generate bone cancer vs. inflammatory pain. *Pain* **99,** 397–406.

Mao, J., Price, D. D., Hayes, R. L., Lu, J., and Mayer, D. J. (1992). Differential roles of NMDA and non-NMDA receptor activation in induction and maintenance of thermal hyperalgesia in rats with painful peripheral mononeuropathy. *Brain Res.* **598,** 271–278.

Mao, J., Price, D. D., and Mayer, D. J. (1995a). Experimental mononeuropathy reduces the antinociceptive effects of morphine: Implications for common intracellular mechanisms involved in morphine tolerance and neuropathic pain. *Pain* **61,** 353–364.

Mao, J., Price, D. D., Phillips, L. L., Lu, J., and Mayer, D. J. (1995b). Increases in protein kinase C gamma immunoreactivity in the spinal cord dorsal horn of rats with painful mononeuropathy. *Neurosci. Lett.* **198,** 75–78.

Max, M. B. (1990). Towards physiologically based treatment of patients with neuropathic pain. *Pain* **42,** 131–137.

Mizoguchi, H., Watanabe, H., Hayashi, T., Sakurada, W., Sawai, T., Fujimura, T., Sakurada, T., and Sakurada, S. (2006). Possible involvement of dynorphin A-(1–17) release via μ_1-opioid receptors in spinal antinociception by endomorphin-2. *J. Pharmacol. Exp. Ther.* **317,** 362–368.

Mizoguchi, H., Watanabe, C., Watanabe, H., Moriyama, K., Sato, B., Ohwada, K., Yonezawa, A., Sakurada, T., and Sakurada, S. (2007). Involvement of endogenous opioid peptides in the antinociception induced by the novel dermorphin tetrapeptide analog amidino-TAPA. *Eur. J. Pharmacol.* **560,** 150–159.

Moulin, D. E., Palma, D., Watling, C., and Schulz, V. (2005). Methadone in the management of intractable neuropathic noncancer pain. *Can. J. Neurol. Sci.* **32,** 340–343.

Narita, M., Yajima, Y., Aoki, T., Ozaki, S., Mizoguchi, H., Tseng, L. F., and Suzuki, T. (2000). Up-regulation of the TrkB receptor in mice injured by the partial ligation of the sciatic nerve. *Eur. J. Pharmacol.* **401,** 187–190.

Narita, M., Kuzumaki, N., Suzuki, M., Narita, M., Oe, K., Yamazaki, M., Yajima, Y., and Suzuki, T. (2004). Increased phosphorylated-mu-opioid receptor immunoreactivity in the mouse spinal cord following sciatic nerve ligation. *Neurosci. Lett.* **354,** 148–152.

Narita, M., Nakajima, M., Miyoshi, K., Narita, M., Nagumo, Y., Miyatake, M., Yajima, Y., Yanagida, K., Yamazaki, M., and Suzuki, T. (2007). Role of spinal voltage-dependent calcium channel alpha 2 delta-1 subunit in the expression of a neuropathic pain-like state in mice. *Life Sci.* **80,** 2015–2024.

Narita, M., Nakamura, A., Ozaki, M., Imai, S., Miyoshi, K., Suzuki, M., and Suzuki, T. (2008). Comparative pharmacological profiles of morphine and oxycodone under a neuropathic pain-like state in mice: Evidence for less sensitivity to morphine. *Neuropsychopharmacology* **33,** 1097–1112.

Nielsen, C. K., Ross, F. B., and Smith, M. T. (2000). Incomplete, asymmetric, and route-dependent cross-tolerance between oxycodone and morphine in the Dark Agouti rat. *J. Pharmacol. Exp. Ther.* **295,** 91–99.

Nielsen, C. K., Ross, F. B., Lotfipour, S., Saini, K. S., Edwards, S. R., and Smith, M. T. (2007). Oxycodone and morphine have distinctly different pharmacological profiles: Radioligand binding and behavioural studies in two rat models of neuropathic pain. *Pain* **132,** 289–300.

Ogawa, T., Miyamae, T., Murayama, K., Okuyama, K., Okayama, T., Hagiwara, M., Sakurada, S., and Morikawa, T. (2002). Synthesis and structure-activity relationships of an orally available and long-acting analgesic peptide, N^{α}-amidino-Tyr-D-Arg-Phe-MeβAla-OH (ADAMB). *J. Med. Chem.* **45,** 5081–5089.

Okon, T. (2007). Ketamine: An introduction for the pain and palliative medicine physician. *Pain Physician* **10,** 493–500.

Okuse, K., Chaplan, S. R., McMahon, S. B., Luo, Z. D., Calcutt, N. A., Scott, B. P., Akopian, A. N., and Wood, J. N. (1997). Regulation of expression of the sensory neuron-specific sodium channel SNS in inflammatory and neuropathic pain. *Mol. Cell. Neurosci.* **10,** 196–207.

Ossipov, M. H., Lopez, Y., Nichols, M. L., Bian, D., and Porreca, F. (1995). Inhibition by spinal morphine of the tail-fick response is attenuated in rats with nerve ligation injury. *Neurosci. Lett.* **199,** 83–86.

Pollack, M. H., Matthews, J., and Scott, E. L. (1998). Gabapentin as a potential treatment for anxiety disorders. *Am. J. Psychiatry* **155,** 992–993.

Porreca, F., Tang, Q. B., Bian, D., Riedl, M., Elde, R., and Lai, J. (1998). Spinal opioid mu receptor expression in lumbar spinal cord of rats following nerve injury. *Brain Res.* **795,** 197–203.

Portenoy, R. K., Foley, K. M., and Inturrisi, C. E. (1990). The nature of opioid responsiveness and its implications for neuropathic pain: New hypotheses derived from studies of opioid infusions. *Pain* **43,** 273–286.

Przewlocka, B., Mika, J., Labuz, D., Toth, G., and Przewlocki, R. (1999). Spinal analgesic action of endomorphins in acute, inflammatory and neuropathic pain in rats. *Eur. J. Pharmacol.* **367,** 189–196.

Rashid, M. H., and Ueda, H. (2002). Nonopioid and neuropathy-specific analgesic action of the nootropic drug nefiracetam in mice. *J. Pharmacol. Exp. Ther.* **303,** 226–231.

Rhodin, A., Grönbladh, L., Nilsson, L. H., and Gordh, T. (2006). Methadone treatment of chronic non-malignant pain and opioid dependence-a long-term follow-up. *Eur. J. Pain* **10,** 271–278.

Roberts, R. E., and McLean, W. G. (1997). Protein kinase C isozyme expression in sciatic nerves and spinal cords of experimentally diabetic rats. *Brain Res.* **754,** 147–156.

Rose, M. A., and Kam, P. C. (2002). Gabapentin: Pharmacology and its use in pain management. *Anaesthesia* **57,** 451–462.

Sakurada, S., Hayashi, T., Yuhki, M., Orito, T., Zadina, J. E., Kastin, A. J., Fujimura, T., Murayama, K., Sakurada, C., Sakurada, T., Narita, M., Suzuki, T., *et al.* (2001). Differential antinociceptive effects induced by intrathecally administered endomorphin-1 and endomorphin-2 in the mouse. *Eur. J. Pharmacol.* **427,** 203–210.

Scott, D. A., Wright, C. E., and Angus, J. A. (2002). Actions of intrathecal omega-conotoxins CVID, GVIA, MVIIA, and morphine in acute and neuropathic pain in the rat. *Eur. J. Pharmacol.* **451,** 279–286.

Shimoyama, M., Tanaka, K., Hasue, F., and Shimoyama, N. (2002). A mouse model of neuropathic cancer pain. *Pain* **99,** 167–174.

Shir, Y., and Seltzer, Z. (1990). A-fibers mediate mechanical hyperesthesia and allodynia and C-fibers mediate thermal hyperalgesia in a new model of causalgiform pain disorders in rats. *Neurosci. Lett.* **115,** 62–67.

Sindrup, S. H., and Jensen, T. S. (1999). Efficacy of pharmacological treatments of neuropathic pain: An update and effect related to mechanism of drug action. *Pain* **83,** 389–400.

Takasaki, I., Andoh, T., Nitta, M., Takahata, H., Nemoto, H., Shiraki, K., Nojima, H., and Kuraishi, Y. (2000a). Pharmacological and immunohistochemical characterization of a mouse model of acute herpetic pain. *Jpn. J. Pharmacol.* **83,** 319–326.

Takasaki, I., Andoh, T., Shiraki, K., and Kuraishi, Y. (2000b). Allodynia and hyperalgesia induced by herpes simplex virus type-1 infection in mice. *Pain* **86,** 95–101.

Taylor, C. P. (2004). The biology and pharmacology of calcium channel α_2-δ proteins. *CNS Drug Rev.* **10,** 183–188.

Trujillo, K. A., and Akil, H. (1991). Inhibition of morphine tolerance and dependence by the NMDA receptor antagonist MK-801. *Science* **251,** 85–87.

Vanderah, T. W., Gardell, L. R., Burgess, S. E., Ibrahim, M., Dogrul, A., Zhong, C. M., Zhang, E. T., Malan, T. P. Jr., Ossipov, M. H., Lai, J., and Porreca, F. (2000). Dynorphin promotes abnormal pain and spinal opioid antinociceptive tolerance. *J. Neurosci.* **20,** 7074–7079.

Wang, Z., Gardell, L. R., Ossipov, M. H., Vanderah, T. W., Brennan, M. B., Hochgeschwender, U., Hruby, V. J., Malan, T. P., Jr., Lai, J., and Porreca, F. (2001). Pronociceptive actions of dynorphin maintain chronic neuropathic pain. *J. Neurosci.* **21,** 1779–1786.

Watson, C. P. N., and Babul, N. (1998). Efficacy of oxycodone in neuropathic pain: A randomized trial in postherpetic neuralgia. *Neurology* **50,** 1837–1841.

Watson, C. P., Moulin, D., Watt-Watson, J., Gordon, A., and Eisenhoffer, J. (2003). Controlled-release oxycodone relieves neuropathic pain: A randomized controlled trial in painful diabetic neuropathy. *Pain* **105,** 71–78.

Wong, E. H., Knight, A. R., and Woodruff, G. N. (1988). [^{3}H]MK-801 labels a site on the *N*-methyl-D-aspartate receptor channel complex in rat brain membranes. *J. Neurochem.* **50,** 274–281.

Yajima, Y., Narita, M., Shimamura, M., Narita, M., Kubota, C., and Suzuki, T. (2003). Differential involvement of spinal protein kinase C and protein kinase A in neuropathic and inflammatory pain in mice. *Brain Res.* **992,** 288–293.

Yajima, Y., Narita, M., Usui, A., Kaneko, C., Miyatake, M., Narita, M., Yamaguchi, T., Tamaki, H., Wachi, H., Seyama, Y., and Suzuki, T. (2005). Direct evidence for the involvement of brain-derived neurotrophic factor in the development of a neuropathic pain-like state in mice. *J. Neurochem.* **93,** 584–594.

Yang, L., Zhang, F. X., Huang, F., Lu, Y. J., Li, G. D., Bao, L., Xiao, H. S., and Zhang, X. (2004). Peripheral nerve injury induces trans-synaptic modification of channels, receptors and signal pathways in rat dorsal spinal cord. *Eur. J. Neurosci.* **19,** 871–883.

Yarnitsky, D., and Eisenberg, E. (1998). Neuropathic pain: Between positive and negative ends. *Pain Forum* **7,** 241–242.

Zhang, W., Slusher, B., Murakawa, Y., Wozniak, K. M., Tsukamoto, T., Jackson, P. F., and Sima, A. A. (2002). GCPII (NAALADase) inhibition prevents long-term diabetic neuropathy in type 1 diabetic BB/Wor rats. *J. Neurol. Sci.* **194,** 21–28.

Zurek, J. R., Nadeson, R., and Goodchild, C. S. (2001). Spinal and supraspinal components of opioid antinociception in streptozotocin induced diabetic neuropathy in rats. *Pain* **90,** 57–63.

REGULATED EXOCYTOSIS FROM ASTROCYTES: PHYSIOLOGICAL AND PATHOLOGICAL RELATED ASPECTS

Corrado Calì, Julie Marchaland, Paola Spagnuolo, Julien Gremion, and Paola Bezzi

Department of Cellular Biology and Morphology (DBCM), Faculty of Medicine, University of Lausanne, rue du Bugnon 9, 1005 Lausanne, Switzerland

Astrocytes have traditionally been considered ancillary, satellite cells of the nervous system. However, it is a very recent acquisition that glial cells generate signaling loops which are integral to the brain circuitry and participate, interactively with neuronal networks, in the processing of information. Such a conceptual breakthrough makes this field of investigation one of the hottest in neuroscience, as it calls for a revision of past theories of brain function as well as for new strategies of experimental exploration of brain function. Glial cells are electrically not excitable, and it was only the use of optical recording techniques together with calcium sensitive dyes, that allowed the chemical excitability of glial cells to become apparent. Studies using these new techniques have shown for the first time that glial cells are activated by surrounding synaptic activity and translate neuronal signals into their own calcium code. Intracellular calcium concentration ($[Ca^{2+}]_i$) elevations in glial cells have then shown to underlie spatial transfer of information in the glial network, accompanied by release of chemical transmitters (gliotransmitters) such as glutamate and back-signaling to neurons. As a consequence, optical imaging techniques applied to cell cultures or intact tissue have become a state-of-the-art technology for studying glial cell signaling.

INTERNATIONAL REVIEW OF NEUROBIOLOGY, VOL. 85
DOI: 10.1016/S0074-7742(09)85020-4

The molecular mechanisms leading to release of "gliotransmitters," especially glutamate, from glia are under debate. Accumulating evidence clearly indicates that astrocytes secrete numerous transmitters by Ca^{2+}-dependent exocytosis. This review will discuss the mechanisms underlying the release of chemical transmitters from astrocytes with a particular emphasis to the regulated exocytosis processes.

I. Introduction

Since the first description of glial cells, Camillo Golgi (1843–1926) and Santiago Ramón y Cajal (1852–1934) recognized that astrocytes are located in strategic positions between neurons and capillaries to act as a conduit for signals between different cells types in the central nervous system (CNS; Ramon y Cajal, 1899). Later on, examination of the nervous system at the ultrastructural level has shown that astrocytes can be intimately associated with synapses, literally enwrapping many pre- and postsynaptic terminals. For instance, in the hippocampus, 57% of the axon–spine interfaces are associated with astrocytes (Ventura and Harris, 1999). It is likely that this close physical relationship provides an opportunity for many functional interactions between astrocytes and neurons.

Despite these important morphological indications, a role for glia in information processing has been neglected for a long time. The most important reason for this is related to the fact that neurophysiologists have taken advantage of the electrical properties of neurons to investigate neuronal signaling *in situ* and *in vitro*. Glial cells are electrically nonexcitable instead and exhibit a complex array of cellular processes making difficult to study the points of interaction with neurons and sites of both release of chemical transmitters and of neuroligand stimulation.

In the late 1980s and early 1990s, electrophysiological and immunohistochemical studies began to appear, which demonstrated that astrocytes *in situ* expressed neuroligand-gated ion channels and specific receptors. Later on (mid-1990s), the discovery of new fluorescence tools for studying intracellular ions in living cells and the application of Ca^{2+}-imaging technology provided the technological capability to make breakthroughs in our understanding of astrocyte functions. During that period many laboratories demonstrated that astrocytes *in situ* express a variety of neuroligand receptors linked to Ca^{2+} mobilization. Thanks to number of studies performed in acute slices, it is now clear that the *in situ* astrocytes exhibit a wide variety of neuroligand receptors linked through G proteins to the mobilization of internal Ca^{2+} stores (Fiacco and McCarthy, 2006; Porter and McCarthy, 1996) and that the astrocytic receptor systems are activated by neurotransmitters released from synaptic terminals during neuronal activity

(Araque *et al.*, 2001; Carmignoto, 2000; Haydon and Carmignoto, 2006; Pasti *et al.*, 1997; Porter and McCarthy, 1996). These results constituted the first major breakthrough that has provided a completely new understanding of the physiological role of astrocytes in the neuronal–glial signaling system. Hence, an increasing number of observations outlined that glial cells, despite generating neither conducting action potentials nor apparently forming synapses with neuronal cells, interact with neurons in a more complex manner than simply providing structural, metabolic, and trophic support. These findings have encouraged more detailed research on glial cells.

In the past 15 years it has become increasingly clear that signaling between neurons and astrocytes may play a crucial role in the information processing that the brain carries out (Araque *et al.*, 2001; Bezzi and Volterra, 2001; Haydon and Carmignoto, 2006; Santello and Volterra, 2008; Volterra and Meldolesi, 2005; Volterra and Steinhauser, 2004). While it is now clear that astrocytes respond to neurotransmitters released during synaptic activity, a still controversial issue in glial biology is whether neuronal activity, by inducing calcium waves in astrocytes, induces secretion of neuroactive substances from astrocytes back onto synapses in a process known as gliotransmission (Barres, 2008). The reason why many neuroscientists are still skeptical about gliotransmission is the lack of a detailed description of the basic cell biological properties of astrocytes; the secretory aspect is a typical example.

II. Are Astrocytes Specialized Secretory Cells?

A. Astrocytes Contain and Release Chemical Transmitters (Gliotransmitters)

The existence of communication systems based on the release of chemical transmitters between astrocytes and other brain cells was first hypothesized at the end of the 1980 supported by the observation that glial cells contain, synthesize, and release a variety of compounds (Martin, 1992). At that time, however, the view of glia as passive elements in the CNS continued to dominate scientific thinking and the whole concept remained largely hypothetical. It is now clear that astrocytes are highly secretive cells (Volterra and Bezzi, 2002); they are able to synthesize and release numbers of different chemical transmitters such as (i) excitatory and inhibitory aminoacids (D-serine, homocysteic acid or quinolinic acid, glutamate, aspartate, GABA, glycine, taurine, etc.), (ii) other classical neurotransmitters (acetylcholine, noradrenaline, dopamine, serotonin, histamine, etc.), (iii) ATP and related nucleotides and nucleosides, (purine nucleotides, ATP, GTP and their diphosphate and monophosphate derivates, adenosine and guanosine, UTP), (iv) eicosanoids and other lipid mediators (cyclooxygenase—COX-products: prostaglandins (PGs), lipooxygenase—LOX-products, epooxygenase—EPOX-products), (v) neuropeptides (proenkephalin, angiotensinogen, endothelins), (vi) neurotrophins (nerve growth factor, neurotrophin-3, brain-derived neurotrophic factor), (vii) cytokines (interleukins (IL), interferons (IFN), tumor necrosis factors (TNF)), and (viii) structurally associated chemokines and growth factors (Bergami *et al.*, 2008; Bezzi and Volterra, 2001; Blum *et al.*, 2008; Fields and Stevens, 2000; Fujita *et al.*, 2009; Hussy *et al.*, 2000; Kang *et al.*, 2008; Liu *et al.*, 2008; Medhora, 2000; Sanzgiri *et al.*, 1999; Snyder and Kim, 2000). For many of these, we do not know the mechanism(s) of release and their dynamic properties yet. For instance, for eicosanoid, we know that all type of glial cells possess the enzymatic machinery for producing arachidonic acid and downstream metabolites. COX, LOX, and EPOX enzymes are constitutively expressed in cultured astrocytes, their activity increase with intracellular Ca^{2+} elevations and their formation is stimulated by endogenous ligands. However, only few studies have reported their dynamics: eicosanoids are rapidly released upon G protein coupled receptor (GPCR) stimulation (Bezzi *et al.*, 1998, 2001) by a pulsatile fashion (Zonta *et al.*, 2003) but the exact mechanism and site(s) of PGs action are not fully elucidated. Similarly, cytokines and chemokines are known to be produced in the brain in response to pathophysiological stimuli and to participate in inflammation. For this reason, whether cytokines play roles in physiological brain processes including rapid communication and synaptic transmission remained largely unexplored. Recently, it has been shown that TNFα can be released from astrocytes and microglia in response to physiological stimuli (SDF-1α; Bezzi *et al.*, 2001; Domercq *et al.*, 2006) with a time course in the order of seconds and therefore compatible with rapid signaling.

During the last 15 years, numbers of laboratories focused their studies on mechanisms of amino acid release from astrocytes (Malarkey and Parpura, 2008); several different mechanisms of release from glia have been documented, including a Ca^{2+}-dependent process, as well as a number of Ca^{2+}-independent processes involving either (i) volume-sensitive organic anion channels (Haskew-Layton *et al.*, 2008; Kimelberg *et al.*, 1990; Mongin and Kimelberg, 2002), (ii) hemichannels (Cotrina *et al.*, 1998; Stout *et al.*, 2002; Ye *et al.*, 2003), (iii) P2X7 receptor channels (Duan *et al.*, 2003; Kukley *et al.*, 2001), (iv) reversed operation of reuptake carriers (Attwell *et al.*, 1993; Longuemare and Swanson, 1997; Re *et al.*, 2006; Rossi *et al.*, 2000; Szatkowski *et al.*, 1990; Volterra *et al.*, 1996), or (v) exchange via the cystine–glutamate antiporter (Allen *et al.*, 2001; Baker *et al.*, 2002; Bender *et al.*, 2000; Moran *et al.*, 2003, 2005; Shanker and Aschner, 2001; Tang and Kalivas, 2003). Swelling of glial cells stimulates release of osmolytes through the volume-sensitive anion channels. Interestingly, these include neuroactive amino acids like taurine, aspartate, and glutamate (Kimelberg *et al.*, 1990). The purpose of their release may be twofold: on one hand, to decrease the volume of the swollen cell; on the other, to signal to neighboring cells. This may occur during physiological processes and serve regulatory functions, or under pathological conditions, including traumatic injuries, ischemia, and hepatic encephalopathy, and contribute to brain damage. The signaling mechanism linking osmotic-dependent volume variation to channel opening and osmolyte release is not fully defined. Most studies indicate that hypo-osmotic release is Ca^{2+}-independent.

The reversed operation of reuptake carriers may occur during pathological conditions as well. In certain conditions, they can run backwards, pumping transmitter out of cells and serving as a Ca^{2+}-independent, nonvesicular mechanism for transmitter release, and intercellular communication (Attwell *et al.*, 1993). Direct evidence of this reversed function exists for GABA and glutamate transporters of both neuronal and glial cells. Glutamate uptake carriers are transporters with the peculiar stoichiometry of 3 Na^+ + 1 H^+ cotransported inside the cells together with one glutamate molecule and 1 K^+ countertransported outside (Danbolt *et al.*, 1992). The pioneering work of Attwell and collaborators, based on recording glutamate uptake currents from Müller glial cells of salamander retina, directly demonstrated that glutamate transport runs backwards when external K^+ rises significantly and enough Na^+ and glutamate are present intracellularly (Szatkowski *et al.*, 1990).

Thus, some of the amino acids are released via multiple pathways, activated under different conditions, at different loci and/or with different modalities and, therefore, implying distinct functional consequences (Vesce *et al.*, 2007): for example, glutamate can be released by any of the five processes listed above. Among the gliotransmitters, glutamate is the best characterized; the pioneering work of Parpura and collaborators in the 1994 led to the first characterization of a Ca^{2+}-dependent mechanism for release of glutamate from astrocytes (Parpura *et al.*, 1994). This work described for the first time a rapid release of glutamate from cultured astrocytes that apparently has features totally distinct from known processes such as the reversed uptake or swelling-induced release (Attwell *et al.*, 1993). Such features included: (a) receptor-mediated activation by endogenous ligands; and (b) dependence on intracellular Ca^{2+} elevations. Subsequent pharmacological studies directly demonstrated the autonomy of this form of release that operates independently of other glutamate release pathways. After this pionieristic study it was essential that similar experiments were performed in more intact preparations in order to determine whether this phenomenon represented a physiological signaling or merely a curiosity of cell culture. Confocal imaging studies together with results from enzymatic glutamate assays have provided the first compelling demonstration of the Ca^{2+}-dependent glutamate release from astrocytes in hippocampal slices (Bezzi *et al.*, 1998, 2001; Pasti *et al.*, 1997). More recently, a wide range of studies from our and other groups have confirmed these initial observations, strongly supporting the theory that the calcium-dependent glutamate pathway is triggered by activation of diverse

GPCRs, such as chemokine receptor CXCR4 (Bezzi *et al.*, 2001, Calì *et al.*, 2008), or purinergic receptor P2Y1 (Domercq *et al.*, 2006) and required release of calcium from internal stores (Jeremic *et al.*, 2001; Kang *et al.*, 2005; Sanzgiri *et al.*, 1999; Takano *et al.*, 2005). In 2001 we discovered that, in addition to classical transmitters, such as glutamate or ATP, the CXCL12 chemokine induces calcium-dependent glutamate release from astrocytes through a direct activation of the GPCR CXCR4 (Bezzi *et al.*, 2001). Interestingly, the chemokine-mediated glutamate release process seemed to involve a long chain of intracellular and extracellular events related to the release of two chemical mediators, the TNFα and the PGs. Interestingly, the release process was sensitive to inhibitors of neuronal exocytosis such as clostridial toxins (tetanus and botulinum B neurotoxin) and bafilomycin A1 (Baf A1), a blocker of vesicular H^+-ATPase (Araque *et al.*, 2001; Bezzi *et al.*, 1998, 2001, 2004; Domercq *et al.*, 2006; Pasti *et al.*, 2001). These last results suggested a strong similarity with neurosecretion. At that time, however, it was not clear whether glutamate was released directly from the cytosol or whether an exocytosis process was involved. In the last decade, several studies have suggested that glutamate and some few other gliotransmitters such as D-serine, ATP, neuropeptide Y (NPY), atrial natriuretic peptide (ANP), pro-BDNF (Bergami *et al.*, 2008; Cali *et al.*, 2008; Marchaland *et al.*, 2008; Martineau *et al.*, 2008; Mothet *et al.*, 2005; Pangrsic *et al.*, 2007; Pryazhnikov and Khiroug, 2008; Ramamoorthy and Whim, 2008; Zhang *et al.*, 2007), can be released by Ca^{2+}-dependent regulated exocytosis. Referring to glutamate, the evidence for a regulated exocytotic mechanism will be discussed in the next paragraph.

B. First Evidence for the Regulated Exocytosis of Glutamate from Astrocytes

Until few years ago, the most direct evidence that Ca^{2+}-dependent glutamate release occurs via exocytosis came from pharmacological experiments using clostridial neurotoxins and other agents that selectively interfere with neuronal exocytosis. For instance, it was shown that clostridial neurotoxins, including tetanus neurotoxin (TeNT) and the seven serotypes of botulinum neurotoxin (BoNT A–G), blocked Ca^{2+}-dependent glutamate release in astrocytes both in cultures and *in situ* (Bezzi *et al.*, 1998, 2001; Jeftinija *et al.*, 1997; Pascual *et al.*, 2001; Pasti *et al.*, 2001). If clostridial toxins blocked release of glutamate, then astrocytes must express proteins that are substrate for these toxins. In neuronal cells, according to the SNARE (an acronym derived from "SNAP and NSF attachment receptors") hypothesis, the core complex responsible for initiation of exocytosis is formed by three proteins: the vesicular protein, synaptobrevin II, and the two terminal membrane proteins, syntaxin I and SNAP-25 (Jahn and Scheller, 2006;

Sudhof, 2004). Astrocytes located in hippocampal dentate gyrus express all these three proteins, in particular, synaptobrevin II together with its homologue cellubrevin (Bezzi *et al.*, 2004; Jourdain *et al.*, 2007), syntaxin I, and SNAP-23, an isoform of SNAP-25 (Andrea Volterra, personal communication). Astrocytes, at least in cultures, would express also other proteins implicated in neuronal exocytosis, such as synaptotagmin I, synapsin I, syntaxin I, and rab3a (Anlauf and Derouiche, 2005; Calegari *et al.*, 1999; Hepp *et al.*, 1999; Jeftinija *et al.*, 1997; Madison *et al.*, 1996; Maienschein *et al.*, 1999; Parpura *et al.*, 1995).

Similarly for clostridial toxins, inhibition of glutamate release process with Baf A1 predicted that cells must possess organelles expressing proton-dependent vesicular glutamate transporter (VGLUT). Baf A1, in fact, interferes with H^+-ATPase leading to alkalinization of vesicular lumen and collapsing the proton gradient necessary for VGLUT to transport glutamate into glutamatergic vesicles. VGLUTs are vesicles membrane proteins responsible for uptaking glutamate into synaptic vesicle (SV) in glutamatergic terminal (Fremeau *et al.*, 2004). Recent results show the expression of VGLUTs 1, 2, and 3 in cultures (Bezzi *et al.*, 2004; Kreft *et al.*, 2004; Montana *et al.*, 2004; Zhang *et al.*, 2004b), freshly isolated (Bezzi *et al.*, 2004; Montana *et al.*, 2004), and *in vivo* astrocytes (Bezzi *et al.*, 2004; Zhang *et al.*, 2004b), strongly suggesting that astrocytes must possess compartments able to uptake and store glutamate. Despite these indications, a conclusive demonstration that glutamate is released via exocytosis was still missing: neither glutamate-containing vesicles were ultrastructurally identified nor the dynamics of their fusions with the plasma membrane were shown. This issue was strongly debated and unresolved until few years ago; when in collaboration with Vidar Gundersen (Univ of Oslo) and Christian Steinhäuser (Univ of Bonn), our laboratory identified by means of electron microscopy the glutamate storing vesicular compartment in astrocytes from adult hippocampus (Bezzi *et al.*, 2004) with similar morphological and molecular properties with SVs in glutamatergic terminal. Vesicles in astrocytes both *in situ* and in cultures (i) are grouped very close to plasma membrane (about 100 nm), (ii) have a clear appearance (they are not electrondense), (iii) have a small diameter (about 30–50 nm of diameter), and (iv) express proteins responsible both for glutamate uptake (VGLUT) and for fusion event (a v-SNARE protein cellubrevin); the overall characteristics allowed to classify astrocytic vesicles as synaptic-like microvesicles (SLMVs, Bergersen and Gundersen, 2008). Unlike synaptic terminals, which have dense vesicle clusters, electron micrographs of astroglial processes do not display obvious vesicle accumulations (Nedergaard *et al.*, 2002). Hippocampal astrocytes contain groups of 10 VGLUT-positive vesicles often opposite to extrasynaptic dendritic membranes carrying NMDA receptors and in close proximity to presynaptically expressed NMDA receptors (Jourdain *et al.*, 2007). In the hippocampus, 57% of the axon–spine interfaces are associated with astrocytes (Ventura and Harris, 1999); therefore, the proximity of astrocytic VGLUT-containing vesicles to neuronal

NMDA receptors in the hippocampus, suggests that astrocytes *in situ* may focally secrete glutamate onto neurons. Furthermore, because a single astrocyte can make contacts with multiple neurons (Bushong *et al.*, 2002), these nonneuronal cells are also strategically located to act as a bridge and transfer information between neighboring neurons, an intriguing potential role that has only been recently appreciated.

To study whether VGLUT-positive SLMVs take part in exocytosis, astrocytes were transiently transfected with a chimerical fluorescent construct VGLUT-EGFP and, 3 days after the expression of the fluorescent protein, stained briefly with acridine orange (AO), a fluorescent marker of fusion events (Jahn *et al.*, 2003; Tsuboi *et al.*, 2000). Single vesicles beneath the plasma membrane were illuminated with total internal reflection fluorescence (TIRF) imaging apparatus. TIRF imaging has recently allowed spectacular visualization, up to molecular resolution and in real time, of physiological processes occurring on or near the surface of single living cells, including vesicle fusions during exocytosis in synaptic terminals (Steyer and Almers, 2001). Thus, this was an ideal approach to study localization and dynamics of the glutamate-secreting vesicles in astrocytes. Exocytosis in cultured astrocytes was evoked with (RS)-3,5-dihydrophenylglycine (DHPG), a specific agonist of group I of mGluR, which is known to stimulate Ca^{2+}-dependent glutamate release from astrocytes (Bezzi *et al.*, 1998). Fusion of vesicles occurred as a burst (0.8–1 s) involving about 30% of the VGLUT-positive vesicles present in the TIRF field, most likely the readily releasable pool. Moreover, fusion events were abolished by both preexposing cells with the intracellular Ca^{2+} chelator BAPTA/AM and with TeNT which cleaves the SNAREs proteins, blocking vesicles fusion at the level of SNARE complex formation. This study, by revealing that rapid and quantal release of glutamate is not an exclusive property of neurons, represents the first compelling demonstration of the secretory nature of astrocytes, a finding that may have major implications for the understanding of brain signaling processes.

Accumulating evidence clearly indicates that astrocytes secrete numerous transmitters by Ca^{2+}-dependent exocytosis (Bergami *et al.*, 2008; Bezzi *et al.*, 2004; Bowser and Khakh, 2007; Chen *et al.*, 2005; Domercq *et al.*, 2006; Li *et al.*, 2008; Marchaland *et al.*, 2008; Nadrigny *et al.*, 2007; Ramamoorthy and Whim, 2008; Zhang *et al.*, 2004a,b). The available data, however, are far from giving a complete and coherent description of the exo-endocytosis processes in astrocytes. As a result, the existence of a regulated pathway of exo-endocytosis in astrocytes is still an argument widely debated. Critical information is missing concerning the spatial–temporal characteristics of exocytosis, endocytosis and recycling and the underlying stimulus-secretion coupling mechanism. Such information is in fact of great relevance for understanding the functional importance of astrocytes. Indeed, in neurons the speed of SVs endocytosis is critical: SVs recycling provides a mechanism to avoid SVs pool depletion during repetitive action potential firing. Thus, in order to consider exocytosis of glutamate from astrocytes as a physiological mechanism participating in fast chemical signaling in the brain, we must know whether astrocytes possess an apparatus for local recycling able to maintain a continuous steady-state level of release. Moreover, further uncertainty is created by the emerging heterogeneity of the secretory pathways in astrocytes. Various types of organelles have been proposed to underlie exocytosis of gliotransmitters: from SLMVs (Bezzi *et al.*, 2004; Crippa *et al.*, 2006; Jourdain *et al.*, 2007) to dense-core granules (DCGs) (Coco *et al.*, 2003; Ramamoorthy and Whim, 2008; Striedinger *et al.*, 2007) and to lysosomes (Li *et al.*, 2008; Zhang *et al.*, 2007) and to extralarge organelles with several μm-long diameter (Xu *et al.*, 2007). Past studies, however, have not taken into account such heterogeneity and by using generic fluorescent markers of exocytosis they do not distinguish among populations (Bezzi *et al.*, 2004; Coco *et al.*, 2003; Xu *et al.*, 2007; Zhang *et al.*, 2007). The simultaneous recording of fusion signals from distinct populations of secretory organelles leads to significant discrepancies in the kinetic values of exocytosis (Bezzi *et al.*, 2004; Bowser and Khakh, 2007; Chen *et al.*, 2005; Jaiswal *et al.*, 2007). As a consequence, their description of astrocytic secretion is probably not fully accurate, mixing contributions by more than one exocytic organelle population (Bowser and Khakh, 2007; Chen *et al.*, 2005; Jaiswal *et al.*, 2007; Nadrigny *et al.*, 2007; Xu *et al.*, 2007).

C. Astrocytes Possess Different Exocytotic Organelles and Pathways for Regulated Chemical Transmitters (Gliotransmitters) Release

Astrocytes, like specialized secretory cells, contain at least the two major classes of secretory vesicles, SLMVs (Bergersen and Gundersen, 2008), the DCGs which store and release distinct cargo (Hannah *et al.*, 1999; Morgan and Burgoyne, 1997) and lysosomes (Jaiswal *et al.*, 2007; Li *et al.*, 2008; Zhang *et al.*, 2007). In neurons and specialized secretory cells, classical transmitters and peptides are located in SVs and DCGs, respectively (Kupfermann, 1991). These organelles are typically found in different regions of the cells; in particular, in neuronal cells SVs are clustered at the active zones, whereas peptide-containing DCGs are diffusively distributed in axons or dendrites (Hartmann *et al.*, 2001; Meldolesi *et al.*, 2004; Pickel *et al.*, 1995). Table I summarizes and compares our current knowledge about the two pathways of regulated secretion in specialized secretory cells and in astrocytes (Fernandez-Chacon and Sudhof, 1999; Gerber and Sudhof, 2002; Sudhof, 1995). As one can notice, in contrast to secretory cells, the molecular, morphological, and dynamic properties of the secretory processes in astrocytes are largely unknown and need to be investigated.

TABLE I

CURRENT KNOWLEDGE ABOUT REGULATED SECRETORY PATHWAYS IN SECRETORY CELLS AND ASTROCYTES

Secretory cells	Astrocytes
Peptide secretion	
Large secretory granules (>100 nm radius)	Large secretory granules (>100 nm radius)
Recycling via Golgi complex	Yes
Slow sustained secretion (60 ms–30 s)	Yes
Small number of predocked vesicles	*Not yet investigated*
Exocytosis targeted to large plasma membrane section	*Not yet investigated*
Complex secretory mixes (e.g., multiple peptides, Catecholamines, nucleotides)	Complex secretory mixes (e.g., SgII, ATP, ANP, NPY) (?)
Neurotransmitter release	
Small synaptic vesicles (<25 nm radius)	Small SLMV (30–50 nm radius)
Local recycling	Yes, two different modalities of exo-endocytosis
Fast, short-lasting secretion (0.1–6 ms)	Short lasting secretion (about 200 ms)
Large number of predocked vesicle	*No conclusive demonstration*
Exocytosis restricted to specialized zones (e.g., synaptic active zone)	*Not yet investigated*
Release of one or two low-molecular weight compounds (?) no conclusive demonstration	Hypothesized release of one or two low-M compounds (glutamate, D-serine)

Table I summarizes the characteristics of the two principal pathways of regulated exocytosis (peptide secretion and neurotransmitter release) in specialized secretory cells and astrocytes. Briefly, the first pathway (peptide secretion) secretes primarily polypeptides (e.g., insulin), which are initially synthesized in the endoplasmic reticulum, modified in the Golgi apparatus, and processed proteolytically in precursor organelles. These organelles mature to DCGs and are exocytosed in response to appropriate stimuli. After exocytosis, DCGs need to recycle via the Golgi complex in order to be refilled with secretory material. The second pathway (neurotransmitter release) secretes primarily low-molecular weight substances (e.g., classical neurotransmitters such as glutamate or GABA) which are synthesized in the cytosol, taken up into the SLMVs and secreted by regulated exocytosis. In contrast with DCGs, SLMVs recycle locally and can be reused independently of the Golgi complex. Most commonly, regulated exocytosis is stimulated both for SLMVs and DCGs by Ca^{2+} influx, but the dynamics of secretion are very different. In secretory cells, the two pathways differ in the mechanisms by which secretory vesicles are filled with secretory material and by which the vesicles recycle after exocytosis for a new round of secretion. In addition, they exhibit different timing of secretion and distinct physiological properties. Information in astrocytes is mostly missing (indicated by "*not yet investigated*") or not conclusive.

1. *Synaptic-Like Microvesicles*

Synaptic-like microvesicles (SLMVs) represent the best characterized vesicular organelles in astrocytes. They strongly resemble SVs of nerve terminals in size and shape (Bergersen and Gundersen, 2008; Bezzi *et al.*, 2004; Crippa *et al.*, 2006;

Jourdain *et al.*, 2007) and are equipped with transport proteins for uptake of transmitters and with proteins participating in vesicles exo- and endocytosis and recycling. In a recent work from our laboratory, to study the characteristics of exo-endocytosis and recycling of glutamatergic SLMVs in astrocytes, we took advantage of chimerical protein VGLUT1 (VGLUT type 1)-pHluorin (Marchaland *et al.*, 2008). VGLUT1-pHluorin combines the SLMV-specific targeting assured by VGLUT1 to the properties of superecliptic pHluorin, a GFP with modified pH sensitivity (Miesenbock *et al.*, 1998; Ryan and Reuter, 2001). When attached to the luminal domain of a vesicle resident protein, pHluorin allows a direct detection of fusion and retrieval of single vesicles (Balaji and Ryan, 2007; Sankaranarayanan and Ryan, 2000).

The intracellular distribution of VGLUT1-pHluorin (Fig. 1) showed that the vast majority of the GFP-labeled vesicles were positive for the SNARE protein cellubrevin (>90%, Fig. 1A; Bezzi *et al.*, 2004; Crippa *et al.*, 2006), contained glutamate (Fig. 1B) and displayed a diameter (about 40 nm) compatible with SLMVs in other secretory cells. VGLUT1-pHluorin partially overlapped with early or recycling endosomes (~20% and 30%, respectively), *trans*-Golgi network (~8%), and late endosomes/lysosomes (~6%, Fig. 2C). The recycling endosomes positive for VGLUT1-pHluorin, which constitute the 35% of the GFP-labeled organelles, contribute only for 9.5% of the total fusion events evoked by DHPG (Coggins *et al.*, 2007; Marchaland *et al.*, 2008; Park *et al.*, 2006).

In order to define the specific characteristics of exo-endocytosis of glutamatergic SLMVs, we applied for the first time in astrocytes a strategy recently developed for studying the dynamics of glutamatergic SVs at synapses (Voglmaier *et al.*, 2006). This consists in the use of VGLUT1-pHluorin in combination with EPI and TIRF illuminations (EPIi and TIRFi). The use of EPIi in real time with TIRFi was aimed at providing a comparative analysis of pHluorin fluorescence changes in the whole cell and in its submembrane compartment. With our first set of experiments, we defined the kinetics of exocytosis, endocytosis and reacidification of SLMVs at the whole-cell level. A major challenge consisted in separating the exocytosis, and the two components of the endocytic process: the movement of the recently fused vesicles from the plasma membrane to the cytoplasm (out of the EW field) and the reacidification process. This was achieved with "alkaline trapping" experiments in which we isolated the exocytosis signal by blocking reacidification with the proton pump inhibitor Baf A1 (Sankaranarayanan and Ryan, 2001). Once Baf A1 is taken up by SVs during exocytosis, it blocks the proton pump and prevents reacidification, trapping VGLUT1-pHluorin-expressing vesicles in the fluorescent state (Ryan, 2001; Fig. 2A). Addition of Baf A1 thereby eliminates all changes of the pHluorin fluorescence signal due to the reacidification component of recycling to reveal pure exocytosis. In the presence of Baf A1, the rising phase produced by DHPG under EPIi represented a pure measure of SLMVs exocytosis (Fig. 2B). Comparison with the EPIi signal for the

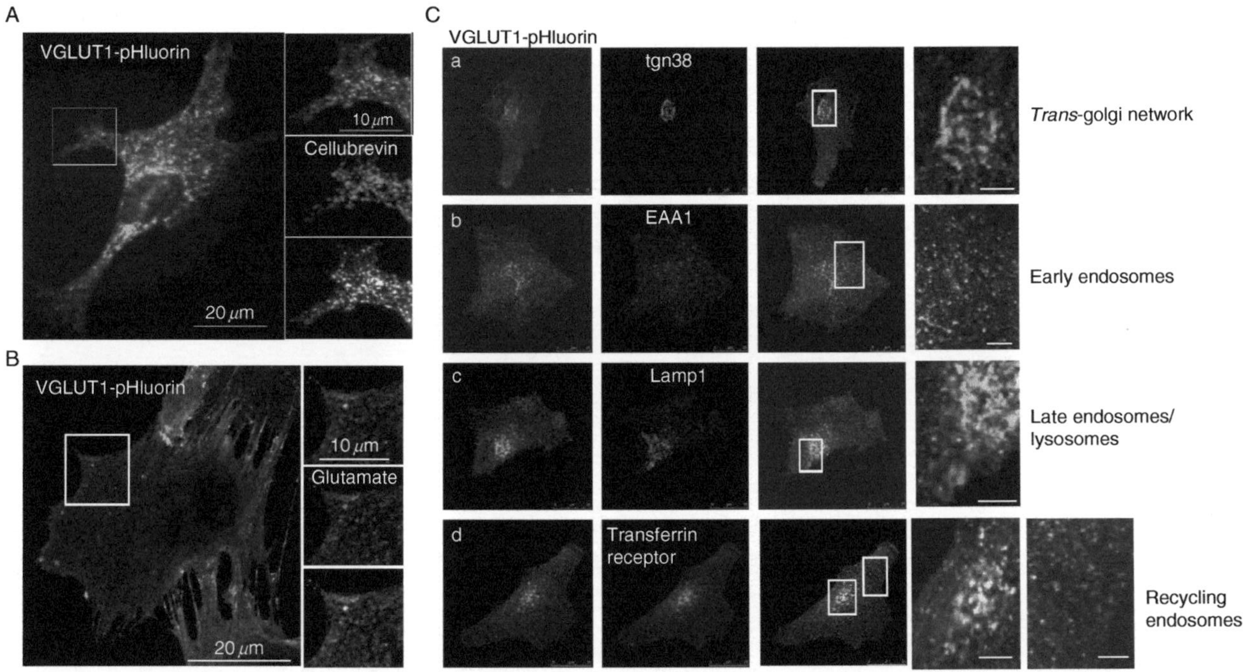

FIG. 1. The intracellular distribution of VGLUT1-pHluorin. (A, B) Images of two astrocytes transfected with VGLUT1-pHluorin (B, green transfected cell). The punctate GFP signal (green dots) and the immunofluorescent signal for endogenous cellubrevin (A, red dots) and for endogenous glutamate (B, red dots) largely colocalize (yellow dots, colocalization VGLUT1 versus cellubrevin: $96 \pm 3\%$; VGLUT1 versus glutamate: $88 \pm 8\%$; $n = 5$). Double immunofluorescent labeling was performed on rat cortical astrocyte cultures fixed in 4% *p*-formaldehyde and subsequently incubated with a guinea pig antibody directed against cellubrevin and with a purified rabbit antibody against glutamate (1:1000, Synaptic Systems GmbH; 1:3000, Vidar Gundersen's lab, batch number 607) or with a rabbit antibody directed against GFP (1:500, Chemicon), visualized with Cy3- or Alexa 488-conjugated secondary antibodies (1:200; Molecular Probes).

same stimulus in the absence of Baf A1 (that represents a balance between exocytosis and the decay of pHluorin fluorescence after fusion pore closure and reacidification) enabled us to determine the cumulative amount of reacidification by calculating the difference between the two; the resulting "difference curve" provided the complete kinetic of reacidification processes during and after the stimulus.

Then, by comparing the normalized EPIi and TIRFi curves in the presence of Baf A1, we calculated the kinetic of the movement of SLMVs out of EW field that we can consider as the first step of the endocytic process. The first derivative of the endocytosis curve shows the evolution of the speed of endocytic events during and after the stimulus; they start at ~600 ms during the stimulus, reach two maxima at 1.2 and 2.8 s and then end after ~6 s. The bimodal distribution indicates the existence of two distinct phases of endocytosis and most probably, of two distinct modes of secretion.

In the second set of experiments, we analyzed exo-endocytosis processes at the single vesicle level with TIRFi. The fast imaging protocol (40 Hz) applied in these sets of experiments provided further kinetic information and added new information on the modalities of exocytosis and recycling. We found that the burst (total duration 1.6–2 s) displayed bimodal distribution, which is in line with the observations previously made for endocytosis and strongly suggests the existence of two components/modes of SLMV exo-endocytosis. By analyzing single fusion events in more detail, we noticed heterogeneity both in the origin of the SLMVs that underwent stimulated exocytosis ("residents" vs. "newcomers"; Zenisek *et al.*, 2000) and in their mode of fusion (kiss-and-run vs. full-collapse type of fusion). Interestingly, the rapid phase of the exocytic burst (0–400 ms) was sustained almost exclusively by "resident" vesicles undergoing kiss-and-run fusion, whereas the slow phase (500 ms–1.6 s) mainly by "newcomers" vesicles undergoing full-collapse fusion (Marchaland *et al.*, 2008). This duality is reminiscent of observations in neurons where only readily releasable SVs, are rapidly recycled and reused (Harata *et al.*, 2006). The functional meaning of the two modalities of

---

Scale bar: 20 μm. (C) In this figure the left panels (green A, B, C, D) show astrocytes transfected with VGLUT1-pHluorin construct revealed by rabbit antibody against GFP (1:500, Chemicon). The middle panels show (red) the markers of the organelles, revealed by mouse antibodies against specific markers of (A) *trans*-Golgi network (tgn38; 1:1000, Alexis Biochemicals); (B) early endosomes (EAA1; 1:100, BD Transduction Lab); (C) late endosomes, multivesicular bodies and lysosomes (lamp1; 1:100, Calbiochem); and (D) recycling endosomes (transferrin receptor; 1:100, Invitrogen). The right panels show the merged images. The boxed regions in the merged images correspond to the high magnifications on the right. Double immunofluorescent labeling was performed on rat cortical astrocyte cultures that were fixed in 4% *p*-formaldehyde and subsequently incubated with antibodies directed against GFP and specific markers, visualized with Cy3- or FITC-conjugated secondary antibodies (1:200; Molecular Probes). Scale bar: 5 μm. Reproduced with permission from Marchaland *et al.* (2008). (See Color Insert.)

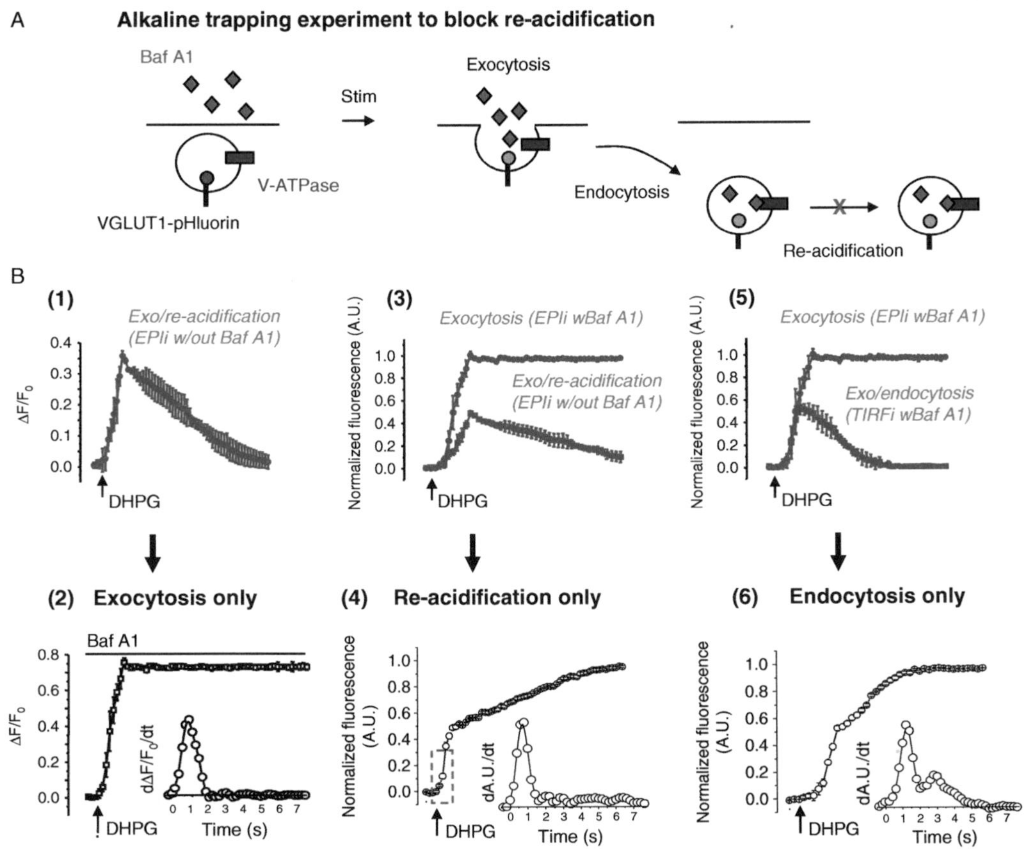

FIG. 2. Kinetics of exocytosis, endocytosis, and reacidification of glutamatergic vesicles in astrocytes. (A) In the presence of Baf A1, a V-type ATPase inhibitor, VGLUT1-pHluorin and vesicle lumens become trapped in the alkaline state after fusion. The signal represents a pure measure of exocytosis because reacidification is blocked (Ryan, 2001). (B) Astrocytes transfected with VGLUT1-pHluorin have been stimulated with DHPG (100 μM) and imaged with TIRFi

secretion could be of great importance in the brain physiology. Studies in synaptic terminals and other secretory cells indicate that full-collapse fusion (Heuser and Reese, 1973) and kiss-and-run (Fesce *et al.*, 1994; Valtorta *et al.*, 2001) coexist and play active roles in exocytic events. In particular, in small nerve terminals, kiss-and-run has an additional role of enabling nerve terminals to respond to high-frequency inputs with a scarce number of vesicles. In astrocytes, it is possible that, like at nerve terminals, the two modalities of fusion can each contribute to maintain cellular communication over a wide range of different stimuli.

2. *Dense-Core Granules*

There is significantly less information concerning DCGs in astrocytes. Proteins belonging to the family of granins (such as chromogranins and secretogranins) are known to be stored in the DCGs of many neuroendocrine cells together with neuropeptides and hormones (Malosio *et al.*, 2004; Meldolesi *et al.*, 2004; Rosa and Gerdes, 1994; Winkler and Fischer-Colbrie, 1992). In view of their typical localization, granins are among the most useful markers to investigate the presence of DCGs in neurons in different areas of the mammalian CNS (Meldolesi *et al.*, 2004; Scammell, 1993; Taupenot, 2007). In 1999, Calegari and colleagues showed for the first time that secretogranin II (SgII) is expressed in cultured hippocampal astrocytes (Calegari *et al.*, 1999). At the ultrastructural level, SgII appeared to be packaged in large DCGs (diameter > 100 nm) located in the Golgi apparatus and near the tubular structure of the *trans*-Golgi networks, where biogenesis of secretory granules is known to take place. Release of intracellular-stored SgII was evoked by treatment with various secretagogues (e.g., ionomycin,

---

and EPIi. (1) The curves of the DHPG-evoked pHluorin fluorescence (pHF) signal represents at all times the net balance of exocytosis (*EXO(t)*) and reacidification. (2) Baf A1, a V-type ATPase inhibitor, blocks their proton pump and prevents reacidification, trapping VGLUT1-pHluorin-expressing vesicles in the fluorescent state. Thus, when Baf A1 is present under EPIi there is only *EXO(t)*(2). We have first analyzed the cumulative curve obtained with Baf A1; by calculating the first derivative of the curve we obtained the temporal distribution of exocytic events (inset). (3–6) In order to obtain information about the kinetics of reacidification and the movement of vesicles after exocytosis, we compared EPI and TIRF curves that have been normalized to the maximum measured with each EPI and TIRF methods, respectively. (3) Superposition of EPIi curves obtained in the presence (*EPIwBafA1*) and in the absence (EPIw/out Baf A1) of Baf A1 (5 μM). By subtracting fluorescence without Baf A1 from that with Baf A1 (*FwBaf(t)–Fw/outBaf(t)* = *ENDO′(t)*) we obtained a curve representing only reacidification. (4) By calculating the first derivative we obtained the temporal distribution of reacidification processes (inset). (5) Superposition of EPIi and TIRFi curves obtained in the presence of Baf A1 (*EPIwBafA1* and *TIRFwBafA1*). The TIRFwBafA1 represents the results of the two competitive processes: exocytosis and movement out of the EW field: *FTIRFwBaf(t)* = *EXO(t)* − *MOV(t)*. Thus, by subtracting *TIRFwBafA1* from that of *EPIwBafA1* (*FwBaf(t)* − *FTIRFwBaf(t)* = *MOV(t)*) we obtained a curve representing only the movement out of the EW field or the first step of endocytosis. (6) Finally, by calculating the first derivative we obtained the temporal distribution of endocytic events (inset). Reproduced with permission from Marchaland *et al.* (2008).

dibutyryl-cAMP, and bradykinin) in a Ca^{2+}-dependent manner. Later on, by means of subcellular fractionation on sucrose equilibrium gradients, ATP was found to be present in the same fractions as SgII, which were distinct from the fractions containing synaptobrevin/VAMP2 (Coco *et al.*, 2003). These findings suggested the possible existence of two distinct classes of secretory vesicles in astrocytes. In keeping, Ca^{2+}-dependent secretion of a peptide, the ANP, was reported in a study that monitored destaining of FM dyes at the whole-cell level (Krzan *et al.*, 2003). Very recently, Ramamoorthy and Whim (2008) reported that cortical astrocytes are able to synthesize both foreign and native neuropeptides, including NPY, a peptide widely distributed throughout the mammalian nervous system (Gray and Morley, 1986). Immunocytochemical studies revealed NPY-immunoreactive puncta that colocalize with markers of DCGs and not with markers of SLMVs (such as VGLUT). Activation of group I of mGluR resulted in a Ca^{2+}-dependent fusion of NPY-containing DCGs with the cell membrane and consequent peptide secretion. This study indicates for the first time that astrocytes, like neurons, have a regulated secretory pathway that is responsible for the release of multiple classes of chemical transmitters. Moreover, because NPY is a potent regulator of synaptic transmission (van den Pol *et al.*, 1996) and NPY receptors are found throughout the CNS (Dumont *et al.*, 2004), astrocytes may be involved in the peptidergic regulation of synaptic transmission.

3. *Lysosomes*

Lysosomes have been considered to be a major storage site of immune-signaling substances, such as proinflammatory cytokines (Andrei *et al.*, 2004) and adenosine (Lukashev *et al.*, 2004; Pisoni and Thoene, 1989), and has been implicated in intercellular communication at the immunological synapse (McNeil and Kirchhausen, 2005). Recent studies have revealed that elevated calcium in astrocytes does induce a special kind of regulated secretion from secretory lysosomes (Jaiswal *et al.*, 2007; Li *et al.*, 2008; Zhang *et al.*, 2007). Secretory lysosomes are enriched in certain cell types such as immune cells and glia. In oligodendrocytes, secretory lysosomes secrete myelin proteins and likely play a critical role in myelination (Trajkovic *et al.*, 2006). In astrocytes, secretory lysosomes release ATP and blockade of ATP release from secretory lysosomes prevents the propagation of calcium waves between neighboring astrocytes. Although these studies have been so far focused on astrocytes in culture, a similar mechanism of release is likely to occur *in vivo*, as acutely isolated astrocytes express some mRNAs of proteins involved in lysosome secretion (Cahoy *et al.*, 2008). ATP release by astrocytes regulates CNS synaptic transmission *in vivo* (Pascual *et al.*, 2005) and it is thus likely that this glial release occurs from secretory lysosomes. However, activated at physiological cytosolic Ca^{2+} concentrations, the observed Ca^{2+}-dependent release of astrocytic secretory lysosomes operates on timescale orders of magnitude slower than neurotransmission; the slow kinetic of release therefore raise the question of the functional importance of the lysosomal exocytosis in the brain.

D. Localized Calcium Microdomains Control Exocytosis of Glutamatergic SLMVs in Astrocytes

In the CNS the concept that localized microdomains of calcium are responsible for triggering vesicle fusion generally refers to neurons (Becherer *et al.*, 2003). Concerning glial cells, although the existence of structural calcium microdomains has been described in the fine profiles of the cerebellar Bergmann glia (Grosche *et al.*, 1999, 2002), very little is known about their physiological implications. Previous works monitoring calcium responses to GPCR activation in these cells focused their observations to the somatic cell region (Bezzi *et al.*, 1998; Fiacco and McCarthy, 2006; Parri *et al.*, 2001; Pasti *et al.*, 1997, 2001; Porter and McCarthy, 1996). The observed somatic calcium elevations in astrocytes upon the GPCR activation showed too slow kinetics (1.5 s in the case of mGluR activation) to underlie the rapid burst of SLMVs exocytosis recently reported (Bezzi *et al.*, 2004; Cali *et al.*, 2008; Marchaland *et al.*, 2008). In order to explain the rapidity of the burst of exocytosis in astrocytes (Bezzi *et al.*, 2004; Cali *et al.*, 2008; Jourdain *et al.*, 2007), we hypothesized the existence of localized calcium microdomains in the sites of exocytosis. Indeed, recently obtained results by our group confirmed our hypothesis (Marchaland *et al.*, 2008); we found that endoplasmic reticulum (ER) tubules are located very close to plasma membrane (~100 nm) and within ~400 nm from the nearest SLMV in the tiny structural domains of femto-litres volume space. By looking ER in astrocytes with TIRF illumination (penetration depth ~100 nm, Fig. 3), we can appreciate the complex spatial organization of the tubules and cisterns approaching the plasma membrane; they form, together with SLMVs and plasma membrane, complex and peculiar structures of submicrometer space that might limit diffusion exchange of signaling molecules, including calcium (Fig. 3). This organization might provide the structural basis for local and possibly global calcium signaling patterns necessary to control diverse intracellular processes simultaneously. Moreover, ER structures and SLMVs lie in the submembrane compartment in tight spatial proximity, with an average distance of 300–500 nm. These results are in line with recent observations where ER tubules have been considered as spatially distinct compartments, structurally and functionally coupled to the plasma membrane via cytoskeletal scaffold proteins (Blaustein and Golovina, 2001; Sala *et al.*, 2005; Tse *et al.*, 1997; Wu *et al.*, 2006). By using fast acquisition rate TIRF imaging, we then investigated the functional implication of the submembrane ER domain; we found that, upon stimulation of group I of mGluR, submicrometer localized calcium

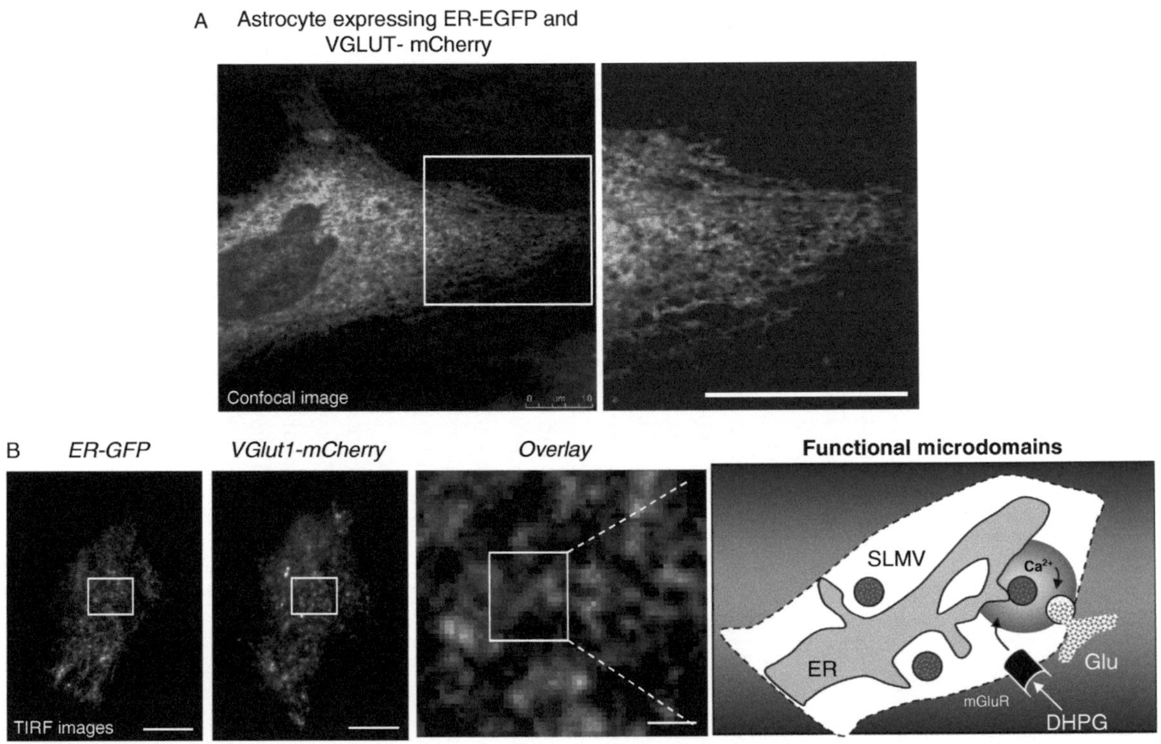

FIG. 3. Submembrane calcium microdomains control exocytosis of glutamate from astrocytes. (A) Confocal images showing an astrocyte double transfected with ER-GFP and VGLUT1-mCherry. ER forms a complex network of tubules and cisterns within the whole cell. (B) TIRF images of an astrocyte transfected with ER-GFP and VGLUT1-mCherry. High magnification image confirms that SLMVs (red dots) within the TIRF illumination field are close to the portion of ER (green structures) approaching plasma membrane. On the right, cartoon showing that spatial localization of mGluR, ER tubules, submembrane calcium events, and glutamatergic SLMVs could define a functional Ca^{2+} microdomain. Reproduced with permission from Marchaland *et al.* (2008). (See Color Insert.)

elevations (hot spots) are generated in the subplasma membrane domains of ER. Interestingly, in most cases the submembrane calcium events occurred at/near sites where SLMVs underwent exocytosis (interdistance: ≤ 280 nm) and were in strict temporal correlation with the fusion events. Indeed, temporal distribution of both submembrane Ca^{2+} and the fusion events displayed similar, biphasic distribution, with the two peaks of calcium events preceding the corresponding peaks of fusion events (Marchaland *et al.*, 2008).

The localized submembrane calcium events in astrocytes are reminiscent of the elementary signals of calcium due to the opening of a spatially restricted group of IP3Rs channels, the so-called "calcium puffs" (Bootman *et al.*, 2001), that have been described in different cell lines (Rizzuto and Pozzan, 2006). The calcium puffs similar to the submembrane calcium events in astrocytes typically consist of a fast elevation of the intracellular calcium (~50 ms) with limited spatial spread (~2–6 μm; Bootman *et al.*, 1997; Thomas *et al.*, 2000). As for the astrocytic signals, whether the localized hot spots of calcium in astrocytes represent a structural organization of IP3R clusters or a convergence of modulatory inputs (for instance the mGluR clusters), remains to be investigated. Interestingly, in several cell types, the calcium puffs act as "pacemaker" and control the frequency of repetitive calcium spiking (the phenomenon called calcium oscillations). Sometimes, the cumulative recruitment of calcium puffs can lead to the initiation and propagation of a global cytosolic calcium signal (Tovey *et al.*, 2001). In astrocytes, we cannot exclude the existence of such a phenomenon; indeed, we found that some local subplasma membrane calcium hot spots propagate from the site where they originate a long distances in the TIRF illumination (up to 14 μm) causing a submembrane calcium wave.

It is likely that the submembrane calcium events merge into propagating waves even in the *z* direction (out of the TIRF illumination) causing the mGluR-mediated global cytosolic calcium signal with the typical slower rise (time to peak about 1.5 s). It is therefore possible that local (submembrane) and global calcium signaling in astrocytes control different calcium-dependent cellular processes.

The findings recently reported can be relevant to understand the functional role of transmitter release from astrocytes in brain function. Interestingly, a recent report shows for the first time *in vivo* the occurrence of fast astrocytic calcium events in response to neuronal activity, peaking with millisecond timescale from sensory stimulation (Winship *et al.*, 2007).

This finding suggests that the fast calcium events we have recently identified are not a peculiarity of astrocytes in cell culture but may correspond to events taking place in astrocytes of the living brain. Establishing whether fast calcium events *in vivo* are associated to transmitter release via SLMV exocytosis becomes, therefore, of the outmost importance in order to define the type of modulatory influence exerted by astrocytes on neighboring neuronal circuits.

III. Calcium-Dependent Glutamate Release from Astrocytes is Deregulated in Pathological Conditions with an Inflammatory Component

The appreciation that brain activity involves interactive signaling between neurons and glia opens new perspectives for understanding the pathogenesis of brain diseases (Bezzi and Volterra, 2001). Affecting neuron–astrocyte interactions might turn out to have a more profound impact on the functionality of the associated neuronal circuits than previously thought and, in some cases, even to be the primary cause of the neuropathology. The brain inflammation is known to cause major morphological and functional changes in glial cells, particularly astrocytes and microglia, broadly defined as "reactive gliosis" (Bezzi and Volterra, 2001; Vesce *et al.*, 2007). The signals exchanged between the two glial cell types during these events are largely unknown, yet their transition from the resting to the activated state appears to be associated with a marked up regulation of several genes and the secretion of factors like cytokines, eicosanoids, reactive oxygen species, nitric oxide and excitatory amino acids (Perry *et al.*, 1995). The inflammation is not only found to produce profound alterations in the structural relations between neurons and astrocytes (Fitch *et al.*, 1999) but also within glial networks. In particular, in the presence of inflammatory cytokines such as interleukin-1β (IL-1β) and TNFα, changes have been observed in the expression of junctional proteins (Duffy *et al.*, 2000; John *et al.*, 1999), in the propagation of intracellular calcium waves (Liu *et al.*, 2000) and the glutamate release (Bezzi *et al.*, 2001). Thus, when local inflammatory reaction is triggered in the brain (as it may occur, for example, in Alzheimer's disease, in the acquired immunodeficiency syndrome (AIDS) neuropathology or after brain ischemia), microglial cells that rapidly migrate to the injury site (Davalos *et al.*, 2005; Nimmerjahn *et al.*, 2005) become activated and start releasing a number of mediators such as IL-1β, IL-6, PGs and TNFα, deeply altering the properties of glial networks (Bezzi and Volterra, 2001). Moreover, in view of the control exerted by prostaglandins and TNFα on calcium-dependent glutamate release from astrocytes (Bezzi *et al.*, 1998, 2001; Domercq *et al.*, 2006), overproduction of these mediators during neuroinflammation might favor an increased and deleterious glutamatergic input from astrocytes to neurons (Vesce *et al.*, 2007). Consequently, the neuron–glial signaling might be perturbed. This hypothesis is substantiated by direct experimental evidence.

A. The Case of HIV-Associated Dementia

Infection with human immunodeficiency virus-1 (HIV-1) can destroy the immune system and lead to AIDS; the virus can induce severe and debilitating neurological problems as well (Ellis *et al.*, 1997; Kaul *et al.*, 2001, 2005;

Power *et al.*, 2002). The syndrome of cognitive and motor dysfunction observed in the presence of HIV-1 has been designated HIV-associated dementia (HAD) or AIDS-dementia complex (Gendelman and Persidsky, 2005; Kaul *et al.*, 2001; McArthur *et al.*, 1993). Although highly active antiretroviral therapy (HAART) has decreased the incidence of HAD, it does not seem to provide complete protection from or reversal of HAD (Dore *et al.*, 1999; Major *et al.*, 2000). Currently, there is no specific treatment for HAD, mainly because of an incomplete understanding of the mechanisms by which HIV infection causes neuronal injury and apoptosis (Kaul *et al.*, 2001). It is widely accepted that HIV entry into CNS occurs via infected monocytes cells (macrophages and microglial cells; Gartner, 2000; Koenig *et al.*, 1986; Persidsky *et al.*, 2001); however, the injury and apoptotic cell death occur only in neurons (Eilbott *et al.*, 1989). Once in the brain, infected macrophages or microglia release viral envelope glycoproteins (gp120), cytokines (e.g., TNFα) and chemokines, which in turn activate uninfected glial cells that start to release neurotoxic substances (Giulian *et al.*, 1996; Wesselingh *et al.*, 1997) such as quinolinic acid and other excitatory amino acids (EAAs, such as glutamate), L-cysteine, arachidonic acid, PAF, free radicals and TNFα (Kaul *et al.*, 2001, 2005). These substances, including the gp120 released from infected cells, induce neuronal injury (Miller and Meucci, 1999), dendritic and synaptic damage and apoptosis (Everall *et al.*, 1999; Masliah *et al.*, 1997) through direct (Toggas *et al.*, 1994) or indirect routes (for instance via release of glutamate from astrocytes, Bezzi *et al.*, 2001). Neuronal death is therefore thought to occur via interactions with infected microglial cells as well as with astrocytes (Kaul *et al.*, 2001; Meucci and Miller, 1996; Miller and Meucci, 1999; Toggas *et al.*, 1996). In support to these findings, the topographic distribution of neuronal apoptosis is correlated with evidence of structural atrophy and closely associated with markers for microglia activation.

Infection of macrophages and lymphocytes by HIV-1 can occur after binding of the viral envelope protein gp120 to one of several possible chemokine receptors in conjunction with CD4 (Kaul *et al.*, 2005; Miller and Meucci, 1999). Generally, T cells are infected via the α-chemokine receptor CXCR4 and/or the β-chemokine receptor CCR5. In contrast, macrophages and microglia are primarily infected via the β-chemokine receptor CCR5 or CCR3, but the α-chemokine receptor CXCR4 may also be involved (He *et al.*, 1997; Michael and Moore, 1999). The HIV coreceptors CCR5 and CXCR4, among other chemokines receptors, are also present on neurons and astrocytes (Rottman *et al.*, 1997; Zhang *et al.*, 1998).

As CXCR4 receptor serves as a HIV coreceptor (Feng *et al.*, 1996) and is expressed on glial and neuronal cells, several *in vitro* studies have suggested that CXCR4 is directly involved in HIV-associated neuronal damage while CCR5 may additionally have a protective role (Hesselgesser *et al.*, 1998; Kaul and Lipton, 1999; Meucci *et al.*, 1998). In cerebrocortical neurons and neuronal cell lines from humans and rodents, picomolar concentrations of HIV-1 gp120 can induce neuronal death via CXCR4 receptors (Chen *et al.*, 2002; Hesselgesser *et al.*, 1998; Ohagen *et al.*, 1999). In mixed neuronal–glial cerebrocortical cultures that mimic the cellular composition of the intact brain, this apoptotic death appears to be predominantly mediated by the release of microglial toxins rather than by direct neuronal damage (Bezzi *et al.*, 2001; Garden *et al.*, 2004).

By using mixed neuronal–glial cerebrocortical cultures, acute brain slices and an *in vivo* model of HAD, we have further investigated the role of CXCR4 chemokine receptor in the neurotoxicity of gp120. The rapid CXCR4-triggered signaling cascade (Fig. 2) displays the involvement of TNFα. Indeed, data obtained not only in cell cultures but also in the TNFα knockout mice demonstrate the essential involvement of the cytokine in the coupling of GPCRs to exocytotic glutamate release from astrocytes (Bezzi *et al.*, 2001; Domercq *et al.*, 2006). In the pathological brain, such as during HAD, astrocytes and microglia often form local foci of reactive cells around the sites of lesion or infection. In a recent paper we mimicked this condition by adding lipopolysaccharide (LPS) activated microglia to astrocyte-pure cultures in about the same ratio (1:10) existing in the brain. Simultaneous stimulation of CXCR4 with its natural ligand CXCL12 or with the glycoprotein gp120IIIB (a T-tropic form of the HIV envelope protein), in reactive microglia and astrocytes resulted in a dramatic amplification of TNFα release from both cells; the TNFα amplification, as a consequence, resulted in strong potentiation of calcium-dependent glutamate release from the astrocytes (Bezzi *et al.*, 2001; Fig. 4). Since shedding of viral proteins in the HIV-1

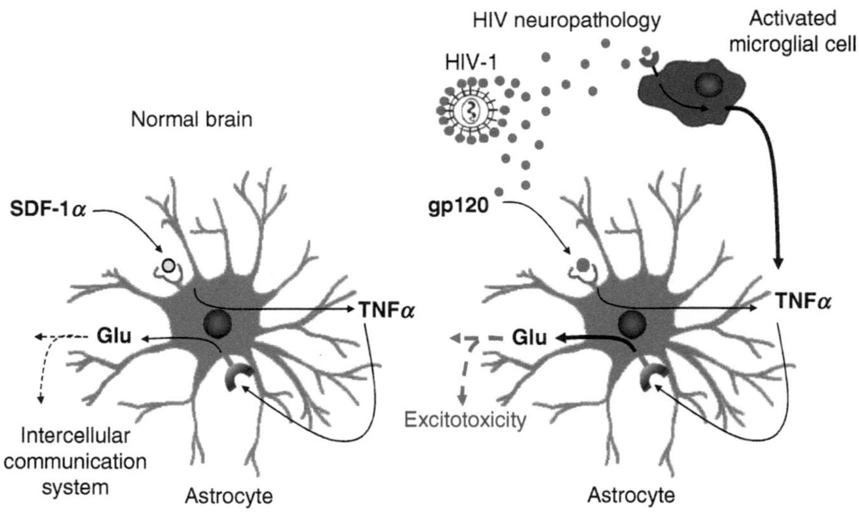

FIG. 4. TNFα-dependent alteration of CXCR4-dependent glutamate release from astrocytes during inflammatory conditions.

infected brain is an uncontrolled process, it is likely that an over stimulation of glial CXCR4 receptor could lead to both an excessive glutamate release from astrocytes and the excitotoxic neuronal cell death. Indeed, in separate set of experiments performed in hippocampal cultures containing neurons, astrocytes and activated microglia, we could demonstrate that the microglia- and TNFα-dependent potentiation of astrocyte glutamate release had neurotoxic consequences, inducing slow apoptotic death of neuronal subpopulations. Finally, experiments done at the whole-animal level confirmed that the identified CXCR4-acitvated cascade is part of the mechanism responsible for the toxic action of gp120 in the brain. In agreement with previous findings obtained by Gigi Bagetta's group (Bagetta *et al.*, 1999), subchronic intracerebroventricular microinfusions of gp120IIIB (100 ng/daily for 7 days) consistently induced early microglial activation and delayed apoptotic cell death in the rat brain neocortex. When the CXCR4 antagonist or anti-TNFα antibodies were coadministered with the viral glycoprotein, the number of apoptotic cells observed was significantly reduced (Bezzi *et al.*, 2001). We therefore discovered that gp120 acts as a CXCR4 agonist in both astrocytes and microglia and, similarly to the endogenous CXCL12, triggers potent TNFα-dependent glutamate release from astrocytes (Fig. 4).

The dramatic amplification of the cytokine-induced response in the presence of activated microglia is consistent with a switch from physiology to excitotoxicity. In view of their broad significance, the TNFα-dependent synergism between reactive microglia and astrocytes leading to excitotoxicity might be operative not only in HAD but also in other neurodegenerative diseases.

During HAD, a local inflammatory reaction is triggered in the brain; this phenomenon induces microglial cells to converge to the site of injury, to become activated and to start releasing a number of mediators, notably inflammatory cytokines such as TNFα, that alter profoundly the properties of astrocytes. HIV-infected microglia become reactive and shed viral particles, including the envelope glycoprotein gp120 which can act as the CXCR4 molecules present in astrocytes and also in the reactive microglia itself, thus inducing release of higher concentrations of TNFα compared to the physiological conditions. The increase of TNFR1 activation in astrocytes results in massive and ultimately excitotoxic release of glutamate.

References

Allen, J. W., Shanker, G., and Aschner, M. (2001). Methylmercury inhibits the *in vitro* uptake of the glutathione precursor, cystine, in astrocytes, but not in neurons. *Brain Res.* **894,** 131–140.

Andrei, C., Margiocco, P., Poggi, A., Lotti, L. V., Torrisi, M. R., and Rubartelli, A. (2004). Phospholipases C and A2 control lysosome-mediated IL-1 beta secretion: Implications for inflammatory processes. *Proc. Natl. Acad. Sci. USA* **101,** 9745–9750.

Anlauf, E., and Derouiche, A. (2005). Astrocytic exocytosis vesicles and glutamate: A high-resolution immunofluorescence study. *Glia* **49,** 96–106.

Araque, A., Carmignoto, G., and Haydon, P. G. (2001). Dynamic signaling between astrocytes and neurons. *Annu. Rev. Physiol.* **63,** 795–813.

Attwell, D., Barbour, B., and Szatkowski, M. (1993). Nonvesicular release of neurotransmitter. *Neuron* **11,** 401–407.

Bagetta, G., Corasaniti, M. T., Berliocchi, L., Nistico, R., Giammarioli, A. M., Malorni, W., Aloe, L., and Finazzi-Agro, A. (1999). Involvement of interleukin-1beta in the mechanism of human immunodeficiency virus type 1 (HIV-1) recombinant protein gp120-induced apoptosis in the neocortex of rat. *Neuroscience* **89,** 1051–1066.

Baker, D. A., Shen, H., and Kalivas, P. W. (2002). Cystine/glutamate exchange serves as the source for extracellular glutamate: Modifications by repeated cocaine administration. *Amino Acids* **23,** 161–162.

Balaji, J., and Ryan, T. A. (2007). Single-vesicle imaging reveals that synaptic vesicle exocytosis and endocytosis are coupled by a single stochastic mode. *Proc. Natl. Acad. Sci. USA* **104,** 20576–20581.

Barres, B. A. (2008). The mystery and magic of glia: A perspective on their roles in health and disease. *Neuron* **60,** 430–440.

Becherer, U., Moser, T., Stuhmer, W., and Oheim, M. (2003). Calcium regulates exocytosis at the level of single vesicles. *Nat. Neurosci.* **6,** 846–853.

Bender, A. S., Reichelt, W., and Norenberg, M. D. (2000). Characterization of cystine uptake in cultured astrocytes. *Neurochem. Int.* **37,** 269–276.

Bergami, M., Santi, S., Formaggio, E., Cagnoli, C., Verderio, C., Blum, R., Berninger, B., Matteoli, M., and Canossa, M. (2008). Uptake and recycling of pro-BDNF for transmitter-induced secretion by cortical astrocytes. *J. Cell Biol.* **183,** 213–221.

Bergersen, L. H., and Gundersen, V. (2009). Morphological evidence for vesicular glutamate release from astrocytes. *Neuroscience* **158,** 260–265.

Bezzi, P., and Volterra, A. (2001). A neuron-glia signalling network in the active brain. *Curr. Opin. Neurobiol.* **11,** 387–394.

Bezzi, P., Carmignoto, G., Pasti, L., Vesce, S., Rossi, D., Rizzini, B. L., Pozzan, T., and Volterra, A. (1998). Prostaglandins stimulate calcium-dependent glutamate release in astrocytes. *Nature* **391,** 281–285.

Bezzi, P., Domercq, M., Brambilla, L., Galli, R., Schols, D., De Clercq, E., Vescovi, A., Bagetta, G., Kollias, G., Meldolesi, J., and Volterra, A. (2001). CXCR4-activated astrocyte glutamate release via TNFalpha: Amplification by microglia triggers neurotoxicity. *Nat. Neurosci.* **4,** 702–710.

Bezzi, P., Gundersen, V., Galbete, J. L., Seifert, G., Steinhauser, C., Pilati, E., and Volterra, A. (2004). Astrocytes contain a vesicular compartment that is competent for regulated exocytosis of glutamate. *Nat. Neurosci.* **7,** 613–620.

Blaustein, M. P., and Golovina, V. A (2001). Structural complexity and functional diversity of endoplasmic reticulum Ca(2+) stores. *Trends Neurosci.* **24,** 602–608.

Blum, A. E., Joseph, S. M., Przybylski, R. J., and Dubyak, G. R. (2008). Rho-family GTPases modulate Ca(2+)-dependent ATP release from astrocytes. *Am. J. Physiol. Cell Physiol.* **295,** C231–C241.

Bootman, M., Niggli, E., Berridge, M., and Lipp, P. (1997). Imaging the hierarchical Ca2+ signalling system in HeLa cells. *J. Physiol.* **499**(Pt. 2), 307–314.

Bootman, M. D., Lipp, P., and Berridge, M. J. (2001). The organisation and functions of local Ca(2+) signals. *J. Cell Sci.* **114,** 2213–2222.

Bowser, D. N., and Khakh, B. S. (2007). Two forms of single-vesicle astrocyte exocytosis imaged with total internal reflection fluorescence microscopy. *Proc. Natl. Acad. Sci. USA* **104,** 4212–4217.

Bushong, E. A., Martone, M. E., Jones, Y. Z., and Ellisman, M. H. (2002). Protoplasmic astrocytes in CA1 stratum radiatum occupy separate anatomical domains. *J. Neurosci.* **22,** 183–192.

Cahoy, J. D., Emery, B., Kaushal, A., Foo, L. C., Zamanian, J. L., Christopherson, K. S., Xing, Y., Lubischer, J. L., Krieg, P. A., Krupenko, S. A., Thompson, W. J., and Barres, B. A. (2008). A transcriptome database for astrocytes, neurons, and oligodendrocytes: A new resource for understanding brain development and function. *J. Neurosci.* **28,** 264–278.

Calegari, F., Coco, S., Taverna, E., Bassetti, M., Verderio, C., Corradi, N., Matteoli, M., and Rosa, P. (1999). A regulated secretory pathway in cultured hippocampal astrocytes. *J. Biol. Chem.* **274,** 22539–22547.

Cali, C., Marchaland, J., Regazzi, R., and Bezzi, P. (2008). SDF 1-alpha (CXCL12) triggers glutamate exocytosis from astrocytes on a millisecond time scale: Imaging analysis at the single-vesicle level with TIRF microscopy. *J. Neuroimmunol.* **198,** 82–91.

Carmignoto, G. (2000). Reciprocal communication systems between astrocytes and neurones. *Prog. Neurobiol.* **62,** 561–581.

Chen, D., Zuleger, C., Chu, Q., Maa, Y. F., Osorio, J., and Payne, L. G. (2002). Epidermal powder immunization with a recombinant HIV gp120 targets Langerhans cells and induces enhanced immune responses. *AIDS Res. Hum. Retroviruses* **18,** 715–722.

Chen, X., Wang, L., Zhou, Y., Zheng, L. H., and Zhou, Z. (2005). "Kiss-and-run" glutamate secretion in cultured and freshly isolated rat hippocampal astrocytes. *J. Neurosci.* **25,** 9236–9243.

Coco, S., Calegari, F., Pravettoni, E., Pozzi, D., Taverna, E., Rosa, P., Matteoli, M., and Verderio, C. (2003). Storage and release of ATP from astrocytes in culture. *J. Biol. Chem.* **278,** 1354–1362.

Coggins, M. R., Grabner, C. P., Almers, W., and Zenisek, D. (2007). Stimulated exocytosis of endosomes in goldfish retinal bipolar neurons. *J. Physiol.* **584,** 853–865.

Cotrina, M. L., Kang, J., Lin, J. H., Bueno, E., Hansen, T. W., He, L., Liu, Y., and Nedergaard, M. (1998). Astrocytic gap junctions remain open during ischemic conditions. *J. Neurosci.* **18,** 2520–2537.

Crippa, D., Schenk, U., Francolini, M., Rosa, P., Verderio, C., Zonta, M., Pozzan, T., Matteoli, M., and Carmignoto, G. (2006). Synaptobrevin2-expressing vesicles in rat astrocytes: Insights into molecular characterization, dynamics and exocytosis. *J. Physiol.* **570,** 567–582.

Danbolt, N. C., Storm-Mathisen, J., and Kanner, B. I. (1992). An [Na+ + K+]coupled L-glutamate transporter purified from rat brain is located in glial cell processes. *Neuroscience* **51,** 295–310.

Davalos, D., Grutzendler, J., Yang, G., Kim, J. V., Zuo, Y., Jung, S., Littman, D. R., Dustin, M. L., and Gan, W. B. (2005). ATP mediates rapid microglial response to local brain injury *in vivo*. *Nat. Neurosci.* **8,** 752–758.

Domercq, M., Brambilla, L., Pilati, E., Marchaland, J., Volterra, A., and Bezzi, P. (2006). P2Y1 receptor-evoked glutamate exocytosis from astrocytes: Control by tumor necrosis factor-alpha and prostaglandins. *J. Biol. Chem.* **281,** 30684–30696.

Dore, G. J., Correll, P. K., Li, Y., Kaldor, J. M., Cooper, D. A., and Brew, B. J. (1999). Changes to AIDS dementia complex in the era of highly active antiretroviral therapy. *AIDS* **13,** 1249–1253.

Duan, S., Anderson, C. M., Keung, E. C., Chen, Y., Chen, Y., and Swanson, R. A. (2003). P2X7 receptor-mediated release of excitatory amino acids from astrocytes. *J. Neurosci.* **23,** 1320–1328.

Duffy, H. S., John, G. R., Lee, S. C., Brosnan, C. F., and Spray, D. C. (2000). Reciprocal regulation of the junctional proteins claudin-1 and connexin43 by interleukin-1beta in primary human fetal astrocytes. *J. Neurosci.* **20,** RC114.

Dumont, Y., Chabot, J. G., and Quirion, R. (2004). Receptor autoradiography as mean to explore the possible functional relevance of neuropeptides: Focus on new agonists and antagonists to study natriuretic peptides, neuropeptide Y and calcitonin gene-related peptides. *Peptides* **25,** 365–391.

Eilbott, D. J., Peress, N., Burger, H., LaNeve, D., Orenstein, J., Gendelman, H. E., Seidman, R., and Weiser, B. (1989). Human immunodeficiency virus type 1 in spinal cords of acquired immunodeficiency syndrome patients with myelopathy: Expression and replication in macrophages. *Proc. Natl. Acad. Sci. USA* **86,** 3337–3341.

Ellis, R. J., Deutsch, R., Heaton, R. K., Marcotte, T. D., McCutchan, J. A., Nelson, J. A., Abramson, I., Thal, L. J., Atkinson, J. H., Wallace, M. R., and Grant, I. (1997). Neurocognitive impairment is an independent risk factor for death in HIV infection. San Diego HIV Neurobehavioral Research Center Group. *Arch. Neurol.* **54,** 416–424.

Everall, I. P, Heaton, R. K., Marcotte, T. D., Ellis, R. J., McCutchan, J. A., Atkinson, J. H., Grant, I., Mallory, M., and Masliah, E. (1999). Cortical synaptic density is reduced in mild to moderate human immunodeficiency virus neurocognitive disorder. HNRC Group. HIV Neurobehavioral Research Center. *Brain Pathol.* **9,** 209–217.

Feng, Y., Broder, C. C., Kennedy, P. E., and Berger, E. A. (1996). HIV-1 entry cofactor: Functional cDNA cloning of a seven-transmembrane, G protein-coupled receptor. *Science* **272,** 872–877.

Fernandez-Chacon, R., and Sudhof, T. C. (1999). Genetics of synaptic vesicle function: Toward the complete functional anatomy of an organelle. *Annu. Rev. Physiol.* **61,** 753–776.

Fesce, R., Grohovaz, F., Valtorta, F., and Meldolesi, J. (1994). Neurotransmitter release: Fusion or 'kiss-and-run'? *Trends Cell Biol.* **4,** 1–4.

Fiacco, T. A., and McCarthy, K. D. (2006). Astrocyte calcium elevations: Properties, propagation, and effects on brain signaling. *Glia* **54,** 676–690.

Fields, R. D., and Stevens, B. (2000). ATP: An extracellular signaling molecule between neurons and glia. *Trends Neurosci.* **23,** 625–633.

Fitch, M. T., Doller, C., Combs, C. K., Landreth, G. E., and Silver, J. (1999). Cellular and molecular mechanisms of glial scarring and progressive cavitation: *In vivo* and *in vitro* analysis of inflammation-induced secondary injury after CNS trauma. *J. Neurosci.* **19,** 8182–8198.

Fremeau, R. T., Jr., Voglmaier, S., Seal, R. P., and Edwards, R. H. (2004). VGLUTs define subsets of excitatory neurons and suggest novel roles for glutamate. *Trends Neurosci.* **27,** 98–103.

Fujita, T., Tozaki-Saitoh, H., and Inoue, K. (2009). P2Y1 receptor signaling enhances neuroprotection by astrocytes against oxidative stress via IL-6 release in hippocampal cultures. *Glia* **57,** 244–257.

Garden, G. A., Guo, W., Jayadev, S., Tun, C., Balcaitis, S., Choi, J., Montine, T. J., Moller, T., and Morrison, R. S. (2004). HIV associated neurodegeneration requires p53 in neurons and microglia. *Faseb J.* **18,** 1141–1143.

Gartner, S. (2000). HIV infection and dementia. *Science* **287,** 602–604.

Gendelman, H. E., and Persidsky, Y. (2005). Infections of the nervous system. *Lancet Neurol.* **4,** 12–13.

Gerber, S. H., and Sudhof, T. C. (2002). Molecular determinants of regulated exocytosis. *Diabetes* **51** (Suppl. 1), S3–S11.

Giulian, D., Yu, J., Li, X., Tom, D., Li, J., Wendt, E., Lin, S. N., Schwarcz, R., and Noonan, C. (1996). Study of receptor-mediated neurotoxins released by HIV-1-infected mononuclear phagocytes found in human brain. *J. Neurosci.* **16,** 3139–3153.

Gray, T. S., and Morley, J. E. (1986). Neuropeptide Y: Anatomical distribution and possible function in mammalian nervous system. *Life Sci.* **38,** 389–401.

Grosche, J., Matyash, V., Moller, T., Verkhratsky, A., Reichenbach, A., and Kettenmann, H. (1999). Microdomains for neuron-glia interaction: Parallel fiber signaling to Bergmann glial cells. *Nat. Neurosci.* **2,** 139–143.

Grosche, J., Kettenmann, H., and Reichenbach, A. (2002). Bergmann glial cells form distinct morphological structures to interact with cerebellar neurons. *J. Neurosci. Res.* **68,** 138–149.

Hannah, M. J., Schmidt, A. A., and Huttner, W. B. (1999). Synaptic vesicle biogenesis. *Annu. Rev. Cell Dev. Biol.* **15,** 733–798.

Harata, N. C., Aravanis, A. M., and Tsien, R. W. (2006). Kiss-and-run and full-collapse fusion as modes of exo-endocytosis in neurosecretion. *J. Neurochem.* **97,** 1546–1570.

Hartmann, M., Heumann, R., and Lessmann, V. (2001). Synaptic secretion of BDNF after high-frequency stimulation of glutamatergic synapses. *EMBO J.* **20,** 5887–5897.

Haskew-Layton, R. E., Rudkouskaya, A., Jin, Y., Feustel, P. J., Kimelberg, H. K., and Mongin, A. A. (2008). Two distinct modes of hypoosmotic medium-induced release of excitatory amino acids and taurine in the rat brain *in vivo*. *PLoS ONE* **3,** e3543.

Haydon, P. G., and Carmignoto, G. (2006). Astrocyte control of synaptic transmission and neurovascular coupling. *Physiol. Rev.* **86,** 1009–1031.

He, J., Chen, Y., Farzan, M., Choe, H., Ohagen, A., Gartner, S., Busciglio, J., Yang, X., Hofmann, W., Newman, W., Mackay, C. R., Sodroski, J., *et al.* (1997). CCR3 and CCR5 are co-receptors for HIV-1 infection of microglia. *Nature* **385,** 645–649.

Hepp, R., Perraut, M., Chasserot-Golaz, S., Galli, T., Aunis, D., Langley, K., and Grant, N. J. (1999). Cultured glial cells express the SNAP-25 analogue SNAP-23. *Glia* **27,** 181–187.

Hesselgesser, J., Taub, D., Baskar, P., Greenberg, M., Hoxie, J., Kolson, D. L., and Horuk, R. (1998). Neuronal apoptosis induced by HIV-1 gp120 and the chemokine SDF-1 alpha is mediated by the chemokine receptor CXCR4. *Curr. Biol.* **8,** 595–598.

Heuser, J. E., and Reese, T. S. (1973). Evidence for recycling of synaptic vesicle membrane during transmitter release at the frog neuromuscular junction. *J. Cell Biol.* **57,** 315–344.

Hussy, N., Deleuze, C., Desarmenien, M. G., and Moos, F. C. (2000). Osmotic regulation of neuronal activity: A new role for taurine and glial cells in a hypothalamic neuroendocrine structure. *Prog. Neurobiol.* **62,** 113–134.

Jahn, R., and Scheller, R. H. (2006). SNAREs—Engines for membrane fusion. *Nat. Rev. Mol. Cell Biol.* **7,** 631–643.

Jahn, R., Lang, T., and Sudhof, T. C. (2003). Membrane fusion. *Cell* **112,** 519–533.

Jaiswal, J. K., Fix, M., Takano, T., Nedergaard, M., and Simon, S. M. (2007). Resolving vesicle fusion from lysis to monitor calcium-triggered lysosomal exocytosis in astrocytes. *Proc. Natl. Acad. Sci. USA* **104,** 14151–14156.

Jeftinija, S. D., Jeftinija, K. V., and Stefanovic, G. (1997). Cultured astrocytes express proteins involved in vesicular glutamate release. *Brain Res.* **750,** 41–47.

Jeremic, A., Jeftinija, K., Stevanovic, J., Glavaski, A., and Jeftinija, S. (2001). ATP stimulates calcium-dependent glutamate release from cultured astrocytes. *J. Neurochem.* **77,** 664–675.

John, G. R., Scemes, E., Suadicani, S. O., Liu, J. S., Charles, P. C., Lee, S. C., Spray, D. C., and Brosnan, C. F. (1999). IL-1beta differentially regulates calcium wave propagation between primary human fetal astrocytes via pathways involving P2 receptors and gap junction channels. *Proc. Natl. Acad. Sci. USA* **96,** 11613–11618.

Jourdain, P., Bergersen, L. H., Bhaukaurally, K., Bezzi, P., Santello, M., Domercq, M., Matute, C., Tonello, F., Gundersen, V., and Volterra, A. (2007). Glutamate exocytosis from astrocytes controls synaptic strength. *Nat. Neurosci.* **10,** 331–339.

Kang, N., Xu, J., Xu, Q., Nedergaard, M., and Kang, J. (2005). Astrocytic glutamate release-induced transient depolarization and epileptiform discharges in hippocampal CA1 pyramidal neurons. *J. Neurophysiol.* **94,** 4121–4130.

Kang, J., Kang, N., Lovatt, D., Torres, A., Zhao, Z., Lin, J., and Nedergaard, M. (2008). Connexin 43 hemichannels are permeable to ATP. *J. Neurosci.* **28,** 4702–4711.

Kaul, M., and Lipton, S. A. (1999). Chemokines and activated macrophages in HIV gp120-induced neuronal apoptosis. *Proc. Natl. Acad. Sci. USA* **96,** 8212–8216.

Kaul, M., Garden, G. A., and Lipton, S. A. (2001). Pathways to neuronal injury and apoptosis in HIV-associated dementia. *Nature* **410,** 988–994.

Kaul, M., Zheng, J., Okamoto, S., Gendelman, H. E., and Lipton, S. A. (2005). HIV-1 infection and AIDS: Consequences for the central nervous system. *Cell Death Differ.* **12**(Suppl. 1), 878–892.

Kimelberg, H. K., Goderie, S. K., Higman, S., Pang, S., and Waniewski, R. A. (1990). Swelling-induced release of glutamate, aspartate, and taurine from astrocyte cultures. *J. Neurosci.* **10,** 1583–1591.

Koenig, R. E., Gautier, T., and Levy, J. A. (1986). Unusual intrafamilial transmission of human immunodeficiency virus. *Lancet* **2,** 627.

Kreft, M., Stenovec, M., Rupnik, M., Grilc, S., Krzan, M., Potokar, M., Pangrsic, T., Haydon, P. G., and Zorec, R. (2004). Properties of Ca(2+)-dependent exocytosis in cultured astrocytes. *Glia* **46,** 437–445.

Krzan, M., Stenovec, M., Kreft, M., Pangrsic, T., Grilc, S., Haydon, P. G., and Zorec, R. (2003). Calcium-dependent exocytosis of atrial natriuretic peptide from astrocytes. *J. Neurosci.* **23,** 1580–1583.

Kukley, M., Barden, J. A., Steinhauser, C., and Jabs, R. (2001). Distribution of P2X receptors on astrocytes in juvenile rat hippocampus. *Glia* **36,** 11–21.

Kupfermann, I. (1991). Functional studies of cotransmission. *Physiol. Rev.* **71,** 683–732.

Li, D., Ropert, N., Koulakoff, A., Giaume, C., and Oheim, M. (2008). Lysosomes are the major vesicular compartment undergoing Ca2+-regulated exocytosis from cortical astrocytes. *J. Neurosci.* **28,** 7648–7658.

Liu, J. S., John, G. R., Sikora, A., Lee, S. C., and Brosnan, C. F. (2000). Modulation of interleukin-1beta and tumor necrosis factor alpha signaling by P2 purinergic receptors in human fetal astrocytes. *J. Neurosci.* **20,** 5292–5299.

Liu, X. S., Chopp, M., Santra, M., Hozeska-Solgot, A., Zhang, R. L., Wang, L., Teng, H., Lu, M., and Zhang, Z. G. (2008). Functional response to SDF1 alpha through over-expression of CXCR4 on adult subventricular zone progenitor cells. *Brain Res.* **1226,** 18–26.

Longuemare, M. C., and Swanson, R. A. (1997). Net glutamate release from astrocytes is not induced by extracellular potassium concentrations attainable in brain. *J. Neurochem.* **69,** 879–882.

Lukashev, D., Ohta, A., Apasov, S., Chen, J. F., and Sitkovsky, M. (2004). Cutting edge: Physiologic attenuation of proinflammatory transcription by the Gs protein-coupled A2A adenosine receptor *in vivo*. *J. Immunol.* **173,** 21–24.

Madison, D. L., Kruger, W. H., Kim, T., and Pfeiffer, S. E. (1996). Differential expression of rab3 isoforms in oligodendrocytes and astrocytes. *J. Neurosci. Res.* **45,** 258–268.

Maienschein, V., Marxen, M., Volknandt, W., and Zimmermann, H. (1999). A plethora of presynaptic proteins associated with ATP-storing organelles in cultured astrocytes. *Glia* **26,** 233–244.

Major, E. O., Rausch, D., Marra, C., and Clifford, D. (2000). HIV-associated dementia. *Science* **288,** 440–442.

Malarkey, E. B., and Parpura, V. (2008). Mechanisms of glutamate release from astrocytes. *Neurochem. Int.* **52,** 142–154.

Malosio, M. L., Giordano, T., Laslop, A., and Meldolesi, J. (2004). Dense-core granules: A specific hallmark of the neuronal/neurosecretory cell phenotype. *J. Cell Sci.* **117,** 743–749.

Marchaland, J., Cali, C., Voglmaier, S. M., Li, H., Regazzi, R., Edwards, R. H., and Bezzi, P. (2008). Fast subplasma membrane Ca2+ transients control exo-endocytosis of synaptic-like microvesicles in astrocytes. *J. Neurosci.* **28,** 9122–9132.

Martin, D. L. (1992). Synthesis and release of neuroactive substances by glial cells. *Glia* **5,** 81–94.

Martineau, M., Galli, T., Baux, G., and Mothet, J. P. (2008). Confocal imaging and tracking of the exocytotic routes for D-serine-mediated gliotransmission. *Glia* **56,** 1271–1284.

Masliah, E., Heaton, R. K., Marcotte, T. D., Ellis, R. J., Wiley, C. A., Mallory, M., Achim, C. L., McCutchan, J. A., Nelson, J. A., Atkinson, J. H., and Grant, I. (1997). Dendritic injury is a pathological substrate for human immunodeficiency virus-related cognitive disorders. HNRC Group. The HIV Neurobehavioral Research Center. *Ann. Neurol.* **42,** 963–972.

McArthur, J. C., Hoover, D. R., Bacellar, H., Miller, E. N., Cohen, B. A., Becker, J. T., Graham, N. M., McArthur, J. H., Selnes, O. A., Jacobson, L. P., *et al.* (1993). Dementia in AIDS patients: Incidence and risk factors. Multicenter AIDS Cohort Study. *Neurology* **43,** 2245–2252.

McNeil, P. L., and Kirchhausen, T. (2005). An emergency response team for membrane repair. *Nat. Rev. Mol. Cell Biol.* **6,** 499–505.

Medhora, M. M. (2000). Retinoic acid upregulates beta(1)-integrin in vascular smooth muscle cells and alters adhesion to fibronectin. *Am. J. Physiol. Heart Circ. Physiol.* **279,** H382–H387.

Meldolesi, J., Chieregatti, E., and Luisa Malosio, M. (2004). Requirements for the identification of dense-core granules. *Trends Cell Biol.* **14,** 13–19.

Meucci, O., and Miller, R. J. (1996). gp120-induced neurotoxicity in hippocampal pyramidal neuron cultures: Protective action of TGF-beta1. *J. Neurosci.* **16,** 4080–4088.

Meucci, O., Fatatis, A., Simen, A. A., Bushell, T. J., Gray, P. W., and Miller, R. J. (1998). Chemokines regulate hippocampal neuronal signaling and gp120 neurotoxicity. *Proc. Natl. Acad. Sci. USA* **95,** 14500–14505.

Michael, N. L., and Moore, J. P. (1999). HIV-1 entry inhibitors: Evading the issue. *Nat. Med.* **5,** 740–742.

Miesenbock, G., De Angelis, D. A., and Rothman, J. E. (1998). Visualizing secretion and synaptic transmission with pH-sensitive green fluorescent proteins. *Nature* **394,** 192–195.

Miller, R. J., and Meucci, O. (1999). AIDS and the brain: Is there a chemokine connection? *Trends Neurosci.* **22,** 471–479.

Mongin, A. A., and Kimelberg, H. K. (2002). ATP potently modulates anion channel-mediated excitatory amino acid release from cultured astrocytes. *Am. J. Physiol. Cell Physiol.* **283,** C569–C578.

Montana, V., Ni, Y., Sunjara, V., Hua, X., and Parpura, V. (2004). Vesicular glutamate transporter-dependent glutamate release from astrocytes. *J. Neurosci.* **24,** 2633–2642.

Morgan, A., and Burgoyne, R. D. (1997). Common mechanisms for regulated exocytosis in the chromaffin cell and the synapse. *Semin. Cell Dev. Biol.* **8,** 141–149.

Moran, M. M., Melendez, R., Baker, D., Kalivas, P. W., and Seamans, J. K. (2003). Cystine/glutamate antiporter regulation of vesicular glutamate release. *Ann. NY Acad. Sci.* **1003,** 445–447.

Moran, M. M., McFarland, K., Melendez, R. I., Kalivas, P. W., and Seamans, J. K. (2005). Cystine/glutamate exchange regulates metabotropic glutamate receptor presynaptic inhibition of excitatory transmission and vulnerability to cocaine seeking. *J. Neurosci.* **25,** 6389–6393.

Mothet, J. P., Pollegioni, L., Ouanounou, G., Martineau, M., Fossier, P., and Baux, G. (2005). Glutamate receptor activation triggers a calcium-dependent and SNARE protein-dependent release of the gliotransmitter D-serine. *Proc. Natl. Acad. Sci. USA* **102,** 5606–5611.

Nadrigny, F., Li, D., Kemnitz, K., Ropert, N., Koulakoff, A., Rudolph, S., Vitali, M., Giaume, C., Kirchhoff, F., and Oheim, M. (2007). Systematic colocalization errors between acridine orange and EGFP in astrocyte vesicular organelles. *Biophys. J.* **93,** 969–980.

Nedergaard, M., Takano, T., and Hansen, A. J. (2002). Beyond the role of glutamate as a neurotransmitter. *Nat. Rev. Neurosci.* **3,** 748–755.

Nimmerjahn, A., Kirchhoff, F., and Helmchen, F. (2005). Resting microglial cells are highly dynamic surveillants of brain parenchyma in vivo. *Science* **308,** 1314–1318.

Ohagen, A., Ghosh, S., He, J., Huang, K., Chen, Y., Yuan, M., Osathanondh, R., Gartner, S., Shi, B., Shaw, G., and Gabuzda, D. (1999). Apoptosis induced by infection of primary brain cultures with diverse human immunodeficiency virus type 1 isolates: Evidence for a role of the envelope. *J. Virol.* **73,** 897–906.

Pangrsic, T., Potokar, M., Stenovec, M., Kreft, M., Fabbretti, E., Nistri, A., Pryazhnikov, E., Khiroug, L., Giniatullin, R., and Zorec, R. (2007). Exocytotic release of ATP from cultured astrocytes. *J. Biol. Chem.* **282,** 28749–28758.

Park, M., Salgado, J. M., Ostroff, L., Helton, T. D., Robinson, C. G., Harris, K. M., and Ehlers, M. D. (2006). Plasticity-induced growth of dendritic spines by exocytic trafficking from recycling endosomes. *Neuron* **52,** 817–830.

Parpura, V., Basarsky, T. A., Liu, F., Jeftinija, K., Jeftinija, S., and Haydon, P. G. (1994). Glutamate-mediated astrocyte-neuron signalling. *Nature* **369,** 744–747.

Parpura, V., Fang, Y., Basarsky, T., Jahn, R., and Haydon, P. G. (1995). Expression of synaptobrevin II, cellubrevin and syntaxin but not SNAP-25 in cultured astrocytes. *FEBS Lett.* **377,** 489–492.

Parri, H. R., Gould, T. M., and Crunelli, V. (2001). Spontaneous astrocytic Ca2+ oscillations in situ drive NMDAR-mediated neuronal excitation. *Nat. Neurosci.* **4,** 803–812.

Pascual, M., Climent, E., and Guerri, C. (2001). BDNF induces glutamate release in cerebrocortical nerve terminals and in cortical astrocytes. *Neuroreport* **12,** 2673–2677.

Pascual, O., Casper, K. B., Kubera, C., Zhang, J., Revilla-Sanchez, R., Sul, J. Y., Takano, H., Moss, S. J., McCarthy, K., and Haydon, P. G. (2005). Astrocytic purinergic signaling coordinates synaptic networks. *Science* **310,** 113–116.

Pasti, L., Volterra, A., Pozzan, T., and Carmignoto, G. (1997). Intracellular calcium oscillations in astrocytes: A highly plastic, bidirectional form of communication between neurons and astrocytes in situ. *J. Neurosci.* **17,** 7817–7830.

Pasti, L., Zonta, M., Pozzan, T., Vicini, S., and Carmignoto, G. (2001). Cytosolic calcium oscillations in astrocytes may regulate exocytotic release of glutamate. *J. Neurosci.* **21,** 477–484.

Perry, V.H, Bell, M. D., Brown, H. C., and Matyszak, M. K. (1995). Inflammation in the nervous system. *Curr. Opin. Neurobiol.* **5,** 636–641.

Persidsky, Y., Limoges, J., Rasmussen, J., Zheng, J., Gearing, A., and Gendelman, H. E. (2001). Reduction in glial immunity and neuropathology by a PAF antagonist and an MMP and TNFalpha inhibitor in SCID mice with HIV-1 encephalitis. *J. Neuroimmunol.* **114,** 57–68.

Pickel, V. M, Chan, J., Veznedaroglu, E., and Milner, T. A. (1995). Neuropeptide Y and dynorphin-immunoreactive large dense-core vesicles are strategically localized for presynaptic modulation in the hippocampal formation and substantia nigra. *Synapse* **19,** 160–169.

Pisoni, R. L., and Thoene, J. G. (1989). Detection and characterization of a nucleoside transport system in human fibroblast lysosomes. *J. Biol. Chem.* **264,** 4850–4856.

Porter, J. T., and McCarthy, K. D. (1996). Hippocampal astrocytes in situ respond to glutamate released from synaptic terminals. *J. Neurosci.* **16,** 5073–5081.

Power, C., Gill, M. J., and Johnson, R. T. (2002). Progress in clinical neurosciences: The neuropathogenesis of HIV infection: Host-virus interaction and the impact of therapy. *Can. J. Neurol. Sci.* **29,** 19–32.

Pryazhnikov, E., and Khiroug, L. (2008). Sub-micromolar increase in [Ca(2+)](i) triggers delayed exocytosis of ATP in cultured astrocytes. *Glia* **56,** 38–49.

Ramamoorthy, P., and Whim, M. D. (2008). Trafficking and fusion of neuropeptide Y-containing dense-core granules in astrocytes. *J. Neurosci.* **28,** 13815–13827.

Ramon y Cajal, S. (1899). Contribucion al conocimiento de la neuroglia del cerebro humano. *Trab. Lab. Invest. Biol. XI (1913)* 255–315.

Re, D. B., Nafia, I., Melon, C., Shimamoto, K., Kerkerian-Le Goff, L., and Had-Aissouni, L. (2006). Glutamate leakage from a compartmentalized intracellular metabolic pool and activation of the lipoxygenase pathway mediate oxidative astrocyte death by reversed glutamate transport. *Glia* **54,** 47–57.

Rizzuto, R., and Pozzan, T. (2006). Microdomains of intracellular Ca2+: Molecular determinants and functional consequences. *Physiol. Rev.* **86,** 369–408.

Rosa, P., and Gerdes, H. H. (1994). The granin protein family: Markers for neuroendocrine cells and tools for the diagnosis of neuroendocrine tumors. *J. Endocrinol. Invest.* **17,** 207–225.

Rossi, D. J., Oshima, T., and Attwell, D. (2000). Glutamate release in severe brain ischaemia is mainly by reversed uptake. *Nature* **403,** 316–321.

Rottman, J. B., Ganley, K. P., Williams, K., Wu, L., Mackay, C. R., and Ringler, D. J. (1997). Cellular localization of the chemokine receptor CCR5. Correlation to cellular targets of HIV-1 infection. *Am. J. Pathol.* **151,** 1341–1351.

Ryan, T. A. (2001). Presynaptic imaging techniques. *Curr. Opin. Neurobiol.* **11,** 544–549.

Ryan, T. A., and Reuter, H. (2001). Measurements of vesicle recycling in central neurons. *News Physiol. Sci.* **16,** 10–14.

Sala, C., Roussignol, G., Meldolesi, J., and Fagni, L. (2005). Key role of the postsynaptic density scaffold proteins Shank and Homer in the functional architecture of Ca2+ homeostasis at dendritic spines in hippocampal neurons. *J. Neurosci.* **25,** 4587–4592.

Sankaranarayanan, S., and Ryan, T. A. (2000). Real-time measurements of vesicle-SNARE recycling in synapses of the central nervous system. *Nat. Cell Biol.* **2,** 197–204.

Sankaranarayanan, S., and Ryan, T. A. (2001). Calcium accelerates endocytosis of vSNAREs at hippocampal synapses. *Nat. Neurosci.* **4,** 129–136.

Santello, M., and Volterra, A. (2009). Synaptic modulation by astrocytes via Ca(2+)-dependent glutamate release. *Neuroscience* **158,** 253–259.

Sanzgiri, R. P., Araque, A., and Haydon, P. G. (1999). Prostaglandin E(2) stimulates glutamate receptor-dependent astrocyte neuromodulation in cultured hippocampal cells. *J. Neurobiol.* **41,** 221–229.

Scammell, J. G. (1993). Granins markers of the regulated secretory pathway. *Trends Endocrinol. Metab.* **4,** 14–18.

Shanker, G., and Aschner, M. (2001). Identification and characterization of uptake systems for cystine and cysteine in cultured astrocytes and neurons: Evidence for methylmercury-targeted disruption of astrocyte transport. *J. Neurosci. Res.* **66,** 998–1002.

Snyder, S. H., and Kim, P. M. (2000). D-amino acids as putative neurotransmitters: Focus on D-serine. *Neurochem. Res.* **25,** 553–560.

Steyer, J. A., and Almers, W. (2001). A real-time view of life within 100 nm of the plasma membrane. *Nat. Rev. Mol. Cell Biol.* **2,** 268–275.

Stout, C. E., Costantin, J. L., Naus, C. C., and Charles, A. C. (2002). Intercellular calcium signaling in astrocytes via ATP release through connexin hemichannels. *J. Biol. Chem.* **277,** 10482–10488.

Striedinger, K., Meda, P., and Scemes, E. (2007). Exocytosis of ATP from astrocyte progenitors modulates spontaneous Ca2+ oscillations and cell migration. *Glia* **55,** 652–662.

Sudhof, T. C. (1995). The synaptic vesicle cycle: A cascade of protein-protein interactions. *Nature* **375,** 645–653.

Sudhof, T. C. (2004). The synaptic vesicle cycle. *Annu. Rev. Neurosci.* **27,** 509–547.

Szatkowski, M., Barbour, B., and Attwell, D. (1990). Non-vesicular release of glutamate from glial cells by reversed electrogenic glutamate uptake. *Nature* **348,** 443–446.

Takano, T., Kang, J., Jaiswal, J. K., Simon, S. M., Lin, J. H., Yu, Y., Li, Y., Yang, J., Dienel, G., Zielke, H. R., and Nedergaard, M. (2005). Receptor-mediated glutamate release from volume sensitive channels in astrocytes. *Proc. Natl. Acad. Sci. USA* **102,** 16466–16471.

Tang, X. C., and Kalivas, P. W. (2003). Bidirectional modulation of cystine/glutamate exchanger activity in cultured cortical astrocytes. *Ann. NY Acad. Sci.* **1003,** 472–475.

Taupenot, L. (2007). Analysis of regulated secretion using PC12 cells. *Curr. Protoc. Cell Biol.* **36,** 15.12.1–15.12.13.

Thomas, D., Lipp, P., Tovey, S. C., Berridge, M. J., Li, W., Tsien, R. Y., and Bootman, M. D. (2000). Microscopic properties of elementary Ca2+ release sites in non-excitable cells. *Curr. Biol.* **10,** 8–15.

Toggas, S. M., Masliah, E., Rockenstein, E. M., Rall, G. F., Abraham, C. R., and Mucke, L. (1994). Central nervous system damage produced by expression of the HIV-1 coat protein gp120 in transgenic mice. *Nature* **367,** 188–193.

Toggas, S. M., Masliah, E., and Mucke, L. (1996). Prevention of HIV-1 gp120-induced neuronal damage in the central nervous system of transgenic mice by the NMDA receptor antagonist memantine. *Brain Res.* **706,** 303–307.

Tovey, S. C., de Smet, P., Lipp, P., Thomas, D., Young, K. W., Missiaen, L., De Smedt, H., Parys, J. B., Berridge, M. J., Thuring, J., Holmes, A., and Bootman, M. D. (2001). Calcium puffs are generic InsP(3)-activated elementary calcium signals and are downregulated by prolonged hormonal stimulation to inhibit cellular calcium responses. *J. Cell Sci.* **114,** 3979–3989.

Trajkovic, K., Dhaunchak, A. S., Goncalves, J. T., Wenzel, D., Schneider, A., Bunt, G., Nave, K. A., and Simons, M. (2006). Neuron to glia signaling triggers myelin membrane exocytosis from endosomal storage sites. *J. Cell Biol.* **172,** 937–948.

Tse, F. W., Tse, A., Hille, B., Horstmann, H., and Almers, W. (1997). Local Ca2+ release from internal stores controls exocytosis in pituitary gonadotrophs. *Neuron* **18,** 121–132.

Tsuboi, T., Zhao, C., Terakawa, S., and Rutter, G. A. (2000). Simultaneous evanescent wave imaging of insulin vesicle membrane and cargo during a single exocytotic event. *Curr. Biol.* **10,** 1307–1310.

Valtorta, F., Meldolesi, J., and Fesce, R. (2001). Synaptic vesicles: Is kissing a matter of competence? *Trends Cell Biol.* **11,** 324–328.

van den Pol, A. N., Obrietan, K., Chen, G., and Belousov, A. B. (1996). Neuropeptide Y-mediated long-term depression of excitatory activity in suprachiasmatic nucleus neurons. *J. Neurosci.* **16,** 5883–5895.

Ventura, R., and Harris, K. M. (1999). Three-dimensional relationships between hippocampal synapses and astrocytes. *J. Neurosci.* **19,** 6897–6906.

Vesce, S., Rossi, D., Brambilla, L., and Volterra, A. (2007). Glutamate release from astrocytes in physiological conditions and in neurodegenerative disorders characterized by neuroinflammation. *Int. Rev. Neurobiol.* **82,** 57–71.

Voglmaier, S. M., Kam, K., Yang, H., Fortin, D. L., Hua, Z., Nicoll, R. A., and Edwards, R. H. (2006). Distinct endocytic pathways control the rate and extent of synaptic vesicle protein recycling. *Neuron* **51,** 71–84.

Volterra, A., and Bezzi, P. (2002). Release of transmitters from glial cells. *In* "The Tripartite Synapse: Glia in Synaptic Transmission" (A. Volterra, P. J. Magistretti, and P. G. Haydon, Eds.), pp. 164–182. Oxford University Press, UK.

Volterra, A., and Meldolesi, J. (2005). Astrocytes, from brain glue to communication elements: The revolution continues. *Nat. Rev. Neurosci.* **6,** 626–640.

Volterra, A., and Steinhauser, C. (2004). Glial modulation of synaptic transmission in the hippocampus. *Glia* **47,** 249–257.

Volterra, A., Bezzi, P., Rizzini, B. L., Trotti, D., Ullensvang, K., Danbolt, N. C., and Racagni, G. (1996). The competitive transport inhibitor L-trans-pyrrolidine-2, 4-dicarboxylate triggers excitotoxicity in rat cortical neuron-astrocyte co-cultures via glutamate release rather than uptake inhibition. *Eur. J. Neurosci.* **8,** 2019–2028.

Wesselingh, S. L., Takahashi, K., Glass, J. D., McArthur, J. C., Griffin, J. W., and Griffin, D. E. (1997). Cellular localization of tumor necrosis factor mRNA in neurological tissue from HIV-infected patients by combined reverse transcriptase/polymerase chain reaction in situ hybridization and immunohistochemistry. *J. Neuroimmunol.* **74,** 1–8.

Winkler, H., and Fischer-Colbrie, R. (1992). The chromogranins A and B: The first 25 years and future perspectives. *Neuroscience* **49,** 497–528.

Winship, I.R, Plaa, N., and Murphy, T. H. (2007). Rapid astrocyte calcium signals correlate with neuronal activity and onset of the hemodynamic response *in vivo*. *J. Neurosci.* **27,** 6268–6272.

Wu, M. M., Buchanan, J., Luik, R. M., and Lewis, R. S. (2006). Ca2+ store depletion causes STIM1 to accumulate in ER regions closely associated with the plasma membrane. *J. Cell Biol.* **174,** 803–813.

Xu, J., Peng, H., Kang, N., Zhao, Z., Lin, J. H., Stanton, P. K., and Kang, J. (2007). Glutamate-induced exocytosis of glutamate from astrocytes. *J. Biol. Chem.* **282,** 24185–24197.

Ye, Z. C., Wyeth, M. S., Baltan-Tekkok, S., and Ransom, B. R. (2003). Functional hemichannels in astrocytes: A novel mechanism of glutamate release. *J. Neurosci.* **23,** 3588–3596.

Zenisek, D., Steyer, J. A., and Almers, W. (2000). Transport, capture and exocytosis of single synaptic vesicles at active zones. *Nature* **406,** 849–854.

Zhang, L., He, T., Talal, A., Wang, G., Frankel, S. S., and Ho, D. D. (1998). In vivo distribution of the human immunodeficiency virus/simian immunodeficiency virus coreceptors: CXCR4, CCR3, and CCR5. *J. Virol.* **72,** 5035–5045.

Zhang, Q., Fukuda, M., Van Bockstaele, E., Pascual, O., and Haydon, P. G. (2004a). Synaptotagmin IV regulates glial glutamate release. *Proc. Natl. Acad. Sci. USA* **101,** 9441–9446.

Zhang, Q., Pangrsic, T., Kreft, M., Krzan, M., Li, N., Sul, J. Y., Halassa, M., Van Bockstaele, E., Zorec, R., and Haydon, P. G. (2004b). Fusion-related release of glutamate from astrocytes. *J. Biol. Chem.* **279,** 12724–12733.

Zhang, Z., Chen, G., Zhou, W., Song, A., Xu, T., Luo, Q., Wang, W., Gu, X. S., and Duan, S. (2007). Regulated ATP release from astrocytes through lysosome exocytosis. *Nat. Cell Biol.* **9,** 945–953.

Zonta, M., Sebelin, A., Gobbo, S., Fellin, T., Pozzan, T., and Carmignoto, G. (2003). Glutamate-mediated cytosolic calcium oscillations regulate a pulsatile prostaglandin release from cultured rat astrocytes. *J. Physiol.* **553,** 407–414.

GLUTAMATE RELEASE FROM ASTROCYTIC GLIOSOMES UNDER PHYSIOLOGICAL AND PATHOLOGICAL CONDITIONS

Marco Milanese,* Tiziana Bonifacino,* Simona Zappettini,* Cesare Usai,[†]
Carlo Tacchetti,[‡,§] Mario Nobile,[†] and Giambattista Bonanno*,[¶,||]

*Department of Experimental Medicine, Section of Pharmacology and Toxicology,
University of Genoa, 16148 Genoa, Italy
[†]Institute of Biophysics, National Research Council, 16149 Genoa, Italy
[‡]Department of Experimental Medicine, Section of Human Anatomy, University of Genoa,
16132 Genoa, Italy
[§]FIRC Institute of Molecular Oncology (IFOM), 20139 Milan, Italy
[¶]Center of Excellence for Biomedical Research, University of Genoa, 16132 Genoa, Italy
[||]National Institute for Neuroscience (INN), 10125 Turin, Italy

Glial subcellular particles (gliosomes) have been purified from rat cerebral cortex or mouse spinal cord and investigated for their ability to release glutamate. Confocal microscopy showed that gliosomes are enriched with glia-specific proteins, such as GFAP and S-100 but not neuronal proteins, such as PSD-95, MAP-2, and β-tubulin III. Furthermore, gliosomes exhibit labeling neither for integrin-αM nor for myelin basic protein, specific for microglia and oligodendrocytes, respectively. The gliosomal fraction contains proteins of the exocytotic machinery coexisting with GFAP. Consistent with ultrastructural analysis, several nonclustered vesicles are present in the gliosome cytoplasm. Finally, gliosomes represent functional organelles that actively export glutamate when subjected to

INTERNATIONAL REVIEW OF
NEUROBIOLOGY, VOL. 85
DOI: 10.1016/S0074-7742(09)85021-6

releasing stimuli, such as ionomycin, high KCl, veratrine, 4-aminopyridine, AMPA, or ATP by mechanisms involving extracellular Ca^{2+}, Ca^{2+} release from intracellular stores as well as reversal of glutamate transporters. In addition, gliosomes can release glutamate also by a mechanism involving heterologous transporter activation (heterotransporters) located on glutamate-releasing and glutamate transporter-expressing (homotransporters) gliosomes. This glutamate release involves reversal of glutamate transporters and anion channel opening, but not exocytosis. Both the exocytotic and the heterotransporter-mediated glutamate release were more abundant in gliosomes prepared from the spinal cord of transgenic mice, model of amyotrophic lateral sclerosis, than in controls; suggesting the involvement of astrocytic glutamate release in the excitotoxicity proposed as a cause of motor neuron degeneration. The results support the view that gliosomes may represent a viable preparation that allows to study mechanisms of astrocytic transmitter release and its regulation in healthy animals and in animal models of brain diseases.

I. New Perspectives in Astrocyte Function

The impact of astrocytes on CNS function has recently attracted the interest of many investigators, and the numerous outcomes in the field have led to dramatic conceptual changes about the role of these glial cells, formerly thought to provide only structural and trophic support to neurons.

An increasing number of papers suggest that astrocytes share at least some of the features typical of neurons: they possess transporters able to capture neurotransmitters and neuromodulators from the extracellular space (Bergles and Jahr, 1997; Mennerick and Zorumski, 1994), express receptors able to sense signals from the outside of the cell, and synthesize and release gliotransmitters (see Volknandt, 2002 and references therein). Astrocytes, due to their intimate spatial relationship with neuronal synaptic contacts, can directly respond to synaptically released messengers and, in turn, communicate via signaling substances with neurons.

In particular, these abilities have been extensively studied referring to the excitatory transmission (for a review, see Haydon, 2001; Volterra and Meldolesi, 2005). Several lines of evidence suggest that glutamate released from neurons can activate both ionotropic and metabotropic receptors located on astroglial cells, inducing intracellular Ca^{2+} elevation (Dani *et al.*, 1992; Pasti *et al.*, 2001; Porter and McCarthy, 1996), which is associated with glutamate release (Parpura and Haydon, 2000; Pasti *et al.*, 2001). It has been demonstrated that glutamate release can also be observed following other stimuli, including bradykinin (Parpura *et al.*, 1994), prostaglandin (Bezzi *et al.*, 1998), chemokine (Bezzi *et al.*, 2001), endocannabinoid (Navarrete and Araque, 2008), and 5-hydroxytryptamine (Meller *et al.*, 2002) receptor activation. The release of glutamate evoked by these agents is linked to Ca^{2+} delivery from intracellular stores, emphasizing the evidence that exocytotic-like glutamate release may take place in astrocytes.

The disclosure of this active role of glia led to the model of the "tripartite synapse" (Araque *et al.*, 1999; Bezzi and Volterra, 2001), where, besides the important tasks pursued by pre- and postsynaptic neuronal elements, a pivotal role in regulating synaptic function, strength, and plasticity is played by the glial cells surrounding the above-mentioned structures.

II. Gliosomes as a Model to Study Astrocyte Properties

Most of the studies on the release of glutamate from glia have been carried out using cultured astrocytes (Araque *et al.*, 2000; Bezzi *et al.*, 1998, 2001; Parpura *et al.*, 1994); very few work explored other systems such as astroglioma cells (Meller *et al.*, 2002), acutely isolated astrocytes (Rutledge and Kimelberg, 1996), or brain slices, where the astrocytary transmitter release has been isolated from that of neuronal origin (Carmignoto *et al.*, 1998; Navarrete and Araque, 2008). In our laboratory, we studied the possibility of using a glial subcellular particle preparation acutely isolated from the brain of the adult rodent, which we named gliosomes.

Purified gliosomes (and synaptosomes) utilized in the experiments here described have been prepared from rat or mouse brain tissue by homogenization and purification on a discontinuous Percoll® gradient essentially according to Nakamura *et al.* (1993, 1994) with minor modifications (Stigliani *et al.*, 2006). The tissue was homogenized in 14 volumes of 0.32 M sucrose, buffered at pH 7.4 with Tris–HCl, using a glass–teflon tissue grinder (clearance 0.25 mm, 12 up–down strokes in about 1 min). The homogenate was centrifuged (5 min, 1000*g* at 4 °C) to remove nuclei and debris and the supernatant gently stratified on a discontinuous Percoll® gradient (2%, 6%, 10%, and 20% v/v in Tris-buffered sucrose) and centrifuged at 33,500*g* for 5 min at 4 °C. The layers between 2% and 6% Percoll® (gliosomal fraction) and between 10% and 20% Percoll® (synaptosomal fraction) were collected, washed by centrifugation and resuspended in physiological medium.

Several studies have taken advantage of the characteristics of the gliosome preparation to study functional aspects of glial cells. These studies allowed identification of specific cell distribution, function, and molecular mechanisms of a number of transmitter and modulator targets, mainly membrane transporters (see for instance Daniels and Vickroy, 1999; Hirst *et al.*, 1998; Pedrazzi *et al.*, 2006; Raiteri *et al.*, 2005a; Suchak *et al.*, 2003).

In a recent paper (Stigliani *et al.*, 2006), we characterized in detail gliosomes purified from adult rat cerebral cortex, pointing out biochemical and morphological evidence in support to the concept that our gliosomal fraction is largely purified from synaptosomes. Figure 1A shows the results obtained when the presence of glial and neuronal markers was studied by Western blot: the astrocyte markers, glial fibrillary acidic protein (GFAP) and Ca^{2+}-binding protein S-100, were expressed in gliosomes more abundantly than in synaptosomes; while the neuronal markers PSD-95 and β-tubulin III were enriched in the synaptosomal fraction. Confocal microscopy highlights the extensive labeling of particles present in the gliosomal preparation by GFAP (Fig. 1B); about 90% of particles present in this preparation were positive for GFAP using selective antibodies (not shown). Moreover, GFAP-positive gliosomes presented only a very modest positiveness for antibodies raised against the neuronal markers PSD-95, microtubule-associated protein 2 (MAP-2), or β-tubulin III (Fig. 1B), thus supporting the idea that gliosomes represent a preparation with low synaptosomal contamination. Of note, PSD-95, MAP-2, and β-tubulin III extensively marked synaptosomes under the same experimental conditions (not shown). Moreover, GFAP-expressing gliosomal preparation did not exhibit labeling either for integrin-αM or for myelin basic protein (MBP), two proteins selectively expressed in microglia and oligodendrocytes, respectively (not shown). Accordingly, the ultrastructural analysis pointed out that the gliosome fraction displayed morphological differences compared to the synaptosomes. First, the gliosome fraction contained a very much lower number of postsynaptic densities, compared to the synaptosome fraction (Fig. 1C). Second, several vesicles with a diameter of approximately 30 nm scattered within the cytoplasm were present in about 35% of the gliosomes. These vesicles were either uncoated, or clathrin coated, and did not show a clustered configuration, at variance to synaptosomes (Fig. 1D).

Interestingly, similar to synaptosomes and to cultured astrocytes (Montana *et al.*, 2006), the SNARE proteins, synaptobrevin-2 (VAMP-2), syntaxin-1, SNAP-23, and SNAP-25, known to form the core complex, as well as the Ca^{2+} sensor synaptotagmin-1 and the regulatory protein synapsin-1, required to execute exocytotic neurotransmitter release (Südhof, 1995) could be detected in the purified gliosomal fraction (Fig. 2A). A substantial colocalization of the core complex proteins with the GFAP-positive particles could be evidentiated by the confocal experiments reported in Fig. 2B. The analysis of different image couples indicated that about 55% of GFAP-expressing particles coexpress VAMP-2 immunoreactivity and about 70% of GFAP colocalizes with both syntaxin-1 and SNAP-23. The GFAP-expressing gliosomal preparation also showed a significant (about 35%) vesicular glutamate transporter 1 (vGLUT-1) staining. Also, VAMP-2 and vGLUT-1 appear to be coexpressed in gliosomes: about 65% of VAMP-2 colocalizes with the vesicular glutamate transporter. Conversely, almost the totality of vGLUT-1 coexpresses with GFAP or VAMP-2.

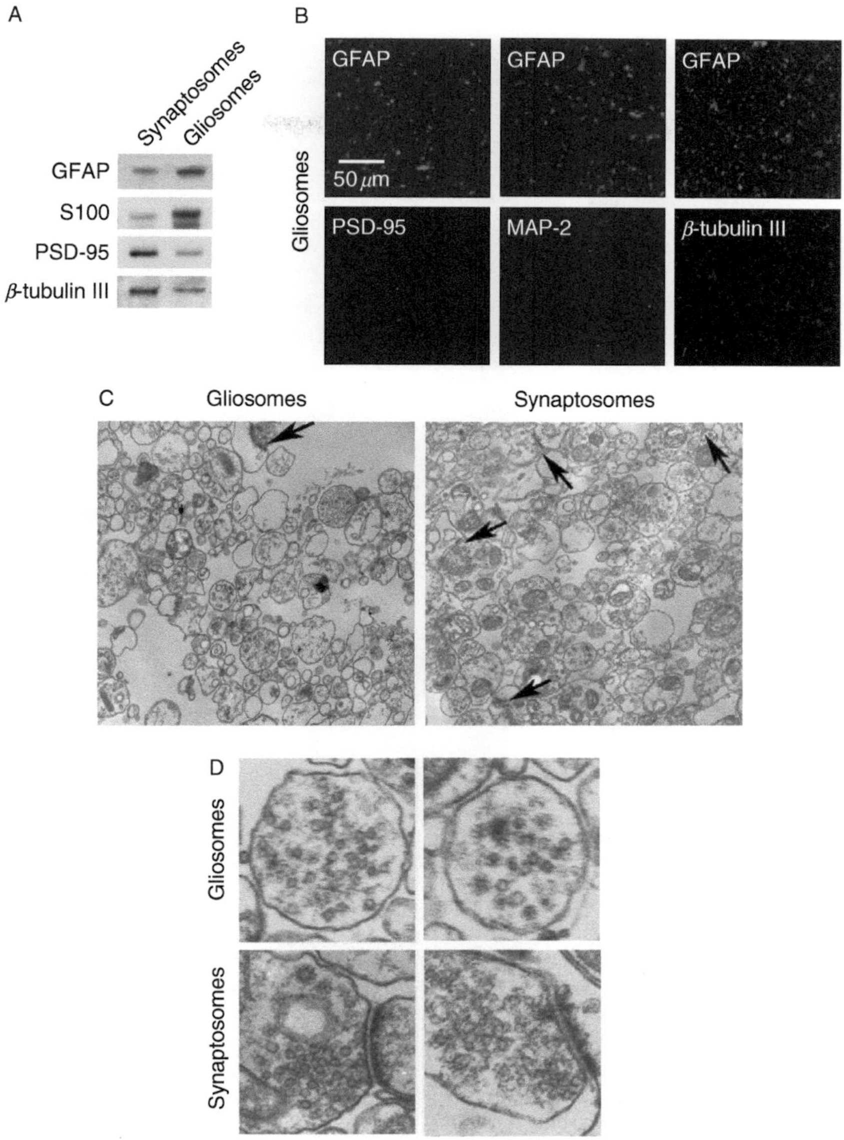

FIG. 1. (A) Expression of glia- and neuron-specific proteins in gliosomes and synaptosomes purified from rat cerebral cortex. Glial fibrillary acidic protein (GFAP), glial Ca^{2+}-binding protein S-100, neuronal postsynaptic density protein of 95 kDa (PSD-95), and neuronal β-tubulin III immunoreactivity in gliosomes and synaptosomes were evaluated by Western blotting. (B) Identification by immunocytochemistry and confocal microscopy of GFAP, PSD-95, neuronal microtubule-associated protein type 2 (MAP-2), and β-tubulin III. (C) Electron micrographs of gliosome and synaptosome fractions, showing the different presence of postsynaptic densities in the two preparations (arrows). (D) Electron micrographs of gliosome and synaptosome fractions, showing the cytosolic vesicle organization. (See Color Insert.)

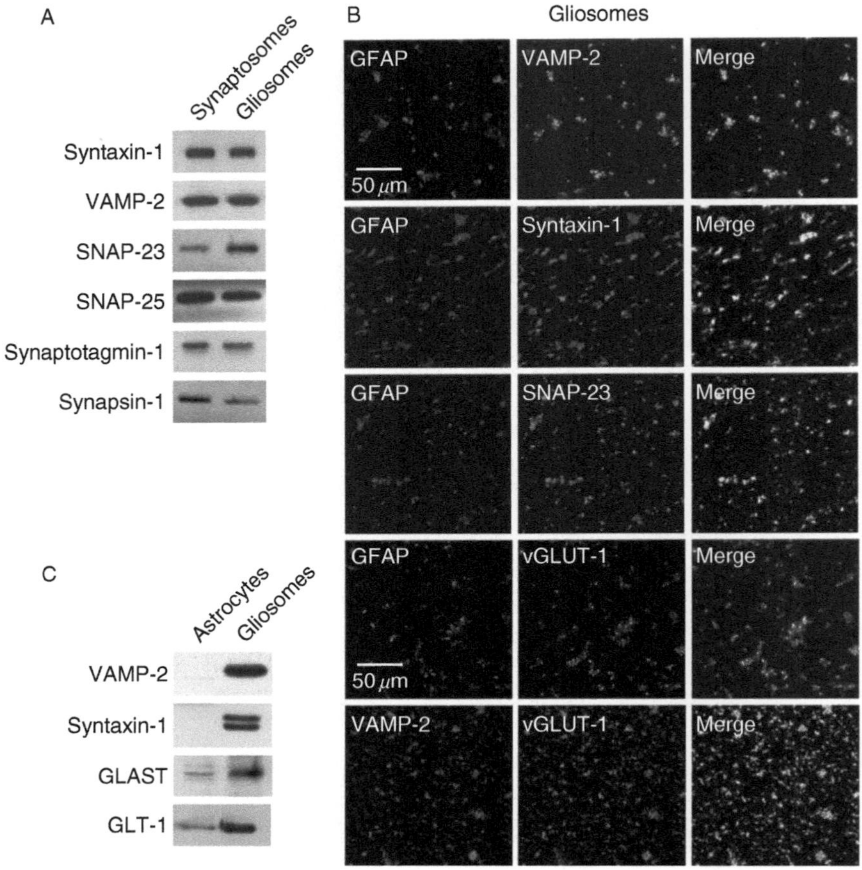

FIG. 2. (A) Expression of proteins of the release machinery in gliosomes and synaptosomes purified from rat cerebral cortex. Syntaxin-1, vesicular-associated membrane protein type 2 (VAMP-2), synaptosome-associate membrane protein of 23 kDa (SNAP-23) or 25 kDa (SNAP-25), synaptotagmin-1, and synapsin-1 immunoreactivity was evaluated by Western blotting. (B) Immunocytochemical identification of glial fibrillary acidic protein (GFAP) and its colocalization with VAMP-2, syntaxin-1, SNAP-23, and vesicular glutamate transporter type (vGLUT-1) immunoreactivity in gliosomes. Immunocytochemical colocalization of VAMP-2 and vGLUT-1. Samples were analyzed by laser confocal microscopy. (C) Expression of VAMP-2, syntaxin-1, glutamate–aspartate transporter (GLAST), and glutamate transporter of type 1 (GLT1) in gliosomes or in neonatal cultured astrocytes. Samples were analyzed by Western blot. (See Color Insert.)

Interestingly, expression of the SNARE complex proteins, VAMP-2 and sintaxin-1, and of the glial-specific glutamate transporters of type 1 (GLT1) and GLAST were much more enriched in gliosomes than in astrocytes in culture, as outlined by the Western blot experiments reported in Fig. 2C. This finding suggests that during brain tissue homogenization, gliosomes are formed by a process similar to that originating synaptosomes: that is, they are "pinched off" particles coming from glia cell arborizations. It has been proposed that astrocytes possess dedicated regions at the processes surrounding the synapses, by which they sense the neuronal messengers for a point-to-point neuron to astrocyte communication. Astrocytes have been suggested to release transmitters from these specialized areas (Araque *et al.*, 1999; Carmignoto, 2000; Grosche *et al.*, 1999). Accordingly, a number of evidences indicate that the vesicular release sites of astrocytes might be situated at the processes rather than all the cell bodies (reviewed by Montana *et al.*, 2006). It could be proposed that the gliosomal preparation is enriched with these specific areas, where the release machinery of the glial cell should be concentrated.

III. Exocytotic Release of Glutamate from Gliosomes

The experiments shown above point out that gliosomes represent a highly purified astrocyte-derived subcellular preparation that possess the machinery to actuate exocytotic gliotransmitter release. We tested this hypothesis by monitoring directly the release of glutamate from gliosomes evoked by stimuli that augment the cytosolic Ca^{2+} concentration $[Ca^{2+}]_i$, by means of different mechanisms, in *in vitro* release experiments.

A. Increasing the Gliosome $[Ca^{2+}]_i$ by Ionomycin Induces Glutamate Release

Functional experiments conducted in rat cerebral cortex showed that purified gliosomes are able to take up and release glutamate when exposed to stimuli able to induce increase of intracellular Ca^{2+}, such as ionomycin.

The release of $[^3H]$-D-Asp or endogenous glutamate was studied, taking advantage of the uniqueness of a superfusion technique that we have used for several years to study neurotransmitter release from synaptosomes (Raiteri and Raiteri, 2000). The system consists of several (up to 24) parallel superfusion chambers thermostated at 37 °C in which very thin layers, mostly monolayers, of synaptosomes (gliosomes in our experiments), plated on microporous filters, are up–down superfused in conditions in which any released compound is immediately removed by the superfusion fluid. Such a rapid removal prevents indirect effects: in particular, the changes of glutamate release observed following exposure to various agents are essentially due to direct actions on excitatory amino acid-releasing particles with minimal or no involvement of neighboring elements. The fast taking away of released glutamate does not allow (i) their reuptake and, therefore, their exchange with cytosolic excitatory amino acid transporters substrates and (ii) their feedback on presynaptic targets like release-regulating receptors. If substrates just released are virtually absent from the particle biophase, release by Ca^{2+}-dependent exocytosis or by Ca^{2+}-independent transporter reversal can be monitored under appropriate conditions.

Ionomycin, a Ca^{2+}-selective ionophore capable to mediate Ca^{2+} influx without voltage-sensitive Ca^{2+} channel (VSCC) activation and previously shown to induce transmitter exocytosis from nerve terminals (Sanchez-Prieto *et al.*, 1987; Verhage *et al.*, 1991), produced a dose-dependent stimulus-evoked release of endogenous glutamate (Fig. 3A; Stigliani *et al.*, 2006). Figure 3B shows that the release induced by ionomycin was entirely dependent on the presence of external Ca^{2+} and significantly decreased by bafilomycin-A1, a vesicle membrane V-type ATPase inhibitor (Bowman *et al.*, 1988; Floor *et al.*, 1990), which is expected to prevent the accumulation of the amino acid into vesicles (Moriyama and Futai, 1990; Roseth *et al.*, 1995). The same pattern was observed when the release was studied after prelabeling with [^{3}H]-D-aspartate, which can be taken up through the glial glutamate transporter and mimic glutamate (not shown). Under our experimental conditions, low concentrations of ionomycin appeared to release even higher amounts of glutamate from gliosomes than from synaptosomes. Conversely, synaptosomes were superior glutamate releasers when higher concentrations of the Ca^{2+} ionophore were applied (not shown).

The dependence on Ca^{2+} and the sensitivity to bafilomycin-A1 of the ionomycin-evoked release of glutamate from gliosomes suggest that the stimulus-induced release is exocytotic in nature. In line, experiments performed with the fluorescent dye acridine orange (Zoccarato *et al.*, 1999) showed that gliosomes were able to accumulate the dye into acidic cytoplasmic organelles and that the application of ionomycin-induced fusion of these organelles with the plasma membrane that was almost totally external Ca^{2+}-dependent (Fig. 3C; Stigliani *et al.*, 2006).

B. Increasing Gliosome $[Ca^{2+}]_i$ by ATP or AMPA Receptor Activation Induces Glutamate Release

To rule out the possibility that the exocytosis-like release of glutamate evoked by ionomycin might be linked to a unique characteristic of this agent, or result from unforeseen damage to gliosome integrity related to the use of the ionophore, we tested the effects of ATP or AMPA, which has been reported to increase $[Ca^{2+}]_i$ and to induce glutamate release from cultured astrocytes by activating the respective membrane receptors (Jeremic *et al.*, 2001; Volterra and Meldolesi, 2005; Zhang *et al.*, 2004). As illustrated in Fig. 3D, ATP induced the release of [^{3}H]-D-aspartate in a concentration-dependent manner. ATP released a comparable

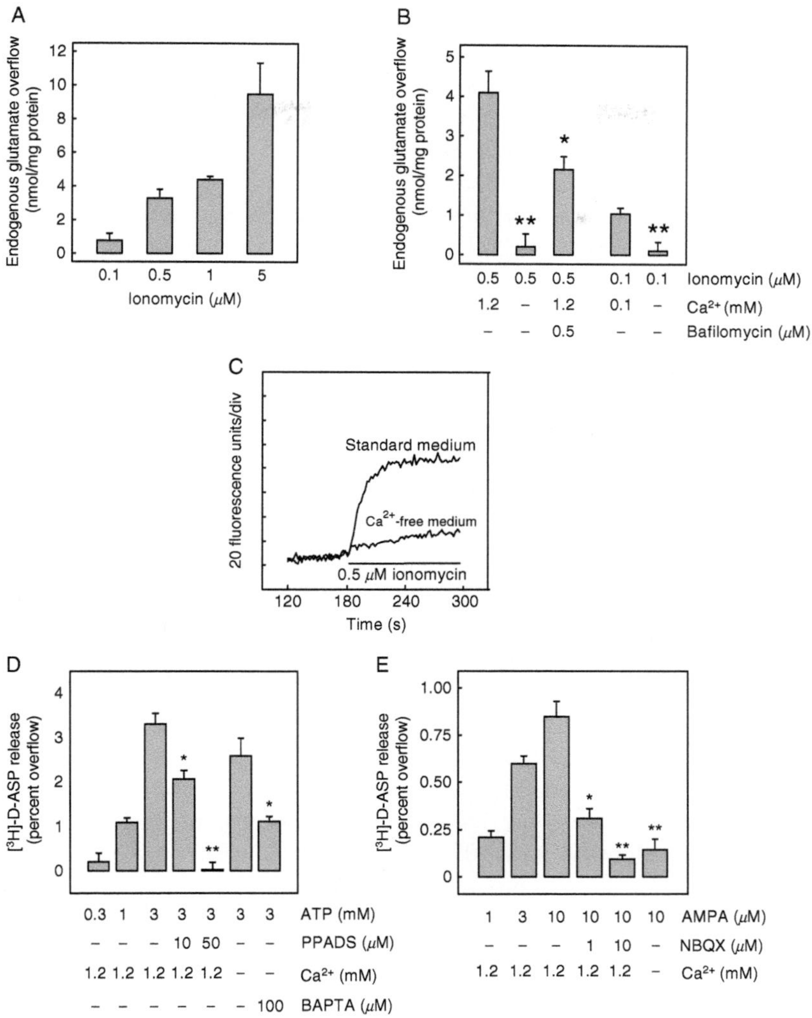

FIG. 3. (A) Effects of ionomycin (0.1–5 μM) on the release of endogenous glutamate from rat cerebral cortex gliosomes. Gliosomes were exposed in superfusion to a 90-s pulse of the Ca^{2+} ionophore. (B) Effects of Ca^{2+} omission and of bafilomycin on the release of endogenous glutamate induced by ionomycin in rat cerebral cortex gliosomes. *$p < 0.01$, **$p < 0.001$ versus the respective ionomycin-induced overflow (two-tailed Student's *t*-test). (C) Effect of ionomycin on vesicle membrane fusion in rat cerebral cortex purified gliosomes measured by means of the fluorescent dye acridine orange. (D) Effects of ATP (0.3–3 mM) on the release of [^{3}H]-D-aspartate from gliosomes purified from rat cerebral cortex. Effects of Ca^{2+} omission and BAPTA and antagonism by pyridoxalphosphate-6-azophenyl-2′,4′-disulfonic acid (PPADS) on the transmitter release induced by ATP. *$p < 0.01$, **$p < 0.001$ versus the respective ATP-induced overflow (two-tailed Student's *t*-test). (E) Effects of AMPA (1–10 μM) on the release of [^{3}H]-D-aspartate from gliosomes purified from rat cerebral cortex. Effects of Ca^{2+} omission and antagonism by NBQX on the transmitter release induced by AMPA. *$p < 0.01$, **$p < 0.001$ versus the respective AMPA-induced overflow (two-tailed Student's *t*-test).

amount of [^{3}H]-D-aspartate when applied to purified synaptosomes (data not shown). The effect of ATP was significantly reduced by the selective P2 receptor antagonist PPADS but it was only minimally affected by omission of extracellular Ca^{2+}. On the contrary, the ATP-induced release of [^{3}H]-D-aspartate was significantly diminished by preloading gliosomes with the Ca^{2+} chelator BAPTA, suggesting the involvement of intracellular Ca^{2+}. Also, AMPA (Fig. 3E) stimulated [^{3}H]-D-aspartate release in a concentration-dependent way from purified gliosomes, an effect abolished by the AMPA receptor antagonist NBQX and by omitting extracellular Ca^{2+}.

C. Increasing Gliosome $[Ca^{2+}]_i$ by Membrane Depolarization Induces Glutamate Release

External Ca^{2+} entry, leading to the increase of $[Ca^{2+}]_i$, has been reported to occur after electrical (MacVicar *et al.*, 1991) or chemical (Jensen and Chiu, 1991) depolarization of cultured astrocytes, as well as after KCl depolarization of astrocytes *in situ* utilizing acute brain slices (Porter and McCarthy, 1995). Accordingly, astrocytes express membrane ion channels, including voltage-sensitive Na^+ and K^+ channels as well as VSCCs, which may represent the molecular substrate of this aptitude (Barres *et al.*, 1990; Verkhratsky and Steinhauser, 2000).

Notwithstanding these indications, very few studies have tried to correlate depolarization and externally driven $[Ca^{2+}]_i$ modifications with transmitter release. In addition, the importance of astrocytic VSCCs in triggering exocytosis has been questioned (Carmignoto *et al.*, 1998). Overall, it is a common faith that depolarization and external Ca^{2+} hardly success in stimulating transmitter release from astrocytes (Montana *et al.*, 2006). As a matter of facts, it has been reported that KCl depolarization can provoke glutamate release from neonatal astrocytes in culture but only at markedly elevated concentrations and by mechanisms involving volume-activated Cl^- channels (Kimelberg *et al.*, 1990; Rutledge and Kimelberg, 1996). We have studied here the release of glutamate induced by membrane depolarization utilizing gliosomes as a model of adult astrocytes *in vitro* and found that indeed depolarization can induce exocytosis in this preparation.

Mild membrane depolarization obtained by 90-s application of 15 or 35 mM KCl increased [^{3}H]-D-aspartate (Fig. 4A) or endogenous glutamate (Fig. 4B) release from gliosomes. Neurotransmitter release was partly dependent on external Ca^{2+} and partly due to reversal of glutamate transporters. The external Ca^{2+} dependency of the gliosomal glutamate release suggests that high KCl can depolarize the glial plasma membrane, leading to Ca^{2+} entry and exocytosis, as in neurons and synaptosomes. This hypothesis was verified by directly measuring gliosomal membrane potential, cytosolic Ca^{2+} concentration $[Ca^{2+}]_i$, and vesicle fusion. KCl increased gliosomal membrane potential (Fig. 4C), cytosolic $[Ca^{2+}]_i$

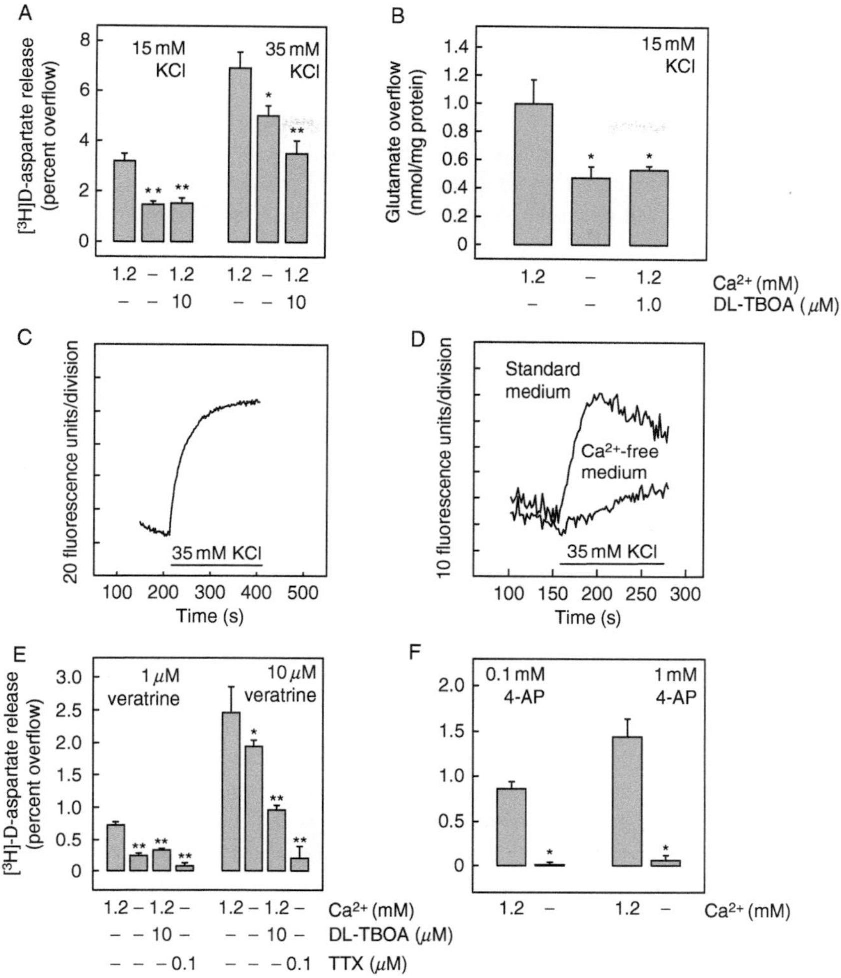

FIG. 4. (A) Effects of KCl on the release of [^{3}H]-D-aspartate from rat cerebral cortex gliosomes. Gliosomes were exposed in superfusion to a 90-s pulse of the depolarizing agent. Effects of Ca^{2+} omission and DL-TBOA on the transmitter release induced by KCl. *$p < 0.05$, **$p < 0.001$ when compared to respective KCl-induced overflow (two-tailed Student's t-test). (B) Effects of KCl on the release of endogenous glutamate from rat cerebral cortex gliosomes. Gliosomes were exposed in superfusion to a 90-s pulse of the depolarizing agent. Effects of Ca^{2+} omission and DL-TBOA on the transmitter release induced by KCl. *$p < 0.01$ when compared to respective KCl-induced overflow (two-tailed Student's t-test). (C) Effect of KCl (35 mM) on membrane potential of gliosomes prepared from rat cerebral cortex measured by means of the fluorescent dye Rhodamine-6G. (D) Effect of KCl on cytosolic vesicle fusion of gliosomes prepared from rat cerebral cortex measured by means of the fluorescent dye acridine orange. (E) Effects of veratrine on the release of [^{3}H]-D-aspartate from rat cerebral cortex gliosomes. Gliosomes were exposed in superfusion to a 90-s pulse of veratrine. Effects of Ca^{2+} omission, DL-TBOA, or TTX on the transmitter release induced by veratrine. *$p < 0.05$,

(not shown), and vesicle fusion rate (Fig. 4D), suggesting the involvement of exocytotic-like processes. Glutamate release from gliosomes was independent from VSCC opening; it was instead abolished by the Na^+/Ca^{2+} exchanger blocker KB-R7943 suggesting a role for this exchanger which, working in reverse mode, would allow Ca^{2+} entry during depolarization (not shown; Paluzzi *et al.*, 2007). Noteworthy, also the KCl-induced $[Ca^{2+}]_i$ increase was insensitive to VSCC activation but was abolished by KB-R7943.

We also investigated the releasing properties of other two depolarizing stimuli: 4-aminopyridine (4-AP), a K^+ channel blocker that leads to pulsatile membrane depolarization and glutamate release (Tibbs *et al.*, 1989), and veratrine, a mixture of alkaloids known to directly activate voltage-dependent Na^+ channels and to cause depolarization of neuronal plasma membranes (Narahashi, 1974). A 90-s pulse of veratrine (1 or 10 μM; Fig. 4E) increased the release of $[^3H]$-D-aspartate from gliosomes, an effect completely prevented by the Na^+ channel blocker tetrodotoxin. The release of $[^3H]$-D-aspartate evoked by 1 μM veratrine in gliosomes was largely (about 60%) dependent on external Ca^{2+} and partly (40%) blocked by DL-TBOA. The 10 μM veratrine-induced gliosomal neurotransmitter release was scarcely (about 25%) Ca^{2+}-dependent and largely (about 75%) carrier mediate. Also, 4-AP concentration dependently evoked $[^3H]$-D-aspartate release from prelabeled gliosomes (Fig. 4F). Neurotransmitter release was completely dependent on external Ca^{2+}. Also, the release of $[^3H]$-D-aspartate induced by veratrine or 4-AP was reduced by the Na^+/Ca^{2+} exchanger blocker KB-R7943.

One possible reason for the discrepancy between the present results with gliosomes and data in the literature, reporting that cultured astrocytes do not actuate glutamate exocytosis when subjected to K^+ depolarization (Kimelberg *et al.*, 1990; Rutledge and Kimelberg, 1996; Szatkowski *et al.*, 1990), may be due to the origin of gliosomes. In fact, they are acutely obtained from astrocytes of the adult rat brain, where they matured in the presence of neurons, while cultured astrocytes are usually prepared from neonatal animals. To investigate on this possibility, we obtained cultures of astrocytes from adult rats and measured $[Ca^{2+}]_i$ modifications and glutamate release induced by high K^+, in comparison with classical neonatal rat-prepared astrocytes. Adult astrocyte cultures were obtained by tissue explants derived from the superficial layer of adult (70–90 days) rat cortices and >95% of cells were astrocytes. Noteworthy, K^+ elicited increase of $[Ca^{2+}]_i$ in

---

**$p < 0.001$ versus the respective veratrine-induced overflow (two-tailed Student's *t*-test). (F) Effects of 4-aminopyridine (4-AP) on the release of $[^3H]$-D-aspartate from rat cerebral cortex gliosomes. Gliosomes were exposed in superfusion to a 90-s pulse of 4-AP. Effects of Ca^{2+} omission on the transmitter release induced by 4-AP. *$p < 0.001$ versus the respective 4-AP-induced overflow (two-tailed Student's *t*-test).

adult, not in neonatal astrocytes in culture (Fig. 5A). This cytosolic [Ca^{2+}] augmentation resulted in Ca^{2+}-dependent endogenous glutamate release (Fig. 5B). Glutamate release was even more marked in *in vitro* neuron-conditioned adult astrocytes. The endogenous glutamate release from adult astrocyte was almost abolished by omitting Ca^{2+} from the extracellular milieu and by incubating for 24-h cells with botulinum toxin C1, which cleaves the core complex proteins syntaxin-1 and SNAP-25 (Foran *et al.*, 1996; Schiavo *et al.*, 1995). As in the case of gliosomes, KCl-induced Ca^{2+} influx and glutamate release were abolished by KB-R7943 also in cultured adult astrocytes (not shown; Paluzzi *et al.*, 2007). These

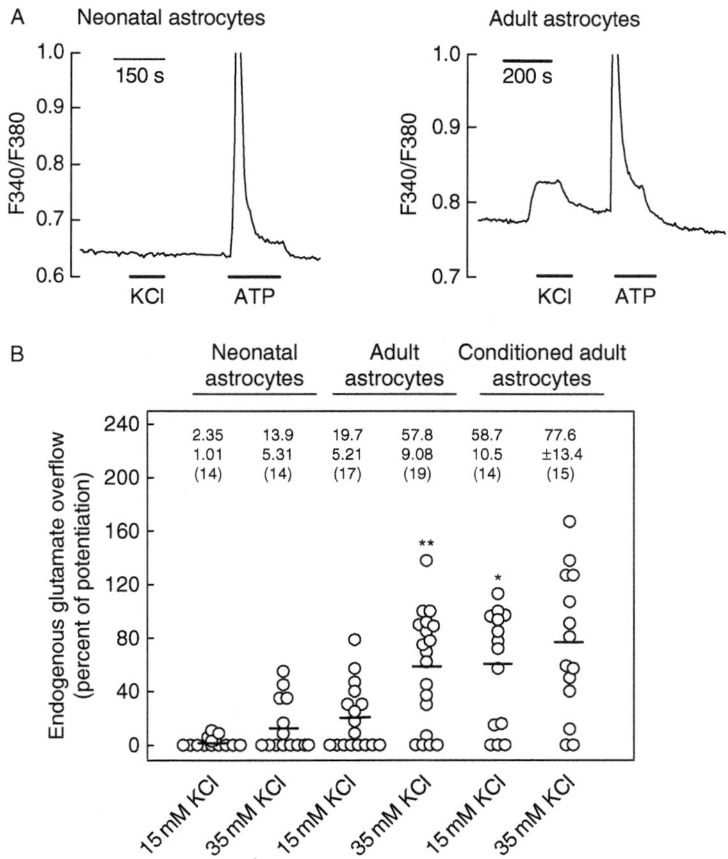

Fig. 5. (A) Effect of KCl (35 mM) on the cytosolic Ca^{2+} concentration in cultured astrocytes prepared from the cerebral cortex of neonatal (2-day-old) or adult (70–90-day-old) rats measured by means of the fluorescent dye FURA-2. (B) Effects of KCl on the release of endogenous glutamate from cultured astrocytes prepared from the cerebral cortex of neonatal or adult rats.

data reveal that depolarization-triggered glutamate exocytosis may occur *in vitro* from *in situ* matured adult astrocytes.

IV. Glutamate Release Induced by Heterotransporter Activation

Our laboratory has found that transporters for different transmitters often coexist on the same axon terminal, that is, a transporter which recaptures the endogenous transmitter just released and under some circumstances can release it through a carrier-mediated process (homotransporters) and transporters which recognize and take up transmitters coming from adjacent structures (heterotransporters; see Fig. 6A for a scheme). Activation of a heterotransporter invariably elicits release of the transmitter taken up previously through the coexisting homotransporter or endogenously synthesized. The release caused by heterotransporter activation takes place through multiple mechanisms including exocytosis, either dependent on external Ca^{2+} or dependent on Ca^{2+} mobilized from intraterminal stores, homotransporter reversal, anion channel opening (Bonanno and Raiteri, 1994; Raiteri *et al.*, 2002).

We decided to exploit the characteristics of the gliosome preparation to clarify whether the heterotransporter phenomenon could also exist on astrocyte, which have been reported to express transporters for different transmitters (Theodosis *et al.*, 2008). In particular, we investigated on the possible coexistence of glycine and GABA transporters and of GABA and glutamate transporters on the same gliosome in mouse spinal cord and on their ability to modulate the release of gliotransmitters.

A. Glycine Heterotransporter-Induced GABA Release

Gliosomes accumulated [^{3}H]GABA through GAT1 transporters and, when exposed to glycine in superfusion, they released the radioactive amino acid in a concentration-dependent manner (Fig. 6B). The effect of glycine at purified gliosomes is similar to that previously determined in synaptosomes, where glycine stimulated [^{3}H]GABA release via heterotransporter activation (Raiteri *et al.*, 2001, 2008). Studying the mechanism by which glycine evokes GABA release in gliosomes, we found that the effect of glycine was insensitive to strychnine and to 5,7-dichlorokynurenate, indicating that the amino acid does not act at its classical glycine-operated Cl^- channels or at the glycine coagonist site on NMDA receptors (Fig. 6C). The figure also shows that the glycine-evoked release of [^{3}H]GABA from gliosomes was abolished by GDA, a compound found to inhibit the uptake of [^{3}H]glycine into rat cortex synaptosomes (Javitt and Frusciante, 1997),

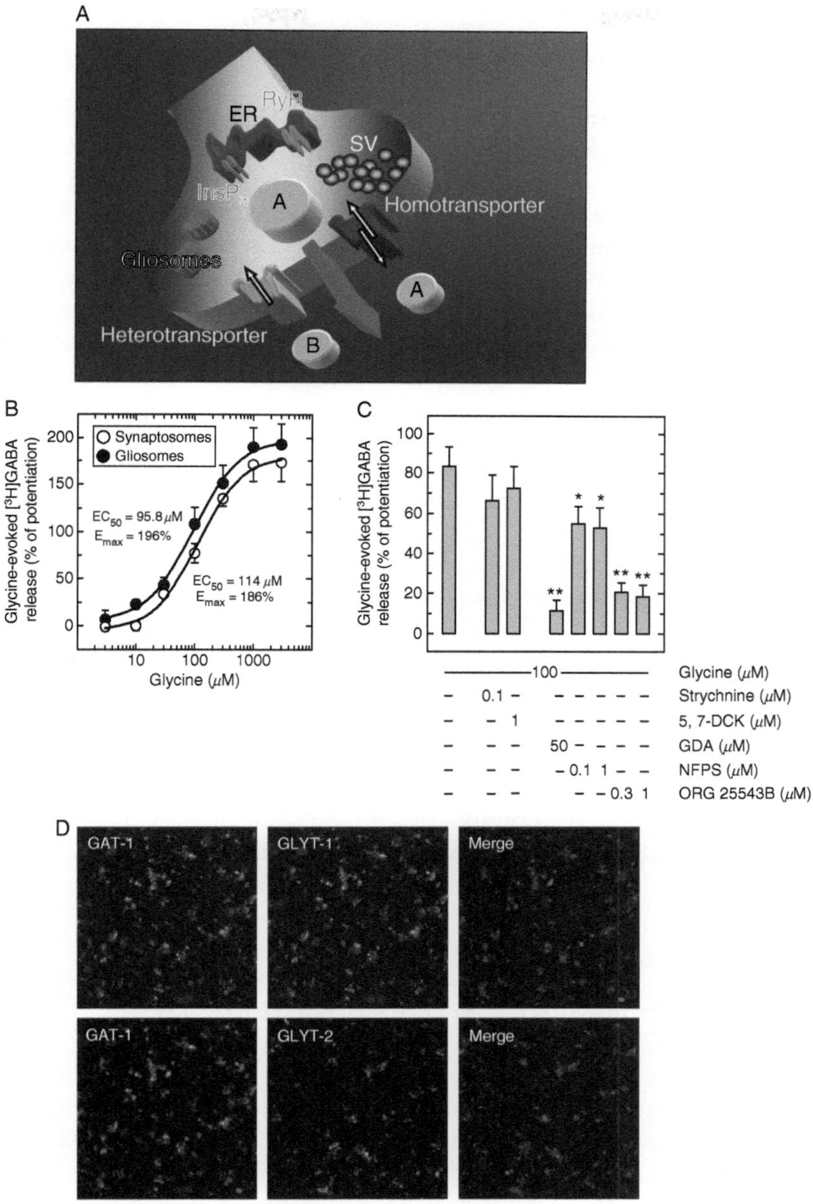

FIG. 6. (A) Representative scheme illustrating the heterotransporter mechanisms. (B) Effects of glycine (3–3000 μM) on the release of [^{3}H]GABA from gliosomes and synaptosomes prepared from rat cerebral cortex. (C) Effects of strychnine, 5,7-dichlorokynurenic acid (5,7-DCK), glycyldodecylamide (GDA), *N*-[3-(4′-fluorophenyl)-3-(4′-phenylphenoxy)propyl]sarcosine (NFPS), and 4-benzoyl-3,5-dimethoxy-*N*-[1-(dimethylaminociclopenthyl)methyl]benzamide (ORG25543B) on [^{3}H]GABA release

compatible with the idea that release of GABA takes place as a consequence of glycine penetration through its selective transporters into GABA-releasing particles.

Glycine transporters exist as two types, termed GLYT1 and GLYT2; being the first manly expressed in neurons and the second in astrocytes. We found therefore of interest to investigate the relative contribution of GLYT1 and GLYT2 to the glycine-evoked release of [^{3}H]GABA from purified gliosomes. Figure 6C shows that the releasing effect of glycine was significantly more sensitive to the selective GLYT2 inhibitor ORG25543 (Caulfield *et al.*, 2001) than to the selective GLYT1 inhibitor NFPS (Atkinson *et al.*, 2001). Surprisingly, it can be argued from these findings that the neuronal GLYT2 contributed more efficiently than the glial GLYT1 to mediate glycine potentiation in [^{3}H]GABA-releasing gliosomes. These functional results were largely supported by confocal microscopy analysis, showing that indeed GLYT1 are more abundantly expressed in gliosomes and GLYT2 in synaptosomes (not shown) but that coexpression of GAT1 and GLYT2 in gliosomes was more copious than GAT1 and GLYT1 coexistence (Fig. 6D).

We also investigated the mode of exit of [^{3}H]GABA from gliosomes exposed in superfusion to glycine: the effects of glycine were insensitive to the removal of external Ca^{2+} ions, and it did not decrease when external Ca^{2+} was removed and cytosolic Ca^{2+} was inactivated by entrapping BAPTA into gliosomes (Raiteri *et al.*, 2000). On the contrary, the presence in the superfusion medium of the GABA GAT1 transporter blocker SKF89976A (Larsson *et al.*, 1988) inhibited in a concentration-dependent manner the glycine-evoked release of [^{3}H]GABA, suggesting that Exocytotic processes are not involved and that [^{3}H]GABA is released by glycine through reversal of the GAT1 homotransporters leading to carrier-mediated transmitter release (Levi and Raiteri, 1993).

B. GABA Heterotransporter-Induced Glutamate Release

Similar results have been obtained when the effect of GABA on the release of glutamate was studied in spinal cord gliosomes. In this experiment, purified gliosomes were labeled with [^{3}H]-D-aspartate and exposed in superfusion to GABA.

induced by glycine. *$p < 0.01$, **$p < 0.001$ versus the respective GABA-induced overflow (two-tailed Student's *t*-test). (D) Immunocytochemical identification of glycine transporters type 1 and 2 (GLYT1, GLYT2) and their colocalization with the GABA transporter type 1 (GAT1). Samples were analyzed by laser confocal microscopy. (See Color Insert.)

GABA concentration dependently evoked the release of [^{3}H]-D-aspartate (maximal effect about 120% potentiation; $EC_{50} = 15.7\ \mu M$). The effect of GABA was prevented neither by the $GABA_A$ receptor antagonist SR95531 nor by the $GABA_B$ receptor antagonist CGP52432, excluding receptor involvement. The GABA-induced release of the excitatory amino acid was prevented by the GABA transport inhibitor SKF89976A, suggesting involvement of GABA transporters of the GAT1 type placed on glutamate-releasing astrocytes. Indeed, confocal microscopy showed that GAT1 is coexpressed with the glutamate transporters EAAT1 and EAAT2 in the majority of glial particles. As to the mode of exit of [^{3}H]-D-aspartate, the GABA effect was external Ca^{2+} independent and was not decreased when cytosolic Ca^{2+} ions were chelated by BAPTA. The release was almost completely reduced by the anion channel blockers niflumic acid and NPPB, suggesting that the release of glutamate was due to the opening of these nonspecific channels that, among other anions, are also permeable by glutamate (manuscript in preparation).

V. Glutamate Release from Gliosomes in a Mouse Model of Amyotrophic Lateral Sclerosis

A. Heterotransporter-Mediated Glutamate Release

Among the different hypotheses to explain motor neurons death in amyotrophic lateral sclerosis (ALS), glutamate-mediated excitotoxicity may play a mayor role (Morrison and Morrison, 1999). Abnormalities in glutamate transport, mainly a reduced expression and function of GLT1, were observed in synaptic preparations of motor cortex and spinal cord in ALS. It has been suggested that this GLT1 activity reduction could explain the higher levels of glutamate in ALS patients and in animal models of the disease (Gruzman *et al.*, 2007; Pardo *et al.*, 2006). Since GLT1 is mainly localized in astroglial processes, it was hypothesized that glial cells play a role in the development of the disease. Alternatively, elevated extracellular concentrations of glutamate may well be due to augmentation of neuronal glutamate release rather than to the astrocyte-localized inhibition of reuptake.

Studying the heterotransporter-mediated release modulation in synaptosomes purified from mouse spinal cord, we have found that glycine and GABA can be taken up by selective heterotransporters into nerve terminals endowed with glutamate homotransporters, thus causing release of glutamate (Raiteri *et al.*, 2005a,b). The glutamate release was in part due to homotransporter reversal and largely to anion channel opening. Furthermore, we have recently found that the ability of GABA and glycine heterotransporters to elicit release of glutamate from mouse spinal cord synaptosomes is dramatically enhanced in a transgenic mouse model of ALS (Raiteri *et al.*, 2003, 2004), possibly contributing to the reported augmented availability of glutamate in the extracellular fluids of these animal and ALS patients and to excitotoxicity. We applied the study of heterotransporter paradigm to gliosomes purified from the spinal cord of these transgenic mice to reveal if this mechanism could be enhanced also in glial cell, thus contributing jointly with neurons to the excessive release of the excitatory amino acid in the synaptic cleft.

B6SJL–TgN SOD1–G93A(+)1Gur mice expressing high copy number of mutant human SOD1 with a Gly^{93}Ala substitution [SOD1–G93A(+)] and B6SJL–TgN (SOD1)2Gur mice expressing wild-type human SOD1 [SOD1(+)] (Dal Canto and Gurney, 1994; Gurney *et al.*, 1994), obtained from Jackson Laboratories (Bar Harbor, ME), were used. Nontransgenic littermates of SOD1–G93A(+) and SOD1(+) mice were used as controls [SOD1(−)]. SOD1/G93A(+) mice develop the first symptoms around day 60 and reach the end-stage disease 8–11 weeks later. Death usually occurs between 120 and 140 days of life. This period was chosen for the experiments at the end stage of the disease; experiments have also been conducted at the 30-day presymptomatic stage.

The release of $[^3H]$-D-aspartate was concentration dependently enhanced by GABA in SOD1(+) control mice (maximal effect about 90% potentiation; $EC_{50} = 18.7\ \mu M$). These figures were comparable to those obtained in nontransgenic littermates belonging to the B6SJL strain (maximal effect about 100; $EC_{50} = 16.1$). Interestingly, the GABA-induced potentiation of $[^3H]$-D-aspartate release was significantly enhanced (maximal effect about 135%) in SOD1(+)/G93A(+) mice at the late stage of the pathology while the EC_{50} (14.5 μM) was unmodified. The effects of GABA were almost completely blocked by 30 μM of the GAT1 blocker SKF89976A. The GABA-induced $[^3H]$-D-ASP release was largely reduced by the anion channel blocker niflumic acid both in SOD1(+)/G93A(+) and in SOD1(+) mice. As a consequence, it can be assumed that the surplus of $[^3H]$-D-Asp release measured in the transgenic mutant mice is triggered by the same mechanisms taking place in SOD1(+) mice wild-type animals (manuscript in preparation).

To ascertain whether the potentiation of $[^3H]$-D-Asp release observed in gliosomes from spinal cord of symptomatic mice is already present in presymptomatic animals, experiments were performed with 30-day-old SOD1(+)/G93A(+) mice. The results show that the release of $[^3H]$-D-aspartate elicited by varying concentrations of GABA was already enhanced in 30 days SOD1(+)/G93A(+) respect to controls.

B. Depolarization-Evoked Glutamate Release

Very recently, we have also studied the release of $[^3H]$-D-aspartate and of endogenous glutamate induced by depolarizing and nondepolarizing stimuli, known to induce exocytotic neurotransmitter release, in synaptosomes from the spinal cord of SOD1/G93A(+) mice. Exposure to 15 or 25 mM KCl or to 0.3 or 1 μM ionomycin provoked an almost complete Ca^{2+}-dependent release of glutamate. The exocytotic release induced by KCl or ionomycin was dramatically increased in symptomatic SOD1(+)/G93A(+) mice than in controls. The higher glutamate release in mutant animals was already present in early-symptomatic 70–90 and presymptomatic 30–40-day-old mice. Noticeably, both the stimulus-evoked release of [^{3}H]GABA in spinal cord and of [^{3}H]-D-aspartate in motor cortex of SOD1(+)/G93A(+) mice did not differ from controls. Modification of phosphorylative pathways of synapsin-1 seems to be at the basis of the excessive glutamate release observed (manuscript in preparation). The results indicate that spinal cord glutamatergic nerve terminals of SOD1/G93A(+) mice undergoes to some presynaptic modifications which may sustain the increased glutamate exocytosis.

Paralleling the experiments of the preceding paragraph with heterotransporters, we tested whether mouse spinal cord gliosomes are capable to release glutamate exocytotically when subjected to stimuli that raise $[Ca^{2+}]_i$, as shown above in rat cerebral cortex, and whether this release was augmented in SOD1/G93A(+) mice, respect to controls. The results collected, although preliminary, shows that spinal cord gliosomes react to KCl depolarization producing glutamate exocytosis and that the 15 mM KCl-evoked [^{3}H]-D-aspartate release was greatly augmented in the transgenic mouse model of ALS.

VI. Concluding Remarks

Purified astrocyte-derived organelles isolated from the adult rat brain, referred as to gliosomes, are able to take up and release glutamate when subjected to a variety of stimuli. In particular, we have here shown that gliosomes are able to release glutamate in an exocytotic mode when subjected to stimuli able to increase $[Ca^{2+}]_i$, both allowing entering from the extracellular space and mobilization from cytosolic stores. In particular, purified gliosomes release glutamate when exposed to stimuli known to induce membrane depolarization and exocytotic release in neurons. Depolarization conditions such as elevated KCl, veratrine, or 4-AP can trigger glutamate release by two major mechanisms: vesicular exocytosis, involving extracellular Ca^{2+} entry, and reversal of the glutamate transporters, both thermodynamically linked to the collapse of the sodium gradients following membrane depolarization. The mechanism allowing Ca^{2+} entry is not linked to VSCC activation but to the Na^{+}/Ca^{2+} exchanger, working in the reverse mode due to Na^{+} accumulation into gliosomes during depolarization.

The reasons of the discrepancy between the failure of cultured neonatal astrocytes to release exocytotically glutamate by depolarization, reported in the literature, and our own results can be due to the *in situ* maturation of gliosome-producing astrocytes. The hypothesis that gliosomes may resemble mature astrocytes is strengthened by the observation that they are better glutamate releasers than adult astrocyte in culture and that adult astrocytes release glutamate even more efficiently if they have been conditioned in culture with neurons before experiments, a situation which more closely mimics the *in vivo* maturation of astrocytes, from which gliosomes originate.

Gliosomes hold also another mechanism of release previously described in synaptosomes: the heterotransporter-mediated gliotransmitters release. In particular, two heterotransporters, selective for glycine and GABA, respectively, have been described. Activation of glycine heterotransporters releases GABA and activation of GABA heterotransporters releases glutamate from spinal cord gliosomes.

Interestingly, both the exocytotic and the GABA heterotransporter-induced glutamate release were most pronounced in gliosomes prepared from the spinal cord of SOD1–G93A(+) mouse, a transgenic animal model of ALS, suggesting that astrocytic release may play a role in excitotoxicity, proposed as a cause of motor neuron degeneration.

To conclude, gliosomes may represent a viable preparation that allows to study mechanisms of transmitter release and its regulation in adult astrocytes. In this respect, gliosomes may have a number of advantages when compared to cultured astrocytes: they can be rapidly prepared and, most important, they originate directly from mature brain astrocytes. Due to their characteristics, gliosomes can be obtained from animals acutely or chronically treated with drugs, from knockout or knockdown animals, from animals that are models of brain diseases and from fresh human brain samples of surgical origin.

References

Araque, A., Parpura, V., Sanzgiri, R. P., and Haydon, P. G. (1999). Tripartite synapses: Glia, the unacknowledged partner. *Trends Neurosci.* **22,** 208–215.

Araque, A., Li, N., Doyle, R. T., and Haydon, P. G. (2000). SNARE protein-dependent glutamate release from astrocytes. *J. Neurosci.* **20,** 666–673.

Atkinson, B. N., Bell, S. C., De Vivo, M., Kowalski, L. R., Lechner, S. M., Ognyanov, V. I., Tham, C. S., Tsai, C., Jia, J., Ashton, D., and Klitenick, M. A. (2001). ALX 5407: A potent, selective inhibitor of the hGlyT1 glycine transporter. *Mol. Pharmacol.* **60,** 1414–1420.

Barres, B. A., Koroshetz, W. J., Chun, L. L., and Corey, D. P. (1990). Ion channel expression by white matter glia: The type-1 astrocyte. *Neuron* **5,** 527–544.

Bergles, D. E., and Jahr, C. E. (1997). Synaptic activation of glutamate transporters in hippocampal astrocytes. *Neuron* **19,** 1297–1308.

Bezzi, P., and Volterra, A. (2001). A neuron–glia signalling network in the active brain. *Curr. Opin. Neurobiol.* **11,** 387–394.

Bezzi, P., Carmignoto, G., Pasti, L., Vesce, S., Rossi, D., Rizzini, B. L., Pozzan, T., and Volterra, A. (1998). Prostaglandins stimulate calcium-dependent glutamate release in astrocytes. *Nature* **391,** 281–285.

Bezzi, P., Domercq, M., Brambilla, L., Galli, R., Schols, D., De Clercq, E., Vescovi, A., Bagetta, G., Kollias, G., Meldolesi, J., and Volterra, A. (2001). CXCR4-activated astrocyte glutamate release via TNFa: Amplification by microglia triggers neurotoxicity. *Nat. Neurosci.* **4,** 702–710.

Bonanno, G., and Raiteri, M. (1994). Release-regulating presynaptic heterocarriers. *Prog. Neurobiol.* **44,** 451–462.

Bowman, E. J., Siebers, A., and Altendorf, K. (1988). Bafilomycins: A class of inhibitors of membrane ATPase from microorganisms, animal cells and plant cells. *Biochemistry* **85,** 7972–7976.

Carmignoto, G. (2000). Reciprocal communication systems between astrocytes and neurones. *Prog. Neurobiol.* **62,** 561–581.

Carmignoto, G., Pasti, L., and Pozzan, T. (1998). On the role of voltage-dependent calcium channels in calcium signaling of astrocytes *in situ*. *J. Neurosci.* **18,** 4637–4645.

Caulfield, W. L., Collie, I. T., Dickins, R. S., Epemolu, O., McGuire, R., Hill, D. R., McVey, G., Morphy, J. R., Rankovic, Z., and Sundaram, H. (2001). The first potent and selective inhibitors of the glycine transporter type 2. *J. Med. Chem.* **44,** 2679–2682.

Dal Canto, M. C., and Gurney, M. E. (1994). Development of central nervous system pathology in a murine transgenic model of human amyotrophic lateral sclerosis. *Am. J. Pathol.* **145,** 1271–1279.

Daniels, K. K., and Vickroy, T. W. (1999). Reversible activation of glutamate transport in rat brain glia by protein kinase C and an okadaic acid-sensitive phosphoprotein phosphatase. *Neurochem. Res.* **24,** 1017–1025.

Dani, J. W., Chernjavsky, A., and Smith, S. J. (1992). Neuronal activity triggers calcium waves in hippocampal astrocyte networks. *Neuron* **8,** 429–440.

Floor, E., Leventhal, P. S., and Schaeffer, S. F. (1990). Partial purification and characterization of the vacuolar H^+-ATPase of mammalian synaptic vesicles. *J. Neurochem.* **55,** 1663–1670.

Foran, P., Lawrence, G. W., Shone, C. C., Foster, K. A., and Dolly, J. O. (1996). Botulinum neurotoxin C1 cleaves both syntaxin and SNAP-25 in intact and permeabilized chromaffin cells: Correlation with its blockade of catecholamine release. *Biochemistry* **25,** 2630–2636.

Grosche, J., Matyash, V., Moller, T., Verkhratsky, A., Reichenbach, A., and Kettenmann, H. (1999). Microdomains for neuron–glia interaction: Parallel fiber signaling to Bergmann glial cells. *Nat. Neurosci.* **2,** 139–143.

Gruzman, A., Wood, W. L., Alpert, E., Prasad, M. D., Miller, R. G., Rothstein, J. D., Bowser, R., Hamilton, R., Wood, T. D., Cleveland, D. W., Lingappa, V. R., and Jian Liu, J. (2007). Common molecular signature in SOD1 for both sporadic and familial amyotrophic lateral sclerosis. *Proc. Natl. Acad. Sci. USA* **104,** 12524–12529.

Gurney, M. E., Pu, H., Chiu, A. Y., Dal Canto, M. C., Polchow, C. Y., Alexander, D. D., Caliendo, J., Hentati, A., Kwon, Y. W., Deng, H. X., Chen, W., and Zhai, P. (1994). Motor neuron degeneration in mice that express a human Cu,Zn superoxide dismutase mutation. *Science* **264,** 1772–1775.

Haydon, P. G. (2001). Glia: Listening and talking to the synapse. *Nat. Rev. Neurosci.* **2,** 185–193.

Hirst, W. D., Price, G. W., Rattray, M., and Wilkin, G. P. (1998). Serotonin transporters in adult rat brain astrocytes revealed by [^{3}H]5-HT uptake into glial plasmalemmal vesicles. *Neurochem. Int.* **33,** 11–22.

Javitt, D. C., and Frusciante, M. (1997). Glycyldodecylamide, a phencyclidine behavioral antagonist, blocks cortical glycine uptake: Implications for schizophrenia and substance abuse. *Psychopharmacology* **129,** 96–98.

Jensen, A. M., and Chiu, S. Y. (1991). Differential intracellular responses to glutamate in type 1 and type 2 cultured brain astrocytes. *J. Neurosci.* **11,** 1674–1684.

Jeremic, A., Jeftinija, K., Stevsanovic, J., Glavavaski, A., and Jeftinija, S. (2001). ATP stimulated calcium-dependent glutamate release from cultured astrocytes. *J. Neurochem.* **77,** 664–675.

Kimelberg, H. K., Goderie, S. K., Igman, S., Pang, S., and Wanlewski, R. A. (1990). Swelling-induced release of glutamate, aspartate, and taurine from astrocyte cultures. *J. Neurosci.* **10,** 1583–1591.

Larsson, O. M., Falch, E., Krogsgaard-Larsen, P., and Schousboe, A. (1988). Kinetic characterization of inhibition of γ-aminobutyric acid uptake into cultured neurons and astrocytes by 4,4-diphenyl-3-butenyl derivatives of nipecotic acid and guvacine. *J. Neurochem.* **50,** 818–823.

Levi, G., and Raiteri, M. (1993). Carrier-mediated release of neurotransmitters. *Trends Neurosci.* **16,** 415–419.

MacVicar, B. A., Hochman, D., Dealy, M. J., and Weiss, S. (1991). Modulation of intracellular Ca^{2+} in cultured astrocytes by influx through voltage activated Ca^{2+} channels. *Glia* **4,** 448–455.

Meller, R., Harrison, P. J., and Elliott, J. M. (2002). *In vitro* evidence that 5-hydroxytryptamine increases efflux of glial glutamate via 5-HT_{2A} receptor activation. *J. Neurosci. Res.* **67,** 399–405.

Mennerick, S., and Zorumski, C. F. (1994). Glial contributions to excitatory neurotransmission in cultured hippocampal cells. *Nature* **368,** 59–62.

Montana, V., Malarkey, E. B., Verderio, C., Matteoli, M., and Parpura, V. (2006). Vesicular transmitter release from astrocytes. *Glia* **54,** 700–715.

Moriyama, Y., and Futai, M. (1990). H(+)-ATPase, a primary pump for accumulation of neurotransmitters, is a major constituent of brain synaptic vesicles. *Biochem. Biophys. Res. Commun.* **173,** 443–448.

Morrison, B. M., and Morrison, J. H. (1999). Amyotrophic lateral sclerosis associated with mutations in superoxide dismutase: A putative mechanism of degeneration. *Brain Res. Rev.* **29,** 121–135.

Nakamura, Y., Iga, K., Shibata, T., Shudo, M., and Kataoka, K. (1993). Glial plasmalemmal vesicles: A subcellular fraction from rat hippocampal homogenate distinct from synaptosomes. *Glia* **9,** 48–56.

Nakamura, Y., Kubo, H., and Kataoka, K. (1994). Uptake of transmitter amino acids by glial plasmalemmal vesicles from different regions of rat central nervous system. *Neurochem. Res.* **19,** 1145–1150.

Narahashi, T. (1974). Chemicals as tool in the study of excitable membranes. *Physiol. Rev.* **54,** 813–889.

Navarrete, M., and Araque, A. (2008). Endocannabinoids mediate neuron–astrocyte communication. *Neuron* **57,** 883–893.

Paluzzi, S., Alloisio, S., Zappettini, S., Milanese, M., Raiteri, L., Nobile, M., and Bonanno, G. (2007). Adult astroglia is competent for Na^+/Ca^{2+} exchanger-operated exocytotic glutamate release triggered by mild depolarization. *J. Neurochem.* **103,** 1196–1207.

Pardo, A. C., Wong, V., Benson, L. M., Dykes, M., Tanaka, K., Rothstein, J. D., and Maragakis, N. J. (2006). Loss of the astrocyte glutamate transporter GLT1 modifies disease in SOD1(G93A) mice. *Exp. Neurol.* **201,** 120–130.

Parpura, V., and Haydon, P. G. (2000). Physiological astrocytic calcium levels stimulate glutamate release to modulate adjacent neurons. *Proc. Natl. Acad. Sci. USA* **97,** 8629–8634.

Parpura, V., Basarasky, T. A., Liu, F., Jeftinija, K., Jeftinija, S., and Haydon, P. G. (1994). Glutamate-mediated astrocyte–neuron signalling. *Nature* **369,** 744–747.

Pasti, L., Zonta, M., Pozzan, T., Vicini, S., and Carmignoto, G. (2001). Cytosolic calcium oscillations in astrocytes may regulate exocytotic release of glutamate. *J. Neurosci.* **21,** 477–484.

Pedrazzi, M., Raiteri, L., Bonanno, G., Patrone, M., Ledda, S., Passalacqua, M., Milanese, M., Melloni, E., Raiteri, M., Pontremoli, S., and Sparatore, B. (2006). Stimulation of excitatory amino acid release from adult mouse brain glia subcellular particles by high mobility group box 1 protein. *J. Neurochem.* **99,** 827–838.

Porter, J. T., and McCarthy, K. D. (1995). GFAP-positive hippocampal astrocytes *in situ* respond to glutamatergic neuroligand with increased in $[Ca^{2+}]_i$. *Glia* **13,** 101–112.

Porter, J. T., and McCarthy, K. D. (1996). Hippocampal astrocytes *in situ* respond to glutamate released from synaptic terminals. *J. Neurosci.* **16,** 5073–5081.

Raiteri, L., and Raiteri, M. (2000). Synaptosomes still viable after 25 years of superfusion. *Neurochem. Res.* **25,** 1265–1274.

Raiteri, M., Sala, R., Fassio, A., Rossetto, O., and Bonanno, G. (2000). Entrapping of impermeant probes of different size into nonpermeabilized synaptosomes as a method to study presynaptic mechanisms. *J. Neurochem.* **74,** 423–431.

Raiteri, L., Raiteri, M., and Bonanno, G. (2001). Glycine is taken up through GLYT1 and GLYT2 transporters into mouse spinal cord axon terminals and causes vesicular release of its proposed cotransmitter GABA. *J. Neurochem.* **76,** 1823–1832.

Raiteri, L., Raiteri, M., and Bonanno, G. (2002). Coexistence and function of different neurotransmitter transporters in the plasma membrane of CNS neurons. *Prog. Neurobiol.* **68,** 287–309.

Raiteri, L., Paolucci, E., Prisco, S., Raiteri, M., and Bonanno, G. (2003). Activation of a glycine transporter on spinal cord neurons causes enhanced glutamate release in a mouse model of amyotrophic lateral sclerosis. *Br. J. Pharmacol.* **138,** 1021–1025.

Raiteri, L., Stigliani, S., Zappettini, S., Mercuri, N. B., Raiteri, M., and Bonanno, G. (2004). Excessive and precocious glutamate release in a mouse model of amyotrophic lateral sclerosis. *Neuropharmacology* **46,** 782–792.

Raiteri, L., Stigliani, S., Patti, L., Usai, C., Bucci, G., Diaspro, A., Raiteri, M., and Bonanno, G. (2005a). Activation of GABA GAT-1 transporters on glutamatergic terminals of mouse spinal cord mediates glutamate release through anion channels and by transporter reversal. *J. Neurosci. Res.* **80,** 424–433.

Raiteri, L., Stigliani, S., Siri, A., Passalacqua, M., Melloni, E., Raiteri, M., and Bonanno, G. (2005b). Glycine taken up through GLYT1 and GLYT2 heterotransporters into glutamatergic axon terminals of mouse spinal cord elicits release of glutamate by homotransporter reversal and through anion channels. *Biochem. Pharmacol.* **69,** 159–168.

Raiteri, L., Stigliani, S., Usai, C., Diaspro, A., Paluzzi, S., Raiteri, M., and Bonanno, G. (2008). Functional expression of release-regulating glycine transporters GLYT1 on GABAergic neurons and GLYT2 on astrocytes in mouse spinal cord. *Neurochem. Int.* **52,** 103–112.

Roseth, S., Fykse, E. M., and Fonnum, F. (1995). Uptake of L-glutamate into rat brain synaptic vesicles: Effect of inhibitors that bind specifically to the glutamate transporter. *J. Neurochem.* **65,** 96–103.

Rutledge, E. M., and Kimelberg, H. K. (1996). Release of [^{3}H]-D-aspartate from primary astrocyte cultures in response to raised external potassium. *J. Neurosci.* **16,** 7803–7811.

Sanchez-Prieto, J., Sihra, T. S., Evans, D., Ashton, A., Dolly, J. O., and Nicholls, D. G. (1987). Botulinum toxin A blocks glutamate exocytosis from guinea-pig cerebral cortical synaptosomes. *Eur. J. Biochem.* **165,** 675–681.

Schiavo, G., Shone, C. C., Bennett, M. K., Scheller, R. H., and Montecucco, C. (1995). Botulinum neurotoxin type C cleaves a single Lys–Ala bond within the carboxyl-terminal region of syntaxins. *J. Biol. Chem.* **270,** 10566–10570.

Stigliani, S., Zappettini, S., Raiteri, L., Passalacqua, M., Melloni, E., Venturi, C., Tacchetti, C., Diaspro, A., Usai, C., and Bonanno, G. (2006). Glia re-sealed particles freshly prepared from adult rat brain are competent for exocytotic release of glutamate. *J. Neurochem.* **96,** 656–668.

Suchak, S. K., Baloyianni, N. V., Perkinton, M. S., Williams, R. J., Meldrum, B. S., and Rattray, M. (2003). The "glial" glutamate transporter, EAAT2 (Glt-1) accounts for high affinity glutamate uptake into adult rodent nerve endings. *J. Neurochem.* **84,** 522–532.

Südhof, T. C. (1995). The synaptic vesicle cycle: A cascade of protein-protein interactions. *Nature* **375,** 645–653.

Szatkowski, M., Barbour, B., and Attwell, D. (1990). Non-vesicular release of glutamate from glial cells by reversed electrogenic glutamate uptake. *Nature* **348,** 443–446.

Theodosis, D. T., Poulain, D. A., and Oliet, S. H. R. (2008). Activity-dependent structural and functional plasticity of astrocyte–neuron interactions. *Physiol. Rev.* **88,** 983–1008.
Tibbs, G. R., Barrie, A. P., Van-Mieghen, F., McMahon, H. T., and Nicholls, D. G. (1989). Repetitive action potentials in isolated nerve terminals in the presence of 4-aminopyridine: Effects on cytosolic free Ca^{2+} and glutamate release. *J. Neurochem.* **53,** 1693–1699.
Verhage, M., McMahon, H. T., Ghijsen, W. E. J. M., Boomsma, F., Scholten, G., Wiegant, V. M., and Nicholls, D. G. (1991). Differential release of amino acids, neuropeptides, and catecholamines from isolated nerve terminals. *Neuron* **6,** 517–524.
Verkhratsky, A., and Steinhauser, C. (2000). Ion channels in glial cells. *Brain Res. Rev.* **32,** 380–412.
Volknandt, W. (2002). Vesicular release mechanisms in astrocytic signalling. *Neurochem. Int.* **41,** 301–306.
Volterra, A., and Meldolesi, J. (2005). Astrocytes, from brain glue to communication elements: The revolution continues. *Nat. Rev. Neurosci.* **6,** 626–640.
Zhang, Q., Pangrsic, T., Kreft, M., Krzan, M., Li, N., Sul, J., Salassa, M., Van Bockstaele, E., Zorec, R., and Haudon, P. G. (2004). Fusion-related release of glutamate from astrocytes. *J. Biol. Chem.* **279,** 12724–12733.
Zoccarato, F., Cavallini, L., and Alexandre, A. (1999). The pH-sensitive dye acridine orange as a tool to monitor exocytosis/endocytosis in synaptosomes. *J. Neurochem.* **72,** 625–633.

NEUROTROPHIC AND NEUROPROTECTIVE ACTIONS OF AN ENHANCER OF GANGLIOSIDE BIOSYNTHESIS

Jin-ichi Inokuchi

Division of Glycopathology, Institute of Molecular Biomembranes and Glycobiology, Tohoku Pharmaceutical University, 4-4-1, komatsushima, Aoba-ku, Sendai 981-8558, Miyagi, Japan

To address the role of brain gangliosides in synaptic plasticity, the synthetic ceramide analog, 1-phenyl-2-decanoylamino-3-morpholino-1-propanol (PDMP) was used to manipulate the biosynthesis of gangliosides in cultured cortical neurons. Spontaneous synchronized oscillatory activity of intracellular Ca^{2+} between the neurons, which represents synapse formation, was suppressed by the depletion of endogenous gangliosides by D-*threo*-PDMP, an inhibitor of glucosylceramide synthase. On the other hand, the enantiomer of inhibitor, L-*threo*-PDMP, could elevate cellular levels of gangliosides by upregulating several glycosyltransferases responsible for ganglioside biosynthesis. This review presents our findings on the neurotrophic actions of L-*threo*-PDMP *in vitro* and *in vivo*. We found that L-PDMP could upregulate neurite outgrowth, and functional synapse formation through activating GM3, GD3, and GQ1b synthases. Simultaneously, the activity of p42 mitogen-activated protein kinase was also facilitated by L-PDMP. To evaluate the efficacy of this drug on long term memory, rats were trained for 2 weeks using an 8-arm radial maze task, and then forebrain ischemia was induced by four-vessel occlusion. Repeated treatment of L-PDMP starting 24 h after the ischemia, improved the deficit of the well-learned spatial memory and prevented the ischemia-induced apoptosis in hippocampus, demonstrating the potential therapeutic use of the ceramide analog for treatment of neurodegenerative disorders.

INTERNATIONAL REVIEW OF NEUROBIOLOGY, VOL. 85
DOI: 10.1016/S0074-7742(09)85022-8

I. Introduction

Gangliosides, a family of sialic acid-containing glycosphingolipids (GSLs), are abundant in CNS (Fig. 1). It has often been reported that exogenous gangliosides can elicit neurite outgrowth and neural repair *in vitro* and *in vivo* (Pepeu *et al.*, 1994; Tsuji *et al.*, 1992; Wu *et al.*, 1991). In particular, GM1 and GQ1b have been found not only to enhance nerve growth factor (NGF)-induced neurite outgrowth, but also to display NGF-like activities themselves (Ferrari *et al.*, 1993; Maysinger *et al.*, 1993; Mutoh *et al.*, 1998; Tsuji *et al.*, 1983). The development of GM2/GD2 synthase knockout mice revealed that a lack of complex gangliosides induces

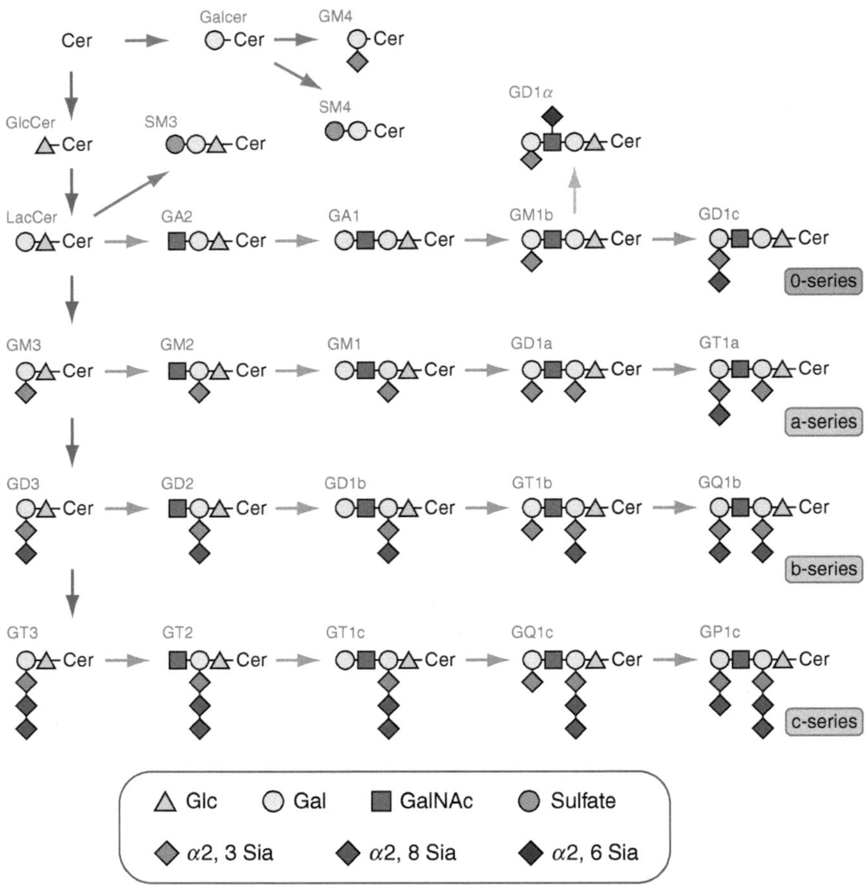

FIG. 1. The ganglioside biosynthetic pathway mainly expressed in brain. (See Color Insert.)

abnormal conduction velocity in somatosensory nerves (Takamiya *et al.*, 1996). The mice also exhibited decreased central myelination and axonal degeneration (Sheikh *et al.*, 1999). Moreover, GM1 reduced glutamate, aspartate, γ-amino-*n*-butyric acid (GABA), and glycine efflux from the cerebral cortex after transient cerebral ischemia in rats (Phillis and O'Regan, 1995), and has been developed clinically for the treatment of neuronal dysfunction.

Various studies have been conducted on excitatory amino acids and their receptors to explain the neuronal cell death (necrosis) after cerebral ischemia. The mechanism of ischemia-induced neuronal damage and the efficacies of antagonists were reviewed by Hara *et al.* (1994). Recently, using caspase-1 knockout or transgenic mice (Friedlander *et al.*, 1997) and Bcl-2 transgenic mice (Martinou *et al.*, 1994), it was shown that the neuronal cell death induced by cerebral ischemia includes apoptosis.

GM1 prevented apoptotic cell death by enhancing TrkA dimerization and consequent autophosphorylation in PC12 cells (Ferrari and Greene, 1998) and decreased the severity of ischemic brain lesions in experimental models (Frontczak-Baniewicz *et al.*, 2000; Hicks *et al.*, 1998). Also in neuron-rich cortical cultures, GM1 and other gangliosides attenuated serum deprivation-induced neuronal apoptosis (Ryu *et al.*, 1999).

We demonstrated that the glucosylceramide synthase inhibitor, D-*threo*-PDMP (D-PDMP; Inokuchi and Radin, 1987; Inokuchi *et al.*, 1989, 1990), inhibited functional synapse formation (Mizutani *et al.*, 1996), neurite outgrowth, autophosphorylation of Trk, and Trk-initiated intracellular protein kinase cascades (Mutoh *et al.*, 1998). These suppressive effects were specifically reversed by the addition of GQ1b or GM1. Conversely, it was found that the enantiomeric form of D-PDMP, L-*threo*-PDMP (L-PDMP), increased the cellular content and biosynthesis of gangliosides in B16 melanoma cells (Inokuchi *et al.*, 1989, 1995) and stimulated neurite outgrowth (Usuki *et al.*, 1996) and functional synapse formation with a concomitant increase of ganglioside biosynthesis in primary cultured rat embryonic cortical neurons (Inokuchi *et al.*, 1997). Thus, L-PDMP or its analogs might be valuable for clinical use in the postischemic treatment of cerebro-vascular diseases.

In this review, I summarize our approaches to explore a new therapeutic intervention for the treatment of neurological disorders by enhancing the ganglioside biosynthesis.

II. Development of a Ceramide Analog PDMP

1-Phenyl-2-decanoylamino-3-morpholino-1-propanol (PDMP) is a synthetic analog of ceramide and possesses two chiral centers at the Cl and C2 positions, to which hydroxyl and aminoacyl groups are attached. It thus forms four

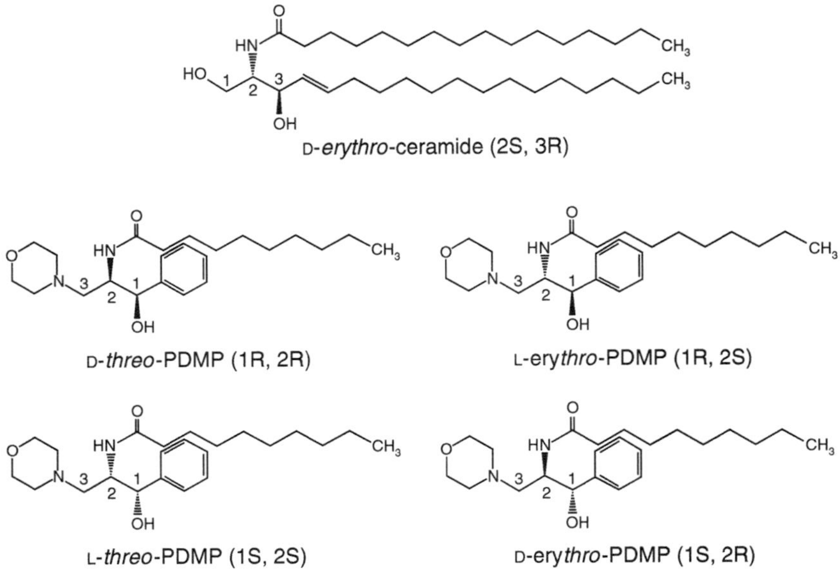

FIG. 2. Structure of ceramide and four stereoisomers of PDMP.

isomers (Fig. 2). The stereospecific action of PDMP isomers on UDP-glucose: *N*-acylsphingosine glucosyltransferase [glucosylceramide (GlcCer) synthase] has been clearly demonstrated, as only D-PDMP was able to inhibit GlcCer synthase and the other isomers, including the L-*threo* and *erythro* forms, could not affect this enzyme activity at all *in vitro* (Inokuchi and Radin, 1987). D-PDMP leads to extensive depletion of endogenous GSLs, including gangliosides biosynthesized from GlcCer, and causes accumulation of ceramide, and it has proved useful as a tool for studying various functional roles of endogenous GSLs (Radin *et al.*, 1993). Unexpectedly, we found that when B16 melanoma cells were incubated with L-PDMP, the cellular levels of the major GSLs—GlcCer, lactosylceramide (LacCer), and GM3—became significantly elevated (Inokuchi *et al.*, 1989). In addition, it was demonstrated that L-PDMP could enhance the activities of glycosyltransferases, including GlcCer synthase, LacCer synthase, and GM3 synthase, and led to elevate GSL levels (Chatterjee, 1991; Inokuchi *et al.*, 1995). We observed that exposure of cultured B 16 cells to cycloheximide for 4 h caused them to lose 50% of LacCer synthase activity (Inokuchi *et al.*, 1995). In the presence of L-PDMP, the loss of enzyme activity was not as great; however, D-PDMP exhibited no protective effect. These results may signify that L-PDMP binds to GSL synthesizing enzymes, stabilizing them against the normal catabolic inactivation process.

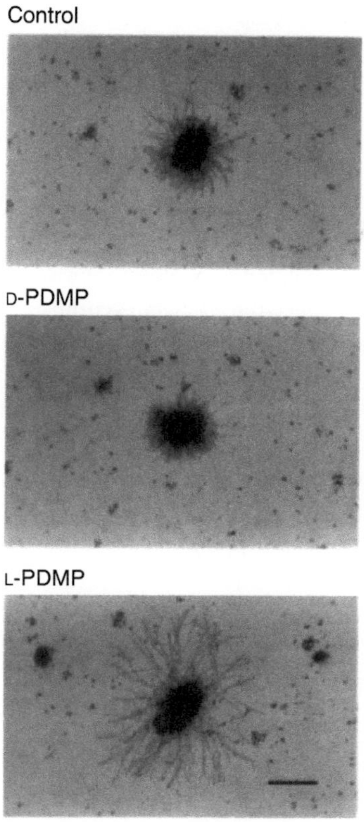

FIG. 3. Effects of PDMP isomers on neurite extension of rat neocortical explants. The explants were incubated for 2 days (top) without or (middle) with 15 μM D-PDMP or (bottom) 15 μM L-PDMP. The procedures were described in Usuki *et al.* (1996).

III. Effects of L- and D-PDMP on Neurite Extension

When explants from embryonic neocortical neurons were cultured with L-PDMP for 2 days in serum-free medium, marked increases in dense neurite outgrowth and the numbers of neurite branches were observed (Usuki *et al.*, 1996). On the other hand, treatment with D-PDMP resulted in an inhibition of neurite outgrowth. A typical microphotograph is presented in Fig. 3. D-PDMP showed a dose-dependent inhibitory effect from 5 to 20 μM. On the other hand, L-PDMP enhanced the neurite outgrowth over the same concentration range.

IV. Facilitation of Functional Synapse Formation and Ganglioside Synthesis by L-PDMP

L-PDMP at 20 and 40 μM in primary culture of cortical neurons increased the frequency of spontaneous synchronous oscillations between the neurons on day 9 of *in vitro* culture (DIV), a 50% increase being achieved compared to controls with no L-PDMP added (Fig. 4A). D-PDMP, which blocks GSL biosynthesis, showed the opposite effect as reported by Mizutani *et al.* (1996). The decreased functional synapse formation was normalized by supplementation of GQ1b but not by the other gangliosides, suggesting that *de novo* synthesis of ganglioside GQ1b is essential for the synaptic activity. The time course of the stimulatory effect of 20 μM L-PDMP indicates that a prolonged exposure to L-PDMP for at least 8 days is required to achieve this effect (Fig. 4B). The change in *de novo* synthesis of gangliosides produced by this treatment was analyzed by metabolic labeling of GSLs with [^{14}C]galactose for 6 h before each harvest (Fig. 5A). The slow but clear elevation of *de novo* synthesis of GSLs became clearly evident by 6 days of exposure to L-PDMP. Metabolic labeling of gangliosides in cortical neurons cultured with 20 μM L-PDMP for 8 days pointed to an elevation in the activity of biosynthetic pathways leading to the gangliosides, particularly GM3, GD3, and GQ 1b

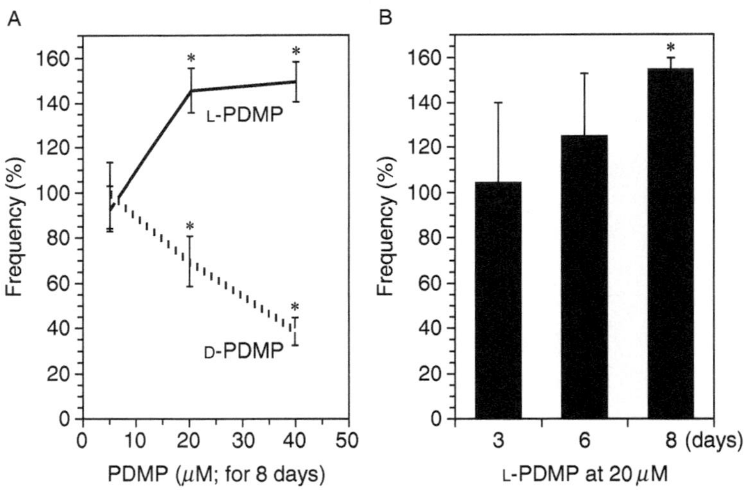

FIG. 4. A synthetic ceramide analog, L-PDMP, facilitates functional synapse formation. (A) The dose-dependent effects of PDMP isomers on the frequencies of synchronous oscillations of the cortical neurons. The cells were treated with PDMP isomers at the concentrations of 5 and 40 mM for 8 days (1–9 DIV). (B) Time dependency of the stimulatory effect of L-PDMP on functional synapse formation. The cells were treated with 20 mM L-PDMP for 1 (8–9 DIV), 3 (6–9 DIV), 6 (3–9 DIV), and 8 days (1–9 DIV). *$P < 0.01$, significant difference from the values in the absence of PDMP at each point.

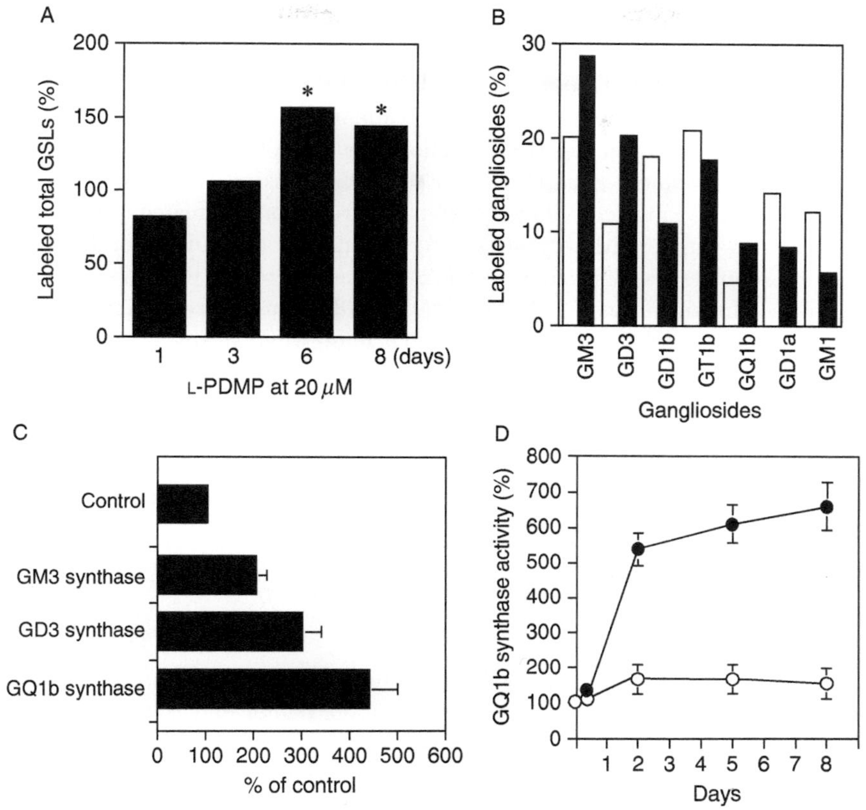

FIG. 5. Stimulation of ganglioside biosynthesis by L-PDMP. (A) Effect of L-PDMP on *de novo* synthesis of total GSLs. The cortical cells were treated with L-PDMP for different intervals. These cells were metabolically labeled with [^{14}C]galactose for 6 h before harvest and total labeled GSLs per mg protein were counted after purification as described previously (Usuki *et al.*, 1996). (B) Selective acceleration of *de novo* synthesis of gangliosides, GM3, GD3, and GQ1b by L-PDMP. The cells were treated with or without 20 μM L-PDMP for 8 days (1–9 DIV) and metabolically labeled. Equal amounts of radioactivities of total GSLs (12,000 dpm) in control (open bar) and the L-PDMP treated cells (closed bar) were separated by a silica gel HPTLC using chloroform/methanol/0.22% $CaCl_2$ (60:35:8, v/v/v). The radioactivity of each ganglioside species was detected with the BAS 2000 imaging analyzer. (C) Activation of ganglioside synthases in intact cortical cells by L-PDMP. The cells were treated with or without 20 μM L-PDMP for 8 days (1–9 DIV). GM3 synthase and GD3 synthase were assayed as described previously (Usuki *et al.*, 1996). (D) The activation profile of GQ1b synthase by 20 μM L-PDMP. Cells were treated in the absence (open circle) or presence (closed circle) of PDMP for the indicated intervals, and then the GQ1b synthase activities of the cell lysates of each culture were measured. Data are mean ± S.D. values of triplicate determinations.

(Fig. 5B). The elevation of ganglioside synthesis was confirmed to be due, in part at least, to increased activity of three gangliosides synthases, as assayed with lysates prepared from the cells pretreated with L-PDMP as described above.

GM3 synthase was increased by 100%, GD3 synthase by 200%, and GQ1b synthase by 340% (Fig. 5C). Analysis of the time course of GQ1b enzymatic synthesis, showed that the treatment with 20 μM L-PDMP had no effect by 10 h but by 2 days the activity had already risen about 265% over the control activity (Fig. 5D). This delayed type of activation was also observed in our previous experiments (Inokuchi *et al.*, 1989, 1990, 1995; Usuki *et al.*, 1996). In other assay conditions in which cell lysates prepared from non-pretreated cortical cells were treated with 20 μM L-PDMP during the enzyme assays, none of the above enzyme activities was affected (not shown) and similar results were also obtained previously (Inokuchi *et al.*, 1989, 1990, 1995, 1998). It has been also demonstrated that, when intracellular distribution of the fluorescent analog of PDMP was examined, the fluorescence could be detected at Golgi apparatus within 30 min after the exogenous addition of this analog (Rosenwald *et al.*, 1992). These results suggest that L-PDMP does not activate the transferase directly but through interaction with some other factors.

V. Upregulation of p42 MAPK Activity

Activation of p42 mitogen-activated protein kinase (MAPK) was found to occur in response to glutamate agonist stimulation (Bading and Greenberg, 1991; English and Sweatt, 1996) and was correlated with the spontaneous synaptic activity in cortical neurons (Fiore *et al.*, 1993). Since the cortical cell culture forms glutamatergic synapses (Robinson *et al.*, 1993), we measured the content and activity of p42 MAPK in the cortical cells treated with or without 20 μM L-PDMP (Inokuchi *et al.*, 1997). As shown in Fig. 6, L-PDMP had almost no effect on the content of p42 MAPK, however, this kinase activity was elevated in a slow but long-lasting manner. L-PDMP did not affect the MAPK activity in a short-term experiment (1–60 min) in the absence of serum. These results suggest that the activation profile of p42 MAPK by L-PDMP could be correlated with both increased ganglioside biosynthesis and synaptic activity.

VI. Improvement of the Spatial Memory Deficit and the Apoptotic Neuronal Death in Ischemic Rats

In the course of trials to evaluate the efficacy of L-PDMP on memory *in vivo*, we have tested the effect of the two PDMP isomers on the deficit of previously acquired special memory after transient forebrain ischemia in rats (Inokuchi *et al.*,

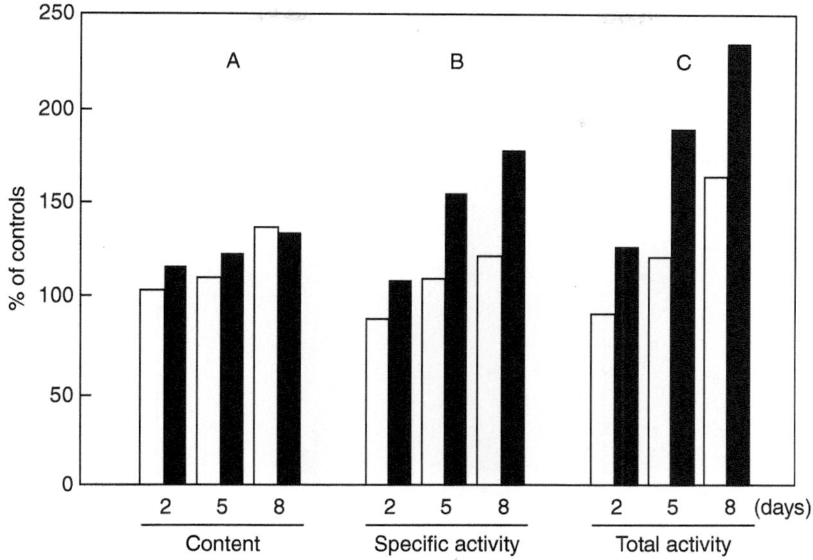

FIG. 6. Effect of L-PDMP on the content and activity of p42 MAPK in the cortical cells. Cells were treated in the absence (open bar) or presence (closed bar) of 20 μM L-PDMP for 2 (1–3 DIV), 5 (1–6 DIV), and 8 days (1–9 DIV). (A) p42 MAPK content of the cell extract was measured by Western blotting as described in methods. (B) Specific activity of p42 MAPK was shown. Immunoprecipitated MAPK activity was measured using the in-gel kinase assay and normalized for MAPK content measured by Western blotting. (C) To compare the total activity of p42 MAPK in the cortical cells, the specific activity was multiplied by MAPK content. Data are expressed as a mean percentage (duplicate) of MAPK content or activity of the cells taking each control value at 1 DIV as 100% (immediately before the addition of L-PDMP).

1997; Yamagishi *et al.*, 2003). The experimental protocol is shown in Fig. 7A. Vehicle-treated repeated cerebral ischemia rats showed a significant deficit in spatial cognition in the 8-arm radial maze task on the 7th day after reperfusion. On the other hand, a 6-day regimen with L-PDMP (40 mg/kg, twice a day) treatment starting at 24 h after ischemia (Group 1) resulted in a significant increase in correct choices and a decrease in errors in the radial maze task (Fig. 7B). A 4-day regimen with L-PDMP (40 mg/kg, twice a day) treatment starting at 72 h after ischemia (Group 2) did not show significant improvement (Fig. 7B). Neither of the two PDMP isomers had any effect on heart rates, blood pressure, or body temperature.

After behavioral evaluation in the 8-arm radial maze task, the rat's brain was fixed and apoptosis was identified in the hippocampal CA1 by TUNEL assay. Apoptotic cells were found in vehicle-treated rats on the 7th day after ischemia (mean $F \pm$ S.E.M.: 78.4 F 5.7 TUNEL-positive cells/mm^2). As shown in Fig. 8,

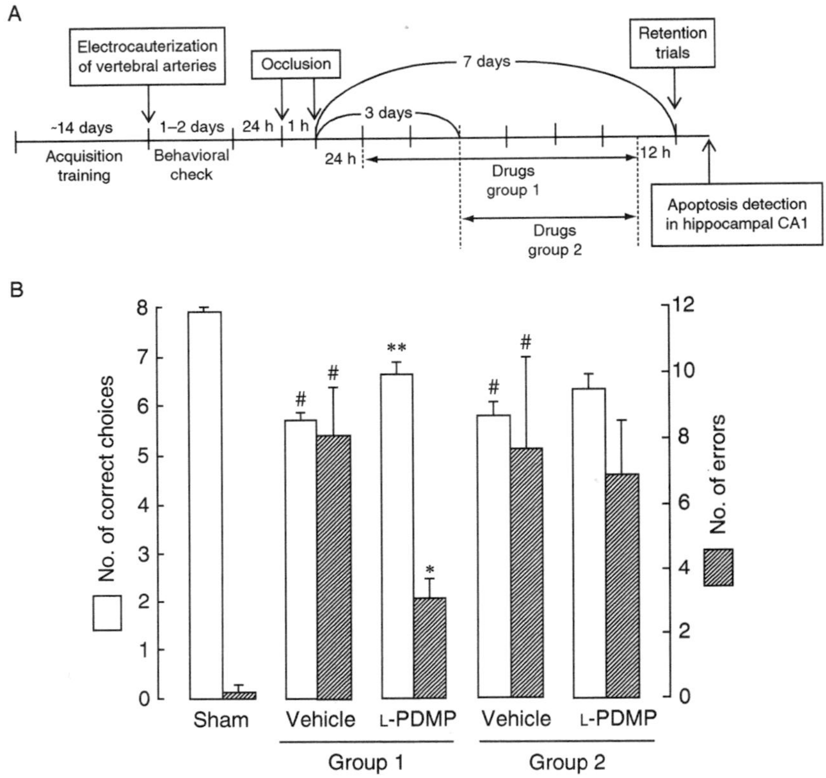

FIG. 7. Effects of L-PDMP on spatial cognition deficit induced by repeated cerebral ischemia. (A) Protocol for behavioral experiments to evaluate the efficacy of L-PDMP against spatial cognition deficit in rats with cerebral ischemia. (B) Sham (sham-operated rats, $n = 11$), Group 1-vehicle, L-PDMP (i.p. injections of vehicle or 40 mg/kg L-PDMP twice a day for 6 days from 24 h after ischemia, $n = 13$), Group 2-vehicle, L-PDMP (i.p. injection of vehicle or 40 mg/kg L-PDMP twice a day for 4 days from 3 days after ischemia, $n = 9$ and $n = 10$, respectively). $^{\#}P < 0.001$ versus sham-operated rats, $^{*}P < 0.05$ and $^{**}P < 0.01$ versus vehicle-treated rats (Wilcoxon's rank sum test).

continuous treatment with L-PDMP tended to attenuate apoptotic cell death in the hippocampal CA1. In Group 1, apoptotic cells were fewer than $25/\text{mm}^2$ in three of five animals (Group 1: 48.6 *F* 24.4 TUNEL-positive cells/mm^2). In Group 2, apoptotic cells were fewer than $25/\text{mm}^2$ in one of five animals (Group 2: 61.2 *F* 13.2 TUNEL-positive cells/mm^2).

Thus, we confirmed the amelioration by L-PDMP of the ischemia-induced deficit of spatial cognition and apoptotic cell death in CA1 even when treatment was started 24 h after repeated cerebral ischemia.

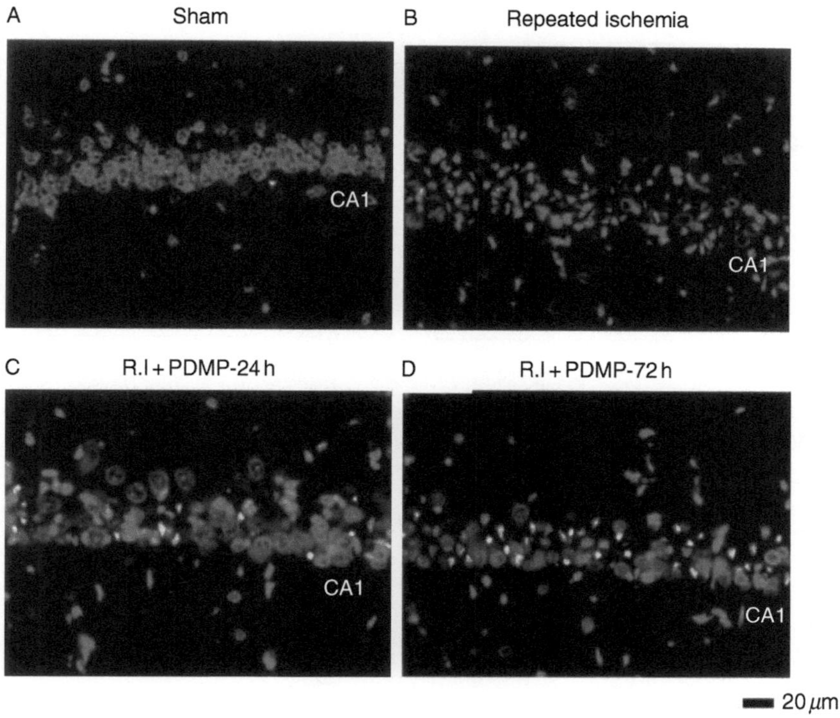

FIG. 8. Hippocampal CA1 in rats with repeated cerebral ischemia 7 days after reperfusion. Green fluorescence shows fragmented DNA in an apoptotic cell. Typical photomicrographs of sham-operated rats (A), repeated ischemia (R.I.) + vehicle (Group 1) (B), R.I. + L-PDMP (Group 1) (C), and R.I. + L-PDMP (Group 2) (D). Group 1: i.p. injections of vehicle or 40 mg/kg L-PDMP twice a day for 6 days from 24 h after ischemia, Group 2: i.p. injections of 40 mg/kg L-PDMP twice a day for 4 days from 3 days after ischemia. (See Color Insert.)

VII. Effect of L-PDMP on Biosynthesis of Cortical Gangliosides after Repeated Cerebral Ischemia

The successful results of L-PDMP treatment in ischemic rats encouraged us to investigate whether L-PDMP is able to stimulate ganglioside biosynthesis *in vivo* under a similar drug administration schedule (Yamagishi *et al.*, 2003). As a result of metabolic radiolabeling of gangliosides in the ischemic brain, we found that L-PDMP stimulated the biosynthesis of major gangliosides involving GM3 at the top of the biosynthetic pathway (Fig. 9). In particular, the biosynthesis of b-series gangliosides was significantly increased compared with that in vehicle-treated rats on days 3 and 5 after ischemia. These results correlate with the acceleration of GM3, GD3, and GQ1b synthases and functional synapse formation by L-PDMP

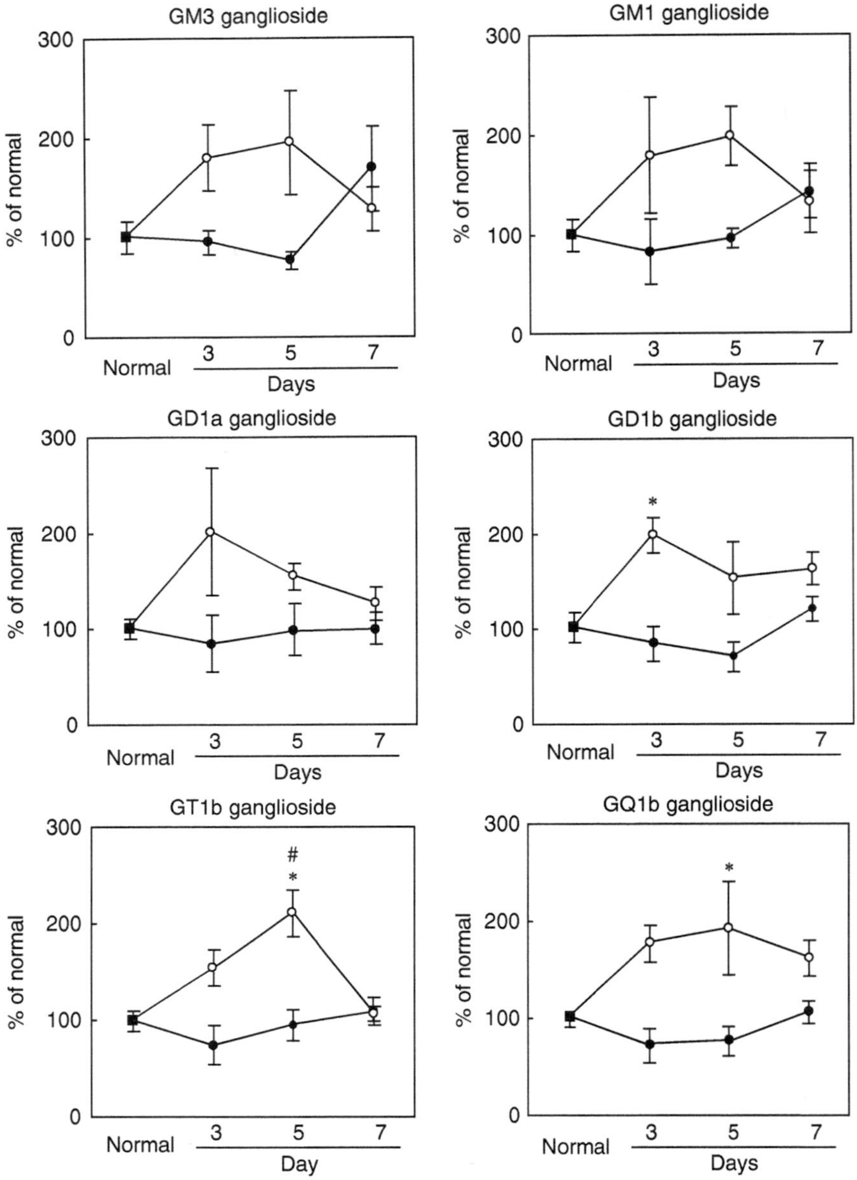

FIG. 9. Effects of L-PDMP on biosynthesis of cortical gangliosides in rats with cerebral ischemia. Normal rats: ($n = 6$), vehicle group (●, i.p. injections of vehicle twice a day from 24 h after ischemia): day 3 ($n = 5$), day 5 ($n = 6$), day 7 ($n = 4$), L-PDMP group (o, i.p. injections of 40 mg/kg L-PDMP twice a day from 24 h after ischemia): day 3 ($n = 4$), day 5 ($n = 6$), day 7 ($n = 4$). *$P < 0.05$ versus vehicle group, $^{\#}P < 0.05$ versus normal rats (Tukey's test).

in primary cultured cortical neurons (Inokuchi *et al.*, 1997). Since the upregulation of L-PDMP returned to the normal level by day 7, this compound might have a positive effect in the acute ischemic period.

VIII. Discussion

It has been shown that the biosynthesis of polysialogangliosides and their expression are correlated with differentiation and synaptogenesis of neuronal cells (Ledeen, 1985; Rosner *et al.*, 1992). We demonstrated that a synthetic ceramide analog, L-PDMP, upregulates the biosynthesis of b-series gangliosides in a long-term primary culture of cortical neurons by activating GM3, GD3, and GQ1b synthases (Inokuchi *et al.*, 1997). Under the same culture conditions, L-PDMP facilitates neurite outgrowth (Usuki *et al.*, 1996) and synchronous oscillatory activity between neurons (formation of functional synapses) in a long-lasting manner (Inokuchi *et al.*, 1997).

The importance of endogenous gangliosides in neural functions was directly evidenced by the observation that transfection of GD3 synthase gene into neuroblastoma cells induced the expression of b-series gangliosides, including GD3 and GQ1b, and also caused differentiation into cholinergic neuron-like cells (Kojima *et al.*, 1994). Nagai and Tsuji (1994) and Nagai (1995) originally demonstrated a highly specific neuritegenic effect of exogenous GQ1b on human neuroblastoma cells. Interestingly, they observed the remarkable neuritegenic effect of GQ1b only in the GQ1b-deficient neuroblastoma cells (not in the GQ1b-expressing cell lines). Their results coincided with our observation that the effect of exogenous GQ1b on functional synapse formation could be observed only in the cortical neurons depleted of endogenous gangliosides by D-PDMP, suggesting that the preexisting endogenous GQ1b might be actually involved in neuronal functions (Mizutani *et al.*, 1996). On the other hand, the ability of gangliosides, especially GM1, to exert trophic effects on neurons when applied as exogenous agents has been demonstrated with many *in vitro* and *in vivo* systems. However, exogenous GM1 did not facilitate functional synapse formation in our culture system (Mizutani *et al.*, 1996). Thus, it is suggested that exogenous gangliosides and L-PDMP differ in their mode of action and the induction of *de novo* synthesis of gangliosides can activate neuronal functions.

Synchronization among neuronal responses occurs within cortical regions as diverse as the visual (Gray *et al.*, 1989), sensorimotor (Murthy and Fetz, 1992), and prefrontal cortex (Vaadia *et al.*, 1995). The strength of synchronous coupling between these wide variety spaced cortical neurons changes dynamically during task performance (Bressler *et al.*, 1993). Thus, such synaptic plasticity of cortical neurons could be a part of the mechanism in cortical memory processing

(Fukai, 1994). Kuroda (1989) proposed that the plasticity of synaptic contacts in human association cortex constitutes the cellular mechanism of long-term memory in a "tracing circit" model. Accordingly, the ability of L-PDMP to induce long-lasting upregulation of synchronous oscillatory activity in cultured cortical neurons could be expected to have a beneficial effect on learning and memory *in vivo*. We have tested the effect of L- and D-PDMP on the retention of acquired special memory after transient forebrain ischemia, since it has been reported that a well-learned special memory of a maze task in rats stored mainly in the cortex area (McNaughton *et al.*, 1986; Okada *et al.*, 1995). It is notable that L-PDMP (40 mg/kg), which was administered i.p. as late as 24 h after the ischemic episode, could ameliorate the memory deficit in agreement with our *in vitro* experiments data (Inokuchi *et al.*, 1997, Yamagishi *et al.*, 2003).

Cerebral ischemia-induced deficit of spatial cognition is associated with apoptosis in hippocampal CA1 pyramidal cells, and that the key mechanism of promoting apoptosis starts within 3 days of reperfusion (Iwasaki *et al.*, 1998). In fact, it is reported that the expression of apoptosis-regulating molecules such as Bax and caspase-3 in CA1 neurons is seen in the period from 24 to 72 h following cerebral ischemia (Hara *et al.*, 1996; Ni *et al.*, 1998). In this study, L-PDMP tended to prevent apoptosis in hippocampal CA1 on day 7 after ischemia, when treatment was started 24 h after repeated cerebral ischemia (Fig. 8). Since L-PDMP stimulated cortical ganglioside biosynthesis on day 3 after ischemia when the drug treatment was started 24 h after ischemia (Fig. 8), it is strongly suggested that the enhancement of ganglioside biosynthesis restricts the cascade of neuronal cell death during the period in which cell death after ischemia shifts from necrosis to apoptosis. In addition, L-PDMP treatment caused an increase in brain GM1 levels in Parkinson models produced by subacute 1-methy-4-phenyl-1,2,3,6-tetrahydropyridine (MPTP) administration and resulted in a partial sparing of striatal dopamine levels, suggesting the potential strategy for Parkinson's disease (Schneider *et al.*, 2006).

Several lines of evidence point to the involvement of p42 MAPK in synaptic transmission. First of all, p42 MAPK and its substrates were localized at postsynaptic density (Suzuki *et al.*, 1995). Second, p42MAPK but not p44 MAPK was activated in response to synaptic stimuli, such as glutamate (Bading and Greenberg, 1991), *N*-methyl-d-aspartate, and electroconvulsive stimulation inducing long term potentiation (English and Sweatt, 1996). In addition, it has also been reported that p42 MAPK activity was increased simultaneously with upregulation of endogenous (spontaneous) synaptic activity in cultured cortical neurons (Fiore *et al.*, 1993). Considering these observations, we investigated the content and activity of p42 MAPK in cortical culture forming glutamatergic synapses. The slow but long-lasting facilitation of the synaptic activity by L-PDMP (Fig. 4) was paralleled by the activation profile of p42 MAPK (Fig. 6). Thus, the slow activation of MAPK by L-PDMP may reflect multiple intermediate steps.

Alternatively, the delayed response may reflect indirect activation of p42 MAPK by release of neurotrophins or other factors. At present, the cellular and molecular bases underlying the complex process of ganglioside biosynthesis and regulation are still being elucidated. Therefore, investigation of the possible interaction between ganglioside biosynthesis and signal transduction system including p42 MAPK in relation to their stimulation by L-PDMP is a matter of importance.

In conclusion, we have successfully demonstrated the usefulness of a new approach for treatment of memory deficit by upregulating *de novo* synthesis of gangliosides and synaptic function by applying the synthetic ceramide analog. Our results summarized here open up the possibility of a new therapeutic strategy for neurodegenerative disorders.

Acknowledgments

The author thanks Dr Norman Radin (Emeritus Professor University of Michigan) for continuous encouragement and many colleagues cited this review, whose contributions are indispensable for this study.

References

Bading, H., and Greenberg, M. E. (1991). Stimulation of protein tyrosine phosphorylation by NMDA receptor activation. *Science* **253,** 912–914.

Bressler, S. L., Coppola, R., and Nakamura, R. (1993). Episodic multiregional cortical coherence at multiple frequencies during visual task performance. *Nature* **366,** 153–156.

Chatterjee, S. (1991). Lactosylceramide stimulates aortic smooth muscle cell proliferation. *Biochem. Biophys. Res. Commun.* **181,** 554–561.

English, J. D., and Sweatt, J. D. (1996). Activation of p42 mitogen-activated protein kinase in hippocampal long term potentiation. *J. Biol. Chem.* **271,** 24329–24332.

Fiore, R. S., Murphy, T. H., Sanghera, J. S., Pelech, S. L., and Baraban, J. M. (1993). Activation of p42 mitogen-activated protein kinase by glutamate receptor stimulation in rat primary cortical cultures. *J. Neurochem.* **61,** 1626–1633.

Ferrari, G., and Greene, L. A. (1998). Promotion of neuronal survival by GM1 ganglioside. *Ann. N.Y. Acad. Sci.* **845,** 263–273.

Ferrari, G., Batistatou, A., and Greene, L. A. (1993). Gangliosides rescue neuronal cells from death after trophic factor deprivation. *J. Neurosci.* **13,** 1879–1887.

Friedlander, R. M., Gagliurdini, V., Hara, H., Fink, K. B., Li, W., MacDon-old, G., Fishman, M. C., Greenberg, A. H., Moskowitz, M. A., Yuan, J., (1997). Expression of a dominant negative mutant of interleukin-1β converting enzyme in transgenic mice prevents neuronal cell death induced by trophic factor withdrawal and ischemic brain injury. *J. Exp. Med.* **185,** 933–940.

Frontczak-Baniewicz, M., Gadamski, R., Barskov, I., and Gajkowska, B. (2000). Beneficial effects of GM1 ganglioside on photochemically-induced microvascular injury in cerebral cortex and hypophysis in rat. *Exp. Toxicol. Pathol.* **52,** 111–118.

Fukai, T. (1994). Synchronization of neural activity is a promising mechanism of memory information processing in networks of columns. *Biol. Cybern.* **71,** 215–226.

Gray, C. M., König, P., Engel, A. K., and Singer, W. (1989). Oscillatory responses in cat visual cortex exhibit inter-columnar synchronization which reflects global stimulus properties. *Nature* **338,** 334–337.

Hara, H., Kato, H., Sukamoto, T., Tsukamoto, G., and Kogure, K. (1994). Pharmacological prevention of ischemia-induced brain damage. *Drugs Today* **30,** 123–144.

Hara, A., Iwai, T., Niwa, M., Uematsu, T., Yoshimi, N., Tanaka, T., and Mori, H. (1996). Immunohistochemical detection of Bax and Bcl-2 proteins in gerbil hippocampus following transient forebrain ischemia. *Brain Res.* **711,** 249–253.

Hicks, D., Heidinger, V., Mohand-Said, S., Sahel, J., and Dreyfus, H. (1998). Growth factors and gangliosides as neuroprotective agents in excitotoxicity and ischemia. *Gen. Pharmacol.* **30,** 265–273.

Inokuchi, J., and Radin, N. S. (1987). Preparation of the active isomer of 1-phenyl-2-decanoylamino-3-morpholino-1-propanol, inhibitor of murine glucocerebroside synthetase. *J. Lipid Res.* **28,** 565–571.

Inokuchi, J., Momosaki, K., Shimeno, H., Nagamatsu, A., and Radin, N. S. (1989). Effects of D-*threo*-PDMP, an inhibitor of glucosylceramide synthetase, on expression of cell surface glycolipid antigen and binding to adhesive proteins by B16 melanoma cells. *J. Cell. Physiol.* **141,** 573–583.

Inokuchi, J., Jimbo, M., Momosaki, K., Shimeno, H., Nagamatsu, A., and Radin, N. S. (1990). Inhibition of experimental metastasis of murine Lewis lung carcinoma by an inhibitor of glucosylceramide synthase and its possible mechanism of action. *Cancer Res.* **50,** 6731–6737.

Inokuchi, J., Usuki, S., and Jimbo, M. (1995). Stimulation of glycosphingolipid biosynthesis by L-*threo*-1-phenyl-2-decanoylamino-1-propanol and its homologs in B16 melanoma cells. *J. Biochem.* **117,** 766–773.

Inokuchi, J., Mizutani, A., Jimbo, M., Usuki, S., Yamagishi, K., Mochizuki, H., Muramoto, K., Kobayashi, K., Kuroda, Y., Iwasaki, K., Ohgami, Y., and Fujiwara, M. (1997). Up-regulation of ganglioside biosynthesis, functional synapse formation, and memory retention by a synthetic ceramide analog (l-PDMP). *Biochem. Biophys. Res. Commun.* **237,** 595–600.

Inokuchi, J., Kuroda, Y., Kosaka, S., and Fujiwara, M. (1998). L-*threo*-1-phenyl-2-decanoylamino-3-morpholino-1-propanol stimulates ganglioside biosynthesis, neurite outgrowth and synapse formation in cultured cortical neurons, and ameliorates memory deficits in ischemic rats. *Acta Biochim. Pol.* **45,** 479–492.

Iwasaki, K., Chung, E. H., and Fujiwara, M. (1998). Apoptosis in the repeated cerebral ischemia—behavioral and histochemical study. *Folia Pharmacol. Jpn.* **112**(Suppl. 1), 88–92.

Kojima, N., Kurosawa, N., Nishi, T., Hanai, N., and Tsuji, S. (1994). Induction of cholinergic differentiation with neurite sprouting by *de novo* biosynthesis and expression of GD3 and b-series gangliosides in Neuro2a cells. *J. Biol. Chem.* **269,** 30451–30456.

Kuroda, Y. (1989). Tracing circuit model for the memory process in human brain; Roles of ATP and adenosine derivatives for dynamic change of synaptic connections. *Neurochem. Int.* **14,** 309–319.

Ledeen, R. (1985). Gangliosides of the neurons. *Trends Neurosci.* **8,** 169–174.

Martinou, J.-C., Dubois-Dauphin, M., Staple, J. K., Rodriguez, I., Frankowski, H., Missotten, M., Albertini, P., Talabot, D., Catsicas, S., Pietra, C., and Huarte, J. (1994). Overexpression of BCL-2 in transgenic mice protects neurons from naturally occurring cell death and experimental ischemia. *Neuron* **13,** 1017–1030.

Maysinger, D., Filipovic-Grcic, J., and Cuello, A. C. (1993). Effects of coencapsulated NGF and GM1 in rats with cortical lesions. *NeuroReport* **4,** 971–974.

McNaughton, B. L., Barnes, C. A., Rao, G., Baldwin, J., and Rasmussen, M. (1986). Long-term enhancement of hippocampal synaptic transmission and the acquisition of spatial information. *J. Neurosci.* **6,** 563–571.

Mizutani, A., Kuroda, Y., Muramoto, K., Kobayashi, K., Yamagishi, K., and Inokuchi, J. (1996). Effects of glucosylceramide synthase inhibitor and ganglioside GQ1b on synchronous oscillations of intracellular Ca^{2+} in cultured cortical neurons. *Biochem. Biophys. Res. Commun.* **222,** 494–498.

Murthy, V. N., and Fetz, E. E. (1992). Coherent 25- to 35-Hz oscillations in the sensorimotor cortex of awake behaving monkeys. *Proc. Natl. Acad. Sci. USA* **89,** 5670–5674.

Mutoh, T., Tokuda, A., Inokuchi, J., and Kuriyama, M. (1998). Glucosylceramide synthase inhibitor inhibits the action of nerve growth factor in PC12 cells. *J. Biol. Chem.* **273,** 26001–26007.

Nagai, Y. (1995). Functional roles of gangliosides in bio-signaling. *Behav. Brain Res.* **66,** 99–104.

Nagai, Y., and Tsuji, S. (1994). Significance of ganglioside-mediated glycosignal transduction in neuronal differentiation and development. *Prog. Brain Res.* **101,** 119–126.

Ni, B., Wu, X., Su, Y., Stephenson, D., Smalstig, E. B., and Clemens, J. (1998). Transient global forebrain ischemia induces a prolong expression of the caspase-3 mRNA in rat hippocampal pyramidal neurons. *J. Cereb. Blood Flow Metab.* **18,** 248–256.

Okada, M., Nakanishi, H., Tamura, A., Urae, A., Mine, K., Yamamoto, K., and Fujiwara, M. (1995). Long-term spatial cognitive impairment after middle cerebral artery occlusion in rats: No involvement of the hippocampus. *J. Cereb. Blood Flow Metab.* **15,** 1012–1021.

Pepeu, G., Oderfeld-Nowak, B., and Casamenti, F. (1994). CNS pharmacology of gangliosides. *Prog. Brain Res.* **101,** 327–335.

Phillis, J. W., and O'Regan, M. H. (1995). GM1 ganglioside inhibits ischemia release of amino acid neurotransmitters from rat cortex. *NeuroReport* **6,** 2010–2012.

Pulsinelli, W. A., and Brierley, J. B. (1979). A new model of bilateral hemispheric ischemia in the unanesthetized rat. *Stroke* **10,** 267–272.

Radin, N. S., Shayman, J. A., and Inokuchi, J. (1993). Metabolic effects of inhibiting glucosylceramide synthesis with PDMP and other substances. *Adv. Lipid Res.* **26,** 183–213.

Robinson, H. P., Kawahara, M., Jimbo, Y., Torimitsu, K., Kuroda, Y., and Kawana, A. (1993). Periodic synchronized bursting and intracellular calcium transients elicited by low magnesium in cultured cortical neurons. *J. Neurophysiol.* **70,** 1606–1616.

Rosenwald, A. G., Machamer, C. E., and Pagano, R. E. (1992). Effects of a sphingolipid synthesis inhibitor on membrane transport through the secretory pathway. *Biochemistry* **31,** 3581–3590.

Rösner, H., al-Aqtum, M., and Rahmann, H. (1992). Gangliosides and neuronal differentiation. *Neurochem. Int.* **20,** 339–351.

Ryu, B. R., Choi, D. W., Hartley, D. M., Costa, E., Jou, I., and Gwag, B. J. (1999). Attenuation of cortical neuronal apoptosis by gangliosides. *J. Pharmacol. Exp. Ther.* **290,** 811–816.

Sheardown, M. J., Suzak, P. D., and Nordolm, L. (1993). AMPA but not NMDA, receptor antagonism is neuroprotective in gerbil ischemia, even when delayed 24 h. *Eur. J. Pharmacol.* **236,** 347–353.

Sheikh, K. A., Sun, J., Liu, Y., Kawai, H., Crawford, T. O., Proia, R. L., Griffin, J. W., and Schnaar, R. L. (1999). Mice lacking complex gangliosides develop Wallerian degeneration and myelination defects. *Proc. Natl. Acad. Sci. USA* **96,** 7532–7537.

Schneider, J. S., Bradbury, K. A., Anada, Y., Inokuchi, J., and Anderson, D. W. (2006). The synthetic ceramide analog l-PDMP partially protects striatal dopamine levels but does not promote dopamine neuron survival in murine models of parkinsonism. *Brain Res.* **1099,** 199–205.

Suzuki, T., Okumura-Noji, K., and Nishida, E. (1995). ERK2-type mitogen-activated protein kinase (MAPK) and its substrates in postsynaptic density fractions from the rat brain. *Neurosci. Res.* **22,** 277–285.

Takamiya, K., Yamamoto, A., Furukawa, K., Yamashiro, S., Shin, M., Okada, M., Fukumoto, S., Haraguchi, M., Takeda, N., Fujimura, K., Sakae, M., Kishikawa, M., *et al.* (1996). Mice with disrupted GM2/GD2 synthase gene lack complex gangliosides but exhibit only subtle defects in their nervous system. *Proc. Natl. Acad. Sci. USA* **93,** 10662–10667.

Tsuji, S., Arita, M., and Nagai, Y. (1983). GQ1b, a bioactive ganglioside that exhibits novel nerve growth factor (NGF)-like activities in two neuroblastoma cell lines. *J. Biochem.* **94,** 303–306.

Tsuji, S., Yamashita, T., Matsuda, Y., and Nagai, Y. (1992). A novel glycosignaling system: GQ1b-dependent neuritogenesis of human neuroblastoma cell line, GOTO, is closely associated with GQ1b-dependent ecto-type protein phosphorylation. *Neurochem. Int.* **21,** 549–554.

Usuki, S., Hamanoue, M., Kohsaka, S., and Inokuchi, J. (1996). Induction of ganglioside biosynthesis and neurite outgrowth of primary cultured neurons by L-*threo*-1-phenyl-2-decanoylamino-3-morpholino-1-propanol. *J. Neurochem.* **67,** 1821–1830.

Vaadia, E., Haalman, I., Abeles, M., Bergman, H., Prut, Y., Slovin, H., and Aertsen, A. (1995). Dynamics of neuronal interactions in monkey cortex in relation to behavioural events. *Nature* **373,** 515–518.

Wu, G., Lu, Z., and Ledeen, R. W. (1991). Correlation of gangliotetraose gangliosides with neurite forming potential of neuroblastoma cells. *Dev. Brain Res.* **61,** 217–228.

Yamagishi, K., Mishima, K., Ohgami, Y., Iwasaki, K., Jimbo, M., Matsuda, H., Fujiwara, M., Igarashi, Y., and Inokuchi, J. (2003). A synthetic ceramide analog ameliorates spatial cognition deficit and stimulates biosynthesis of brain gangliosides in rats wtih cerebral ischema. *Eur. J. Pharmacol.* **462,** 53–60.

INVOLVEMENT OF ENDOCANNABINOID SIGNALING IN THE NEUROPROTECTIVE EFFECTS OF SUBTYPE 1 METABOTROPIC GLUTAMATE RECEPTOR ANTAGONISTS IN MODELS OF CEREBRAL ISCHEMIA

Elisa Landucci, Francesca Boscia,* Elisabetta Gerace, Tania Scartabelli, Andrea Cozzi, Flavio Moroni, Guido Mannaioni, and Domenico E. Pellegrini-Giampietro

Dipartimento di Farmacologia Preclinica e Clinica, Università di Firenze, Firenze 50139, Italy
*Dipartimento di Neuroscienze, Università di Napoli "Federico II", Napoli 80131, Italy

Experimental evidence indicates that metabotropic glutamate (mGlu) receptors of the mGlu1 and mGlu5 subtypes play a differential role in models of cerebral ischemia and that only mGlu1 receptors are implicated in the pathways leading to postischemic neuronal injury. The localization of mGlu1 receptors in GABA-containing interneurons rather than in hippocampal CA1 pyramidal cells that are vulnerable to ischemia has prompted experimental studies that have demonstrated mGlu1 receptor antagonist agents attenuate postischemic injury by enhancing GABA-mediated neurotransmission, thus providing a new viewpoint on the neuroprotective mechanism of these pharmacological agents. In view of the recent discovery of a functional interaction between group I mGlu receptors and the cannabinoid system in the modulation of synaptic transmission, we propose a novel mechanism that predicts that the neuroprotective effects of

INTERNATIONAL REVIEW OF NEUROBIOLOGY, VOL. 85
DOI: 10.1016/S0074-7742(09)85023-X

mGlu1 receptor antagonists on CA1 pyramidal cells are mediated by a mechanism that overcomes the "synaptic circuit break" operated by endocannabinoids on GABAergic transmission.

I. Introduction

Most synapses in the central nervous system use glutamate as an excitatory neurotransmitter. Besides its physiological role in normal synaptic transmission and in mechanisms that underlie neuronal plasticity, glutamate is responsible for apoptotic and necrotic neuronal death, a process known as "excitotoxicity," in a number of acute and chronic neurodegenerative diseases (Choi, 1996; Martin *et al.*, 1998). It is nowadays established that the physiological and potentially pathological actions of glutamate are mediated by its interaction with at least two classes of receptors: the inotropic glutamate receptors (NMDA, AMPA, and kainate receptors) that belong to the superfamily of ion channel-linked receptors, and the metabotropic glutamate (mGlu) receptors, coupled via G-proteins to the production of second messengers.

The eight mGlu receptors cloned to date have been subdivided into three groups based on their structural homology, transduction mechanism, and pharmacological profile (Conn and Pin, 1997). Whereas mGlu receptors of the I and II groups are negatively coupled to adenylate cyclase, mGlu receptors of the I group (mGlu1 and mGlu5) are coupled to Gq proteins and hence to phosphoinositide hydrolysis: their activation leads to the production of inositol-trisphosphate and diacylglycerol and to intracellular free Ca^{2+} mobilization. Because these events are known to exert a profound influence upon synaptic plasticity and neuronal death mechanisms, many research groups in the past few years have addressed the role of group I mGlu receptors in experimental models of neuronal plasticity (Riedel and Reymann, 1996) and neurological disorders (Bordi and Ugolini, 1999; Nicoletti *et al.*, 1999). In our laboratory, we have identified a differential role for these receptor subtypes both in potentiating the NMDA receptor response (Mannaioni *et al.*, 1996) and in the regulation of hippocampal CA1 pyramidal cell function (Mannaioni *et al.*, 2001). In addition, we have characterized a series of selective mGlu1 receptor antagonists, such as 1-aminoindan-1,5-dicarboxylic acid (AIDA) (Moroni *et al.*, 1997), (*S*)-(+)-2-(3′-carboxybicyclo[1.1.1]pentyl)-glycine (CBPG) (Pellicciari *et al.*, 1996), and 3-methyl-aminothiophene dicarboxylic acid (3-MATIDA) (Moroni *et al.*, 2002), and used them to evaluate their protective effects against oxygen-glucose deprivation (OGD) *in vitro* or focal or global ischemic injury *in vivo* (Pellegrini-Giampietro, 2003).

II. Role of mGlu Receptors in CA1 Hippocampal Postischemic Neuronal Death

Stroke is the third leading cause of death after heart attack and cancer and the most common cause of disability in the Western world. Unfortunately, clinical trials for stroke with neuroprotective drugs have been generally unsuccessful so far. Along with the development of more appropriate experimental animal models, drugs with a better therapeutic index and aimed at alternative targets in the excitotoxic cascade are required for the development of new and effective neuroprotective therapies for cerebral ischemia. Group I mGlu receptors represent one of these attractive new targets, and as such they have been thoroughly investigated in a variety of experimental models of cerebral ischemia.

A. Antagonists of mGlu1 but not mGlu5 Receptors are Neuroprotective in Models of Cerebral Ischemia

The use of agents that stimulate group I mGlu receptors has failed to produce clear-cut information on their role in the mechanisms that are responsible for postischemic neuronal death (Nicoletti *et al.*, 1999; Pellegrini-Giampietro, 2003). The poor selectivity of ligands, the possible coupling of these receptors to alternative signaling pathways, and the observed neuroprotective effects of elevated cytosolic free Ca^{2+} and PKC activation under certain conditions are all factors that could explain the conflicting results observed with mGlu1 and/or mGlu5 receptor agonists.

On the other hand, the use of mGlu1 receptor antagonists has consistently generated neuroprotective results, indicating that the endogenous activation of this receptor subtype is likely to play a major role in neurodegeneration. We have shown that competitive and noncompetitive antagonists displaying increasing degrees of selectivity for mGlu1 receptors reduce neuronal injury in cultured neocortical cells and in hippocampal slices exposed to OGD (Moroni *et al.*, 2002; Pellegrini-Giampietro *et al.*, 1999a,b). In these *in vitro* models, the neuroprotective effects are evident even when mGlu1 receptor antagonists are added to the incubation medium up to 60 min after OGD. In addition, different laboratories have reported that activation of mGlu1 receptors might also contribute to CA1 pyramidal cell death in the transient two-vessel occlusion gerbil model of global ischemia (Bruno *et al.*, 1999; Henrich-Noack *et al.*, 1998; Pellegrini-Giampietro *et al.*, 1999a) and to the size and volume of the infarct in models of focal ischemia (De Vry *et al.*, 2001; Rauca *et al.*, 1998; Moroni *et al.*, 2002).

Although mGlu1 and mGlu5 receptors share a high degree of sequence homology and virtually identical transduction pathways, the results obtained with selective mGlu5 antagonists in models of cerebral ischemia are somewhat different and not particularly encouraging. The initial observation that the mGlu5 receptor-selective noncompetitive antagonist 2-methyl-6-(phenylethynyl)-pyridine (MPEP) was more effective than AIDA in reducing CA1 pyramidal cell loss following transient global ischemia in the gerbil (Muralikrishna Rao *et al.*, 2000) could not be confirmed in a subsequent study (Bao *et al.*, 2001). Similarly, MPEP was not neuroprotective following OGD *in vitro* (Meli *et al.*, 2002) nor after permanent middle cerebral artery occlusion (MCAO) in the rat (Gasparini *et al.*, 2002). In rats subjected to transient focal ischemia, MPEP was able to reduce the size of the ischemic infarct when administered i.c.v. at a high dose, but the protective effect was ascribed to a noncompetitive antagonism of NMDA receptors, rather than to an interaction with mGlu5 receptors (Bao *et al.*, 2001).

B. Expression of mGlu1 Receptors in CA1 Hippocampal Neurons

Group I mGlu receptors have been shown to display a characteristic perisynaptic localization at the postsynaptic membrane of glutamatergic synapses (Luján *et al.*, 1996), where they can regulate neuronal excitability by modulating a variety of K^+ channels and NMDA or AMPA receptors. Activation of mGlu1 receptors may hence contribute to postischemic neuronal injury through multiple noxious mechanisms, including an increase in intracellular free Ca^{2+} or the potentiation of ionotropic glutamate receptor responses (Nicoletti *et al.*, 1999; Pellegrini-Giampietro, 2003). However, the peculiar localization of this receptor subtype has prompted numerous studies in the past few years that have provided a new viewpoint on the neuroprotective mechanisms of mGlu1 receptor antagonists.

In the CA1 hippocampal subregion, mGlu1 and mGlu5 receptors display a complementary distribution: whereas mGlu5 is prominently expressed in dendritic fields of vulnerable pyramidal cells, the mGlu1α isoform, which is the main alternatively spliced variant of the mGlu1 gene in this area, is notably expressed in distinct classes of interneurons (Ferraguti *et al.*, 2004) and, in particular, in somatostatin-positive GABA-containing interneurons of the stratum oriens–alveus (Baude *et al.*, 1993) that appear to be resistant to global ischemia (Blasco-Ibáñez and Freund, 1995). mGlu1α receptor immunoreactivity has also been identified in other types of interneurons that are located in various CA1 strata and target different pyramidal cell dendritic domains (Ferraguti *et al.*, 2004), and also in GABA-containing interneurons in the neocortex (Stinehelfer *et al.*, 2000), in the striatum (Tallaksen-Greene *et al.*, 1998) and in the substantia nigra pars reticulata (Marino *et al.*, 2001). Because potentiation of GABA-mediated transmission is known to exert a neuroprotective effect against postischemic injury (Schwartz-Bloom and Sah, 2001), these anatomical observations suggest that neuroprotection by mGlu1 receptor antagonists might be induced by changes in the function of these GABAergic interneurons.

C. The Protective Effects of mGlu1 Receptor Antagonists are Mediated by an Increase in GABAergic Neurotransmission in CA1

The hypothesis that mGlu1 receptor antagonists attenuate postischemic injury by enhancing GABA-mediated neurotransmission was first proposed by studies showing that perfusion with neuroprotective doses of AIDA increased the concentrations of GABA in the hippocampal dialyzate of gerbils subjected to global ischemia (Pellegrini-Giampietro *et al.*, 1999b). In hippocampal slices exposed to OGD, the attenuation of CA1 damage observed with another mGlu1 receptor antagonist, 3-MATIDA, was mimicked by $GABA_A$ and $GABA_B$ receptor agonists and reduced by GABA receptor antagonists (Cozzi *et al.*, 2002). Likewise, LY367385 and CPCCOEt produced an increase in the extracellular levels of GABA in the corpus striatum of freely moving rats that was associated with a reduction in NMDA neurotoxicity (Battaglia *et al.*, 2001). In cultured neuronal cells exposed to NMDA, the neuroprotective effects of mGlu1 receptor antagonists were occluded by the previous application of GABA and SKF89976A (a GABA transporter inhibitor) and prevented by $GABA_A$ and $GABA_B$ receptor antagonists (Battaglia *et al.*, 2001). It is interesting to note that the reduction of spontaneous epileptiform activity induced by 3-MATIDA was similarly reproduced by a GABA receptor agonist and reverted by an antagonist in mouse cortical wedges (Cozzi *et al.*, 2002). Taken together, these data suggest that a common GABA-mediated mechanism, which involves the release of GABA and stimulation of GABA receptors, might contribute to the neuroprotective effects of mGlu1 receptor antagonists.

This hypothesis implies that mGlu1 receptors are located presynaptically and that their activation inhibits the release of GABA. Indeed, functional data support the existence of presynaptic mGlu1 receptors modulating neurotransmitter release in the hippocampus (Manahan-Vaughan *et al.*, 1999) and neocortex (Moroni *et al.*, 1998). Other studies have shown that stimulation of group I mGlu receptors in the hippocampal CA1 and other brain areas leads to increased principal cell excitability via the presynaptic inhibition of GABA release from interneurons (Gereau and Conn, 1995; Mannaioni *et al.*, 2001; Marino *et al.*, 2001; Morishita *et al.*, 1998). It has been proposed that presynaptic inhibition of neurotransmitter release by mGlu1 receptors could be mediated by suppression of Ca^{2+} currents through N- or P/Q-type channels (Choi and Lovinger, 1996) or by activation of a Ca^{2+}-dependent K^{+} conductance (Fiorillo and Williams, 1998). To date, only group I receptors of the mGlu5 subtype have been detected at a presynaptic level (Romano *et al.*, 1995), but two reports provide electron microscopy evidence for mGlu1α and mGlu5 staining in GABAergic presynaptic terminals and preterminal GABAergic axons in the substantia nigra (Hubert *et al.*, 2001; Marino *et al.*, 2001). Activation of these presynaptic receptors appears to be responsible for the decrease in inhibitory transmission observed in this area (Marino *et al.*, 2001) and a similar mechanism may be operative in the hippocampus. Hence, we proposed a hypothetical model in which mGlu1 receptors exert a negative control upon the release of GABA from GABAergic interneurons (Cozzi *et al.*, 2002). Blockade of mGlu1 receptors may provide neuroprotection in this model by increasing the release of GABA and promoting the activation of postsynaptic $GABA_A$ receptors that hyperpolarize vulnerable principal neurons.

III. Interactions Between mGlu1 Receptors and Endocannabinoids in the CA1 Hippocampal Region

Although anatomical studies have essentially demonstrated that mGlu1α receptors, at least in the hippocampus, are expressed almost exclusively in non-principal cells, functional evidence exists for the presence of mGlu1 together with mGlu5 receptors in CA1 pyramidal cells (Mannaioni *et al.*, 2001; Rae and Irving, 2004). Moreover, mGlu1β and mGlu1d splice variants have been shown to be expressed in hippocampal principal neurons, albeit not abundantly in CA1 (Berthele *et al.*, 1998; Ferraguti *et al.*, 1998). Hence, it is also possible that a presynaptic decrease in inhibitory transmission can be indirectly mediated by mGlu1 receptors located postsynaptically in principal neurons.

A. Postsynaptic mGlu1 Receptors may Indirectly Inhibit the Release of GABA via the Formation of Endocannabinoids

Emerging studies from several laboratories have revealed that some of the effects of group I mGlu receptors in the CNS are indirectly mediated by a novel signaling mechanism which entails a functional interaction with the endocannabinoid system (see for reviews: Alger, 2002; Chevaleyre *et al.*, 2006; Doherty and Dingledine, 2003; Freund *et al.*, 2003; Kano *et al.*, 2002). Specifically, activation of group I mGlu receptors has been shown to promote in postsynaptic cells the production of endogenous cannabinoids that retrogradely diffuse and suppress the release of GABA by activation of CB1 cannabinoid receptors located on the presynaptic terminals of interneurons (Ohno-Shosaku *et al.*, 2001; Wilson and Nicoll, 2001). This mechanism appears to be responsible for the above-mentioned finding that group I mGlu receptor agonists depress (rather than stimulate) inhibitory postsynaptic potentials in the CA1 hippocampal region, given that

CB1 receptor antagonists are able to prevent this effect (Neu *et al.*, 2007; Ohno-Shosaku *et al.*, 2002). Similarly, a group I mGlu receptor-mediated retrograde inhibition of GABA release through endocannabinoids has been described in CA1 for both short-(Varma *et al.*, 2001) and long-term (Chevaleyre and Castillo, 2003) forms of synaptic plasticity, as well as for the persistent effects of experimental febrile seizures on the enhancement of hippocampal excitability (Chen *et al.*, 2007).

Preliminary results from our laboratory suggest that a similar mechanism may be involved in mediating the GABAergic neuroprotective effects of mGlu1 receptors. To address this point we used organotypic hippocampal slices exposed to 30 min OGD (Fig. 1). As reported (Pellegrini-Giampietro *et al.*, 1999a), the OGD-induced injury observed in the CA1 subregion was significantly prevented by

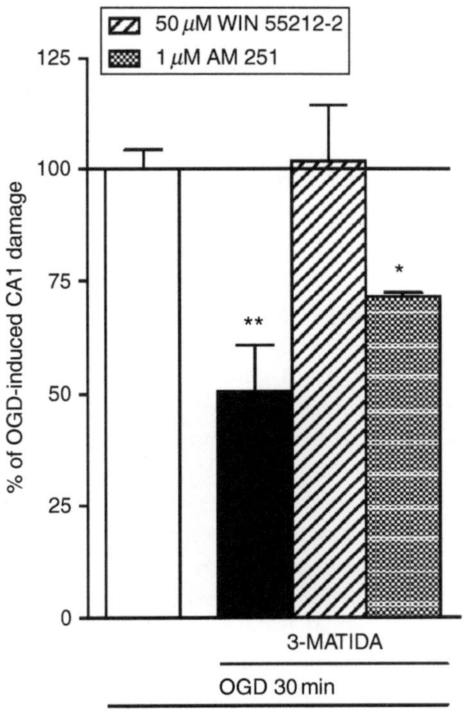

FIG. 1. The CB1 receptor agonist WIN 55212-2, but not the CB1 receptor antagonist AM 251, reverts the neuroprotective effects of the mGlu1 antagonist 3-MATIDA in rat organotypic hippocampal slices exposed to OGD. Drugs were present in the incubation medium during the 30 min OGD exposure and the subsequent 24-h recovery period. Bars represent the mean ± SEM of at least six experiments. $^{**}P < 0.01$ and $^{**}P < 0.05$ versus OGD alone.

adding the mGlu1 antagonist 3-MATIDA to the incubation medium. Interestingly, the CB1 receptor agonist WIN 55212-2, but not the antagonist AM 251, completely reverted this effect, suggesting that the neuroprotection afforded by the mGlu1 antagonist is mediated by CB1 receptors.

B. Colocalization of mGlu1α and CB1 Receptors in a Subset of Stratum Radiatum Hippocampal Interneurons

In the CA1 hippocampal region, both mGlu1α and CB1 receptors are characteristically expressed in GABAergic interneurons. As described, mGlu1α receptors are enriched in interneurons of the stratum oriens–alveus that contain somatostatin, but are also present in interneurons expressing vasoactive intestinal peptide (VIP) and/or calretinin and in a subpopulation of cholecystokinin (CCK)-immunopositive interneurons (Ferraguti *et al.*, 2004). On the other hand, CB1 receptors are primarily expressed in CCK-immunoreactive basket cells of the hippocampus (Katona *et al.*, 1999), but they have also been described to be partially colocalized with neurons containing the calcium binding proteins calretinin and calbindin in the same area (Tsou *et al.*, 1999).

In a recent study (Boscia *et al.*, 2008), we examined the distribution of mGlu1α and CB1 receptors in the CA1 region of rat organotypic hippocampal slice cultures. Because these two receptor types appeared to be expressed in distinct, but at least partially overlapping classes of nonprincipal cells, we further investigated their coexistence in hippocampal interneurons of the CA1 subregion, by performing a double-labeling confocal fluorescence analysis with specific mGlu1α and CB1 receptor antibodies both in rat organotypic hippocampal slice cultures and in hippocampal sections from adult rat brain. Our results showed that a subset of interneurons, mainly located in the stratum radiatum, was double-labeled for both mGlu1α and CB1 receptors. High magnification of individual neurons immunopositive for both mGlu1α and CB1 in the stratum radiatum confirmed the peculiar perikaryal distribution of the two proteins (Fig. 2). In fact, mGlu1α immunoreactivity was mainly confined to the somatic plasma membrane, whereas a punctate staining pattern of the cytoplasm was observed with the anti-CB1 antibodies, possibly corresponding to the cytoplasmic organelle localization previously described for this receptor (Bodor *et al.*, 2005; Katona *et al.*, 1999). By using the "mirror technique" (Kosaka *et al.*, 1985) in adjacent sections, we observed that the double-labeled cells for mGlu1α and CB1 receptors were also immunopositive for the CCK peptide. Quantitative analysis revealed that in the stratum radiatum the majority (92%) of the CB1-positive cells and 19% of the mGlu1α-positive cells expressed both receptors. Triple immunofluorescence staining showed partial colabeling of mGlu1α- and CB1-immunopositive cells with the vesicular glutamate transporter 3 and calbindin, two molecular

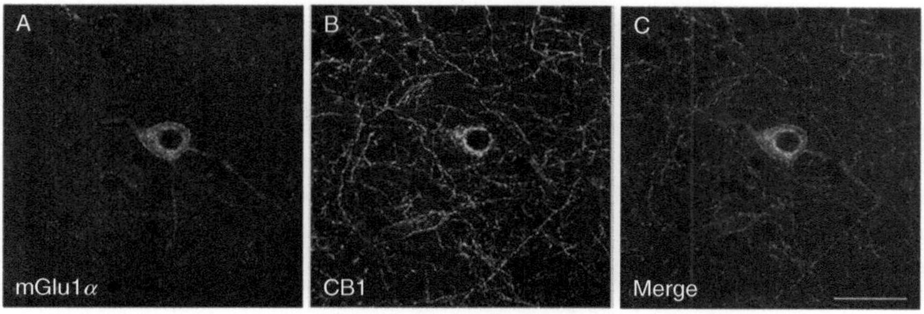

FIG. 2. Confocal microscopic images of an organotypic hippocampal slice depicting an interneuron displaying both CB1- and mGlu1α-like immunoreactivity in the stratum radiatum of the CA1 region. High magnification displays a characteristic membrane staining for mGlu1α (A) and a typical punctate labeling pattern for CB1 (B). Scale bar: 50 μm. (See Color Insert.)

markers that are known to be coexpressed with CCK in interneurons (Cope *et al.*, 2002; Somogyi *et al.*, 2004), which may suggest that the cells coexpressing both mGlu1α and CB1 receptors are Schaffer collateral-associated interneurons.

IV. Conclusions

Altogether, the results we have discussed in this review point out to a cooperation between mGlu1 receptors and the endocannabinoid system in the mechanisms that lead to postischemic neuronal death. The pieces of evidence supporting this view include the following: (i) mGlu1 but not mGlu5 receptor antagonists attenuate ischemic and OGD injury, (ii) mGlu1 receptor antagonists enhance the release of GABA at doses that reduce postischemic damage, (iii) CB1 agonists prevent the neuroprotective effects of mGlu1 receptor antagonists against OGD injury, and (iv) mGlu1α and CB1 receptors are coexpressed in a subpopulation of hippocampal cells that is suggestive of Schaffer collateral-associated interneurons. Hence, it appears as though the protective effects of mGlu1 receptor antagonists are mediated by a mechanism that overcomes the "synaptic circuit break" operated by endocannabinoids on GABAergic transmission (Hajos *et al.*, 2000; Katona *et al.*, 1999).

We would like to propose three hypothetic models providing a possible explanation for the neuroprotective effects of mGlu1 receptors antagonists against postischemic neuronal injury in the hippocampus. (1) A *presynaptic mGlu1 and CB1 receptor model*, in which mGlu1 and CB1 receptors are located presynaptically in the same interneuron terminals and the release of transmitter is negatively controlled by both receptors: mGlu1 antagonists will enhance the release of

GABA and thus provide neuroprotection, CB1 agonists will reduce the release and prevent the neuroprotective effect of mGlu1 receptor antagonists. (2) A *postsynaptic mGlu1 receptor model*, in which mGlu1 receptors are located postsynaptically in pyramidal cells and their activation promotes the formation of endocannabinoids that can diffuse the membrane and act as retrograde transmitters to activate CB1 receptors in presynaptic interneuron terminals: mGlu1 antagonists will indirectly inhibit CB1 receptors and thus increase the release of GABA and provide neuroprotection, CB1 antagonists will directly prevent this effect. (3) *A polysynaptic GABAergic disinhibition model*, in which mGlu1 are postsynaptic and CB1 receptors are presynaptic in different interneuron populations connected in series (Fig. 3). Activation of mGlu1 receptors in this model will increase the firing of the first interneuron and lead to inhibition of the second one (possibly a basket cell) innervating the perisomatic region of pyramidal cells: mGlu1 antagonists will thus increase the net output of GABA upon

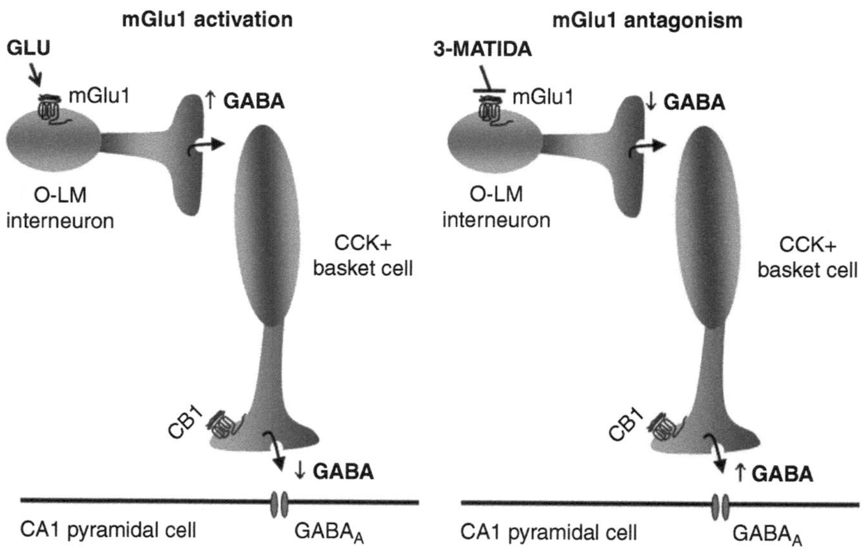

FIG. 3. Polysynaptic GABAergic disinhibition model, providing a possible explanation for the neuroprotective effects of mGlu1 receptors antagonists against postischemic neuronal injury in the hippocampus. mGlu1 receptors are expressed in dendrites and somata of oriens–lacunosum molecular (O–LM) interneurons, which are connected in series with CCK^+ basket cells that express presynaptic CB1 receptors. Activation of mGlu1 receptors in this model will increase the firing of the O–LM interneuron and lead to inhibition of the basket cell innervating the perisomatic region of pyramidal cells: mGlu1 antagonists like 3-MATIDA will thus increase the net output of GABA upon pyramidal cells and provide neuroprotection, CB1 agonists are expected to reduce the output from basket cell terminals and prevent this effect.

pyramidal cells and provide neuroprotection, CB1 agonists will reduce the output from basket cell terminals and prevent this effect.

Further studies are required to determine the validity of these hypotheses. The clarification of these mechanisms is expected to provide new insight into the possible targets for new therapeutic interventions for stroke and other ischemia-related syndromes.

Acknowledgments

This work was supported by grants from the Italian Ministry of University and Research (MIUR, PRIN 2006 project), from the Ente Cassa di Risparmio di Firenze, and from the University of Florence.

References

Alger, B. E. (2002). Retrograde signaling in the regulation of synaptic transmission: Focus on endocannabinoids. *Prog. Neurobiol.* **68,** 247–286.

Bao, W. L., Williams, A. J., Faden, A. I., and Tortella, F. C. (2001). Selective mGluR5 receptor antagonist or agonist provides neuroprotection in a rat model of focal cerebral ischemia. *Brain Res.* **922,** 173–179.

Battaglia, G., Bruno, V., Pisani, A., Centonze, D., Catania, M. V., Calabresi, P., and Nicoletti, F. (2001). Selective blockade of type-1 metabotropic glutamate receptors induces neuroprotection by enhancing gabaergic transmission. *Mol. Cell. Neurosci.* **17,** 1071–1083.

Baude, A., Nusser, Z., Roberts, J. D. B., Mulvihill, E., McIlhinney, R. A. J., and Somogyi, P. (1993). The metabotropic glutamate receptor (mGluR1α) is concentrated at perisynaptic membrane of neuronal subpopulations as detected by immunogold reaction. *Neuron* **11,** 771–787.

Berthele, A., Laurie, D. J., Platzer, S., Zieglgansberger, W., Tolle, T. R., and Sommer, B. (1998). Differential expression of rat and human type I metabotropic glutamate receptor splice variant messenger RNAs. *Neuroscience* **85,** 733–749.

Blasco-Ibáñez, J. M., and Freund, T. F. (1995). Synaptic input of horizontal interneurons in stratum oriens of the hippocampal CA1 subfield: Structural basis of feed-back activation. *Eur. J. Neurosci.* **7,** 2170–2180.

Bodor, A. L., Katona, I., Nyiri, G., Mackie, K., Ledent, C., Hajos, N., and Freund, T. F. (2005). Endocannabinoid signaling in rat somatosensory cortex: Laminar differences and involvement of specific interneuron types. *J. Neurosci.* **25,** 6845–6856.

Bordi, F., and Ugolini, A. (1999). Group I metabotropic glutamate receptors: Implications for brain diseases. *Prog. Neurobiol.* **59,** 55–79.

Boscia, F., Ferraguti, F., Annunziato, L., Moroni, F., and Pellegrini-Giampietro, D. E. (2008). mGlu1α receptors are co-expressed with CB1 receptors in a subset of interneurons in the CA1 region of organotypic hippocampal slice cultures and adult rat brain. *Neuropharmacology* **55,** 428–439.

Bruno, V., Battaglia, G., Kingston, A. E., O'Neill, M. J., Catania, M. V., Di Grezla, R., and Nicoletti, F. (1999). Neuroprotective activity of the potent and selective mGlu1a metabotropic glutamate receptor antagonist, (+)-2-methyl-4-carboxyphenylglycine (LY367385): Comparison with LY357366, a broader spectrum antagonist with equal affinity for mGlu1a and mGlu5 receptors. *Neuropharmacology* **38,** 199–207.

Chen, K., Neu, A., Howard, A. L., Foldy, C., Echegoyen, J., Hilgenberg, L., Smith, M., Mackie, K., and Soltesz, I. (2007). Prevention of plasticity of endocannabinoid signaling inhibits persistent limbic hyperexcitability caused by developmental seizures. *J. Neurosci.* **27,** 46–58.

Chevaleyre, V., and Castillo, P. E. (2003). Heterosynaptic LTD of hippocampal GABAergic synapses. A novel role of endocannabinoids in regulating excitability. *Neuron* **38,** 461–472.

Chevaleyre, V., Takahashi, K. A., and Castillo, P. E. (2006). Endocannabinoid-mediated synaptic plasticity in the CNS. *Annu. Rev. Neurosci.* **29,** 37–76.

Choi, D. W. (1996). Ischemia-induced neuronal apoptosis. *Curr. Opin. Neurobiol.* **6,** 667–672.

Choi, S., and Lovinger, D. M. (1996). Metabotropic glutamate receptor modulation of voltage-gated Ca^{2+} -channels involves multiple receptor subtypes in cortical neurons. *J. Neurosci.* **16,** 36–45.

Conn, P. J., and Pin, J.-P. (1997). Pharmacology and functions of metabotropic glutamate receptors. *Annu. Rev. Pharmacol. Toxicol.* **37,** 205–237.

Cope, D. W., Maccaferri, G., Marton, L. F., Roberts, J. D., Cobden, P. M., and Somogyi, P. (2002). Cholecystokinin-immunopositive basket and Schaffer collateral-associated interneurones target different domains of pyramidal cells in the CA1 area of the rat hippocampus. *Neuroscience* **109,** 63–80.

Cozzi, A., Meli, E., Carlà, V., Moroni, F., and Pellegrini-Giampietro, D. E. (2002). Metabotropic glutamate 1 (mGlu1) receptor antagonists enhance GABAergic neurotransmission: A mechanism for the attenuation of post-ischemic injury and epileptiform activity?. *Neuropharmacology* **43,** 119–130.

De Vry, J., Horvath, E., and Schreiber, R. (2001). Neuroprotective and behavioral effects of the selective metabotropic glutamate mGlu(1) receptor antagonist BAY 36–7620. *Eur. J. Pharmacol.* **428,** 203–214.

Doherty, J., and Dingledine, R. (2003). Functional interactions between cannabinoid and metabotropic glutamate receptors in the central nervous system. *Curr. Opin. Pharmacol.* **3,** 46–53.

Ferraguti, F., Conquet, F., Corti, C., Grandes, P., Kuhn, R., and Knoepfel, T. (1998). Immunohistochemical localization of the mGluR1β metabotropic glutamate receptor in the adult rodent forebrain: Evidence for a different distribution of mGluR1 splice variants. *J. Comp. Neurol.* **400,** 391–407.

Ferraguti, F., Cobden, P., Pollard, M., Cope, D., Shigemoto, R., Watanabe, M., and Somogyi, P. (2004). Immunolocalization of metabotropic glutamate receptor 1α (mGluR1α) in distinct classes of interneurons in the CA1 region of rat hippocampus. *Hippocampus* **14,** 193–215.

Fiorillo, C. D., and Williams, J. T. (1998). Glutamate mediates an inhibitory postsynaptic potential in dopamine neurons. *Nature* **394,** 78–82.

Freund, T. F., Katona, I., and Piomelli, D. (2003). Role of endogenous cannabinoids in synaptic signaling. *Physiol. Rev.* **83,** 1017–1066.

Gasparini, F., Kuhn, R., and Pin, J. P. (2002). Allosteric modulators of group I metabotropic glutamate receptors: novel subtype-selective ligands and therapeutic perspectives. *Curr. Opin. Pharmacol.* **2,** 43–49.

Gereau, R. W. IV, and Conn, P. J. (1995). Multiple presynaptic metabotropic glutamate receptors modulate excitatory and inhibitory synaptic transmission in hippocampal area CA1. *J. Neurosci.* **15,** 6879–6889.

Hajos, N., Katona, I., Naiem, S. S., Mackie, K., Ledent, C., Mody, I., and Freund, T. F. (2000). Cannabinoids inhibit hippocampal GABAergic transmission and network oscillations. *Eur. J. Neurosci.* **12,** 3239–3249.

Henrich-Noack, P., Hatton, C. D., and Reymann, K. G. (1998). The mGlu receptor ligand (S)-4C3HPG protects neurons after global ischaemia in gerbils. *NeuroReport* **9,** 985–988.

Hubert, G. W., Paquet, M., and Smith, Y. (2001). Differential subcellular localization of mGluR1a and mGluR5 in the rat and monkey substantia nigra. *J. Neurosci.* **21,** 1838–1847.

Kano, M., Ohno-Shosaku, T., and Maejima, T. (2002). Retrograde signaling at central synapses via endogenous cannabinoids. *Mol. Psychiatry* **7,** 234–235.

Katona, I., Sperlagh, B., Sik, A., Kafalvi, A., Vizi, E. S., Mackie, K., and Freund, T. F. (1999). Presynaptically located CB1 cannabinoid receptors regulate GABA release from axon terminals of specific hippocampal interneurons. *J. Neurosci.* **19,** 4544–4558.

Kosaka, T., Kosaka, K., Tateishi, K., Hamaoka, Y., Yanaihara, N., Wu, J. Y., and Hama, K. (1985). GABAergic neurons containing CCK-8-like and/or VIP-like immunoreactivities in the rat hippocampus and dentate gyrus. *J. Comp. Neurol.* **239,** 420–430.

Luján, R., Nusser, Z., Roberts, J. D. B., Shigemoto, R., and Somogyi, P. (1996). Perisynaptic localization of metabotropic glutamate receptors mGluR1 and mGluR5 on dendrites and dendritic spines in the rat hippocampus. *Eur. J. Neurosci.* **8,** 1488–1500.

Manahan-Vaughan, D., Herrero, I., Reymann, K. G., and Sánchez-Prieto, J. (1999). Presynaptic group 1 metabotropic glutamate receptors may contribute to the expression of long-term potentiation in the hippocampal CA1 region. *Neuroscience* **94,** 71–82.

Mannaioni, G., Carlà, V., and Moroni, F. (1996). Pharmacological characterization of metabotropic glutamate receptors potentiating NMDA responses in mouse cortical wedge preparations. *Br. J. Pharmacol.* **118,** 1530–1536.

Mannaioni, G., Marino, M. J., Valenti, O., Traynelis, S. F., and Conn, P. J. (2001). Metabotropic glutamate receptors 1 and 5 differentially regulate CA1 pyramidal cell function. *J. Neurosci.* **21,** 5925–5934.

Marino, M. J., Wittmann, M., Bradley, S. R., Hubert, G. W., Smith, Y., and Conn, P. J. (2001). Activation of group I metabotropic glutamate receptors produces a direct excitation and disinhibition of GABAergic projection neurons in the substantia nigra pars reticulata. *J. Neurosci.* **15,** 7001–7012.

Martin, L. J., Al-Abdulla, N. A., Brambrink, A. M., Kirsch, J. R., Sieber, F. E., and Portera-Cailliau, C. (1998). Neurodegeneration in excitotoxicity, global cerebral ischemia, and target deprivation: A perspective on the contributions of apoptosis and necrosis. *Brain Res. Bull.* **46,** 281–309.

Meli, E., Picca, R., Attucci, S., Cozzi, A., Peruginelli, F., Moroni, F., and Pellegrini-Giampietro, D. E. (2002). Activation of mGlu1 but not mGlu5 metabotropic glutamate receptors contributes to post-ischemic neuronal injury *in vitro* and *in vivo*. *Pharmacol. Biochem. Behav.* **73,** 439–446.

Morishita, W., Kirov, S. A., and Alger, B. E. (1998). Evidence for metabotropic glutamate receptor activation in the induction of depolarization-induced suppression of inhibition in hippocampal CA1. *J. Neurosci.* **18,** 4870–4882.

Moroni, F., Lombardi, G., Thomsen, C., Leonardi, P., Attucci, S., Peruginelli, F., Albani Torregrossa, S., Pellegrini-Giampietro, D. E., Luneia, R., and Pellicciari, R. (1997). Pharmacological characterization of 1-aminoindan-1,5-dicarboxylic acid, a potent mGluR1 antagonist. *J. Pharmacol. Exp. Ther.* **281,** 721–729.

Moroni, F., Cozzi, A., Lombardi, G., Sourtcheva, S., Leonardi, P., Carfì, M., and Pellicciari, R. (1998). Presynaptic $mGlu_1$ type receptors potentiate transmitter output in the rat cortex. *Eur. J. Pharmacol.* **347,** 189–195.

Moroni, F., Attucci, S., Cozzi, A., Meli, E., Picca, R., Scheideler, M. A., Pellicciari, R., Noe, C., Sarichelou, I., and Pellegrini-Giampietro, D. E. (2002). The novel and systemically active metabotropic glutamate 1 (mGlu1) receptor antagonist 3-MATIDA reduces post-ischemic neuronal death. *Neuropharmacology* **42,** 741–751.

Muralikrishna Rao, A., Hatcher, J. F., and Dempsey, R. J. (2000). Neuroprotection by group I metabotropic glutamate receptor antagonists in forebrain ischemia of gerbil. *Neurosci. Lett.* **293,** 1–4.

Neu, A., Foldy, C., and Soltesz, I. (2007). Postsynaptic origin of CB1-dependent tonic inhibition of GABA release at cholecystokinin-positive basket cell to pyramidal cell synapses in the CA1 region of the rat hippocampus. *J. Physiol.* **578,** 233–247.

Nicoletti, F., Bruno, V., Catania, M. V., Battaglia, G., Copani, A., Barbagallo, G., Ceña, V., Sánchez-Prieto, J., Spano, P. F., and Pizzi, M. (1999). Group-I metabotropic glutamate receptors: Hypotheses to explain their dual role in neurotoxicity and neuroprotection. *Neuropharmacology* **38,** 1477–1484.

Ohno-Shosaku, T., Maejima, T., and Kano, M. (2001). Endogenous cannabinoids mediate retrograde signals from depolarized postsynaptic neurons to presynaptic terminals. *Neuron* **29,** 729–738.

Ohno-Shosaku, T., Shosaku, J., Tsubokawa, H., and Kano, M. (2002). Cooperative endocannabinoid production by neuronal depolarization and group I metabotropic glutamate receptor activation. *Eur. J. Neurosci.* **15,** 953–961.

Pellegrini-Giampietro, D. E. (2003). The distinct role of mGlu1 receptors in post-ischemic neuronal death. *Trends Pharmacol. Sci.* **24,** 461–470.

Pellegrini-Giampietro, D. E., Cozzi, A., Peruginelli, F., Leonardi, P., Meli, E., Pellicciari, R., and Moroni, F. (1999a). 1-Aminoindan-1,5-dicarboxylic acid and (S)-(+)-2-(3′-carboxybicyclo[1.1.1]pentyl)-glycine, two mGlu1 receptor-preferring antagonists, reduce neuronal death in *in vitro* and *in vivo* models of cerebral ischemia. *Eur. J. Neurosci.* **11,** 3637–3647.

Pellegrini-Giampietro, D. E., Peruginelli, F., Meli, E., Cozzi, A., Albani Torregrossa, S., Pellicciari, R., and Moroni, F. (1999b). Protection with metabotropic glutamate 1 receptor antagonists in models of ischemic neuronal death: Time-course and mechanisms. *Neuropharmacology* **38,** 1607–1619.

Pellicciari, R., Raimondo, M., Marinozzi, M., Natalini, B., Costantino, G., and Thomsen, C. (1996). (*S*)-(+)-2-(3′-Carboxybicyclo[1.1.1]pentyl)-glycine, a structurally new group I metabotropic glutamate receptor antagonist. *J. Med. Chem.* **39,** 2874–2876.

Rae, M. G., and Irving, A. J. (2004). Both mGluR1 and mGluR5 mediate Ca2+ release and inward currents in hippocampal CA1 pyramidal neurons. *Neuropharmacology* **46,** 1057–1069.

Rauca, C., Henrich-Noack, P., Schäfer, K., Höllt, V., and Reymann, K. G. (1998). (*S*)-4C3HPG reduces infarct size after focal cerebral ischemia. *Neuropharmacology* **37,** 1649–1652.

Riedel, G., and Reymann, K. G. (1996). Metabotropic glutamate receptors in hippocampal long-term potentiation and learning and memory. *Acta Physiol. Scand.* **157,** 1–19.

Romano, C., Sesma, M. A., McDonald, C. T., O'Malley, K., van den Pol, A. N., and Olney, J. W. (1995). Distribution of metabotropic glutamate receptor mGluR5 immunoreactivity in rat brain. *J. Comp. Neurol.* **355,** 455–469.

Schwartz-Bloom, R. D., and Sah, R. (2001). γ-Aminobutyric acidA neurotransmission and cerebral ischemia. *J. Neurochem.* **77,** 353–371.

Somogyi, J., Baude, A., Omori, Y., Shimizu, H., El Mestikawy, S., Fukaya, M., Shigemoto, R., Watanabe, M., and Somogyi, P. (2004). GABAergic basket cells expressing cholecystokinin contain vesicular glutamate transporter type 3 (VGLUT3) in their synaptic terminals in hippocampus and isocortex of the rat. *Eur. J. Neurosci.* **19,** 552–569.

Stinehelfer, S., Vruwink, M., and Burette, A. (2000). Immunolocalization of mGluR1alpha in specific populations of local circuit neurons in the cerebral cortex. *Brain Res.* **861,** 37–44.

Tallaksen-Greene, S. J., Kaatz, K. W., Romano, C., and Albin, R. L. (1998). Localization of mGluR1a-like immunoreactivity and mGluR5-like immunoreactivity in identified populations of striatal neurons. *Brain Res.* **780,** 210–217.

Tsou, K., Mackie, K., Sañudo-Peña, M. C., and Walker, J. M. (1999). Cannabinoid CB1 receptors are localized primarily on cholecystokinin-containing gabaergic interneurons in the rat hippocampal formation. *Neuroscience* **93,** 969–975.

Varma, N., Carlson, G. C., Ledent, C., and Alger, B. E. (2001). Metabotropic glutamate receptors drive the endocannabinoid system in hippocampus. *J. Neurosci.* **21,** RC188.

Wilson, R. I., and Nicoll, R. A. (2001). Endogenous cannabinoids mediate retrograde signalling at hippocampal synapses. *Nature* **410,** 588–592.

NF-kappaB DIMERS IN THE REGULATION OF NEURONAL SURVIVAL

Ilenia Sarnico,* Annamaria Lanzillotta,* Marina Benarese,* Manuela Alghisi,* Cristina Baiguera,* Leontino Battistin,[†] PierFranco Spano,*[,†,‡] and Marina Pizzi*[,‡]

*Division of Pharmacology and Experimental Therapeutics, Department of Biomedical Sciences and Biotechnologies, School of Medicine, University of Brescia, Brescia 25123, Italy

[†]IRCCS San Camillo Hospital, 30100 Venice, Italy

[‡]National Institute of Neuroscience, 10125 Turin, Italy

Nuclear factor-kappaB (NF-κB) is a dimeric transcription factor composed of five members, p50, RelA/p65, c-Rel, RelB, and p52 that can diversely combine to form the active transcriptional dimer. NF-κB controls the expression of genes that regulate a broad range of biological processes in the central nervous system such as synaptic plasticity, neurogenesis, and differentiation. Although NF-κB is essential for neuron survival and its activation may protect neurons against oxidative-stresses or ischemia-induced neurodegeneration, NF-κB activation can contribute to inflammatory reactions and apoptotic cell death after brain injury and stroke. It was proposed that the death or survival of neurons might depend on the cell type and the timing of NF-κB activation.

We here discuss recent evidence suggesting that within the same neuronal cell, activation of diverse NF-κB dimers drives opposite effects on neuronal survival. Unbalanced activation of NF-κB p50/RelA dimer over c-Rel-containing complexes contributes to cell death secondary to the ischemic insult. While p50/RelA acts as transcriptional inducer of Bcl-2 family proapoptotic *Bim* and *Noxa* genes, c-Rel dimers specifically promote transcription of antiapototic *Bcl-xL* gene. Changes in the nuclear content of c-Rel dimers strongly affect the threshold of neuron vulnerability to ischemic insult and agents, likewise leptin, activating a NF-κB/c-Rel-dependent transcription elicit neuroprotection in animal models of brain ischemia.

INTERNATIONAL REVIEW OF NEUROBIOLOGY, VOL. 85
DOI: 10.1016/S0074-7742(09)85024-1

Abbreviations

NF-κB, nuclear factor-kappaB; RHD, Rel-homology domain; IκBs, κB inhibitory proteins; IKK, IκB kinase; TNFR, tumor necrosis factor receptor; IL-1, interleukin-1; NMDA, *N*-methyl-D-aspartate; PKC, protein kinase C; PI3K, phosphatidylinositol 3-kinase; MEK, mitogen-activated protein kinase or extracellular signal-regulated kinase; LTD, long-term depression; S100B, S100 calcium binding protein B; MPP^+, 1-methyl-4-phenylpyridinium; MnSOD, manganese superoxide dismutase; Bcl-2, B-cell leukemia/lymphoma 2; MCAO, middle cerebral artery occlusion; OGD, oxygen-glucose deprivation.

I. Nuclear Factor-kappaB (NF-κB) in Brain

In central nervous system (CNS), NF-κB factors act as regulators of growth, differentiation, and adaptive responses to extracellular signals (Kaltschmidt *et al.*, 1993; O'Neill and Kaltschmidt, 1997; West *et al.*, 2002). The NF-κB family transcription factors comprise five members, sharing an N-terminal 300 amino acid Rel-homology domain (RHD), which is identical in 35–61% in all family proteins: RelA (p65), RelB, c-Rel, p50/p105, and p52/p100, encoded by *RELA, RELB, REL, NFKB1*, and *NFKB2*, respectively. The RHD domain allows dimerization, nuclear translocation, and DNA binding. Among the members of the NF-κB family, only RelA, c-Rel, and RelB are directly able to activate the transcription of target genes. The transcriptional capacity of p50 and p52, which are initially synthesized as large precursors called p105 and p100, are dependent on dimerization with RelA, c-Rel, or RelB (Dejardin, 2006; Siebenlist *et al.*, 1994).

Within the CNS, the transcriptionally active form of NF-κB is mostly the p50/RelA heterodimer (Bhakar *et al.*, 2002; Schmidt-Ullrich *et al.*, 1996). NF-κB exists in a latent and a "constitutively" active form in neurons. The "constitutive" NF-κB is within the nucleus, it is transcriptionally active and regulated by synaptic activity (Kaltschmidt *et al.*, 1994; Meffert *et al.*, 2003; Schmidt-Ullrich, *et al.*, 1996). The latent form is sequestered in the cytoplasm through its interaction with an inhibitory protein IκB. Two different intracellular pathways activate NF-κB, the *classic* and the *alternative* pathway, which result in the release of NF-κB from its inhibitors, IκB, and in the nuclear localization of NF-κB (Bonizzi and Karin, 2004). The canonical pathway of NF-κB activation involves an IκB kinase (IKK) complex, composed of two catalytic subunits, IKK1/α and IKK2/β, and a regulatory subunit NF-κB essential modulator (NEMO)/IKKγ. Upon stimulation, IKK2 phosphorylates two N-terminal serines within the IκBs, leading to their ubiquitination and degradation through the proteasome pathway. The alternative pathway, described in the immune system, leads to the processing and cleavage of the p100 precursor to p52 through the phosphorylation of p100 by IKK1. An important component of this pathway is the NF-κB-inducing kinase (NIK) which activates IKK1. In the alternative pathway, p52 mostly dimerizes with RelB.

In CNS the activation of diverse NF-κB factors is developmentally regulated (Bakalkin *et al.*, 1993; Bhakar *et al.*, 2002) and conserves a role in promoting neurogenesis (Denis-Donini *et al.*, 2008). The NF-κB signaling regulates the growth of neuronal processes of maturing neurons (Gutierrez *et al.*, 2005; Koulich *et al.*, 2001; Lezoualc'h *et al.*, 1998; Maggirwar *et al.*, 1998; Middleton *et al.*, 2000), but also cell differentiation and survival (Bhakar *et al.*, 2002; Chiarugi, 2002; Koulich *et al.*, 2001; Maggirwar *et al.*, 1998; Middleton *et al.*, 2000) through regulation of Bcl-2-family antiapoptotic genes (Bhakar *et al.*, 2002; Chiarugi, 2002). NF-κB is present at the synaptic regions where it can act as a signal transducer which transmits transient synaptic signals to the nucleus with a role in behavior, learning, and memory formation (Kaltschmidt *et al.*, 1993; Meberg *et al.*, 1996).

It was demonstrated the involvement of NF-κB in long-term retention of fear memory (Yeh *et al.*, 2002, 2004), in inhibitory avoidance memory (Freudenthal *et al.*, 2005), and spatial long-term memory (Dash *et al.*, 2005). Either RelA or c-Rel and p50 NF-κB factors were found to be involved in mechanisms of cognition (Kassed *et al.*, 2002; Levenson *et al.*, 2004; Meffert and Baltimore, 2005; Meffert *et al.*, 2003; O'Neill and Kaltschmidt, 1997). In forebrain neuronal conditional NF-κB-deficient mice, it has been shown that the loss of NF-κB impairs synaptic transmission, spatial memory formation, and plasticity (Kaltschmidt *et al.*, 2006). The *TNFR/RelA* double knockout mice, in which the deletion of TNFR (tumor necrosis factor receptor) receptor rescues lethality, show a selective learning deficit in the spatial version of the radial arm maze, suggesting a physiological role for NF-κB in adult brain function (Meffert *et al.*, 2003). The *c-Rel*$^{-/-}$ mice display significant deficits of hippocampus-dependent memory formation, as revealed by freezing behavior 24 h after training for contextual fear conditioning (Ahn *et al.*, 2008; Levenson *et al.*, 2004). *c-Rel*$^{-/-}$ mice perform normally in short-term contextual fear conditioning test and, contrary to wild-type littermates, fail to show a preference for a new object (Ahn *et al.*, 2008). Electrophysiological studies show that c-Rel is necessary for either long-term synaptic potentiation (Ahn *et al.*, 2008) and depression (O'Riordan *et al.*, 2006) in the hippocampus. Absence of p50 reduces basal learning abilities, as shown by the deficit in acquisition of p50$^{-/-}$ mice in an active avoidance test (Kassed *et al.*, 2002). The p50$^{-/-}$ mice display a selective defect in short-term spatial memory performance (Denis-Donini *et al.*, 2008) and reduced anxiety-like behaviors in tests of exploratory drive and anxiety (Kassed and Herkenham, 2004).

II. p50/RelA and c-Rel-Containing Dimers Elicit Opposite Regulation of Neuron Vulnerability

In CNS, NF-κB factors participate either in physiological phenomena as differentiation, synaptic transmission, and plasticity or in neurodegenerative processes associated with Alzheimer's disease (Kaltschmidt *et al.*, 1997), Parkinson's disease (Ghosh *et al.*, 2007; Hunot *et al.*, 1997), Huntington's disease (Khoshnan *et al.*, 2004), trauma (Bethea *et al.*, 1998), and ischemia (Clemens *et al.*, 1997; Inta *et al.*, 2006; Nurmi *et al.*, 2004; Schneider *et al.*,1999). The dual function of NF-κB as activator of neuroprotective programs and inducer of neurodegenerative processes has been widely discussed in the last decade (Mattson and Camandola, 2001; Mattson and Meffert, 2006; Pizzi and Spano, 2006). Diverse studies support a role for NF-κB in neuroprotection and mechanisms of brain tolerance (Blondeau *et al.*, 2001; Fridmacher *et al.*, 2003; Kaltschmidt *et al.*, 1999; Mattson *et al.*, 1997; Tamatani *et al.*, 1999; Yu *et al.*, 1999) and others demonstrate the involvement of NF-κB pathway in the apoptotic cell death of injured neurons (Aleyasin *et al.*, 2004; Chen *et al.*, 2005; Grilli *et al.*, 1996; Nakai *et al.*, 2000; Pannaccione *et al.*, 2005; Pizzi *et al.*, 2002, 2005; Valerio *et al.*, 2006). Under neurotoxic stimulation, the inhibition of IκBα phosphorylation or IKK activity can prevent brain damage (Ghosh *et al.*, 2007; Goffi *et al.*, 2005; Sarnico *et al.*, 2008a).

In order to understand which determinants can switch NF-κB activity from a neuroprotective to a neurotoxic mode, recent studies demonstrated that neuronal response to external stimuli relies on a differential activation of NF-κB dimers. Neurotoxic stimuli, such as glutamate, aberrantly induce p50/RelA dimers and their activation promotes a proapoptotic gene expression and neuronal cell death. On the contrary, the proinflammatory cytokine Interleukin-1β (IL-1β) promotes activation of c-Rel-containing dimers, antiapototic gene expression, and neuron survival (Goffi *et al.*, 2005; Pizzi *et al.*, 2002). The selective targeting of RelA prevents toxic effect of glutamate, without modifying prosurvival effects of IL-1β. The selective targeting of c-Rel switches the prosurvival effects of IL-1β into a neurotoxic response (Pizzi *et al.*, 2002). In support of those results, it was found that the oxidative stress in HT22 immortalized hippocampal cells, as well as the proapoptotic β-amyloid peptide in primary neuronal cells, induce p50/RelA dimers (Ishige *et al.*, 2005; Valerio *et al.*, 2006). Moreover, the activation of c-Rel/RelA and c-Rel/p50 dimers by S100B or metabotropic glutamate receptor agonists protects neurons against *N*-methyl-D-aspartate (NMDA)- (Kögel *et al.*, 2004), β-amyloid- (Pizzi *et al.*, 2005), and MPP^+-induced toxicity (Sarnico *et al.*, 2008b) by promoting the expression of MnSOD and Bcl-xL antiapoptotic proteins (Pizzi *et al.*, 2005).

In brain ischemia, NF-κB activation rapidly occurs in neurons and glial cells, and has been proposed to be involved in the pathogenesis of the postischemic injury (Clemens *et al.*, 1997; Herrmann *et al.*, 2005; Nurmi *et al.*, 2004; Schneider *et al.*, 1999; Ueno *et al.*, 2001). Nuclear translocation of NF-κB subunits is induced in the ischemic hemisphere of mice exposed to middle cerebral artery occlusion (MCAO) with a prevalent activation of RelA and p50 subunits (Crack *et al.*, 2006; Inta *et al.*, 2006). With the aim to investigate the specific assembly of diverse NF-κ B factors to form the active dimers during brain ischemia, Sarnico *et al.* (2009) demonstrate that a similar pattern of NF-κB activation occurs in brain ischemic tissues of mice exposed to permanent occlusion of the middle cerebral artery (MCAO) and in primary cortical neurons exposed to oxygen-glucose deprivation (OGD) (Fig. 1). In particular, a major activation of the p50/RelA dimer is associated with inhibition of c-Rel-containing complexes in both *in vivo* and *in vitro* experimental settings. Overexpression of c-Rel or RelA reduces or increases, respectively, the cell susceptibility to anoxia (Fig. 2A). Conversely, targeting c-Rel and RelA, by small interfering RNAs, shows that c-Rel knock-down increases cell vulnerability to OGD, while the RelA silencing significantly reduces the neuronal cell death (Fig. 2B). This in line with evidence that infarct

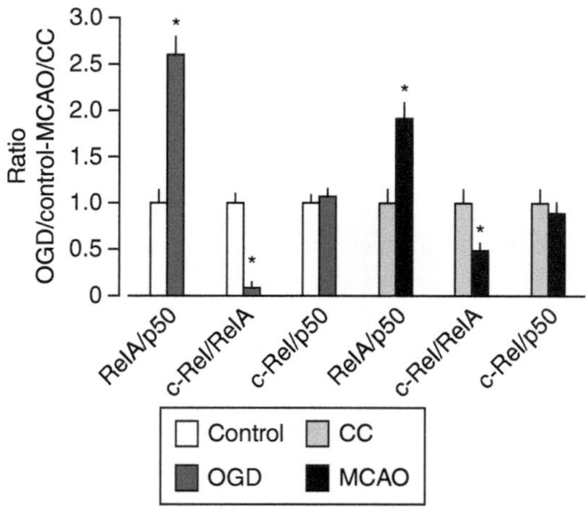

FIG. 1. Densitometry analysis of coimmunoprecipitation studies of NF-κB dimers in nuclear extracts of primary cortical neurons exposed to OGD or cortices from mice subjected to MCAO. Either OGD or MCAO induces nuclear translocation of p50/RelA dimers. The nuclear content of RelA/c-Rel complexes decreased while p50/c-Rel remained unchanged. Data are expressed as ratio of OGD to controls or ratio of MCAO to controlateral cortex. Values are expressed as means ± S.E.M. (* $p \leq 0.05$ vs. the corresponding control value). For methods and details, see Sarnico *et al.* (2009).

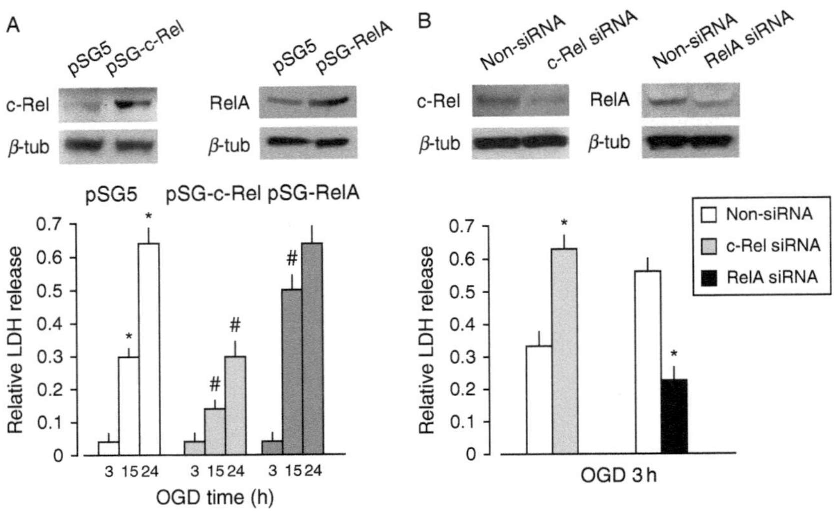

FIG. 2. Opposing regulation of neuronal vulnerability by c-Rel and RelA proteins. (A) Neuronal SK-N-SH cells were transfected with pSG-c-Rel and pSG-RelA plasmids or with a pSG5 vector, and exposed to 3-, 15-, or 24-h OGD. OGD toxicity was prevented by c-Rel overexpression, while overexpressing RelA significantly increased cell loss. Similar results were obtained in three separate experiments run in triplicate (*$p \leq 0.05$ vs. the corresponding control value, $^{\#}p \leq 0.05$ vs. corresponding pSG5-treated cells). (B) Primary cortical neurons were transfected with siRNA cognate to c-Rel gene (c-Rel siRNA), RelA gene (RelA siRNA), or with a control nonspecific siRNA (non-siRNA). c-Rel knockdown made cells more vulnerable to the ischemic insult, while RelA silencing significantly reduced cortical neuron vulnerability to OGD. Values are mean ± S.E.M. of three experiments run in triplicate (*$p \leq 0.05$ vs. the corresponding non-siRNA-treated cells). For methods and details, see Sarnico *et al.* (2009).

size after MCAO is reduced in mice with a brain-conditional deletion of RelA, as well as in p50 knockout mice (Schneider *et al.*, 1999), while no difference is observed in p52 or c-Rel null mice (Inta *et al.*, 2006). The Bcl-2 family proteins involved in brain ischemia can be responsible for the regulation of neuron survival played by different NF-κB factors (Cao *et al.*, 2002; Graham and Chen, 2001; Inta *et al.*, 2006). Among those, the Bcl-2 antiapoptotic member Bcl-xL (Pizzi *et al.*, 2005) and the BH3-only proapoptotic *Bim* and *Noxa* genes (Cao *et al.*, 2002). Bim and Noxa are significantly induced after MCAO and both have been found to be under the transcriptional control of RelA (Inta *et al.*, 2006). Antiapoptotic Bcl-xL has been demonstrated to be transcriptionally regulated by c-Rel in non-neuronal cells (Banerjee *et al.*, 2008; Chen *et al.*, 2000; Kirito *et al.*, 2002) and in neurons (Pizzi *et al.*, 2005). Sarnico *et al.* (2009) show that endogenous Bcl-xL level decreases in cells exposed to OGD. The Bcl-xL content further lowers in c-Rel silenced neurons, but it recovers in c-Rel-overexpressing cells. Analysis of the Bcl-xL promoter activity in neuronal cells reveals that the activation of c-Rel-containing dimers, c-Rel/c-Rel, RelA/c-Rel, and p50/c-Rel, but not p50/RelA activation, induces the Bcl-xL luciferase reporter plasmid in a κB-dependent manner (Fig. 3A). The unresponsiveness of Bcl-xL promoter to p50/RelA in neuronal cells is confirmed by evidence that Bcl-xL reporter plasmid is not induced in cortical neurons exposed to OGD, where endogenous activation of p50/RelA prevails, while the Bim reporter plasmid is highly and κB-specifically activated (Fig. 3B). Thus, the decrease of endogenous Bcl-xL protein observed after OGD may very well be due to activation of an indirect via of degradation of the protein mediated by newly expressed Bim. Bim, by activating Bax-Bak proapoptotic pathway, contributes to the release of cytochrome *c* and activation of caspase-9 cascade (Kuwana *et al.*, 2005), which can be ultimately responsible for the cleavage of Bcl-xL to form its proapoptotic fragments (Chen *et al.*, 2007; Miyawaki *et al.*, 2008). Thus, reducing p50/RelA nuclear translocation, by limiting Bim expression and cytochrome *c* release, could also limit Bcl-xL

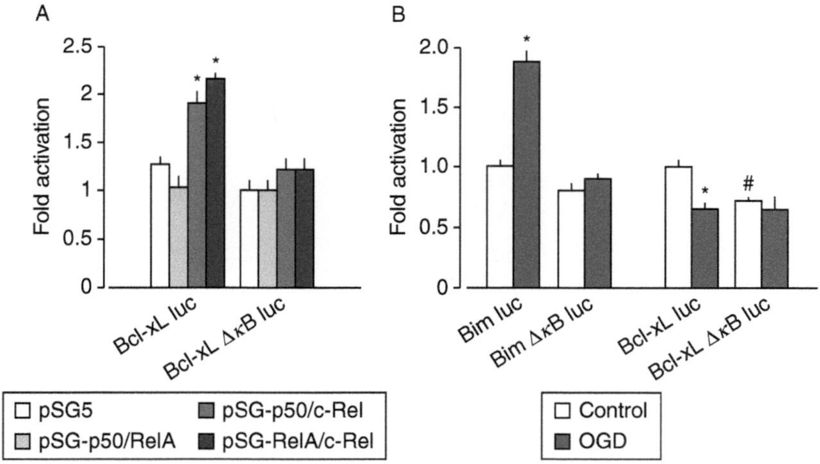

FIG. 3. (A) SK-N-SH neuronal cells were cotransfected with Bcl-xL or Bcl-xL ΔκB luciferase reporter plasmids together with pSG-c-Rel and pSG-RelA, or a combination of pSG-p50, pSG-RelA, and pSG-c-Rel expression plasmids. Control cells were transfected with empty pSG5 vector. Twenty-four hours later the luciferase activity was measured. Mutation of NF-κB binding site blocked the luciferase expression. The c-Rel-containing dimers, p50/c-Rel and RelA/c-Rel, but not p50/RelA complex, were able to activate Bcl-xL promoter. Values are mean ± S.E.M. of three experiments run in triplicate (*$p \leq 0.05$ vs. the corresponding control value). (B) OGD-induced NF-κB activation promotes Bim but not Bcl-xL transcription. Primary cortical neurons were transfected with Bim or Bcl-xL luciferase reporter plasmid or with Bim and Bcl-xL ΔκB luciferase reporter plasmids and then exposed to OGD. Four hours later, luciferase activity was measured. The OGD exposure significantly induced Bim and decreased Bcl-xL promoter activity. Mutation of NF-κB binding sites reduced the luciferase expression. Values are mean ± S.E.M. of three experiments run in triplicate (*$p \leq 0.05$ vs. the corresponding control value; #$p \leq 0.05$ vs. corresponding wild-type luciferase reporter plasmid). For methods and details, see Sarnico *et al.* (2009).

degradation after OGD, as suggested by the recovery of Bcl-xL in RelA-silenced neurons (Sarnico *et al.*, 2009).

In conclusion, the latest acquisition on NF-κB dichotomy shows that within the same neuronal cell unbalanced activation of NF-κB p50/RelA dimer over c-Rel-containing complexes contributes to cell death secondary to the ischemic insult. Opposite effects of NF-κB complexes depend on activation of diverse genes, Bim and Noxa by p50/RelA and Bcl-xL by c-Rel dimers, with opposite functions in the regulation of cell survival. A validation of this model rises from evidence that the adipocyte-derived hormone leptin by activating NF-κB c-Rel-containing dimers c-Rel/p50 and c-Rel/RelA, but not p50/RelA, induces the expression of Bcl-xL and reduces the infarct size in mice exposed to brain ischemia (Valerio *et al.*, 2008). NF-κB activation by leptin involves PI3K, MEK, and PKC pathways, that is the signaling cascade that is also involved in c-Rel-mediated-long-term maintenance of hippocampal LTD (O'Riordan *et al.*, 2006). The knockout of c-Rel abolishes the Bcl-xL expression and neuroprotective effect of leptin in brain ischemia, confirming the fundamental role of c-Rel in mediating neuroprotection (Valerio *et al.*, 2008).

Likewise drugs acting as selective inducers of c-Rel dimers, also specific inhibitors of aberrantly activated p50/RelA would bring a great benefit in the treatment of neurodegenerative conditions. Understanding possible epigenetic mechanisms responsible for diverse p50/RelA functions in physiology and in pathology will reveal new important molecular targets for development of innovative therapies.

Acknowledgments

This work was supported by grants from Italian Ministry of Education, University and Scientific Research-PRIN 2004 and 2005, 2006; Centre of Study and Research on Aging, Brescia; MIUR Center of Excellence for Innovative Diagnostics and Therapeutics (IDET) of Brescia University.

References

Ahn, H. J., Hernandez, C. M., Levenson, J. M., Lubin, F. D., Liou, H. C., and Sweatt, J. D. (2008). c-Rel, an NF-kappaB family transcription factor, is required for hippocampal long-term synaptic plasticity and memory formation. *Learn Mem.* **15,** 539–549.

Aleyasin, H., Cregan, S. P., Iyirhiaro, G., O'Hare, M. J., Callaghan, S. M., Slack, R. S., and Park, D. S. (2004). Nuclear factor-(kappa)B modulates the p53 response in neurons exposed to DNA damage. *J. Neurosci.* **24,** 2963–2973.

Bakalkin, G. Y., Yakovleva, T., and Terenius, L. (1993). NF-kappa B-like factors in the murine brain. Developmentally-regulated and tissue-specific expression. *Brain Res. Mol. Brain Res.* **20,** 137–146.

Banerjee, A., Grumont, R., Gugasyan, R., White, C., Strasser, A., and Gerondakis, S. (2008). NF-{kappa}B1 and c-Rel cooperate to promote the survival of TLR4 activated B cells by neutralizing Bim via distinct mechanisms. *Blood* **112,** 5063–5073.

Bethea, J. R., Castro, M., Keane, R. W., Lee, T. T., Dietrich, W. D., and Yezierski, R. P. (1998). Traumatic spinal cord injury induces nuclear factor-kappaB activation. *J. Neurosci.* **18,** 3251–3260.

Bhakar, A. L., Tannis, L. L., Zeindler, C., Russo, M. P., Jobin, C., Park, D. S., MacPherson, S., and Barker, P. A. (2002). Constitutive nuclear factor-κB activity is required for central neuron survival. *J. Neurosci.* **22,** 8466–8475.

Blondeau, N., Widmann, C., Lazdunski, M., and Heurteaux, C. (2001). Activation of the nuclear factor-kappaB is a key event in brain tolerance. *J. Neurosci.* **21,** 4668–4677.

Bonizzi, G., and Karin, M. (2004). The two NF-kappaB activation pathways and their role in innate and adaptive immunity. *Trends Immunol.* **25,** 280–288.

Cao, G., Pei, W., Ge, H., Liang, Q., Luo, Y., Sharp, F. R., Lu, A., Ran, R., Graham, S. H., and Chen, J. (2002). *In vivo* delivery of a Bcl-xL fusion protein containing the TAT protein transduction domain protects against ischemic brain injury and neuronal apoptosis. *J. Neurosci.* **22,** 5423–5431.

Chen, C., Edelstein, L. C., and Gelinas, C. (2000). The Rel/NF-kappaB family directly activates expression of the apoptosis inhibitor Bcl-x(L). *Mol. Cell. Biol.* **20,** 2687–2695.

Chen, J., Zhou, Y., Mueller-Steiner, S., Chen, L. F., Kwon, H., Yi, S., Mucke, L., and Gan, L. (2005). SIRT1 protects against microglia-dependent amyloid-beta toxicity through inhibiting NF-kappaB signaling. *J. Biol. Chem.* **280,** 40364–40374.

Chen, M., Guerrero, A. D., Huang, L., Shabier, Z., Pan, M., Tan, T. H., and Wang, J. (2007). Caspase-9-induced mitochondrial disruption through cleavage of anti-apoptotic BCL-2 family members. *J. Biol. Chem.* **282,** 33888–33895.

Chiarugi, A. (2002). Characterization of the molecular events following impairment of NF-kappaB-driven transcription in neurons. *Brain Res. Mol. Brain Res.* **109,** 179–188.

Clemens, J. A., Stephenson, D. T., Smalstig, E. B., Dixon, E. P., and Little, S. P. (1997). Global ischemia activates nuclear factor-kappa B in forebrain neurons of rats. *Stroke* **28,** 1073–1080.

Crack, P. J., Taylor, J. M., Ali, U., Mansell, A., and Hertzog, P. J. (2006). Potential contribution of NF-kappaB in neuronal cell death in the glutathione peroxidase-1 knockout mouse in response to ischemia-reperfusion injury. *Stroke* **37,** 1533–1538.

Dash, P. K., Orsi, S. A., and Moore, A. N. (2005). Sequestration of serum response factor in the hippocampus impairs long-term spatial memory. *J. Neurochem.* **93,** 269–278.

Dejardin, E. (2006). The alternative NF-kappaB pathway from biochemistry to biology: Pitfalls and promises for future drug development. *Biochem. Pharmacol.* **72,** 1161–1179.

Denis-Donini, S., Dellarole, A., Crociara, P., Francese, M. T., Bortolotto, V., Quadrato, G., Canonico, P. L., Orsetti, M., Ghi, P., Memo, M., Bonini, S. A., Ferrari-Toninelli, G., *et al.* (2008). Impaired adult neurogenesis associated with short-term memory defects in NF-kappaB p50-deficient mice. *J. Neurosci.* **28,** 3911–3919.

Freudenthal, R., Boccia, M. M., Acosta, G. B., Blake, M. G., Merlo, E., Baratti, C. M., and Romano, A. (2005). NF-kappaB transcription factor is required for inhibitory avoidance long-term memory in mice. *Eur. J. Neurosci.* **21,** 2845–2852.

Fridmacher, V., Kaltschmidt, B., Goudeau, B., Ndiaye, D., Rossi, F. M., Pfeiffer, J., Kaltschmidt, C., Israel, A., and Memet, S. (2003). Forebrain-specific neuronal inhibition of nuclear factor-kappaB activity leads to loss of neuroprotection. *J. Neurosci.* **23,** 9403–9408.

Ghosh, A., Roy, A., Liu, X., Kordower, J. H., Mufson, E. J., Hartley, D. M., Ghosh, S., Mosley, R. L., Gendelman, H. E., and Pahan, K. (2007). Selective inhibition of NF-kappaB activation prevents dopaminergic neuronal loss in a mouse model of Parkinson's disease. *Proc. Natl. Acad. Sci. USA* **104,** 18754–18759.

Goffi, F., Boroni, F., Benarese, M., Sarnico, I., Benetti, A., Spano, P. F., and Pizzi, M. (2005). The inhibitor of I kappa B alpha phosphorylation BAY 11–7082 prevents NMDA neurotoxicity in mouse hippocampal slices. *Neurosci. Lett.* **377,** 147–151.

Graham, S. H., and Chen, J. (2001). Programmed cell death in cerebral ischemia. *J. Cereb. Blood Flow Metab.* **21,** 99–109. (Review).

Grilli, M., Pizzi, M., Memo, M., and Spano, P. (1996). Neuroprotection by aspirin and sodium salicylate through blockade of NF-kappaB activation. *Science* **274,** 1383–1385.

Gutierrez, H., Hale, V.A, Dolcet, X., and Davies, A. (2005). NF-kappaB signalling regulates the growth of neural processes in the developing PNS and CNS. *Development* **132,** 1713–1726.

Herrmann, O., Baumann, B., de Lorenzi, R., Muhammad, S., Zhang, W., Kleesiek, J., Malfertheiner, M., Köhrmann, M., Potrovita, I., Maegele, I., Beyer, C., Burke, J. R., *et al.* (2005). IKK mediates ischemia-induced neuronal death. *Nat. Med.* **11,** 1322–1329.

Hunot, S., Brugg, B., Ricard, D., Michel, P. P., Muriel, M. P., Ruberg, M., Faucheux, B. A., Agid, Y., and Hirsch, E. C. (1997). Nuclear traslocation of NF-κB is increased in dopaminergic neurons of patients with Parkinson's disease. *Proc. Natl. Acad. Sci. USA* **94,** 7531–7536.

Inta, I., Paxian, S., Zhang, W., Pizzi, M., Sarnico, I., Spano, P., Muhammad, S., Herrmann, O., Liou, H. C., Schmid, R. M., and Schwaninger, M. (2006). Bim and Noxa are candidates to mediate the deleterious effect of the NF-κB subunit RelA in cerebral ischemia. *J. Neurosci.* **26,** 12896–12903.

Ishige, K., Tanaka, M., Arakawa, M., Saito, H., and Ito, Y. (2005). Distinct nuclear factor-kappaB/Rel proteins have opposing modulatory effects in glutamate-induced cell death in HT22 cells. *Neurochem. Int.* **47,** 545–555.

Kaltschmidt, C., Kaltschmidt, B., and Baeuerle, P. A. (1993). Brain synapses contain inducible forms of the transcription factor NF-kappa B. *Mech. Dev.* **43,** 135–147.

Kaltschmidt, C., Kaltschmidt, B., Neumann, H., Wekerle, H., and Baeuerle, P. A. (1994). Constitutive NF-kappaB activity in neurons. *Mol. Cell. Biol.* **14,** 3981–3992.

Kaltschmidt, B., Uherek, M., Volk, B., Baeuerle, P. A., and Kaltschmidt, C. (1997). Transcription factor NF-κB is activated in primary neurons by amyloid β peptides and in neurons surrounding early plaques from patients with Alzheimer's disease. *Proc. Natl. Acad. Sci. USA* **94,** 2642–2647.

Kaltschmidt, B., Uherek, M., Wellmann, H., Volk, B., and Kaltschmidt, C. (1999). Inhibition of NF-kappaB potentiates amyloid beta-mediated neuronal apoptosis. *Proc. Natl. Acad. Sci. USA* **96,** 9409–9414.

Kaltschmidt, B., Ndiaye, D., Korte, M., Pothion, S., Arbibe, L., Prullage, M., Pfeiffer, J., Lindecke, A., Staiger, V., Israël, A., Kaltschmidt, C., and Mémet, S. (2006). NF-kappaB regulates spatial memory formation and synaptic plasticity through protein kinase A/CREB signalling. *Mol. Cell. Biol.* **26,** 2936–2946.

Kassed, C. A., and Herkenham, M. (2004). NF-kappaB p50-deficient mice show reduced anxiety-like behaviors in tests of exploratory drive and anxiety. *Behav. Brain Res.* **154,** 577–584.

Kassed, C. A., Willing, A. E., Garbuzova-Davis, S., Sanberg, P. R., and Pennypacker, K. R. (2002). Lack of NF-kappaB p50 exacerbates degeneration of hippocampal neurons after chemical exposure and impairs learning. *Exp. Neurol.* **176,** 277–288.

Khoshnan, A., Ko, J., Watkin, E. E., Paige, L. A., Reinhart, P. H., and Patterson, P. H. (2004). Activation of the IkappaB kinase complex and nuclear factor-kappaB contributes to mutant huntingtin neurotoxicity. *J. Neurosci.* **24,** 7999–8008.

Kirito, K., Watanabe, T., Sawada, K., Endo, H., Ozawa, K., and Komatsu, N. (2002). Thrombopoietin regulates Bcl-xL gene expression through Stat5 and phosphatidylinositol 3-kinase activation pathways. *J. Biol. Chem.* **277,** 8329–8337.

Kögel, D., Peters, M., König, H. G., Hashemi, S. M., Bui, N. T., Arolt, V., Rothermundt, M., and Prehn, J.H (2004). S100B potently activates p65/c-Rel transcriptional complexes in hippocampal neurons: Clinical implications for the role of S100B in excitotoxic brain injury. *Neuroscience* **127,** 913–920.

Koulich, E., Nguyen, T., Johnson, K., Giardina, C., and D'mello, S. (2001). NF-kappaB is involved in the survival of cerebellar granule neurons: Association of IkappaBbeta [correction of Ikappabeta] phosphorylation with cell survival. *J. Neurochem.* **76,** 1188–1198.

Kuwana, T., Bouchier-Hayes, L., Chipuk, J. E., Bonzon, C., Sullivan, B. A., Green, D. R., and Newmeyer, D. D. (2005). BH3 domains of BH3-only proteins differentially regulate Bax-mediated mitochondrial membrane permeabilization both directly and indirectly. *Mol. Cell* **17,** 525–535.

Levenson, J. M., Choi, S., Lee, S. Y., Cao, Y. A., Ahn, H. J., Worley, K. C., Pizzi, M., Liou, H. C., and Sweatt, J. D. (2004). A bioinformatics analysis of memory consolidation reveals involvement of the transcription factor c-rel. *J. Neurosci.* **24,** 3933–3943.

Lezoualc'h, F., Sagara, Y., Holsboer, F., and Behl, C. (1998). High costitutive NF-κB activity mediates resistance to oxidative stress in neuronal cells. *J. Neurosci.* **18,** 3224–3232.

Maggirwar, S. B., Sarmiere, P. D., Dewhurst, S., and Freeman, R. S. (1998). Nerve growth factor-dependent activation of NF-kappaB contributes to survival of sympathetic neurons. *J. Neurosci.* **18,** 10356–10365.

Mattson, M. P., and Camandola, S. (2001). NF-kappaB in neuronal plasticity and neurodegenerative disorders. *J. Clin. Invest.* **107,** 247–254.

Mattson, M. P., and Meffert, M. K. (2006). Roles for NF-kappaB in nerve cell survival, plasticity, and disease. *Cell Death Differ.* **13,** 852–860.

Mattson, M. P., Goodman, Y., Luo, H., Fu, W., and Furukawa, K. (1997). Activation of NF-kappaB protects hippocampal neurons against oxidative stress-induced apoptosis: Evidence for induction of manganese superoxide dismutase and suppression of peroxynitrite production and protein tyrosine nitration. *J. Neurosci. Res.* **49,** 681–697.

Meberg, P. J., Kinney, W. R., Valcourt, E. G., and Routtenberg, A. (1996). Gene expression of the transcription factor NF-kappa B in hippocampus: regulation by synaptic activity. *Brain Res. Mol. Brain Res.* **38,** 179–190.

Meffert, M. K., and Baltimore, D. (2005). Physiological functions for brain NF-κB. *Trends Neurosci.* **28,** 37–43.

Meffert, M. K., Chang, J. M., Wiltgen, B. J., Fanselow, M. S., and Baltimore, D. (2003). NF-kappa B functions in synaptic signaling and behavior. *Nat. Neurosci.* **6,** 1072–1078.

Middleton, G., Hamanoue, M., Enokido, Y., Wyatt, S., Pennica, D., Jaffray, E., Hay, R. T., and Davies, A. M. (2000). Cytokine-induced nuclear factor kappa B activation promotes the survival of developing neurons. *J. Cell. Biol.* **148,** 325–332.

Miyawaki, T., Mashiko, T., Ofengeim, D., Flannery, R. J., Noh, K. M., Fujisawa, S., Bonanni, L., Bennett, M. V., Zukin, R. S., and Jonas, E. A. (2008). Ischemic preconditioning blocks BAD translocation, Bcl-xL cleavage, and large channel activity in mitochondria of postischemic hippocampal neurons. *Proc. Natl. Acad. Sci. USA* **105,** 4892–4897.

Nakai, M., Qin, Z., Wang, Y., and Chase, T. N. (2000). NMDA and non-NMDA receptor-stimulated IkappaB-alpha degradation: Differential effects of the caspase-3 inhibitor DEVD.CHO, ethanol and free radical scavenger OPC-14117. *Brain Res.* **859,** 207–216.

Nurmi, A., Lindsberg, P. J., Koistinaho, M., Zhang, W., Juettler, E., Karjalainen-Lindsberg, M. L., Weih, F., Frank, N., Schwaninger, M., and Koistinaho, J. (2004). Nuclear factor-kappaB contributes to infarction after permanent focal ischemia. *Stroke* **35,** 987–991.

O'Neill, L. A., and Kaltschmidt, C. (1997). NF-kappa B: A crucial transcription factor for glial and neuronal cell function. *Trends Neurosci.* **20,** 252–258.

O'Riordan, K. J., Huang, I. C., Pizzi, M., Spano, P., Boroni, F., Egli, R., Desai, P., Fitch, O., Malone, L., Ahn, H. J., Liou, H. C., Sweatt, J. D., *et al.* (2006). Regulation of nuclear factor kappaB in the hippocampus by group I metabotropic glutamate receptors. *J. Neurosci.* **26,** 4870–4879.

Pannaccione, A., Secondo, A., Scorziello, A., Cali, G., Taglialatela, M., and Annunziato, L. (2005). Nuclear factor-kappaB activation by reactive oxygen species mediates voltage-gated K+ current enhancement by neurotoxic beta-amyloid peptides in nerve growth factor-differentiated PC-12 cells and hippocampal neurones. *J. Neurochem.* **94,** 572–586.

Pizzi, M., and Spano, P. (2006). Distinct roles of diverse nuclear factor-kappaB complexes in neuropathological mechanisms. *Eur. J. Pharmacol.* **545,** 22–28. [Review].

Pizzi, M., Goffi, F., Boroni, F., Benarese, M., Perkins, S. E., Liou, H. C., and Spano, P. (2002). Opposing roles for NF-kappa B/Rel factors p65 and c-Rel in the modulation of neuron survival elicited by glutamate and interleukin-1beta. *J. Biol. Chem.* **277,** 20717–20723.

Pizzi, M., Sarnico, I., Boroni, F., Benarese, M., Steimberg, N., Mazzoleni, G., Dietz, G. P., Bähr, M., Liou, H. C., and Spano, P. F. (2005). NF-κB factor c-Rel mediates neuroprotection elicited by mGlu5 receptor agonists against amyloid β-peptide toxicity. *Cell Death Differ.* **12,** 761–772.

Sarnico, I., Boroni, F., Benarese, M., Alghisi, M., Valerio, A., Battistin, L., Spano, P., and Pizzi, M. (2008a). Targeting IKK2 by pharmacological inhibitor AS602868 prevents excitotoxic injury to neurons and oligodendrocytes. *J. Neural Transm.* **115,** 693–701.

Sarnico, I., Boroni, F., Benarese, M., Sigala, S., Lanzillotta, A., Battistin, L., Spano, P., and Pizzi, M. (2008b). Activation of NF-kappaB p65/c-Rel dimer is associated with neuroprotection elicited by mGlu5 receptor agonists against MPP(+) toxicity in SK-N-SH cells. *J. Neural Transm.* **115,** 669–676.

Sarnico, I., Lanzillotta, A., Boroni, F., Benarese, M., Alghisi, M., Schwaninger, M., Inta, I., Battistin, L., Spano, P., and Pizzi, M. (2009). NF-κB p50/RelA and c-Rel-containing dimers: Opposite regulators of neuron vulnerability to ischemia. *J. Neurochem* **108,** 475–485.

Schmidt-Ullrich, R., Mémet, S., Lilienbaum, A., Feuillard, J., Raphael, M., and Israel, A. (1996). NF-kappaB activity in transgenic mice: Developmental regulation and tissue specificity. *Development* **122,** 2117–2128.

Schneider, A., Martin-Villalba, A., Weih, F., Vogel, J., Wirth, T., and Schwaninger, M. (1999). NF-kappaB is activated and promotes cell death in focal cerebral ischemia. *Nat. Med.* **5,** 554–559.

Siebenlist, U., Franzoso, G., and Brown, K. (1994). Structure, regulation and function of NF-kappa B. *Annu. Rev. Cell Biol.* **10,** 405–455.

Tamatani, M., Che, Y. H., Matsuzaki, H., Ogawa, S., Okado, H., Miyake, S., Mizuno, T., and Tohyama, M. (1999). Tumor necrosis factor induces Bcl-2 and Bcl-x expression through NF-kappaB activation in primary hippocampal neurons. *J. Biol. Chem.* **274,** 8531–8538.

Ueno, T., Sawa, Y., Kitagawa-Sakakida, S., Nishimura, M., Morishita, R., Kaneda, Y., Kohmura, E., Yoshimine, T., and Matsuda, H. (2001). Nuclear factor-kappa B decoy attenuates neuronal damage after global brain ischemia: A future strategy for brain protection during circulatory arrest. *J. Thorac. Cardiovasc. Surg.* **122,** 720–727.

Valerio, A., Boroni, F., Benarese, M., Sarnico, I., Ghisi, V., Bresciani, L. G., Ferrario, M., Borsani, G., Spano, P. F., and Pizzi, M. (2006). NF-κB pathway: A target for preventing β-amyloid (Aβ)-induced neuronal damage and Aβ42 production. *Eur. J. Neurosi.* **23,** 1711–1720.

Valerio, A., Dossena, M., Bertolotti, P., Boroni, F., Sarnico, I., Faraco, G., Chiarugi, A., Frontini, A., Giordano, A., Liou, H. C., De Simoni, M. G., Spano, P. F., *et al.* (2008). Leptin is induced in ischemic cerebral cortex and exerts neuroprotection via NF-κB/c-Rel-dependent transcription. *Stroke* **40,** 610–617.

West, A. E., Griffith, E. C., and Greenberg, M. E. (2002). Regulation of transcription factors by neuronal activity. *Nat. Rev. Neurosci.* **3,** 921–931.

Yeh, S. H., Lin, C. H., Lee, C. F., and Gean, P. W. (2002). A requirement of nuclear factor-kappaB activation in fear-potentiated startle. *J. Biol. Chem.* **277,** 46720–46729.

Yeh, S. H., Lin, C. H., and Gean, P. W. (2004). Acetylation of nuclear factor-kappaB in rat amygdala improves long-term but not short-term retention of fear memory. *Mol. Pharmacol.* **65,** 1286–1292.

Yu, Z., Zhou, D., Bruce-Keller, A. J., Kindy, M. S., and Mattson, M. P. (1999). Lack of the p50 subunit of nuclear factor-kappaB increases the vulnerability of hippocampal neurons to excitotoxic injury. *J. Neurosci.* **19,** 8856–8865.

OXIDATIVE STRESS IN STROKE PATHOPHYSIOLOGY: VALIDATION OF HYDROGEN PEROXIDE METABOLISM AS A PHARMACOLOGICAL TARGET TO AFFORD NEUROPROTECTION

Diana Amantea,* Maria Cristina Marrone,† Robert Nisticò,*,†,‡ Mauro Federici,† Giacinto Bagetta,*,‡ Giorgio Bernardi,†,§ and Nicola Biagio Mercuri†,§

*Department of Pharmacobiology, University of Calabria, Rende (CS), Italy
†C.E.R.C.—Santa Lucia Foundation IRCCS, Rome, Italy
‡Section of Neuropharmacology of Normal and Pathological Neuronal Plasticity, University Centre for Adaptive Disorders and Headache (UCHAD), University of Calabria, Rende (CS), Italy
§Neurological Clinic, Department of Neuroscience, "Tor Vergata" University, Rome, Italy

Reactive oxygen species (ROS) accumulation has been described in the brain following an ischemic insult. Superoxide anion is converted by superoxide dismutase into hydrogen peroxide (H_2O_2), and the latter is then transformed into the toxic hydroxyl radical, through the Haber–Weiss reaction, converted to water by glutathione peroxidase (GPx) or dismuted to water and oxygen through catalase. Accumulation of H_2O_2 has been suggested to exert neurotoxic effects, although recent *in vitro* studies have demonstrated either physiological or protective roles of this molecule in the brain. In particular, oxidative stress is critically involved in brain damage induced by transient cerebral ischemia. Here, we demonstrate that inhibition of GPx by systemic (i.p.) administration of mercaptosuccinate (MS, 1.5–150 mg/kg) dose-dependently reduces brain infarct damage produced by transient (2 h) middle cerebral artery occlusion (MCAo) in rat. Neuroprotection was observed when the drug was administered 15 min before the ischemic insult,

DOI: 10.1016/S0074-7742(09)85025-3

0074-7742/09 $35.00

whereas no effect was detected when the drug was injected 1 h before MCAo or upon reperfusion. Furthermore, application of MS (1 mM) to corticostriatal slices limited the irreversible functional derangement of field potentials caused by a prolonged (12 min) oxygen-glucose deprivation. This effect was reverted by concomitant bath application of the catalase inhibitor 3-aminotriazole (20 mM), suggesting the involvement of catalase in mediating the neuroprotective effects of MS.

Thus, our findings demonstrate that MS is neuroprotective in both *in vivo* and *in vitro* ischemic conditions, through a mechanism which may involve increased endogenous levels of H_2O_2 and its consequent conversion to molecular oxygen by catalase.

I. Introduction

Cerebral ischemia is characterized by complex spatial and temporal events evolving over minutes or even days, leading to tissue damage in the regions supplied by the occluded vessel. Two major mechanisms involved in cellular damage following brain ischemia include amino acid excitotoxicity and oxidative stress produced by free radicals during reperfusion injury (Lo *et al.*, 2003; Warner *et al.*, 2004). Oxidative stress can be traced primarily to formation of superoxide and nitric oxide. Dramatic accumulation of reactive oxygen species (ROS) in the ischemic brain tissue triggers molecular pathways leading to necrosis, apoptosis, and neuroinflammation with subsequent neuronal loss and serious memory and/or motor disturbances (Dirnagl *et al.*, 1999). Principal sources of superoxide include electron leak during mitochondrial electron transport, perturbed mitochondrial metabolism, and inflammatory responses to injury (Warner *et al.*, 2004). Being highly susceptible to oxidative stress, the brain possesses potent defenses against superoxide accumulation, such as free radical scavengers and, most notably, enzymatic antioxidants. Superoxide dismutase (SOD) catalyses dismutation of superoxide to hydrogen peroxide (H_2O_2) (Fridovich, 1995). Overexpression of SOD, as well as administration of SOD mimetics, provides significant neuroprotection in animal models of cerebral ischemia/reperfusion (see Warner *et al.*, 2004).

H_2O_2 can freely cross cell membranes and, although it has modest oxidative potential, it can be metabolized to produce potentially toxic free radicals, such as the hydroxyl radical (OH·), through the Haber–Weiss reaction (Halliwell, 1992). Alternatively, H_2O_2 can be converted to water by glutathione peroxidase (GPx) or dismuted to water and oxygen through catalase (Brannan *et al.*, 1981; De Marchesa *et al.*, 1974). Transgenic mice overexpressing GPx are protected against transient focal brain ischemia damage (Weisbrot-Lefkowitz *et al.*, 1998); whereas increased infarct size and exacerbated apoptosis is observed in GPx knockout mice (Crack *et al.*, 2001), possibly due to accumulation of H_2O_2 in the ischemic/reperfused brain tissue. Interestingly, in addition to possible damaging effects, it has been suggested that H_2O_2 generates sufficient molecular oxygen within the rodent spinal cord and in rat hippocampal slices to support synaptic transmission during hypoxia (Fowler, 1997; Walton and Fulton, 1983). Moreover, we have recently demonstrated that the neuroprotective effect of H_2O_2 against oxygen glucose deprivation (OGD) in rat substantia nigra or hippocampal slices is due to production of molecular oxygen through catalase (Geracitano *et al.*, 2005; Nisticò *et al.*, 2008). Thus, in conditions of reduced oxygen supply, H_2O_2 may exert a protective role through its metabolic degradation to O_2. However, to date, there is no information on whether H_2O_2 may contribute to neuroprotection against brain ischemia *in vivo*.

Here, we demonstrate that systemic administration of mercaptosuccinate (MS), a GPx inhibitor, significantly reduces brain infarct damage produced by transient middle cerebral artery occlusion (MCAo) in rat. Neuroprotection is also observed in corticostriatal slices subjected to OGD, where it is inhibited by the catalase inhibitor 3-aminotriazole (3-AT). Thus, our findings suggest that increased endogenous levels of H_2O_2 during an ischemic insult may provide protection via production of molecular oxygen through catalase.

II. Experimental Procedures

A. Animals and Drug Treatments

Adult male Wistar rats (Charles River, Calco, Como, Italy) were housed under controlled environmental conditions with ambient temperature of 22 °C, relative humidity of 65%, and 12 h light:12 h dark cycle, with free access to food and water.

Mercaptosuccinic acid (1.5–150 mg/kg, Sigma-Aldrich, Milan, Italy) or vehicle (0.01 M phosphate buffered saline (PBS), 1 ml/kg) were administered i.p. 15 min or 1 h before MCAo, or at the onset of reperfusion.

All the experimental procedures were carried out in accordance with the European Community Council Directive on 24 November, 1986 (86/609/EEC), included in the D.M. 116/1992 of the Italian Ministry of Health. All efforts were made to minimize the number of animals used and their suffering.

B. Focal Cerebral Ischemia

Brain ischemia was induced by occlusion of the middle cerebral artery in rats weighing 280–320 g by intraluminal filament, using the relatively noninvasive technique previously described by Longa *et al.* (1989). Briefly, rats were anaesthetized with 5% isoflurane in air, and were maintained with the lowest acceptable concentration of the anaesthetic (1.5–2%). Body temperature was measured with a rectal probe and was kept at 37 °C during the surgical procedure with a heating pad. Under an operating microscope, the external and internal right carotid arteries were exposed through a neck incision. The external carotid artery was cut approximately 3 mm above the common carotid artery bifurcation and a silk suture was tied loosely around the external carotid stump. A silicone-coated nylon filament (diameter: 0.37 mm, Doccol Corporation, Redlands, CA, USA) was then inserted into the external carotid artery and gently advanced into the internal carotid artery, approximately 18 mm from the carotid bifurcation, until mild resistance was felt, thereby indicating occlusion of the origin of the middle cerebral artery in the Willis circle. The silk suture was tightened around the intraluminal filament to prevent bleeding. The wound was then sutured and anaesthesia discontinued.

To allow reperfusion, rats were briefly reanaesthetized with isoflurane, and the nylon filament was withdrawn 2 h after MCAo. After the discontinuation of isoflurane and wound closure, the animals were allowed to awake and were kept in their cages with free access to food and water.

Cerebral blood flow (CBF) was monitored in the cerebral cortex of the ischemic hemisphere corresponding to the supply territory of the middle cerebral artery by laser-doppler flowmetry (DRT4, Moor Instruments, Devon, UK). To this aim, a rectangular bent laser-doppler probe was glued onto the parietal bone (2 mm posterior and 5 mm lateral from bregma) and local CBF was continuously measured from 20 min before the onset of ischemia until 10 min after reperfusion, keeping the animal under isoflurane anaesthesia. Flow values were collected every 5 min before MCAo and after reperfusion; whereas data were collected at 10 min intervals during occlusion.

C. Neuropathology and Quantification of Ischemic Damage

Cerebral infarct volume was evaluated 22 h after reperfusion in rats subjected to 2 h MCAo. Animals were sacrificed by decapitation and the brains were rapidly removed. Eight serial sections from each brain were cut at 2-mm intervals from the frontal pole using a rat brain matrix (Harvard Apparatus, MA, USA). To measure ischemic damage, brain slices were stained in a solution containing 2% 2,3,5-triphenyltetrazolium chloride (TTC) in saline, at 37 °C. After 10 min incubation, the slices were transferred to 10% neutral buffered formaldehyde and stored at 4 °C prior to analysis. Images of TTC-stained sections were captured using a digital scanner and analyzed using an image analysis software (ImageJ, version 1.30). Infarct volume (mm^3) was calculated by summing the infarcted area (unstained) of the eight sections and multiplying by the interval thickness between sections (Li *et al.*, 2000).

D. Electrophysiology

Male Wistar rats (280–320 g of weight) were deeply anaesthetized with halothane and subsequently decapitated. Brain slices were prepared according to a previously described procedure (Geracitano *et al.*, 2003). Briefly, the brain was rapidly removed and corticostriatal coronal slices (300 μm) were cut from a tissue block with the use of a vibratome (at 20–25 °C) in artificial cerebrospinal fluid solution (ACSF).

Slices containing the neostriatum and the neocortex were incubated in a reservoir (1 h, 33 °C) and separately transferred into a recording chamber, completely submerged in a continuously flowing ACSF (32.5–33 °C, 2.5–3 ml/min) and gassed with a 95% O_2–5% CO_2 mixture. The composition of this ACSF was (in mM): NaCl, 126; NaH_2PO_4, 1.2; $MgCl_2$, 1.2; $CaCl_2$, 2.4; KCl, 2.5; $NaHCO_3$, 18; glucose, 10. To study the effects of *in vitro* ischemia, slices were deprived of both oxygen and glucose by omitting glucose from standard ACSF and saturating it with a gas mixture of 95% N_2 and 5% CO_2.

Field potentials were recorded using extracellular glass microelectrodes filled with ACSF (5–10 MΩ) and placed within the dorsal striatum. Signals were fed to an Iso-DAM8 amplifier (WPI), filtered at 1 kHz, acquired and analyzed with the "LTP program" (Anderson and Collingridge, 2001). A bipolar concentric NiCr-insulated stimulating electrode was placed in the white matter between the cortex and the striatum to activate corticostriatal fibers. Test stimuli were delivered every 60 s, at half-maximal intensity. The field potential amplitude was defined as the mean amplitude of the peak negativity, measured from the peak of the early and the late positivity (Calabresi *et al.*, 1992).

MS and 3-amino-1,2,4-triazole (3-AT), obtained from Sigma-Aldrich (Milan, Italy), were dissolved to their final concentrations in ACSF.

E. Statistical Analysis

Data are reported as means ± S.E.M. and statistical analysis was performed by the Student's *t*-test or by ANOVA followed by Dunnett's *post hoc* test, as appropriate. Experimental data were elaborated by means of Prism 3 program (GraphPAD Software for Science, San Diego, CA, USA), and differences were considered statistically significant for $P < 0.05$.

III. Results

A. Neuroprotection by MS Against Transient MCAo-Induced Brain Damage

Systemic (i.p.) administration of the GPx inhibitor MS (1.5–150 mg/kg) dose-dependently reduced brain infarct area and volume produced by 2 h MCAo, as assessed by TTC staining 22 h after reperfusion (Fig. 1A–C). A representative image of the infarcted (pale) areas throughout the brain of rats treated with MS (150 mg/kg) or vehicle (PBS, 1 ml/kg), administered i.p. 15 min before MCAo, is

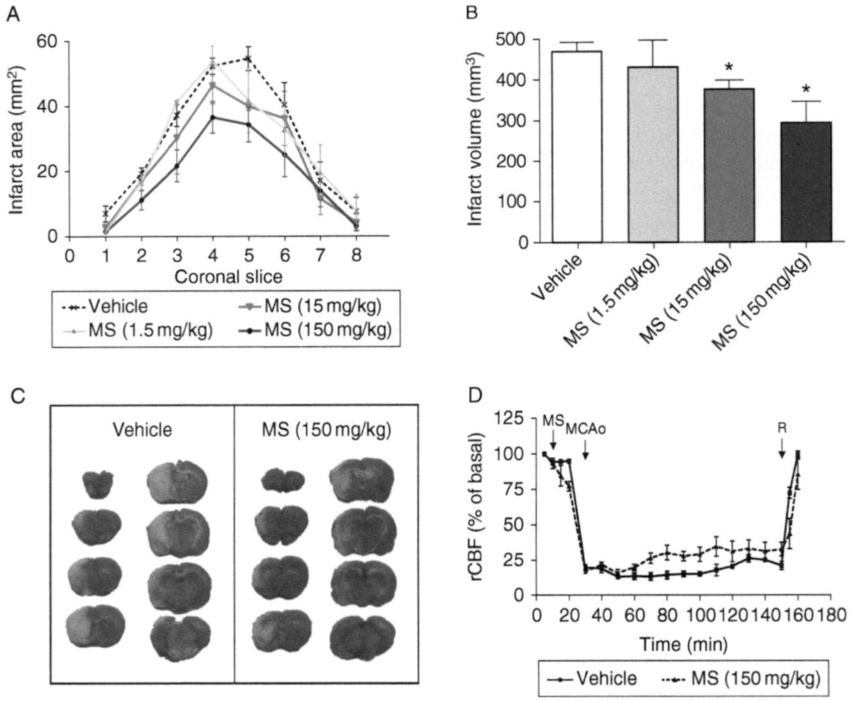

FIG. 1. Systemic administration of MS dose-dependently reduces brain infarct damage produced by transient MCAo in rat. Brain infarct areas (A) and volumes (B) were measured in TTC-stained brain coronal sections from rats treated with mercaptosuccinate (MS), i.p., 15 min before transient (2 h) MCAo. Brain damage was evaluated after 22 h of reperfusion. Values are expressed as mean ± S.E.M.; * indicates $P < 0.05$ versus vehicle (one-way ANOVA followed by Dunnett's posttest; $n = 4$–6 rats per experimental group). (C) Representative brain coronal sections (2 mm thick), stained with TTC, showing the infarct area (unstained) in rats treated with MS (150 mg/kg) or vehicle (PBS, 1 ml/kg) i.p., 15 min before transient (2 h) MCAo followed by 22 h reperfusion. Compared to vehicle-treated animals, MS

shown in Fig. 1C. Ischemic damage in vehicle treated animals involved brain regions supplied by the middle cerebral artery, namely the striatum and the frontoparietal cortex; penumbral regions corresponding to the medial striatum and the motor cortex were rescued as a result of MS treatment (Fig. 1C). Moreover, at the highest neuroprotective dose tested (150 mg/kg), the drug did not significantly affect CBF assessed by laser-doppler flowmetry in the frontoparietal cortex (Fig. 1D).

Neuroprotection was only observed when MS was administered 15 min before MCAo. In fact, administration of the drug 1 h before the induction of ischemia or at the beginning of reperfusion resulted in an infarct volume which was not significantly different from vehicle treated animals (Fig. 2). This suggests that a limited time-window exists to provide reduction of brain infarct damage by GPx inhibition.

B. *In Vitro* Neuroprotection by MS is Mediated by Catalase

Electrophysiological experiments showed that when corticostriatal slices were bathed in an ischemic medium, there was a decrease in the field excitatory post synaptic potential (fEPSP) response that was dependent on the duration of the oxygen and glucose deprivation. The effects of different periods of OGD are shown in Fig. 3A: after 7 min of OGD the fEPSP depression was transient and fully reversible ($n = 4$) upon reperfusion with normal oxygenated ACSF; conversely, synaptic transmission failed to recover when 12 min OGD was applied ($n = 11$). A significant recovery of the fEPSP was observed (61.22 ± 1%, paired Student's *t*-test $P < 0.05$, Fig. 3B, $n = 11$) following superfusion with MS (1 mM), applied 15 min before and during OGD. Longer administration times or higher doses of MS were not able to ameliorate the percentage of fEPSP recovery (data not shown).

In order to confirm and extend previous observations (Geracitano *et al.*, 2005; Nisticò *et al.*, 2008), suggesting a potential involvement of the catalase pathway in mediating the neuroprotective effect of MS, we applied the catalase inhibitor 3-AT. The simultaneous administration of 3-AT (20 mM) with MS (1 mM) completely reverted the recovery promoted by the administration of MS alone ($n = 5$, Fig. 3C). To exclude the possibility that 3-AT could per se contribute to the suppression of synaptic transmission during OGD, we show full recovery (Fig. 3D,

---

administration produced a significant reduction of brain infarct damage in penumbral areas including the medial striatum and the motor cortex. (D) Regional CBF was measured by laser-doppler flowmetry over the ischemic parietal cortex. For each rat, CBF decreased to approximately 20% of the baseline value during the 2-h period of ischemia (MCAo) and recovered to baseline during reperfusion (R). There were no significant differences in regional CBF between vehicle-treated controls ($n = 3$) and MS (150 mg/kg)-treated animals ($n = 3$). Values are expressed as mean ± S.E.M. (See Color Insert.)

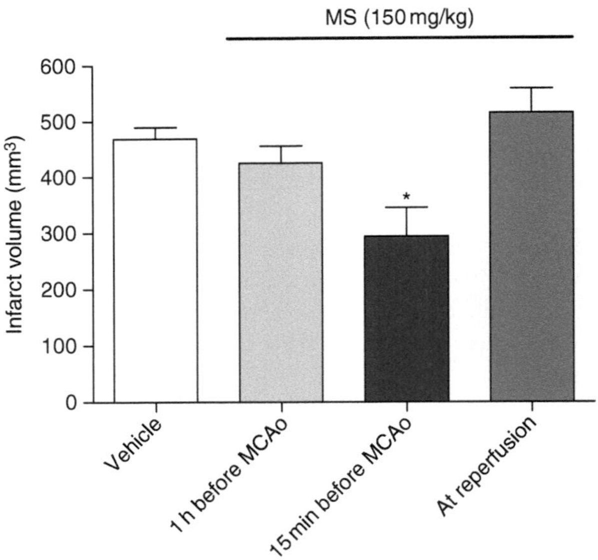

FIG. 2. Time-window for the neuroprotection afforded by MS against MCAo-induced brain damage. Brain infarct volume produced by transient (2 h) MCAo followed by 22 h of reperfusion in rats treated with MS (150 mg/kg) or vehicle (PBS, 1 ml/kg) i.p., 1 h or 15 min before the induction of ischemia, or upon reperfusion. Infarct volume was detected by staining consecutive 2-mm-thick coronal brain slices with TTC as described in the methods section. Data from vehicle-treated animals were pooled together, being the infarct volume values not significantly affected by the treatment schedule. Values are expressed as mean ± S.E.M.; * indicates $P < 0.05$ versus vehicle (one-way ANOVA followed by Dunnett's posttest; $n = 4$–6 rats per experimental group).

$n = 7$) of fEPSPs when 3-AT was superfused during 7 min OGD. This effect was similar to that obtained when 7 min OGD was applied alone (paired Student's *t*-test, $P = 0.22$).

IV. Discussion

The main finding of the present manuscript is that pharmacological inhibition of the GPx activity, reduces the extent of ischemic damage produced by transient MCAo in the rat brain and limits the irreversible functional derangement of field potentials in corticostriatal slices caused by a prolonged (12 min) oxygen-glucose deprivation.

Although there is a general agreement that an ischemic insult facilitates an excessive generation of hydroxyl radicals and, therefore, GPx plays an important role in the defense against H_2O_2-induced damage (Crack *et al.*, 2001; Hoehn *et al.*,

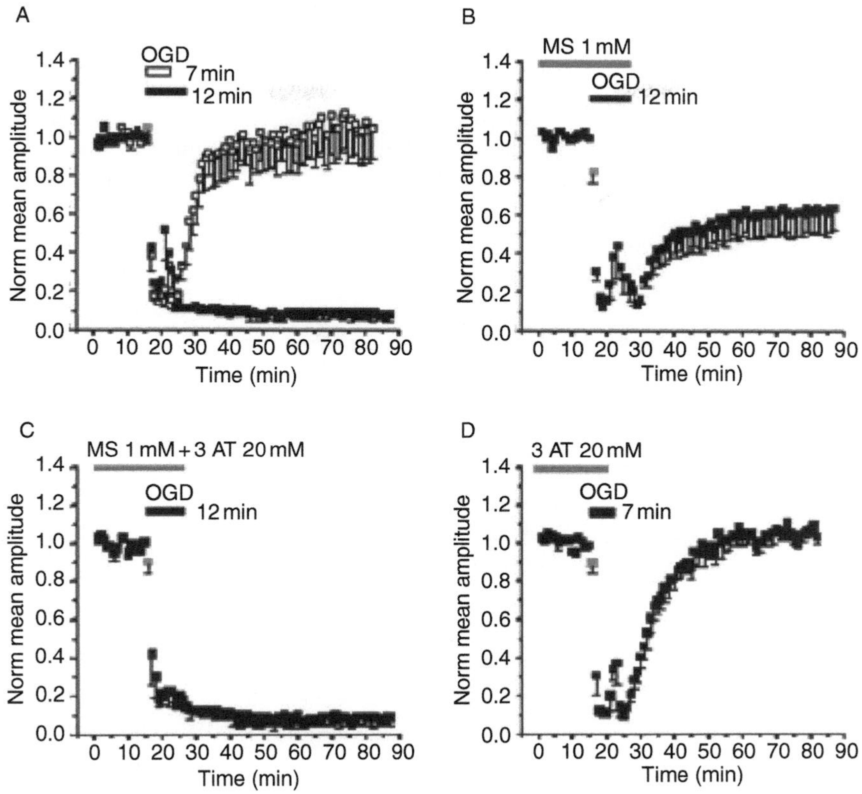

FIG. 3. (A) Exposure to 7 min OGD ($n = 4$, white squares) causes a transient fEPSP depression that is always reversible upon reoxygenation, whereas exposure to 12 min OGD (black squares) causes an irreversible loss of the fEPSP ($n = 1$). (B) Treatment with MS (1 mM), 15 min before and during OGD, protects synaptic responses from the fatal insult ($n = 11$). (C) Administration of 3-AT (20 mM, $n = 5$) reverts the neuroprotection by MS; while (D) it was ineffective on synaptic responses per se ($n = 7$). Bars indicate the time duration of the drug treatment and of the OGD.

2003; Sheldon *et al.*, 2008), there are also experimental data suggesting that an increased production of H_2O_2 may, instead, represent a critical component of the neuroprotective processes that occur during or after an ischemic episode (Fowler, 1997; Geracitano *et al.*, 2005). Here, we have tested the hypothesis that the pharmacological blockade of GPx activity, by reducing the degradation of H_2O_2, might provide more substrate for the action of catalase in converting H_2O_2 into H_2O and O_2. Thus, by using MS, a potent and specific inhibitor of selenium-dependent GPx (Chaudiere *et al.*, 1984), we have observed that this compound protects the brain from the ischemic damage caused under either *in vivo* or *in vitro* experimental conditions.

Systemic administration of MS resulted in a significant reduction of brain infarct volume produced by transient MCAo. The effect of the drug was dose-dependent, although limited to a specific time-window before (15 min) the ischemic event. Since GPx represents a crucial enzyme involved in H_2O_2 elimination in the brain, the net effect of its inhibition by MS is accumulation of H_2O_2 (Warner *et al.*, 2004). Previous studies have suggested that H_2O_2 may reduce release of neurotransmitters, including glutamate, dopamine, and γ-aminobutyric acid through activation of ATP-sensitive K^+ channels (Avshalumov and Rice, 2003; Chen *et al.*, 2001). Although this mechanism has been suggested to underlie the beneficial effects of H_2O_2 during metabolic stress, under our experimental conditions neuroprotection by MS appears to be due to an increased formation of H_2O_2 and, most notably, to its subsequent transformation into O_2 and H_2O by catalase. This hypothesis is supported by the loss of the MS-induced protective effects on the ischemic derangement of the field potentials following a pretreatment with the catalase inhibitor 3-AT.

Accordingly, we have already shown in a model of *in vitro* ischemia (OGD) that a positive modulation of the catalase activity plays an essential role in the H_2O_2-dependent neuroprotection, supplying for the lack of O_2 that occurs in the tissue after an ischemic event (Geracitano *et al.*, 2005; Nisticò *et al.*, 2008).

Our results are in line with those obtained by Vanella *et al.* (1993), showing that a pretreatment with buthionine sulfoximine (BSO), a drug that inhibits GPx activity by reducing glutathione synthesis, prolongs the survival time of rats subjected to 20-min cerebral ischemia. At this point, we have to mention that our results suggest that an acute inhibition of GPx by MS is neuroprotective. In spite of this, data in the literature show that the deleterious effects on cellular metabolism and neuronal survival after a stroke episode, appear when this enzyme is chronically inactivated or genetically ablated (Crack *et al.*, 2001, 2006). Indeed, we have to consider the possibility that various changes in endogenous antioxidant enzymes might occur during cerebral ischemia and reperfusion (Kumari Naga *et al.*, 2007; Ter Horst *et al.*, 1994), giving rise to a complex scenario.

It is worth noting that neurodegenerative processes likely occur after chronic inactivation of GPx associated with ischemia; while under conditions of a transient GPx inhibition, as that produced in our experimental settings, the early toxic events leading to tissue damage after ischemia (Lipton, 1999), are probably inhibited. The lack of toxicity due to transient GPx inhibition is also suggested by the analysis of the field potential that was not modified by the application of MS, demonstrating that basal synaptic transmission and neuronal activity remains viable in spite of the pharmacological inhibition of this enzyme.

Therefore, our results suggest a pattern of an inverse relationship between the susceptibility of neurons to ischemia and the transient (beneficial) and chronic (detrimental) inhibition of GPx activity. Accordingly, the transient and rapid reduction of the activity of this antioxidant enzyme enhances H_2O_2 formation in the ischemia-reoxygenation phase (Lievre *et al.*, 2000; Vanden Hoek *et al.*, 1997), consequently the newly formed H_2O_2 acts as a substrate for catalase in producing the rescuing molecule O_2. On the other hand, a sustained (chronic) inactivation of GPx reduces cellular viability by reducing neuronal resistance to harmful events.

With regard to the protective effects observed under *in vivo* ischemia, the laser-doppler investigation showing no flow modification after treatment with MS rules out that rheological modifications are involved in the neuroprotective action of the drug. This is also confirmed in *in vitro* experiments where an *in situ* rescuing effect of this compound mediated by catalase was demonstrated.

In conclusion, our findings suggest a neuroprotective role for the acute GPx inhibition against transient brain ischemia and are consistent with an involvement of H_2O_2 formation, and its consequent conversion into O_2 by catalase, in mediating the beneficial effect of MS.

References

Anderson, W. W., and Collingridge, G. L. (2001). The LTP program: A data acquisition program for on-line analysis of long-term potentiation and other synaptic events. *J. Neurosci. Methods* **108,** 71–83.

Avshalumov, M. V., and Rice, M. E. (2003). Activation of ATP-sensitive K^+ (K_{ATP}) channels by H_2O_2 underlies glutamate-dependent inhibition of striatal dopamine release. *Proc. Natl. Acad. Sci. USA* **100,** 11729–11734.

Brannan, T. S., Maker, H. S., and Raes, I. P. (1981). Regional distribution of catalase in the adult rat brain. *J. Neurochem.* **36,** 307–309.

Calabresi, P., Maj, R., Pisani, A., Mercuri, N. B., and Bernardi, G. (1992). Long-term synaptic depression in the striatum: Physiological and pharmacological characterization. *J. Neurosci.* **12,** 4224–4233.

Chaudiere, J., Wilhelmsen, E. C., and Tappel, A. L. (1984). Mechanism of selenium-glutathione peroxidase and its inhibition by mercaptocarboxylic acids and other mercaptans. *J. Biol. Chem.* **259,** 1043–1050.

Chen, B. T., Avshalumov, M. V., and Rice, M. E. (2001). H_2O_2 is a novel, endogenous modulator of synaptic dopamine release. *J. Neurophysiol.* **85,** 2468–2476.

Crack, P. J., Taylor, J. M., Flentjar, N. J., de Haan, J., Hertzog, P., Iannello, R. C., and Kola, I. (2001). Increased infarct size and exacerbated apoptosis in the glutathione peroxidase-1 (Gpx-1) knockout mouse brain in response to ischemia/reperfusion injury. *J. Neurochem.* **78,** 1389–1399.

Crack, P. J., Taylor, J. M., Ali, U., Mansell, A., and Hertzog, P. J. (2006). Potential contribution of NF-kappaB in neuronal cell death in the glutathione peroxidase-1 knockout mouse in response to ischemia-reperfusion injury. *Stroke* **37,** 1533–1538.

De Marchesa, O., Guarnirei, M., and McKhann, G. (1974). Glutathione peroxidase levels in brain. *J. Neurochem.* **22,** 773–776.

Dirnagl, U., Iadecola, C., and Moskowitz, M. A. (1999). Pathobiology of ischemic stroke: An integrated view. *Trends Neurosci.* **22,** 391–397.

Fowler, J. C. (1997). Hydrogen peroxide opposes the hypoxic depression of evoked synaptic transmission in rat hippocampal slices. *Brain Res.* **766,** 255–258.

Fridovich, I. (1995). Superoxide radical and superoxide dismutases. *Annu. Rev. Biochem.* **64,** 97–112.

Geracitano, R., Paolucci, E., Prisco, S., Guatteo, E., Zona, C., Longone, P., Ammassari-Teule, M., Bernardi, G., Berretta, N., and Mercuri, N. B. (2003). Altered long-term corticostriatal synaptic plasticity in transgenic mice overexpressing human Cu/Zn superoxide dismutase (GLY (93)→ALA) mutation. *Neuroscience* **118,** 399–408.

Geracitano, R., Tozzi, A., Berretta, N., Florenzano, F., Guatteo, E., Viscomi, M. T., Chiolo, B., Molinari, M., Bernardi, G., and Mercuri, N. B. (2005). Protective role of hydrogen peroxide in oxygen-deprived dopaminergic neurones of the rat substantia nigra. *J. Physiol.* **568,** 97–110.

Halliwell, B. (1992). Reactive oxygen species and the central nervous system. *J. Neurochemistry* **59,** 1609–1623.

Hoehn, B., Yenari, M. A., Sapolsky, R. M., and Steinberg, G. K. (2003). Glutathione peroxidase overexpression inhibits cytochrome C release and proapoptotic mediators to protect neurons from experimental stroke. *Stroke* **34,** 2489–2494.

Kumari Naga, K., Panigrahi, M., and Prakash Babu, P. (2007). Changes in endogenous antioxidant enzymes during cerebral ischemia and reperfusion. *Neurol. Res.* **29,** 877–883.

Li, H., Colbourne, F., Sun, P., Zhao, Z., Buchan, A. M., and Iadecola, C. (2000). Caspase inhibitors reduce neuronal injury after focal but not global cerebral ischemia in rats. *Stroke* **31,** 176–182.

Lievre, V., Becuwe, P., Bianchi, A., Koziel, V., Franck, P., Schroeder, H., Nabet, P., Dauca, M., and Daval, J. L. (2000). Free radical production and changes in superoxide dismutases associated with hypoxia/reoxygenation-induced apoptosis of embryonic rat forebrain neurons in culture. *Free Radic. Biol. Med.* **29,** 1291–1301.

Lipton, P. (1999). Ischemic cell death in brain neurons. *Physiol. Rev.* **79,** 1431–1568.

Lo, E. H., Dalkara, T., and Moskowitz, M. A. (2003). Mechanisms, challenges and opportunities in stroke. *Nat. Rev. Neurosci.* **4,** 399–415.

Longa, E. Z., Weinstein, P. R., Carlson, S., and Cummins, R. (1989). Reversible middle cerebral artery occlusion without craniotomy in rats. *Stroke* **20,** 84–91.

Nisticò, R., Piccirilli, S., Cucchiaroni, M. L., Armogida, M., Guatteo, E., Giampà, C., Fusco, F. R., Bernardi, G., Nisticò, G., and Mercuri, N. B. (2008). Neuroprotective effect of hydrogen peroxide on an *in vitro* model of brain ischemia. *Br. J. Pharmacol.* **153,** 1022–1029.

Sheldon, R. A., Christen, S., and Ferriero, D. M. (2008). Genetic and pharmacologic manipulation of oxidative stress after neonatal hypoxia-ischemia. *Int. J. Dev. Neurosci.* **26,** 87–92.

Ter Horst, G. J., Knollema, S., Stuiver, B., Hom, H., Yoshimura, S., Ruiters, M. H., and Korf, J. (1994). Differential glutathione peroxidase mRNA up-regulations in rat forebrain areas after transient hypoxia-ischemia. *Ann. NY Acad. Sci.* **738,** 329–333.

Vanden Hoek, T. L., Li, C., Shao, Z., Schumacker, P. T., and Becker, L. B. (1997). Significant levels of oxidants are generated by isolated cardiomyocytes during ischemia prior to reperfusion. *J. Mol. Cell. Cardiol.* **29,** 2571–2583.

Vanella, A., Di Giacomo, C., Sorrenti, V., Russo, A., Castorina, C., Campisi, A., Renis, M., and Perez-Polo, J. R. (1993). Free radical scavenger depletion in post-ischemic reperfusion brain damage. *Neurochem. Res.* **18,** 1337–1340.

Walton, K., and Fulton, B. (1983). Hydrogen peroxide as a source of molecular oxygen for in vitro mammalian CNS preparations. *Brain Res.* **278,** 387–393.

Warner, D. S., Sheg, H., and Batinic-Haberle, I. (2004). Oxidants, antioxidants and the ischemic brain. *J. Exp. Biol.* **207,** 3221–3231.

Weisbrot-Lefkowitz, M., Reuhl, K., Perry, B., Chan, P. H., Inouye, M., and Mirochnitchenko, O. (1998). Overexpression of human glutathione peroxidase protects transgenic mice against focal cerebral ischemia/reperfusion damage. *Brain Res. Mol. Brain Res.* **53,** 333–338.

ROLE OF Akt AND ERK SIGNALING IN THE NEUROGENESIS FOLLOWING BRAIN ISCHEMIA

Norifumi Shioda,* Feng Han,† and Kohji Fukunaga*

*Department of Pharmacology, Graduate School of Pharmaceutical Sciences, Tohoku University, Sendai 980-8578, Japan

†Institute of Pharmacology & Toxicology and Biochemical Pharmaceutics, College of Pharmaceutical Sciences, Zhejiang University, Hangzhou 310058, China

Generation of the neural precursors persists throughout life in the forebrain subventricular zone (SVZ) and the hippocampal subgranular zone (SGZ) in rodent and human brains. In addition, newborn granule cells in the hippocampal DG are important for learning and memory formation. Brain injuries such as seizure and trauma could trigger the endogenous programs for neurogenesis in the adult brain. Although brain ischemia also stimulates the proliferation of neural progenitor cells in SVZ and SGZ, the most neural progenitor cells are dead within a few days after generation. In addition, there is no therapeutic agent to promote the neurogenesis following brain injury in the adult brain. We found that intraperitoneal administration of vanadium compounds, a stimulator of phosphatidylinositol 3-kinase (PI3K)/Akt and extracellular signal-regulated kinase (ERK) pathways markedly enhances the brain ischemia-induced neurogenesis and promotes the migration of newborn cells. Thus, vanadium compounds are potential therapeutic agents to enhance the ischemia-induced neurogenesis through PI3K/Akt and ERK activation.

INTERNATIONAL REVIEW OF NEUROBIOLOGY, VOL. 85
DOI: 10.1016/S0074-7742(09)85026-5

0074-7742/09 $35.00

Abbreviations

BDNF, brain-derived neurotrophic factor; BrdU, bromodeoxyuridine; CREB, cyclic AMP-responsive element binding protein; DCX, doublecortin; DG, dentate gyrus; EGF, epidermal growth factor; ERK, extracellular signal regulated kinase; FGF, fibroblast growth factor; GCL, granular cell layer; G-CSF, granulocyte colony-stimulating factor; GSK-3β, glycogen synthesis kinase 3β; HB-EGF, heparin-binding epidermal growth factor-like growth factor; HIF-1α, hypoxia-inducible factor-1α; IGF-1, insulin-like growth factor-1; MSCs, marrow stromal cells; MEK, mitogen-activated protein kinase/ERK kinase; MMP, matrix metalloproteinase; NGF, nerve growth factor; NSPCs, neural stem/progenitor cells; PI3K, phosphatidylinositol 3-kinase; RMS, rostral migratory stream; SGZ, subgranular zone; SVZ, subventricular zone; VEGF, vascular endothelial growth factor; VO(OPT), bis(1-oxy-2-pyridinethiolato)oxovanadium(IV).

I. Introduction

The ability to generate the neural progenitor cells persists throughout life in the forebrain subventricular zone (SVZ) and hippocampal subgranular zone (SGZ) in rodent and human brains (Altman and Das, 1965; Eriksson *et al.*, 1998; Gage *et al.*, 1998; Gould *et al.*, 1999a). Neural progenitor cells form neuroblasts that migrate from the SVZ to the olfactory bulb via tangential chain migration (Lois *et al.*, 1996) in a restricted forebrain pathway named as the rostral migratory stream (RMS). Once the neuroblasts reach the subependymal region of the olfactory bulb, they disperse radially and differentiate into the granular and periglomerular neurons (Lois and Alvarez-Buylla, 1994; Lois *et al.*, 1996; Luskin, 1993; Thomas *et al.*, 1996). The neuronal turnover in the granule cell layer (GCL) of the adult dentate gyrus (DG) also maintains the hippocampal function such as the hippocampus-dependent memory formation (Gould *et al.*, 1999b; Scharfman *et al.*, 2000). Notably, the impaired neurogenesis in the SGZ partly accounts for depressive behaviors and epileptogenesis. For example, suppression of neurogenesis in the DG by irradiation impairs the hippocampus-dependent learning and memory formation in the adult rats (Madsen *et al.*, 2003) and mice (Rola *et al.*, 2004). Similarly, the suppression of neurogenesis by irradiation results in a worse outcome in the water-maze learning following cerebral ischemia (Raber *et al.*, 2004). Furthermore, exposure to chronic stresses and aging causes a reduction in cell proliferation in SGZ (Cameron and McKay, 1999; Gould *et al.*, 1998). By contrast, chronic antidepressant treatments increase cell proliferation in SGZ. For example, chronic fluoxetine administration increases the proliferation of the hippocampal progenitors in mice (Encinas *et al.*, 2006). In addition, chronic administration of fluoxetine or rolipram enhances the survival of newborn neurons. Acute treatments with 5-HT_4 receptor agonists, which are potential antidepressants, also promote the neurogenesis. Likewise, antidepressants are able to prevent or reverse the stress-induced decrease in neurogenesis (Warner-Schmidt and Duman, 2007).

Interestingly, brain injury triggers an endogenous program for neurogenesis. Indeed, seizures (Bengzon *et al.*, 1997; Parent *et al.*, 1997) or trauma (Dash *et al.*, 2001) triggers the hippocampal progenitor cell proliferation. Brain ischemia promotes proliferation of the hippocampal neural progenitor cells both in global (Kee *et al.*, 2001; Liu *et al.*, 1998; Yagita *et al.*, 2001) and focal brain ischemia (Gu *et al.*, 2000; Jin *et al.*, 2001; Takasawa *et al.*, 2002; Zhang *et al.*, 2001). Although increased proliferation (Gu *et al.*, 2000; Jin *et al.*, 2001; Zhang *et al.*, 2001; Takasawa *et al.*, 2002) and migration (Arvidsson *et al.*, 2002; Parent *et al.*, 2002) of newborn neurons have been demonstrated in cerebral ischemia, it is unclear the mechanisms of neurogenesis, and whether both proliferation and migration contribute to the recovery of neurological functions in the injured brain. Furthermore, some of the newly generated neurons survive up to 5 weeks after cerebral ischemia, the majority of them fail to survive for the long period (Arvidsson *et al.*, 2002; Parent *et al.*, 2002). Thus, further extensive studies are required to evaluate neurological functions of the adult neurogenesis following brain ischemia.

II. Stimulation of Endogenous Neural Progenitor Proliferation by Neurotrophic Factors in the Hippocampus

One possible mechanism underlying brain ischemia-induced neurogenesis involves stimulation of receptor tyrosine kinases through induction of growth factors such as fibroblast growth factor (FGF), brain-derived neurotrophic factor (BDNF), and nerve growth factor (NGF). Expression of basic FGF and FGF receptors increase in the DG after transient forebrain ischemia concomitant with increased proliferation of neural progenitors (Endoh *et al.*, 1994). BDNF and NGF are also induced after cerebral ischemia (Lindvall *et al.*, 1992). The proliferation of neural stem cells and differentiation to mature neurons are augmented by intraventricular administration of growth factors such as epidermal growth factor (EGF) (Craig *et al.*, 1996), FGF-2 (Kuhn *et al.*, 1997), or BDNF (Pencea *et al.*, 2001) in normal rodent brain. Intraperitoneal administration of insulin growth factor-1 (IGF-1) increases the hippocampal neurogenesis in normal rodent brain (Aberg *et al.*, 2000, 2003). Likewise, in brain ischemia, intraventricular administration both of FGF-2 and EGF (Nakatomi *et al.*, 2002) or EGF alone (Teramoto *et al.*, 2003) increases the number of hippocampal neural progenitor cells. Nakatomi *et al.* (2002) also reported that growth factor treatment stimulates neurogenesis in the hippocampal CA1 region, thereby ameliorating the electrophysiological disturbances seen in the hippocampus and behavioral deficits following rat ischemic injury. Wang *et al.* (2007) documented that overexpression of vascular endothelial growth factor (VEGF) not only reduces the infarct size but also increases the proliferation of neural precursors in the SVZ using a cerebral ischemic model of VEGF transgenic mice. Although these studies strongly suggest that stimulation of receptor tyrosine kinases by growth factors promotes the hippocampal neurogenesis, the molecular mechanism underlying growth factor-induced neurogenesis remains unclear.

III. Transplantation of Neural Stem Cells and Gene Therapy in the Brain Ischemia

Neural transplantation represents an attractive therapy to ameliorate neurological impairments after brain ischemia. Transplantation of bone marrow cells (Chen *et al.*, 2001a), fetal neural stem/progenitor cells (NSPCs) (Ishibashi *et al.*, 2004), umbilical cord blood cells (Chen *et al.*, 2001b), or embryonic stem cells (Hoehn *et al.*, 2002; Wei *et al.*, 2005; Hayashi *et al.*, 2006) had been used for therapy after ischemic tissue damage in animal models. When bone marrow stromal cells (MSCs) were administered intravenously after brain ischemia, the cells migrated selectively and targeted to damaged brain regions, thereby stimulating angiogenesis. The MSC-induced angiogenesis in turn enhances the neural stem cell proliferation and their migration, thereby promoting functional recovery from brain injury (Chen *et al.*, 2003; Li *et al.*, 2002). The molecular mechanism underlying the MSC effect on neural cell proliferation, migration, and angiogenesis involves an increase in expression levels of IGF-1 and IGF-1 receptor (Zhang *et al.*, 2004). In addition, the administration of granulocyte colony-stimulating factor (G-CSF) also improved angiogenesis, thereby promoting survival of the neural progenitor cells after brain ischemia (Sehara *et al.*, 2007).

On the other hand, in gene therapy, adenovirus-mediated gene transfer of heparin-binding epidermal growth factor-like growth factor (HB-EGF) enhances neurogenesis and angiogenesis after focal cerebral ischemia in rats (Sugiura *et al.*, 2005). In future studies, some of these approaches are applicable for behavioral improvements of the neurological deficits following brain injuries.

IV. Cell Signaling to Promote Neurogenesis in the Adult Brain

The brain ischemia-induced neurogenesis involves stimulation of receptor tyrosine kinases by induction of growth factors including FGF-2, EGF, and HB-EGF, thereby stimulating phosphatidylinositol 3-kinase (PI3K)/Akt and extracellular signal-regulated kinase (ERK) pathways. Both Akt and ERK signaling in neural precursors are crucial for the stimulation of adult neurogenesis. For example, carbachol-induced activation of both PI3K/Akt and mitogen-activated protein kinase/ERK kinase (MEK) pathways via muscarinic receptors stimulates DNA synthesis in basic FGF-treated neural progenitors isolated from rat cortical neuroepithelium. PI3K inhibitors (LY294002 and wortmannin) and MEK inhibitor PD98059 inhibit carbachol-induced increases in DNA synthesis in the progenitor cells (Li *et al.*, 2001). HB-EGF also enhances Akt and ERK phosphorylation and PI3K and MEK inhibitors (PD98059 and U0126) reduce HB-EGF-induced bromodeoxyuridine (BrdU) incorporation into cultured cortical progenitor cells (Jin *et al.*, 2005). Likewise in hypoxia/reoxygenation conditions, the proliferation of cultured mouse neural progenitors decreased by treatment with LY294002 and U0126. Thus, Akt and ERK pathways likely account for proliferation in neural progenitor cells (Sung *et al.*, 2007).

Importantly, chronic treatment with a mood stabilizer, sodium valproate, increases the numbers of BrdU positive cells via ERK activation in the DG of adult mice (Hao *et al.*, 2004). The migration of neuroblasts is also mediated by Akt and ERK signaling. Wang *et al.* (2006) reported that erythropoietin-activated endothelial cells enhance neural progenitor migration by secreting matrix metalloproteinase (MMP)-2 and MMP-9, in which PI3K/Akt and ERK signalings are required. In addition, BrdU/doublecortin (DCX) double positive cells from the SVZ express MMP-9 during the 2 weeks recovery period following transient focal cerebral ischemia in mice and MMP inhibitors significantly inhibit migration of neuroblasts (Lee *et al.*, 2006).

V. Vanadium Compounds are Attractive Therapeutics to Promote Neurogenesis in Neurodegenerative Disorders

Although these neurotrophic factors stimulate neurogenesis after brain ischemia, chronic intraventricular administration of peptides or proteins is inappropriate for human therapies because of insufficient penetration into brain tissue. Since various receptor tyrosine kinases, including EGF, FGF-2, BDNF, and VEGF receptors, are negatively regulated by protein tyrosine phosphatases (Ostman and Bohmer, 2001), inhibitors for protein tyrosine phosphatases possibly promote the receptor tyrosine kinases, which are critical for survival and differentiation of neural progenitor cells. Vanadate (V^{5+}) and vanadyl (V^{4+}) compounds strongly inhibit protein tyrosine phosphatases such as protein tyrosine phosphatase 1B (Lu *et al.*, 2001). We first reported neuroprotective effects of sodium orthovanadate in brain ischemia (Kawano *et al.*, 2001, 2002). Sodium orthovanadate treatment restored both Akt and ERK activities reduced following brain ischemia in gerbils. We also documented that sodium orthovanadate enhances the proliferation of progenitor cells in the adult rat SVZ after focal cerebral ischemia (Matsumoto *et al.*, 2006). Sodium orthovanadate treatment indeed increased both Akt and ERK activities in BrdU-positive neural progenitors (Matsumoto *et al.*, 2006). Sakurai *et al.* (2000) synthesized various vanadyl complexes as VO chelates and found that bis(1-oxy-2-pyridinethiolato)oxovanadium(IV) [VO(OPT)] has a strong insulin-mimetic activity *in vitro* in cultured adipose tissue and elicits insulin-mimetic action *in vivo* following intraperitoneal injection or oral administration. Vanadium compounds inhibit protein tyrosine phosphatase activity, thereby eliciting insulin-like actions through enhancement of tyrosine phosphorylation of insulin receptor β-subunit (Tamura *et al.*, 1984) and insulin receptor substrate (Pandey *et al.*, 1998) that in turn activates PI3K/Akt and ERK pathways. In addition to insulin receptors, pervanadate activates ERK-1/2 through EGF receptor activation by competitive inhibition of protein tyrosine phosphatase activity in HeLa cells (Zhao *et al.*, 1996). Inhibition of nonspecific protein tyrosine phosphatases by vanadium compounds likely leads to activation other tyrosine kinase receptors in addition to insulin and EGF receptors. Further extensive studies are required to define the most susceptible tyrosine kinase receptors for vanadium compounds in the hippocampal neural progenitor cells. Vanadyl (V^{4+}) is less toxic than vanadate (V^{5+}), the form found in sodium orthovanadate (Sakurai *et al.*, 1980, 1990; Nakai *et al.*, 1995). Furthermore, we demonstrated that VO(OPT) is safer than sodium orthovanadate in terms of adverse effects and LD_{50} in mice (Shioda *et al.*, 2007). In this context, we selected VO(OPT) as candidate for therapeutics in the neurogeneration therapy.

We found that intraperitoneal administration of VO(OPT) markedly enhances brain ischemia-induced neurogenesis in the SGZ of the mouse hippocampus (Fig. 1, Shioda *et al.*, 2008). The VO(OPT) treatment enhanced not only the number of proliferating cells but also migration of the neuroblasts. The VO (OPT)-induced neurogenesis was associated with increased Akt and ERK activities in the neural precursors in SGZ. Likewise, the VO(OPT)-induced neurogenesis was blocked by treatment with both PI3K/Akt and MEK/ERK inhibitors. Furthermore, the VO(OPT)-induced neurogenesis was associated with an amelioration of cognitive dysfunction following brain ischemia. As mechanism underlying VO(OPT)-induced neurogeneration, we defined that VO(OPT) treatment restores decreased phosphorylation of glycogen synthesis kinase 3β (GSK-3β) at

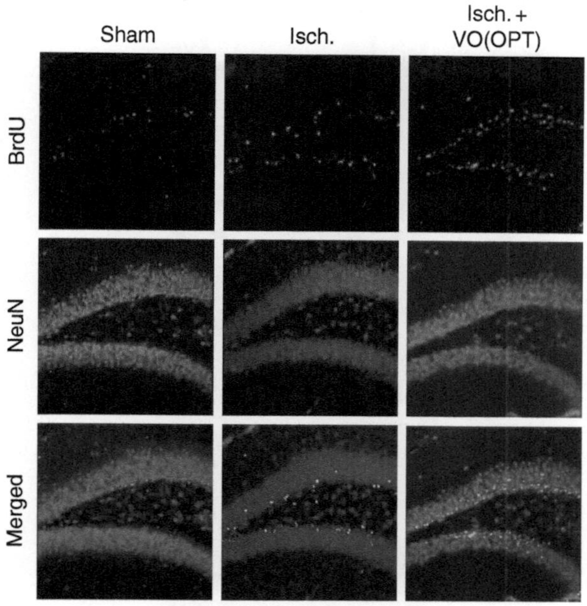

FIG. 1. VO(OPT) treatment promotes ischemia-induced neurogenesis in the hippocampal dentate gyrus. Confocal microscopy images of double staining for BrdU (green), NeuN (red), and merged images were shown. The number of BrdU positive neural progenitor cells increased in the dentate gyrus. The VO(OPT) treatment also stimulated migration of neural progenitor cells from the subgranular zone to the granular cell layer. (See Color Insert.)

Ser-9. GSK-3β inhibition by phosphorylation promotes cytoplasmic accumulation of β-catenin, thereby enhancing not only progenitor cell proliferation but also migration through interaction with cadherins (Fu *et al.*, 2006). The GSK-3β downstream target β-catenin is highly expressed in hippocampal neural precursors (Seki *et al.*, 2007). In this context, GSK-3β inhibition by phosphorylation at Ser-9 with VO(OPT) treatment likely upregulates β-catenin levels in the hippocampal neural precursors. The GSK-3β/β-catenin pathway regulates the proliferation and differentiation of neural progenitor cells (Hirabayashi and Gotoh, 2005). In addition to GSK-3β, hypoxia-inducible factor-1α (HIF-1α) and cyclic AMP-responsive element binding protein (CREB) are potential downstream targets for Akt and ERK signaling, respectively, both of which mediate proliferation, migration, and maturation of the neural progenitor cells in the hippocampus (Fig. 2). Taken together, VO(OPT) is potential therapeutics that promote ischemia-induced neurogenesis through Akt and ERK activation, thereby improving not only memory but also neurological deficits following brain ischemia.

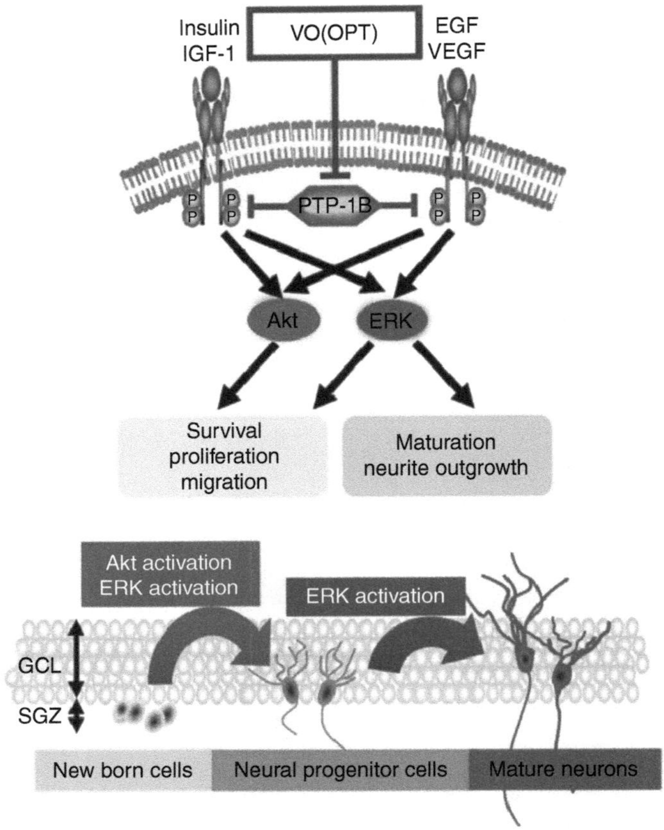

FIG. 2. Mechanisms of VO(OPT)-enhanced neurogenesis after brain ischemia in the hippocampal dentate gyrus. VO(OPT) inhibits protein tyrosine phosphatase 1B, thereby stimulating Akt and ERK signaling. The downstream targets of Akt and ERK include HIF-1α, GSK-3β, and CREB, which are involved in the proliferation, migration, and maturation of the neural progenitor cells. (See Color Insert.)

VI. Conclusion

Adult neurogenesis encourages the development of new strategies to restore neurons in neurodegenerative diseases including brain ischemia. The strategy is also attractive to ameliorate neurological deficits following brain ischemia. We first introduced VO(OPT) as seed compound to promote ischemia-induced neurogenesis and progenitor cell survival through PI3K/Akt and ERK activation, thereby improving not only memory but also neurological deficits following brain ischemia. In transplantation and gene therapy, the promotion of survival by VO

(OPT) is also critical to increase the efficacy of the transplanted neuronal progenitor cells. To define the precise mechanism underlying the VO(OPT)-induced survival, differentiation and migration in the progenitor cells, further extensive studies are required for identification of each factor involved in survival, differentiation, and migration.

References

Aberg, M. A., Aberg, N. D., Hedbacker, H., Oscarsson, J., and Eriksson, P. S. (2000). Peripheral infusion of IGF-I selectively induces neurogenesis in the adult rat hippocampus. *J. Neurosci.* **20,** 2896–2903.

Aberg, M. A., Aberg, N. D., Palmer, T. D., Alborn, A., Carlsson-Skwirut, C., Bang, P., Rosengren, L. E., Olsson, T., Cage, F. H., and Eriksson, P. S. (2003). IGF-1 has a direct proliferative effect in adult hippocampal progenitor cells. *Mol. Cell. Neurosci.* **24,** 23–40.

Altman, J., and Das, G. D. (1965). Autoradiographic and histological evidence of postnatal hippocampal neurogenesis in rats. *J. Comp. Neurol.* **124,** 319–335.

Arvidsson, A., Collin, T., Kirik, D., Kokaia, Z., and Lindvall, O. (2002). Neuronal replacement from endogenous precursors in the adult brain after stroke. *Nat. Med.* **8,** 963–970.

Bengzon, J., Kokaia, Z., Elmer, E., Nanobashvili, A., Kokaia, M., and Lindvall, O. (1997). Apoptosis and proliferation of dentate gyrus neurons after single and intermittent limbic seizures. *Proc. Natl. Acad. Sci. USA* **94,** 10432–10437.

Cameron, H. A., and McKay, R. D. (1999). Restoring production of hippocampal neurons in old age. *Nat. Neurosci.* **2,** 894–897.

Chen, J., Li, Y., Wang, L., Zhang, Z., Lu, D., Lu, M., and Chopp, M. (2001a). Therapeutic benefit of intravenous administration of bone marrow stromal cells after cerebral ischemia in rats. *Stroke* **32,** 1005–1011.

Chen, J., Sanberg, P. R., Li, Y., Wang, L., Lu, M., Willing, A. E., Sanchez-Ramos, J., and Chopp, M. (2001b). Intravenous administration of human umbilical cord blood reduces behavioral deficits after stroke in rats. *Stroke* **32,** 2682–2688.

Chen, J., Zhang, Z. G., Li, Y., Wang, L., Xu, Y. X., Gautam, S. C., Lu, M., Zhu, Z., and Chopp, M. (2003). Intravenous administration of human bone marrow stromal cells induces angiogenesis in the ischemic boundary zone after stroke in rats. *Circ. Res.* **92,** 692–699.

Craig, C. G., Tropepe, V., Morshead, C. M., Reynolds, B. A., Weiss, S., and van der Kooy, D. (1996). *In vivo* growth factor expansion of endogenous subependymal neural precursor cell populations in the adult mouse brain. *J. Neurosci.* **16,** 2649–2658.

Dash, P. K., Mach, S. A., and Moore, A. N. (2001). Enhanced neurogenesis in the rodent hippocampus following traumatic brain injury. *J. Neurosci. Res.* **63,** 313–319.

Encinas, J. M. A., Vaahtokari, A., and Enikolopov, G. (2006). Fluoxetine targets early progenitor cells in the adult brain. *Proc. Natl. Acad. Sci. USA* **103,** 8233–8238.

Endoh, M., Pulsinelli, W. A., and Wagner, J. A. (1994). Transient global ischemia induces dynamic changes in the expression of bFGF and FGF receptor. *Mol. Brain Res.* **22,** 76–88.

Eriksson, P. S., Perfilieva, E., Bjork-Eriksson, T., Alborn, A. M., Nordborg, C., Peterson, D. A., and Gage, F. H. (1998). Neurogenesis in the adult human hippocampus. *Nat. Med.* **4,** 1313–1317.

Fu, X., Sun, H., Klein, W. H., and Mu, X. (2006). Beta-catenin is essential for lamination but not neurogenesis in mouse retinal development. *Dev. Biol.* **299,** 424–437.

Gage, F. H., Kempermann, G., Palmer, T. D., Peterson, D. A., and Ray, J. (1998). Multipotent progenitor cells in the adult dentate gyrus. *J. Neurobiol.* **36,** 249–266.

Gould, E., Tanapat, P., McEwen, B. S., Flügge, G., and Fuchs, E. (1998). Proliferation of granule cell precursors in the dentate gyrus of adult monkeys is diminished by stress. *Proc. Natl. Acad. Sci. USA* **95,** 3168–3171.

Gould, E., Reeves, A. J., Fallah, M., Tanapat, P., Gross, C. G., and Fuchs, E. (1999a). Hippocampal neurogenesis in adult Old World primates. *Proc. Natl. Acad. Sci. USA* **96,** 5263–5267.

Gould, E., Beylin, A., Tanapat, P., Reeves, A., and Shors, T. J. (1999b). Learning enhances adult neurogenesis in the hippocampal formation. *Nat. Neurosci.* **2,** 260–265.

Gu, W., Brannstrom, T., and Wester, P. (2000). Cortical neurogenesis in adult rats after reversible photo thrombotic stroke. *J. Cereb. Blood Flow Metab.* **20,** 1166–1173.

Hao, Y., Creson, T., Zhang, L., Li, P., Du, F., Yuan, P., Gould, T. D., Manji, H. K., and Chen, G. (2004). Mood stabilizer valproate promotes ERK pathway-dependent cortical neuronal growth and neurogenesis. *J. Neurosci.* **24,** 6590–6599.

Hayashi, J., Takagi, Y., Fukuda, H., Imazato, T., Nishimura, M., Fujimoto, M., Takahashi, J., Hashimoto, N., and Nozaki, K. (2006). Primate embryonic stem cell-derived neuronal progenitors transplanted into ischemic brain. *J. Cereb. Blood Flow Metab.* **26,** 906–914.

Hirabayashi, Y., and Gotoh, Y. (2005). Stage-dependent fate determination of neural precursor cells in mouse forebrain. *Neurosci. Res.* **51,** 331–336.

Hoehn, M., Küstermann, E., Blunk, J., Wiedermann, D., Trapp, T., Wecker, S., Föcking, M., Arnold, H., Hescheler, J., Fleischmann, B. K., Schwindt, W., and Bührle, C. (2002). Monitoring of implanted stem cell migration *in vivo*: A highly resolved *in vivo* magnetic resonance imaging investigation of experimental stroke in rat. *Proc. Natl. Acad. Sci. USA* **99,** 16267–16272.

Ishibashi, S., Sakaguchi, M., Kuroiwa, T., Yamasaki, M., Kanemura, Y., Shizuko, I., Shimazaki, T., Onodera, M., Okano, H., and Mizusawa, H. (2004). Human neural stem/progenitor cells, expanded in long-term neurosphere culture, promote functional recovery after focal ischemia in Mongolian gerbils. *J. Neurosci. Res.* **78,** 215–223.

Jin, K., Minami, M., Lan, J. Q., Mao, X. O., Batteur, S., Simon, R. P., and Greenberg, D. A. (2001). Neurogenesis in dentate subgranular zone and rostral subventricular zone after focal cerebral ischemia in the rat. *Proc. Natl. Acad. Sci. USA* **98,** 4710–4715.

Jin, K., Mao, X. O., Del Rio Guerra, G., Jin, L., and Greenberg, D. A. (2005). Heparin-binding epidermal growth factor-like growth factor stimulates cell proliferation in cerebral cortical cultures through phosphatidylinositol 3′-kinase and mitogen-activated protein kinase. *J. Neurosci. Res.* **81,** 497–505.

Kawano, T., Fukunaga, K., Takeuchi, Y., Morioka, M., Yano, S., Hamada, J., Ushio, Y., and Miyamoto, E. (2001). Neuroprotective effect of sodium orthovanadate on delayed neuronal death after transient forebrain ischemia in gerbil hippocampus. *J. Cereb. Blood Flow Metab.* **21,** 1268–1280.

Kawano, T., Morioka, M., Yano, S., Hamada, J., Ushio, Y., Miyamoto, E., and Fukunaga, K. (2002). Decreased Akt activity is associated with activation of Forkhead transcription factor after transient forebrain ischemia in gerbil hippocampus. *J. Cereb. Blood Flow Metab.* **22,** 926–934.

Kee, N. J., Preston, E., and Wojtowicz, J. M. (2001). Enhanced neurogenesis after transient global ischemia in the dentate gyrus of the rat. *Exp. Brain Res.* **136,** 313–320.

Kuhn, H. G., Winkler, J., Kempermann, G., Thal, L. J., and Gage, F. H. (1997). Epidermal growth factor and fibroblast growth factor-2 have different effects on neural progenitors in the adult rat brain. *J. Neurosci.* **17,** 5820–5829.

Lee, S. R., Kim, H. Y., Rogowska, J., Zhao, B. Q., Bhide, P., Parent, J. M., and Lo, E. H. (2006). Involvement of matrix metalloproteinase in neuroblast cell migration from the subventricular zone after stroke. *J. Neurosci* **26,** 3491–3495.

Li, B. S., Ma, W., Zhang, L., Barker, J. L., Stenger, D. A., and Pant, H. C. (2001). Activation of phosphatidylinositol-3 kinase (PI-3K) and extracellular regulated kinases (Erk1/2) is involved in muscarinic receptor-mediated DNA synthesis in neural progenitor cells. *J. Neurosci.* **21,** 1569–1579.

Li, Y., Chen, J., Chen, X. G., Wang, L., Gautam, S. C., Xu, Y. X., Katakowski, M., Zhang, L. J., Lu, M., Janakiraman, N., and Chopp, M. (2002). Human marrow stromal cell therapy for stroke in rat: Neurotrophins and functional recovery. *Neurology* **59,** 514–523.

Lindvall, O., Ernfors, P., Bengzon, J., Kokaia, Z., Smith, M. L., Siesjo, B. K., and Persson, H. (1992). Differential regulation of mRNAs for nerve growth factor, brain-derived neurotrophic factor, and neurotrophin 3 in the adult rat brain following cerebral ischemia and hypoglycemic coma. *Proc. Natl. Acad. Sci. USA* **89,** 648–652.

Liu, J., Solway, K., Messing, R. O., and Sharp, F. R. (1998). Increased neurogenesis in the dentate gyrus after transient global ischemia in gerbils. *J. Neurosci.* **18,** 7768–7778.

Lois, C., and Alvarez-Buylla, A. (1994). Long-distance neuronal migration in the adult mammalian brain. *Science* **264,** 1145–1147.

Lois, C., Garcia-Verdugo, J. M., and Alvarez-Buylla, A. (1996). Chain migration of neuronal precursors. *Science* **271,** 978–981.

Lu, B., Ennis, D., Lai, R., Bogdanovic, E., Nikolov, R., Salamon, L., Fantus, C., Le-Tien, H., and Fantus, I. G. (2001). Enhanced sensitivity of insulin-resistant adipocytes to vanadate is associated with oxidative stress and decreased reduction of vanadate (+5) to vanadyl (+4). *J. Biol. Chem.* **276,** 35589–35598.

Luskin, M. B. (1993). Restricted proliferation and migration of postnatally generated neurons derived from the forebrain subventricular zone. *Neuron* **11,** 173–189.

Madsen, T. M., Kristjansen, P. E., Bolwig, T. G., and Wortwein, G. (2003). Arrested neuronal proliferation and impaired hippocampal function following fractionated brain irradiation in the adult rat. *Neuroscience* **119,** 635–642.

Matsumoto, J., Morioka, M., Hasegawa, Y., Kawano, T., Yoshinaga, Y., Maeda, T., Yano, S., Kai, Y., Fukunaga, K., and Kuratsu, J. I. (2006). Sodium orthovanadate enhances proliferation of progenitor cells in the adult rat subventricular zone after focal cerebral ischemia. *J. Pharmacol. Exp. Ther.* **318,** 982–991.

Nakai, M., Watanabe, H., Fujiwara, C., Kakegawa, H., Satoh, T., Takada, J., Matsushita, R., and Sakurai, H. (1995). Mechanism on insulin-like action of vanadyl sulfate: Studies on interaction between rat adipocytes and vanadium compounds. *Biol. Pharm. Bull.* **18,** 719.

Nakatomi, H., Kuriu, T., Okabe, S., Yamamoto, S., Hatano, O., Kawahara, N., Tamura, A., Kirino, T., and Nakafuku, M. (2002). Regeneration of hippocampal pyramidal neurons after ischemic brain injury by recruitment of endogenous neural progenitors. *Cell* **110,** 429–441.

Ostman, A., and Bohmer, F. D. (2001). Regulation of receptor tyrosine kinase signaling by protein tyrosine phosphatases. *Trends Cell Biol.* **11,** 258–266.

Pandey, S. K., Anand-Srivastava, M. B., and Srivastava, A. K. (1998). Vanadyl sulfate-stimulated glycogen synthesis is associated with activation of phosphatidylinositol 3-kinase and is independent of insulin receptor tyrosine phosphorylation. *Biochemistry* **37,** 7006–7014.

Parent, J. M., Yu, T. W., Leibowitz, R. T., Geschwind, D. H., Sloviter, R. S., and Lowenstein, D. H. (1997). Dentate granule cell neurogenesis is increased by seizures and contributes to aberrant network reorganization in the adult rat hippocampus. *J. Neurosci.* **17,** 3727–3738.

Parent, J. M., Vexler, Z. S., Gong, C., Derugin, N., and Ferriero, D. M. (2002). Rat forebrain neurogenesis and striatal neuron replacement after focal stroke. *Ann. Neurol.* **52,** 802–813.

Pencea, V., Bingaman, K. D., Wiegand, S. J., and Luskin, M. B. (2001). Infusion of brain-derived neurotrophic factor into the lateral ventricle of the adult rat leads to new neurons in the parenchyma of the striatum, septum, thalamus, and hypothalamus. *J. Neurosci.* **21,** 6706–6717.

Raber, J., Fan, Y., Matsumori, Y., Liu, Z., Weinstein, P. R., Fike, J. R., and Liu, J. (2004). Irradiation attenuates neurogenesis and exacerbates ischemia-induced deficits. *Ann. Neurol.* **55,** 381–389.

Rola, R., Raber, J., Rizk, A., Otsuka, S., VandenBerg, S. R., Morhardt, D. R., and Fike, J. R. (2004). Radiation-induced impairment of hippocampal neurogenesis is associated with cognitive deficits in young mice. *Exp. Neurol.* **188,** 316–330.

Sakurai, H., Tsuchiya, K., Nukatsuka, M., Sofue, M., and Kawada, J. (1990). Insulin-like effect of vanadyl ion on streptozotocin-induced diabetic rats. *J. Endocrinol.* **126,** 451.

Sakurai, H., Shimomura, S., Fukuzawa, K., and Ishizu, K. (1980). Detection of oxovanadium(IV) and characterization of its ligand environment in subcellular fractions of the liver of rats treated with pentavalent vanadium(V). *Biochem. Biophys. Res. Commun.* **96,** 293.

Sakurai, H., Sano, H., Takino, T., and Yasui, H. (2000). An orally active antidiabetic vanadyl complex, bis(1-oxy-2-pyridinethiolato)oxovanadium(IV), with VO(S2O2) coordination mode; *in vitro* and *in vivo* evaluations in rats. *J. Inorg. Biochem.* **80,** 99–105.

Scharfman, H. E., Goodman, J. H., and Sollas, A. L. (2000). Granule-like neurons at the hilar/CA3 border after status epilepticus and their synchrony with area CA3 pyramidal cells: Functional implications of seizure-induced neurogenesis. *J. Neurosci.* **20,** 6144–6158.

Seki, T., Namba, T., Mochizuki, H., and Onodera, M. (2007). Clustering, migration, and neurite formation of neural precursor cells in the adult rat hippocampus. *J. Comp. Neurol.* **502,** 275–290.

Sehara, Y., Hayashi, T., Deguchi, K., Zhang, H., Tsuchiya, A., Yamashita, T., Lukic, V., Nagai, M., Kamiya, T., and Abe, K. (2007). G-CSF enhances stem cell proliferation in rat hippocampus after transient middle cerebral artery occlusion. *Neurosci. Lett.* **418,** 248–252.

Shioda, N., Ishigami, T., Han, F., Moriguchi, S., Shibuya, M., Iwabuchi, Y., and Fukunaga, K. (2007). Activation of phosphatidylinositol 3-kinase/protein kinase B pathway by a vanadyl compound mediates its neuroprotective effect in mouse brain ischemia. *Neuroscience* **148,** 221–229.

Shioda, N., Han, F., Morioka, M., and Fukunaga, K. (2008). Bis(1-oxy-2-pyridinethiolato)oxovanadium(IV) enhances neurogenesis via phosphatidylinositol 3-kinase/Akt and extracellular signal regulated kinase activation in the hippocampal subgranular zone after mouse focal cerebral ischemia. *Neuroscience* **155,** 876–887.

Sugiura, S., Kitagawa, K., Tanaka, S., Todo, K., Omura-Matsuoka, E., Sasaki, T., Mabuchi, T., Matsushita, K., Yagita, Y., and Hori, M. (2005). Adenovirus-mediated gene transfer of heparin-binding epidermal growth factor-like growth factor enhances neurogenesis and angiogenesis after focal cerebral ischemia in rats. *Stroke* **36,** 859–864.

Sung, S. M., Jung, D. S., Kwon, C. H., Park, J. Y., Kang, S. K., and Kim, Y. K. (2007). Hypoxia/reoxygenation stimulates proliferation through PKC-dependent activation of ERK and Akt in mouse neural progenitor cells. *Neurochem. Res.* **32,** 1932–1939.

Takasawa, K., Kitagawa, K., Yagita, Y., Sasaki, T., Tanaka, S., Matsushita, K., Ohstuki, T., Miyata, T., Okano, H., Hori, M., and Matsumoto, M. (2002). Increased proliferation of neural progenitor cells but reduced survival of newborn cells in the contralateral hippocampus after focal cerebral ischemia in rats. *J. Cereb. Blood Flow Metab.* **22,** 299–307.

Tamura, S., Brown, T. A., Whipple, J. H., Fujita-Yamaguchi, Y., Dubler, R. E., Cheng, K., and Larner, J. (1984). A novel mechanism for the insulin-like effect of vanadate on glycogen synthase in rat adipocytes. *J. Biol. Chem.* **259,** 6650–6658.

Teramoto, T., Qiu, J., Plumier, J. C., and Moskowitz, M. A. (2003). EGF amplifies the replacement of parvalbumin-expressing striatal interneurons after ischemia. *J. Clin. Invest.* **111,** 1125–1132.

Thomas, L. B., Gates, M. A., and Steindler, D. A. (1996). Young neurons from the adult subependymal zone proliferate and migrate along an astrocyte, extracellular matrix-rich pathway. *Glia* **17,** 1–14.

Wang, L., Zhang, Z. G., Zhang, R. L., Gregg, S. R., Hozeska-Solgot, A., LeTourneau, Y., Wang, Y., and Chopp, M. (2006). Matrix metalloproteinase 2 (MMP2) and MMP9 secreted by erythropoietin-activated endothelial cells promote neural progenitor cell migration. *J. Neurosci.* **26,** 5996–6003.

Wang, Y., Jin, K., Mao, X. O., Xie, L., Banwait, S., Marti, H. H., and Greenberg, D. A. (2007). VEGF-overexpressing transgenic mice show enhanced post-ischemic neurogenesis and neuromigration. *J. Neurosci. Res.* **85,** 740–747.

Warner-Schmidt, J. L., and Duman, R. S. (2007). VEGF is an essential mediator of the neurogenic and behavioral actions of antidepressants. *Proc. Natl. Acad. Sci. USA* **104,** 4647–4652.

Wei, L., Cui, L., Snider, B. J., Rivkin, M., Yu, S. S., Lee, C. S., Adams, L. D., Gottlieb, D. I., Johnson, Jr., E. M., Yu, S. P., and Choi, D. W. (2005). Transplantation of embryonic stem cells overexpressing Bcl-2 promotes functional recovery after transient cerebral ischemia. *Neurobiol. Dis.* **19,** 183–193.

Yagita, Y., Kitagawa, K., Ohtsuki, T., Takasawa, K.-I., Miyata, T., Okano, H., Hori, M., and Matsumoto, M. (2001). Neurogenesis by progenitor cells in the ischemic adult rat hippocampus. *Stroke* **32,** 1890–1896.

Zhang, R. L., Zhang, Z. G., Zhang, L, and Chopp, M. (2001). Proliferation and differentiation of progenitor cells in the cortex and the subventricular zone in the adult rat after focal cerebral ischemia. *Neuroscience* **105,** 33–41.

Zhang, J., Li, Y., Chen, J., Yang, M., Katakowski, M., Lu, M., and Chopp, M. (2004). Expression of insulin-like growth factor 1 and receptor in ischemic rats treated with human marrow stromal cells. *Brain Res.* **1030,** 19–27.

Zhao, Z., Tan, Z., Diltz, C. D., You, M., and Fischer, E. H. (1996). Activation of mitogen-activated protein (MAP) kinase pathway by pervanadate, a potent inhibitor of tyrosine phosphatases. *J. Biol. Chem.* **271,** 22251–22255.

PREVENTION OF GLUTAMATE ACCUMULATION AND UPREGULATION OF PHOSPHO-AKT MAY ACCOUNT FOR NEUROPROTECTION AFFORDED BY BERGAMOT ESSENTIAL OIL AGAINST BRAIN INJURY INDUCED BY FOCAL CEREBRAL ISCHEMIA IN RAT

Diana Amantea,* Vincenza Fratto,† Simona Maida,† Domenicantonio Rotiroti,† Salvatore Ragusa,† Giuseppe Nappi,‡ Giacinto Bagetta,* and Maria Tiziana Corasaniti†

*Department of Pharmacobiology and Center of Neuropharmacology of Normal and Pathological Neuronal Plasticity, UCADH, University of Calabria, 87036 Cosenza, Italy

†Department of Pharmacobiological Sciences, University "Magna Graecia" of Catanzaro, 88100 Catanzaro, Italy

‡Chair of Neurology, University "La Sapienza" of 00161 Rome and IRCCS "C Mondino Institute of Neurology" Foundation, 27100 Pavia, Italy

The effects of bergamot essential oil (BEO; *Citrus bergamia*, Risso) on brain damage caused by permanent focal cerebral ischemia in rat were investigated. Administration of BEO (0.1–0.5 ml/kg but not 1 ml/kg, given intraperitoneally 1 h before occlusion of the middle cerebral artery, MCAo) significantly reduced infarct size after 24 h permanent MCAo. The most effective dose (0.5 ml/kg) resulted in a significant reduction of infarct extension throughout the brain, especially in the medial striatum and the motor cortex as revealed by TTC staining of tissue slices. Microdialysis experiments show that BEO (0.5 ml/kg) did not affect basal amino acid levels, whereas it significantly reduced excitatory amino acid, namely aspartate and glutamate, efflux in the frontoparietal cortex typically observed following

INTERNATIONAL REVIEW OF NEUROBIOLOGY, VOL. 85
DOI: 10.1016/S0074-7742(09)85027-7

0074-7742/09 $35.00

MCAo. Western blotting experiments demonstrated that these early effects were associated, 24 h after permanent MCAo, to a significant increase in the phosphorylation and activity of the prosurvival kinase, Akt. Indeed, BEO significantly enhanced the phosphorylation of the deleterious downstream kinase, GSK-3β, whose activity is negatively regulated via phosphorylation by Akt.

Abbreviations

BEO, bergamot essential oil; GSK-3β, glycogen synthase kinase-3β; MCAo, middle cerebral artery occlusion; PI3-K, phosphatidylinositol 3-kinase

I. Introduction

We recently demonstrated that bergamot essential oil (BEO) concentration dependently minimizes neuronal death induced *in vitro* by excessive stimulation of the *N*-methyl-D-aspartate (NMDA) subtype of glutamate receptor (Corasaniti *et al.*, 2007). In addition to preventing reactive oxygen species accumulation and activation of calpain induced by NMDA, neuroprotection afforded by BEO involves activation of cell survival signals. Specifically, BEO prevents the rapid deactivation of the serine/threonine protein kinase Akt induced by NMDA and the consequent, detrimental, activation of downstream kinase, that is glycogen synthase kinase-3β (GSK-3β) (Corasaniti *et al.*, 2007). These observations are of particular relevance as survival signalings engaged by activation of Akt kinase mediate endogenous prosurvival responses to neuronal stress induced by an ischemic insult (see Fukunaga and Kawano, 2003). Akt functions as a major downstream target of phosphatidylinositol 3-kinase (PI3-K); upon phosphorylation, activated Akt promotes cell survival by phosphorylating and inhibiting several proteins including Bad, caspase-9, and GSK-3β (see Brazil and Hemmings, 2001). Activation of the PI3-K/Akt pathway may rescue neurons of the ischemic penumbra from delayed cell death (see Fukunaga and Kawano, 2003; Zhao *et al.*, 2006). An upregulation of phospho-Akt (p-Akt; Ser473) has been reported in the ischemic penumbra at 1 h after permanent middle cerebral artery occlusion (MCAo) (Shibata *et al.*, 2002) and at 1.5–6 h after reperfusion in a transient model of MCAo (Noshita *et al.*, 2001; Shioda *et al.*, 2007; Zhao *et al.*, 2005); the increase in p-Akt levels in the ischemic penumbra is a transient event as a return toward basal levels has been detected at later time points (24–48 h) in both experimental models of brain ischemia (Noshita *et al.*, 2001; Shibata *et al.*,

2002; Shioda *et al.*, 2007; Zhao *et al.*, 2005). Accumulating evidence demonstrates that exogenous growth factors and a number of neuroprotectans, including estradiol and free radical scavengers, reduce ischemic brain damage and upregulate Akt phosphorylation in both transient and permanent models of MCAo (Amantea *et al.*, 2005; Zhao *et al.*, 2006). Interestingly, we demonstrated that BEO activates the PI3-K/Akt survival cascade in neuronal cells undergone trophic factor deprivation (Corasaniti *et al.*, 2007).

These observations led us to hypothesize that administration of BEO could have a neuroprotective effect against ischemic neuronal injury and that neuroprotection by BEO may stem from maintenance in the active state of endogenous survival programs only transiently activated in the ischemic penumbra. Given that BEO for systemic (intraperitoneal, i.p.) administration crosses the blood–brain barrier (BBB) in rat (Morrone *et al.*, 2007), we considered this of special interest in view of the poor penetration into the brain of a number of neuroprotectants (Wu, 2005).

Interestingly, there is evidence that i.p. injection of BEO transiently affects extracellular levels of neurotransmitter amino acids in the hippocampus of rats thus suggesting that as yet unidentified components of BEO reach the brain where they may interfere with mechanisms relevant to both synaptic plasticity and neurodegeneration (Morrone *et al.*, 2007).

Here we investigate the neuroprotective potential of BEO *in vivo* by using an experimental model of focal brain ischemia in rats. Moreover, we evaluate whether neuroprotection is associated with altered levels of excitatory neurotransmitters and with a modulation of PI3-K/Akt pathway in the ischemic cortex. Our results indicate that BEO indeed protects against ischemic injury in an *in vivo* model of permanent focal brain ischemia in rat and that neuroprotection is associated with reduced excitatory amino acid efflux induced by MCAo in the ischemic cortex and elevation of p-Akt and phospho-GSK-3β (p-GSK-3β) levels in the ischemic penumbra. A preliminary account of this *in vivo* study has been communicated to the British Pharmacologic Society (Morrone *et al.*, 2006).

II. Materials and Methods

A. Animals

Male Wistar rats (250–300 g; Charles River, Calco, Italy) were used. Animals were housed at constant temperature (22 ± 1 °C) and relative humidity (50%) under a regular light-dark schedule (lights on 7 a.m. to 7 p.m.). Food and water were freely available. All experiments were carried out in accordance with the European Community Council Directive of 24 November 1986 (86/609/EEC). All efforts were made to minimize animal suffering and to use only the number of animals necessary to produce reliable results.

B. Permanent MCAo in Rat

Focal brain ischemia was induced in rats by MCAo with an intraluminal filament, using a relatively noninvasive technique previously described by Longa *et al.* (1989). Male Wistar rats (280–320 g body weight), housed in a temperature (22 °C)- and humidity (65%)-controlled colony room, were anesthetized with 5% isoflurane in air and maintained with the lowest acceptable concentration of the anesthetic (1.5–2%) throughout the surgical procedure. Body temperature was measured via a rectal probe and kept at 37 °C with a heating pad. Under an operating microscope, the external and internal right carotid arteries were exposed through a neck incision. The external carotid artery was cut approximately 3 mm above the common carotid artery bifurcation and a silk suture was tied loosely around the external carotid stump. A silicone-coated nylon filament (0.28 mm diameter) was then inserted into the external carotid artery and gently advanced into the internal carotid artery, approximately 18 mm from the carotid bifurcation, until mild resistance was felt, thereby indicating occlusion of the origin of the middle cerebral artery in the Willis circle. The silk suture was tightened around the intraluminal filament to prevent bleeding. The wound was then sutured and anesthesia discontinued. Sham-operated animals underwent the same surgical procedure without insertion of the silicone-coated nylon filament into the internal carotid artery. Rats received i.p. administration of 0.05, 0.1, 0.5, and 1.0 ml/kg of BEO 1 h before permanent MCAo. BEO at 0.5 and 1.0 ml/kg was administered undiluted. Lower doses of BEO (0.05 and 0.1 ml/kg) were administered after diluting the essential oil 1:10 (for the dose 0.05 ml/kg) or 1:5 (for the dose 0.1 ml/kg) in vegetable oil; vehicle-treated animals received, by the same route, 0.45 ml/kg of vegetable oil, corresponding to the maximal volume of vehicle used, but this had no measurable effect (data not shown). Because BEO at 0.5 and 1.0 ml/kg was administered as an undiluted solution, control animals received by the same route an equal volume of saline 1 h before MCAo. No effect of vegetable oil or of saline on infarct size after permanent MCAo was observed (data not shown). The experimental protocol was in accordance to the guidelines of the Italian Ministry of Health for animal care (D.M. 116/1992).

C. Neuropathology and Quantification of Ischemic Damage

Rats were sacrificed by decapitation 24 h following MCA occlusion. The brains were rapidly removed and eight consecutive coronal sections, 2 mm thick, from each brain were cut starting from the frontal pole using a rat brain matrix. To measure ischemic damage, brain slices were stained in a solution containing 2% 2,3,5-triphenyltetrazolium chloride (TTC) in saline, at 37 °C. After 10 min incubation, the slices were transferred to 10% neutral buffered formaldehyde and stored at 4 °C prior to analysis. Images of TTC-stained sections were captured using a digital scanner and analyzed using an image analysis software (ImageJ 1.30v, Research Services Branch, National Institute of Mental Health, Bethesda, MD, USA). The infarct volume (mm^3) was calculated by summing the infarcted area (unstained) of the eight sections and multiplying by the interval thickness between sections (Li *et al.*, 2000).

D. *In Vivo* Microdialysis

Rats were anesthetized with chloral hydrate (0.4 g/kg i.p.) and a microdialysis probe was implanted in the upper frontoparietal cortex (ischemic penumbra). Twenty-four hours after surgery, animals were reanesthetized with urethane (1.5 g/kg i.p.) and the microdialysis probe perfused with artificial CSF (mM: NaCl, 125; KCl, 2.5; $MgCl_2$, 1.18; $CaCl_2$, 1.26; NaH_2PO_4, 0.2; pH 7.0) at a flow rate of 2 ml/min. After 1-h stabilization period, dialysate samples were collected every 10 min for 1 h to establish basal amino acid levels. At this point, BEO or vehicle saline were administered i.p. and 10-min dialysate samples collected for a further 1 h. At the end of collection, pMCAo was performed and sample collection continued for further 3 h. Concentrations of glutamate, aspartate, glycine, GABA, glutamine, taurine, and citrulline were determined by HPLC with fluorescence detection after derivatization with *o*-phthaldialdehyde/mercaptopropionic acid (Morrone *et al.*, 2007; Richards *et al.*, 2000).

E. Preparation of Brain Tissue Homogenates

Individual cortical tissue samples (corresponding to ischemic penumbra) of sham-operated rats and of rats subjected to permanent focal cerebral ischemia were rapidly dissected out and homogenized in a glass homogenizer using 6 volumes of ice-cold lysis buffer containing 20 mM Tris–HCl, pH 7.5, 150 mM NaCl, 2 mM EDTA, 2 mM EGTA, 1% Triton, 1 nM okadaic acid, a cocktail of protease inhibitors (code P8340, Sigma, Milan, Italy) and a cocktail of phosphatase inhibitors (code 524625, Calbiochem, La Jolla, CA, USA); samples were then centrifuged at 20,800×*g* for 15 min at 4 °C. Protein concentration in the supernatant was determined by the DC protein assay (Bio-Rad Laboratories, Milan, Italy).

F. Western Blot Analysis

Proteins were resolved by sodium dodecyl sulfate-polyacrylamide gel electrophoresis (SDS-PAGE; 8%) and electrotransferred to nitrocellulose membranes (Optitran BA-S 83, Schleicher & Schuell Bioscence, Dassel, Germany). Primary antibodies were incubated overnight at 4 °C followed by a horseradish peroxidase-conjugated secondary antibody for 1 h at room temperature. Immunoreactivity was visualized by enhanced chemiluminescent detection (Amersham Biosciences, GE Healthcare, Milan, Italy) and exposure to X-ray films (Hyperfilm ECL, Amersham Biosciences). Autoradiographic films were scanned and densitometric analysis was carried out using Quantiscan software (Biosoft, Cambridge, UK). The following primary antibodies were used: a rabbit polyclonal antibody for Akt at 1:4000 dilution (Cell Signaling Technology, Beverly, MA, USA), a rabbit polyclonal antibody for p-Akt (Ser473) at 1:1000 dilution (Cell Signaling Technology), a rabbit polyclonal antibody against GSK-3β at 1:9000 dilution (Cell Signaling Technology), a rabbit polyclonal antibody for p-GSK-3β (Ser9) at 1:1000 dilutions (Cell Signaling Technology), a rabbit polyclonal antibody for phosphatase and tensin homolog deleted on chromosome 10 (PTEN) phosphorylated at Ser380 (1:1000 dilution; Cell Signaling Technology), a mouse monoclonal antiactin antibody at 1:2000 dilution (clone AC-40; Sigma), a mouse monoclonal anti-α-tubulin antibody at 1:30,000 dilution (clone B-5-1-2; Sigma). Horseradish peroxidase-conjugated goat antirabbit or antimouse IgG (Pierce Biotechnology, Rockford, IL, USA) were used as secondary antibodies.

G. Drugs

The essential oil of bergamot (BEO) was kindly provided by the company "Simone Gatto" (San Pier Niceto, Messina, Italy) together with the certificate of analysis carried out by the "Stazione Sperimentale per le Industrie delle Essenze e dei Derivati dagli Agrumi" (SSEA, Reggio Calabria, Italy). According to the percentages reported in the literature (Mondello *et al.*, 1995; Verzera *et al.*, 1996, 2003), among other substances present in lower percentages, BEO contained: 37.98% D-limonene, 30.02% linalyl acetate, 9.83% linalool, 7.17% γ-terpinene, and 6.15% β-pinene.

H. Statistical Analysis

Data are expressed as the mean ± S.E.M. of the indicated number of independent experiments and evaluated statistically for difference by Student's *t*-test or by one-way analysis of variance (ANOVA) followed by Tukey–Kramer or Dunnett test for multiple comparisons. A value of P less than 0.05 was considered to be significant.

III. Results

Intraperitoneal administration of BEO (0.5 ml/kg, given 1 h before MCAo) robustly reduced infarct size after permanent MCAo (Fig. 1). In fact, the ischemic zone, identified in the right cerebral hemisphere as a distinct pale-stained area in rats subjected to 24 h MCAo (Fig. 1A), was reduced by a pretreatment with BEO (Fig. 1B).

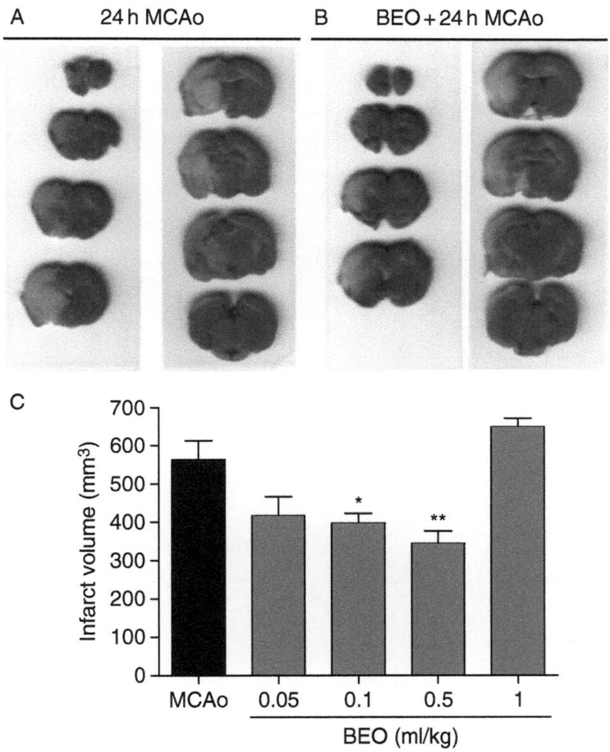

FIG. 1. BEO dose-dependently reduces infarct size after permanent MCAo. Representative images of brain sections from (A) rats (n = 5) sacrificed 24 h after permanent occlusion of middle cerebral artery (24 h MCAo) and (B) BEO-treated rats (n = 5) prior to MCAo; BEO (0.5 ml/kg) was administered i.p. 1 h before MCAo. Brain sections were stained by TTC; the ischemic region appears as a pale-stained area whereas the viable tissue is stained red. (C) Effects of different doses of BEO (0.05–1 ml/kg), administered i.p. 1 h before MCAo, on infarct volume; results are expressed as mean ± S.E.M. (n = 4–6 per group). * and **Denote $P < 0.05$ and $P < 0.01$ versus 24 h MCAo, respectively (ANOVA followed by Dunnett multiple comparisons test). (See Color Insert.)

The infarct volume was significantly ($P < 0.05$ and $P < 0.01$ vs. permanent MCAo, respectively) reduced by 0.1 and 0.5 ml/kg but not by a lower dose of BEO (0.05 ml/kg; Fig. 1C); at 1 ml/kg BEO caused a trend toward an increase of infarct volume as compared to both ischemic- (Fig. 1C) and vehicle (data not shown)-treated animals.

Microdialysis experiments show that BEO (0.5 ml/kg) did not affect basal amino acid levels, whereas it significantly reduced the efflux of excitatory amino acid, namely aspartate and glutamate, in the frontoparietal cortex typically observed following MCAo (Fig. 2). Extracellular levels of glycine, GABA,

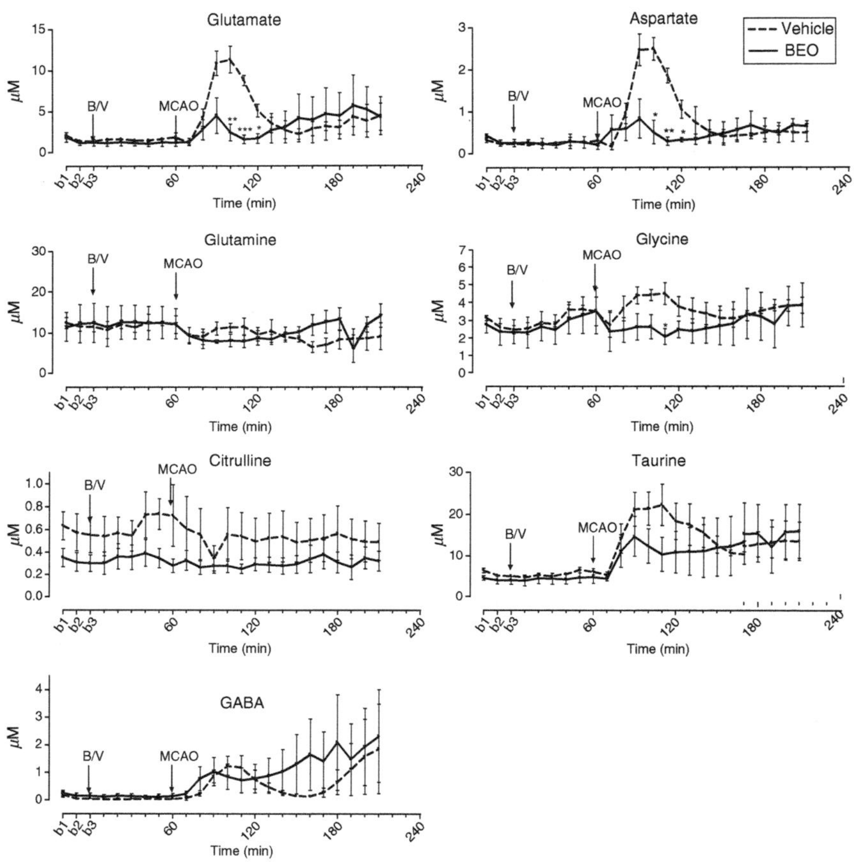

FIG. 2. BEO significantly reduces excitatory amino acid (Asp and Glu) efflux in the frontoparietal cortex typically observed following MCAo. Effect of BEO (B; solid line) or vehicle (V; dotted line) administration (B/V, 0.5 ml/kg, i.p., 1 h before MCAo) on dialysate levels of glutamate, aspartate, glutamine, glycine, citrulline, taurine, and GABA detected in the upper frontoparietal cortex following pMCAo. Data are mean ± S.E.M. (n = 3 per group) $^{*}P < 0.05$, $^{**}P < 0.01$, $^{***}P < 0.001$ versus vehicle, Student's *t*-test.

glutamine, taurine, and citrulline did not change significantly after ischemia nor they were affected by BEO (Fig. 2).

Western blot experiments, performed using brain cortical homogenates of rats sacrificed 24 h after ischemia, show an apparent decrease of p-Akt at Ser473 in the ipsilateral, ischemic, cortex as compared to the contralateral, nonischemic side (Fig. 3A) that, however, did not reach statistical significance (Fig. 3C). In no instance changes in p-Akt were detected in sham-operated animals (Fig. 3A and C). GSK-3β is a downstream kinase negatively regulated by Akt via phosphorylation at Ser9 (Cross *et al.*, 1995; van Weeren *et al.*, 1998). Under our experimental conditions, the apparent deactivation of Akt in the ischemic cortex was associated with a trend toward a reduction of p-GSK-3β (Ser9) (Fig. 3B) that, however, likewise p-Akt, did not reach statistical significance (Fig. 3B).

Next, we investigated whether neuroprotection afforded by BEO was associated with changes in the phosphorylation levels of Akt and GSK-3β. Intraperitoneal administration of BEO (0.5 ml/kg), 1 h before MCAo, significantly ($P <$ 0.05) increases p-Akt expression leaving unaffected the total level of this gene product in the ipsilateral cortex of rats subjected to 24 h occlusion of MCA (Fig. 4). Indeed, a trend toward a reduction of Akt protein expression was observed in ischemic animals and this trend was not affected in rats pretreated with BEO. Besides enhancing the levels of p-Akt (expressed as levels of p-Akt normalized to actin), BEO significantly enhanced ($P < 0.01$) Akt phosphorylation expressed as the ratio of p-Akt/Akt. Interestingly, BEO (0.5 ml/kg) significantly increased the expression of p-GSK-3β (as normalized to tubulin; $P < 0.05$ vs. contralateral side and $P < 0.01$ vs. ipsilateral side of ischemic, untreated animals) and the level of phosphorylation of GSK-3β (as estimated by the ratio of p-GSK-3β/GSK-3β; $P < 0.05$) and this was paralleled by a significant decrease ($P < 0.05$) of the total amount of the protein kinase (Fig. 5). In control animals, i.p. administration of an equal volume of saline or vegetable oil, given 1 h before permanent MCAo, did not affect Akt and GSK-3β phosphorylation in the ischemic cortex (data not shown).

Changes in the expression of p-Akt and p-GSK-3β by BEO are specific as the essential oil did not modify the levels of p-PTEN typically reduced by ischemia (Fig. 6).

IV. Discussion

The results of the present study show that systemic administration of BEO reduces the volume of infarct induced by MCAo in the brain of rat. Neuroprotection afforded by BEO is preceded by minimization of the MCAo-induced increase in excitatory amino acid efflux in the ischemic penumbra. Under ischemic

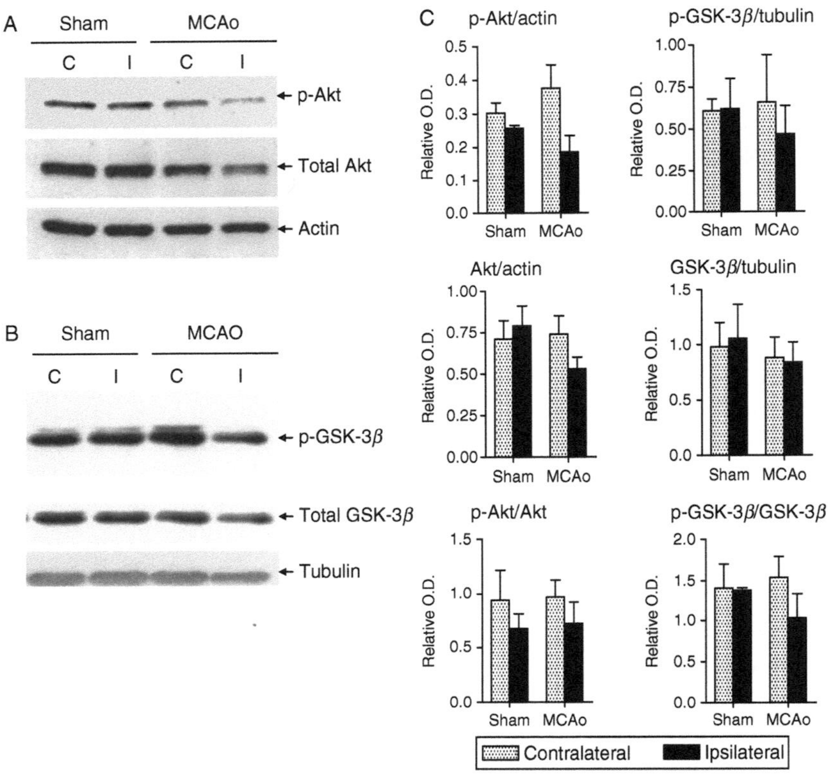

FIG. 3. Phosphorylation level of Akt and GSK-3β following focal cerebral ischemia. (A) Western blot analysis of phospho-Akt (Ser473) (p-Akt) and total Akt performed on brain cortical homogenates from rats sacrificed 24 h after MCAo shows a trend toward a decrease of p-Akt immunoreactivity in the ipsilateral (I), ischemic, cortex as compared to contralateral (C), nonischemic, side. This trend toward a reduction also occurred for total Akt expression, so that phosphorylation level (expressed as the ratio p-Akt/Akt) is not significantly affected by ischemia. The result is representative of three independent experiments. Histograms in (C) show the results of the densitometric analysis of the autoradioraphic bands corresponding to p-Akt, total Akt, and β-actin. p-Akt and Akt levels were normalized to the values yielded by β-actin and Akt phosphorylation was expressed by the ratio of p-Akt/total Akt; data are reported as mean ± S.E.M. ($n = 3$ per group). The same samples were used for the subsequent western blot analysis of phospho-GSK-3β (Ser9) (p-GSK-3β) and total GSK-3β and a representative result of three independent experiments is shown in (B). The results of the densitometric analysis of the bands corresponding to p-GSK-3β, total GSK-3β, and α-tubulin are reported in (C). p-GSK-3β and total GSK-3β were normalized to the values yielded by α-tubulin whereas GSK-3β phosphorylation was calculated from the ratio of p-GSK-3β/total GSK-3β; data are reported as mean ± S.E.M. ($n = 3$ per group).

conditions, altered ionic gradient across the anoxic nerve ending membrane together with the accumulation of reactive oxygen species may cause glutamate transporters to function in a reverse mode leading to accumulation of synaptic

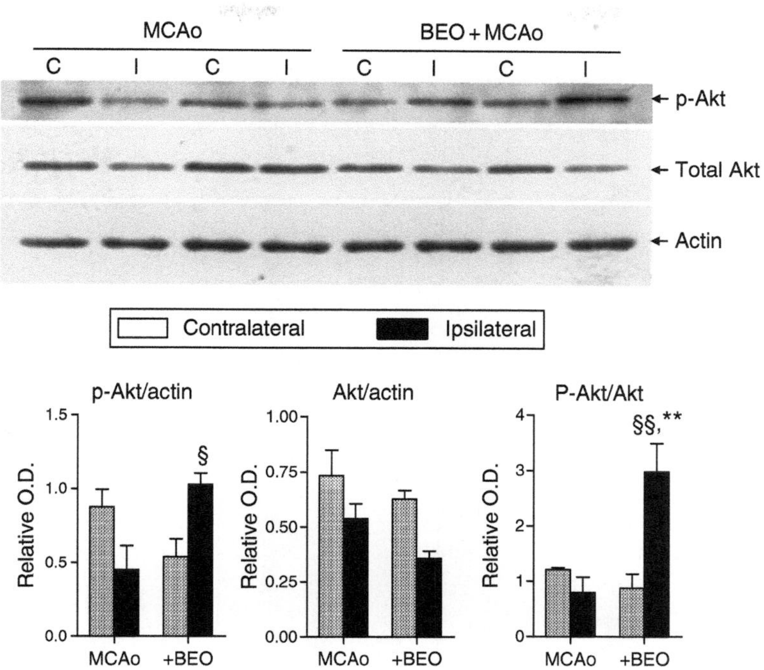

FIG. 4. BEO enhances p-Akt levels in the brain cortical tissue from rats subjected to permanent focal cerebral ischemia. Western blot analysis of phospho-Akt (Ser473) (p-Akt) and total Akt performed using brain cortical homogenates from rats sacrificed 24 h after MCAo shows a trend toward a decrease of p-Akt and total Akt in the ipsilateral (I), ischemic, cortex as compared to contralateral (C), nonischemic, side; intraperitoneal administration of BEO (0.5 ml/kg) 1 h before MCAo enhances p-Akt immunoreactivity in the ischemic cortex without increasing total Akt expression. Histograms show the results of the densitometric analysis of the bands corresponding to p-Akt, total Akt, and β-actin. p-Akt and Akt levels were normalized to the values yielded by β-actin and Akt phosphorylation was expressed by the ratio of p-Akt/total Akt; data are reported as mean ± S.E.M. ($n = 3$ per group). **Denote $P < 0.01$ versus contralateral side; § and §§ denote $P < 0.05$ and $P < 0.01$ versus MCAo, ipsilateral side (ANOVA followed by Tukey–Kramer test for multiple comparisons).

excitatory amino acids (Rossi *et al.*, 2000; Szatkowski and Attwell, 1994; Trotti *et al.*, 1998). Interestingly, here the phytocomplex minimized MCAo-induced increase in excitatory amino acid efflux in the ischemic penumbra at a dose (0.5 ml/kg i.p.) that per se does not affect amino acid efflux under basal conditions suggesting that the underlying mechanism may implicate normalization of the transporter function; whether the latter relates to a free radical scavenging activity (Corasaniti *et al.*, 2007), an inversion of the direction and/or a blockade of excitatory amino acid transport (Morrone *et al.*, 2007) remains to be established. More importantly, minimization of excitatory amino acid accumulation in the

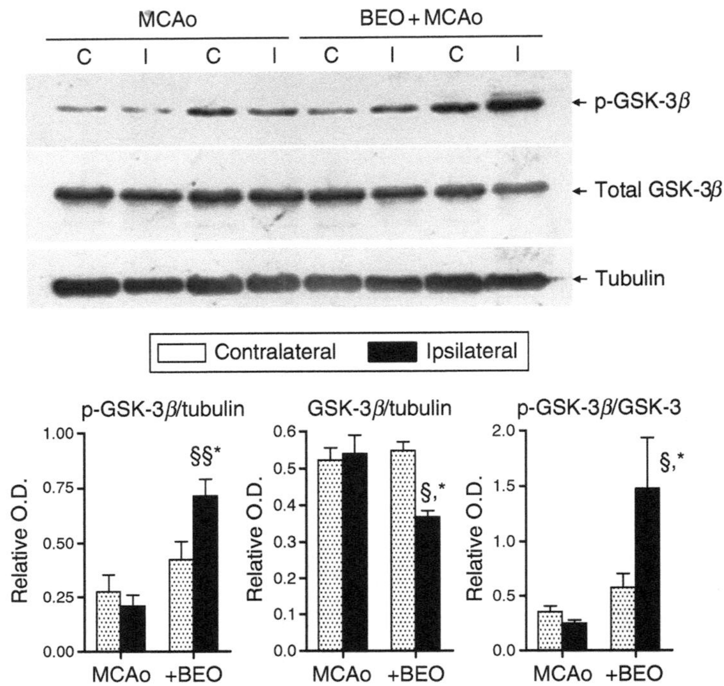

FIG. 5. BEO enhances p-GSK-3β levels in the brain cortical tissue from rats subjected to permanent focal cerebral ischemia. The same samples as in Fig. 4 were used for the subsequent western blot analysis of phospho-GSK-3β (Ser9) (p-GSK-3β) and total GSK-3β. Intraperitoneal administration of BEO (0.5 ml/kg) 1 h before MCAo increases p-GSK-3β immunoreactivity in the ischemic cortex and this is associated with reduced expression of total GSK-3β protein. Histograms show the results of the densitometric analysis of the bands corresponding to p-GSK-3β, total GSK-3β, and α-tubulin; p-GSK-3β and total GSK-3β were normalized to the values yielded by α-tubulin whereas GSK-3β phosphorylation was calculated from the ratio of p-GSK-3β/total GSK-3β. Data are reported as mean $\pm$ S.E.M. ($n = 3$ per group). *Denotes $P < 0.05$ versus contralateral side; § and §§denote $P < 0.05$ and $P < 0.01$, respectively, versus MCAo, ipsilateral side (ANOVA followed by Tukey–Kramer test for multiple comparisons).

ischemic penumbra was accompanied at 24 h by enhanced levels of phosphorylation of the prosurvival gene product Akt and of the downstream kinase GSK-3β. Pospho-Akt (Ser473) immunoreactivity in total protein extracts obtained from the cerebral cortex ipsilateral to the occluded MCA was reduced as compared to contralateral, nonischemic cortex, though this did not reach statistical significance. Phosphorylation is positively linked to Akt activity as the enzyme is activated via phosphorylation at Thr308 and Ser473 (Alessi *et al.*, 1996). The observed apparent reduction in p-Akt levels in the ischemic cortex is associated with a trend toward a reduction in the phosphorylation levels of GSK-3β, a kinase

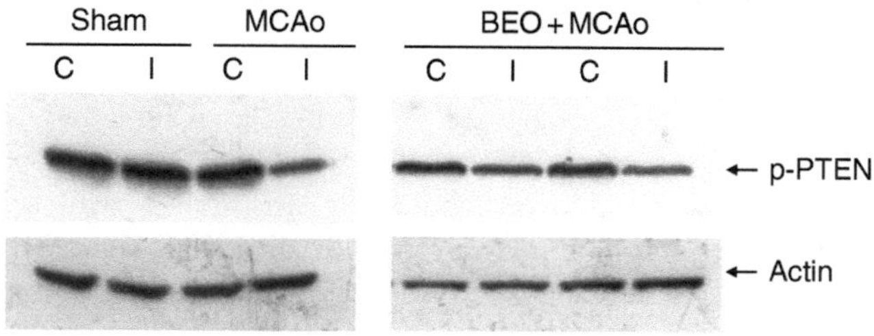

FIG. 6. BEO does not affect decrease in p-PTEN immunoreactivity after stroke. Representative protein bands from western blots of p-PTEN showing reduction in p-PTEN immunoreactivity in brain cortical tissue samples obtained from rats sacrificed 24 h after MCAo. Intraperitoneal administration of BEO (0.5 ml/kg) 1 h before MCAo does not attenuate decrease in p-PTEN induced by stroke. The blots are representative of three animals per experimental group.

negatively regulated by Akt through phosphorylation of the enzyme at Ser9 (Cross *et al.*, 1995; van Weeren *et al.*, 1998). In animals pretreated with BEO and subjected to 24 h occlusion of MCA, a significant increase of p-Akt expression was observed in the ischemic cortex, as compared to the ischemic cortex of control animals. This effect appears to be specific in view of the evidence that BEO did not affect the level of phosphorylated PTEN, typically reduced by ischemia. Activated Akt promotes cell survival by phosphorylating and, thus, inactivating proteins implicated in promoting cell death such as Bad, caspase-9, and GSK-3β, thereby preventing mitochondrial release of cytochrome *c* and caspase-3 activation (Brazil and Hemmings, 2001; Cardone *et al.*, 1998; Grimes and Jope, 2001). Therefore, it can be hypothesized that neuroprotection afforded by BEO may stem from its ability to enhance phosphorylation of Akt. These observations appear of particular interest because neuroprotection against ischemic brain damage is afforded by several agents, including neurotrophic factors such as insulin-like growth factor (IGF-1), hormones, including estrogens and erythropoietin, and this correlates with their ability to upregulate phosphorylation and activity of Akt (see Amantea *et al.*, 2005; Fukunaga and Kawano, 2003; Zhao *et al.*, 2006). We did not perform Akt kinase assay; therefore, to definitely demonstrate that enhancement of Akt phosphorylation by BEO really reflects an increase in kinase activity, we looked for phosphorylation of GSK-3β. BEO preserved phosphorylation of GSK-3β by significantly enhancing levels of p-GSK-3β at Ser9 in the ipsilateral cortex of rats undergone 24 h occlusion of MCA. This observation suggests that changes in Akt phosphorylation induced by BEO are associated with enhanced Akt activity. Changes in the phosphorylation levels of both Akt and GSK-3β induced by BEO were not associated with enhancement of Akt and GSK-3β protein expression. It has been reported that pharmacological inhibition of GSK-3β reduces neuronal death resulting from excitotoxicity *in vitro* (Corasaniti *et al.*, 2007; Facci *et al.*, 2003; Kelly *et al.*, 2004) and cerebral ischemia *in vivo* (Kelly *et al.*, 2004), indicating that GSK-3β inhibition may be beneficial in stroke. Then, it can be suggested that the ability of BEO to upregulate p-GSK-3β levels, and then deleterious activation of the kinase, is instrumental to neuroprotection.

The results obtained in the present study are in line with our previous observations demonstrating that neuroprotection afforded *in vitro* by BEO against NMDA-induced cell death is associated with prevention of injury-induced decrease of Akt and GSK-3β phosphorylation (Corasaniti *et al.*, 2007).

Akt is activated by phosphorylation of Ser473 and Thr308 by a signaling cascade involving PI3-K and 3-phosphoinositide-dependent kinase-1 (PDK-1) (see Brazil and Hemmings, 2001). Receptor tyrosine kinase (RTK) activation by growth factors activates PI3-K which phosphorylates phosphatidylinositol 4,5-biphosphate (PIP2) to phosphatidylinositol 3,4,5-trisphosphate (PIP3) leading to activation of PDK-1 and ultimately to Akt phosphorylation. Conversely, PTEN dephosphorylates PIP3 to PIP2 thereby leading to Akt inactivation (Stambolic *et al.*, 1998; see Vazques and Seller, 2000). In addition, Akt phosphorylation is negatively regulated by protein phosphatase 2A (Andjelkovic *et al.*, 1996). Contrasting data exist relative to p-PTEN expression following ischemic insults (Lee *et al.*, 2004; Omori *et al.*, 2002; Zhao *et al.*, 2005). However, it has been demonstrated that suppressing PTEN expression via administration of siRNApten into the CA1 pyramidal cell layer 3 days before global ischemia protects, in a Akt-dependent manner, against delayed hippocampal CA1 neuronal death in a transient global ischemic model (Ning *et al.*, 2004).

The mechanisms through which BEO preserves Akt phosphorylation in the ischemic penumbra after focal ischemia were not fully investigated in the present work. So, we do not know whether BEO activates PI3-K pathway or whether it directly or indirectly affects the activity of phosphatases. Our present data, however, do allow to exclude a role for PTEN in the level of Akt phosphorylation observed following treatment with BEO. In fact, BEO did not affect expression of p-PTEN (the inactive form of the enzyme) reduced by ischemia. Phosphorylation of three residues (Ser380, T382, and T383) within the C terminus induces PTEN to assume a closed conformation with an inactive phosphatase domain (Rahdar *et al.*, 2009; Vazques *et al.*, 2000, 2001); therefore, reduction of p-PTEN levels in the ischemic cortex might underlie enhanced enzyme activity with PTEN being reducing the levels of PIP3, the positive regulator of Akt activity (Stambolic *et al.*, 1998).

Actually, the mechanism underlying the effects of BEO on Akt phosphorylation were recently investigated in an *in vitro* model of excitotoxic insult, that is exposure of human neuroblastoma cells to NMDA (Corasaniti *et al.*, 2007). We observed that neuroprotection afforded by BEO against NMDA-induced cell death is antagonized in a concentration-dependent fashion by LY294002, a specific PI3-K inhibitor (Vlahos *et al.*, 1994). In addition, LY294002 inhibited the ability of

BEO to restore Akt and GSK-3β phosphorylation reduced by serum deprivation (Corasaniti *et al.*, 2007). Collectively, these observations indicate that a PI3-K-dependent mechanism mediates the effects of BEO on Akt phosphorylation and this may conceivably account for the present *in vivo* observation.

Further experiments are needed to dissect the molecular pathways through which BEO restores Akt phosphorylation in the ischemic cortex of rats subjected to MCAo together with the identification of the active molecules of the essential oil.

Acknowledgments

Partial financial support by the Italian Ministry for University and Scientific Research (PRIN to M.T.C.), Calabria Region (to G.B. and M.T.C) and University of Calabria is gratefully acknowledged.

References

Alessi, D. R., Andjelkovic, M., Caudwell, B., Cron, P., Morrice, N., Cohen, P., and Hemmings, B. A. (1996). Mechanism of activation of protein kinase B by insulin and IGF-1. *EMBO J.* **15,** 6541–6551.

Amantea, D., Russo, R., Bagetta, G., and Corasaniti, M. T. (2005). From clinical evidence to molecular mechanisms underlying neuroprotection afforded by estrogens. *Pharmacol. Res.* **52,** 119–132.

Andjelkovic, M., Jakubowicz, T., Cron, P., Ming, X. F., Han, J. W., and Hemmings, B. A. (1996). Activation and phosphorylation of a pleckstrin homology domain containing protein kinase (RAC-PK/PKB) promoted by serum and protein phosphatase inhibitors. *Proc. Natl. Acad. Sci. USA* **93,** 5699–5704.

Brazil, D. P., and Hemmings, B. A. (2001). Ten years of protein kinase B signalling: A hard Akt to follow. *Trends Biochem. Sci.* **26,** 657–664.

Cardone, M. H., Roy, N., Stennicke, H. R., Salvesen, G. S., Franke, T. F., Stanbridge, E., Frisch, S., and Reed, J. C. (1998). Regulation of cell death protease caspase-9 by phosphorylation. *Science* **282,** 1318–1321.

Corasaniti, M. T., Maiuolo, J., Maida, S., Fratto, V., Navarra, M., Russo, R., Amantea, D., Morrone, L. A., and Bagetta, G. (2007). Cell signaling pathways in the mechanisms of neuroprotection afforded by bergamot essential oil against NMDA-induced cell death *in vitro*. *Br. J. Pharmacol.* **151,** 518–529.

Cross, D. A., Alessi, D. R., Cohen, P., Andjelkovic, M., and Hemmings, B. A. (1995). Inhibition of glycogen synthase kinase-3 by insulin mediated by protein kinase B. *Nature* **378,** 785–789.

Facci, L., Stevens, D. A., and Skaper, S. D. (2003). Glycogen synthase kinase-3 inhibitors protect central neurons against excitotoxicity. *Neuroreport* **14,** 1467–1470.

Fukunaga, K., and Kawano, T. (2003). Akt is a molecular target for signal transduction therapy in brain ischemic insult. *J. Pharmacol. Sci.* **92,** 317–327.

Grimes, C. A., and Jope, R. S. (2001). The multifaceted roles of glycogen synthase kinase 3beta in cellular signaling. *Prog. Neurobiol.* **65,** 391–426.

Kelly, S., Zhao, H., Hua Sun, G., Cheng, D., Qiao, Y., Luo, J., Martin, K., Steinberg, G. K., Harrison, S. D., and Yenari, M. A. (2004). Glycogen synthase kinase 3beta inhibitor Chir025 reduces neuronal death resulting from oxygen-glucose deprivation, glutamate excitotoxicity, and cerebral ischemia. *Exp. Neurol.* **188,** 378–386.

Lee, J. H., Kim, K. J., Lee, Y. K., Park, S. Y., Kim, C. D., Lee, W. S., Rhim, B. Y., and Hong, K. W. (2004). Cilostazol prevents focal cerebral ischemic injury by ebhancing casein kinase 2 phosphorylation and suppression of phosphatase and tensin homolog deleted from chromosome 10 phosphorylation in rats. *JPET* **308,** 896–903.

Li, H., Colbourne, F., Sun, P., Zhao, Z., Buchan, A. M., and Iadecola, C. (2000). Caspase inhibitors reduce neuronal injury after focal but not global cerebral ischemia in rats. *Stroke* **31,** 176–182.

Longa, E. Z., Weinstein, P. R., Carlson, S., and Cummins, R. (1989). Reversible middle cerebral artery occlusion without craniectomy in rats. *Stroke* **20,** 84–91.

Mondello, L., Dugo, P., Bartle, K. D., Dugo, G., and Cotroneo, A. (1995). Automated HPLC-HRGC: A powerful method for essential oils analysis. Part V. Identification of terpene hydrocarbons of bergamot, lemon, mandarin, sweet orange, bitter orange, grapefruit, clementine and mexican lime oils by coupled HPLC-HRGC-MS(ITD). *Flav. Fragr. J.* **10,** 33–42.

Morrone, L. A., Pelle, C., Amantea, D., Rombolà, L., Corasaniti, M. T., and Bagetta, G. (2006). Evidence that systemic administration of the essential oil of bergamot minimizes elevation of glutamate and aspartate in the brain cortex of ischemic rat. *In Proceedings of the British Pharmacological Society at, http://www.pA2online.org/abstracts/Vol4Issue2abst057P.pdf.*

Morrone, L. A., Rombolà, L., Pelle, C., Corasaniti, M. T., Zappettini, S., Paudice, P., Bonanno, G., and Bagetta, G. (2007). The essential oil of bergamot enhances the levels of amino acid neurotransmitters in the hippocampus of rat: Implication of monoterpene hydrocarbons. *Pharmacol. Res.* **55,** 255–262.

Ning, K., Pei, L., Liao, M., Liu, B., Zhang, Y., Jiang, W., Mielke, J. G., Li, L., Chen, Y., El-Hayek, Y. H., Fehlings, M. G., Zhang, X., *et al.* (2004). Dual neuroprotective signaling mediated by downregulating two distinct phosphatase activities of PTEN. *J. Neurosci.* **24,** 4052–4060.

Noshita, N., Lewen, A., Suguwara, T., and Chan, P. H. (2001). Evidence of phosphorylation of Akt and neuronal survival after transient focal cerebral ischemia in mice. *J. Cereb. Blood Flow Metab.* **21,** 1442–1450.

Omori, N., Jin, G., Li, F., Zhang, W. R., Wang, S. J., Hamakawa, Y., Nagano, I., Manabe, Y., Shoji, M., and Abe, K. (2002). Enhanced phosphorylation of PTEN in rat brain after transient middle cerebral artery occlusion. *Brain Res.* **954,** 317–322.

Rahdar, M., Inoue, T., Meyer, T., Zhang, J., Vazquez, F., and Devreotes, P. N. (2009). A phosphorylation-dependent intramolecular interaction regulates the membrane association and activity of the tumor suppressor PTEN. *PNAS* **106,** 480–485.

Richards, D. A., Morrone, L. A., and Bowery, N. G. (2000). Hippocampal extracellular amino acids and EEG spectral analysis in a genetic rat model of absence epilepsy. *Neuropharmacology* **39,** 2433–2441.

Rossi, D. J., Oshima, T., and Attwell, D. (2000). Glutamate release in severe brain ischaemia is mainly by reversed uptake. *Nature* **403,** 316–321.

Shibata, M., Yamawaki, T., Sasaki, T., Hattori, H., Hamada, J., Fukuuchi, Y., Okano, H., and Miura, M. (2002). Upregulation of Akt phosphorylation at the early stage of middle cerebral artery occlusion in mice. *Brain Res.* **942,** 1–10.

Shioda, N., Ishigami, T., Han, F., Moriguchi, S., Shibuya, M., Iwabuchi, Y., and Fukunaga, K. (2007). Activation of phosphatidylinositol 3-kinase/protein kinase B pathway by a vanadyl compound mediates its neuroprotective effect in mouse brain ischemia. *Neuroscience* **148,** 221–229.

Stambolic, V., Suzuki, A., de la Pompa, J. L., Brothers, G. M., Mirtsos, C., Sasaki, T., Ruland, J., Penninger, J. M., Siderovski, D. P., and Mak, T. W. (1998). Negative regulation of PKB/Akt-dependent cell survival by the tumor suppressor PTEN. *Cell* **95,** 29–39.

Szatkowski, M., and Attwell, D. (1994). Triggering and execution of neuronal death in brain ischaemia: Two phases of glutamate release by different mechanisms. *Trends Neurosci.* **17,** 359–365.

Trotti, D., Danbolt, N. C., and Volterra, A. (1998). Glutamate transporters are oxidant-vulnerable: A molecular link between oxidative and excitotoxic neurodegeneration? *Trends Pharmacol. Sci.* **19,** 328–334.

van Weeren, P. C., De Bruyn, K. M., De Vries-Smits, A. M., van Lint, J., and Burgering, B. M. (1998). Essential role for protein kinase B (PKB) in insulin-induced glycogen synthase kinase 3 inactivation. Characterization of dominant-negative mutant of PKB. *J. Biol. Chem.* **273,** 13150–13156.

Vazquez, F., and Sellers, W. R. (2000). The PTEN tumor suppressor protein: An antagonist of phosphoinositide 3-kinase signaling. *Biochim. Biophys. Acta* **1470,** M21–M35.

Vazquez, F., Ramaswamy, S., Nakamura, N., and Sellers, W. R. (2000). Phosphorylation of the PTEN tail regulates protein stability and function. *Mol. Cell. Biol.* **20,** 5010–5018.

Vazquez, F., Grossman, S. R., Takahashi, Y., Rokas, M. V., Nakamura, N., and Sellers, W. R. (2001). Phosphorylation of the PTEN tail acts as an inhibitory switch by preventing its recruitment into a protein complex. *J. Biol. Chem.* **276,** 48627–48630.

Verzera, A., Lamonica, G., Mondello, L., Trozzi, A., and Dugo, G. (1996). The composition of bergamot oil. *Perfumer Flavorist* **21,** 19–34.

Verzera, A., Trozzi, A., Gazea, F., Cicciarello, G., and Cotroneo, A. (2003). Effects of rootstock on the composition of bergamot (Citrus bergamia Risso et Poiteau) essential oil. *J. Agric. Food Chem.* **51,** 206–210.

Vlahos, C. J., Matter, W. F., Hui, K. Y., and Brown, R. F. (1994). A specific inhibitor of phosphatidylinositol 3-kinase, 2-(4-morpholinyl)-8-phenyl-4*H*-1-benzopyran-4-one (LY294002). *J. Biol. Chem.* **269,** 5241–5248.

Wu, D. (2005). Neuroprotection in experimental stroke with targeted neurotrophins. *NeuroRx* **2,** 120–128.

Zhao, H., Shimohata, T., Wang, J. Q., Sun, G., Schaal, D. W., Sapolsky, R. M., and Steinberg, G. K. (2005). Akt contributes to neuroprotection by hypothermia against cerebral ischemia in rats. *J. Neurosci.* **25,** 9794–9806.

Zhao, H., Sapolsky, R. M., and Steinberg, G. K. (2006). Phosphoinositide-3-kinase/Akt survival signal pathways are implicated in neuronal survival after stroke. *Mol. Neurobiol.* **34,** 249–269.

IDENTIFICATION OF NOVEL PHARMACOLOGICAL TARGETS TO MINIMIZE EXCITOTOXIC RETINAL DAMAGE

Rossella Russo,* Domenicantonio Rotiroti,† Cristina Tassorelli,‡ Carlo Nucci,§,¶ Giacinto Bagetta,*,|| Massimo Gilberto Bucci,** Maria Tiziana Corasaniti,† and Luigi Antonio Morrone*,||

*Department of Pharmacobiology, University of Calabria, 87036 Arcavacata di Rende, Italy
†Department of Pharmacobiological Sciences, University "Magna Graecia" of Catanzaro, 88100 Catanzaro, Italy
‡IRCCS "C. Mondino Institute of Neurology" Foundation Department of Neurological Sciences, University of Pavia, 27100 Pavia, Italy
§Physiopathological Optics, Department of Biopathology, University of Rome "Tor Vergata", 00133 Rome, Italy
¶"Mondino-Tor Vergata" Center for Experimental Neurobiology, University of Rome "Tor Vergata", 00133 Rome, Italy
||Section of Neuropharmacology of Normal and Pathological Neuronal Plasticity, Center for Adaptive Disorders and Headache (UCADH), University of Calabria, 87036 Arcavacata di Rende, Italy
**Fondazione G. B. Bietti, IRCCS, 00199 Roma, Italy

Excitotoxic neuronal death is a common feature of neurodegenerative and ischemic diseases of the central nervous system (CNS) and of a variety of ocular diseases, including glaucoma.

Glaucoma, one of the leading causes of blindness in the world, is characterized by a progressive degeneration of retinal ganglion cells (RGCs) and their axons and is often associated with elevated intraocular pressure (IOP). Retinal ischemia/reperfusion induced by experimental elevation of IOP leads to damage and loss of RGCs. Under these conditions, structural, functional, and biochemical changes implicate the accumulation of extracellular glutamate and activation of the excitotoxic cascade.

Beside the activation of associated pathways, death of RGCs is accompanied by impaired endogenous defenses, such as the PI3K/Akt prosurvival pathway.

INTERNATIONAL REVIEW OF NEUROBIOLOGY, VOL. 85
DOI: 10.1016/S0074-7742(09)85028-9

Original neurochemical and pharmacological evidence are discussed here to strengthen the role for excitotoxicity in RGCs death occurring in experimental, angle closure, glaucoma in conjunction with the discovery of novel molecular targets to potentiate endogenous prosurvival defenses in the glaucomatous retina.

I. Introduction

Lucas and Newhouse (1957) firstly reported that subcutaneous injection of glutamate in neonatal rodents had toxic effect on the inner retinal layers. Later on, Sisk and Kuwabara (1985) demonstrated the susceptibility of adult retina to the intravitreal injection of glutamate, showing severe degeneration of the ganglion and inner nuclear layers (INLs).

The observation made by Lucas and Newhouse opened a new lane of research yielding a massive production of experimental observations that, 10 years later, led to the definition of excitotoxicity as the "axon-sparing" neuronal death caused by an excessive or prolonged activation of glutamate receptors (Olney, 1969).

Pathological activation of glutamate receptors is a common feature and one of the primary causes of neuronal death in acute neuronal injury (such as trauma, epilepsy, and brain ischemia) and chronic neurodegenerative diseases (such as Parkinson's disease, Alzheimer diseases, amyotrophic lateral sclerosis, and AIDS dementia) (Choi, 1988; Doble, 1999; Lipton and Rosemberg, 1994). In particular, elevation of extracellular glutamate level is a key factor in the development of neuronal damage under ischemic conditions.

In the mammalian retina, ischemic phenomena are believed to occur in a variety of pathological conditions including glaucoma, anterior ischemic optic neuropathy, retinal and choroidal vessel occlusion, diabetic retinopathy, and traumatic optic neuropathy (Osborne *et al.*, 1999, 2004). The common feature of many of these ocular pathologies is the damage and loss of retinal ganglion cells (RGCs) that, through the optic nerve fibers, send the retinal information to the central areas of the visual pathway.

The partial or complete interruption of the blood flow to the retina, as well as in other tissues, leads to neuronal death through a complex series of biochemical events triggered by the excitotoxic cascade.

The progression of glutamate neurotoxicity can be considered as a process of three sequential steps: the overstimulation of postsynaptic glutamate receptors leading to the accumulation of Ca^{2+}; the amplification of the detrimental signal through the additional Ca^{2+} influx and the release from intracellular stores; the activation of catabolic pathways and the generation of free radicals (Choi, 1990).

Altogether, this knowledge has formed the rational basis of an intense area of pharmacological research that has been devoted to the development of novel neurotherapeutics to minimize excitotoxic RGC damage.

II. Neurochemical and Pharmacological Evidence to Support Excitotoxicity in the Mechanisms of RGC Death in Experimental Glaucoma

Glutamate is the major excitatory neurotransmitter in the retina where it is released by photoreceptors, bipolar, and ganglion cells (Yang, 2004).

Under physiological conditions, glutamate is released from the synaptic vesicles in response to depolarization and acts on postsynaptic *N*-methyl-D-aspartate (NMDA) and non-NMDA ionotropic receptors mediating slow and fast components of the excitatory postsynaptic potentials, respectively (Michaelis, 1998). The neurotransmitter is then rapidly removed from the synaptic cleft by efficient, Na^+-dependent, high-affinity transporters located on both neurons and glia and identified as excitatory amino acid transporters (EAATs) (Danbolt *et al.*, 2001). The efficiency of this process is crucial to terminate the excitatory signal and preventing excitotoxic neuronal damage.

Overactivation of the NMDA subtype of glutamate receptors results in excessive Ca^{2+} influx via the receptor-associated cation channel leading to the activation of calcium-dependent enzymes, such as proteases, endonucleases, and nitric oxide synthases (NOSs) and to the production of nitrogen as well as oxygen free radicals thus contributing to cell death (Choi, 1990; Lipton, 2006; Lipton and Nicotera, 1998; Lynch and Guttmann, 2002).

Several evidence underlie the crucial role of excessive glutamate release under ischemic conditions in the brain (Aarts *et al.*, 2003; Camacho and Massieu, 2006; Dirnagl *et al.*, 1999) and, concurrently, experimental data sustain a comparable role for glutamate under retinal ischemia (Adachi *et al.*, 1998; Louzada-Junior *et al.*, 1992). However, it should be stressed that the mechanisms underlying tolerance to ischemia may differ in brain and in the retina, the latter being more tolerant to ischemia than the former (Iijima *et al.*, 2000; Osborne *et al.*, 2004).

Accumulation of extracellular glutamate during retinal ischemia was first described by Louzada-Junior and collaborators in rabbit (Louzada-Junior *et al.*, 1992) and, subsequently, corroborated by data obtained by Adachi *et al.* (1998) in cat.

Dreyer *et al.* (1996) reported that glutamate levels were elevated in the vitreous of humans and monkeys with glaucoma. However, studies by other investigators failed to confirm this finding (Honkanen *et al.*, 2003; Levkovitch-Verbin *et al.*, 2002) rising doubts about the role of excitotoxicity in glaucoma and suggesting that further studies were needed to fully understand the role of glutamate in the pathogenesis of glaucoma.

Recently, using an intravitreal microdialysis technique, extracellular glutamate has been monitored *in vivo* in the retina of rat undergone transient ischemic insult (Nucci *et al.*, 2005), an experimental setting that models angle closure glaucoma (Buchi *et al.*, 1991; Osborne *et al.*, 2004) and leads to the loss of RGCs (Fig. 1). Under these experimental conditions, elevation of extracellular glutamate is seen during the first 10 min of ischemia, followed by a larger and statistically significant increase during the reperfusion phase (Russo *et al.*, 2008c). The temporal profile of glutamate accumulation seen in rat is similar to that observed in other animal species though the amplitude of the response appears to be smaller (Fig. 2).

The mechanisms of extracellular accumulation of glutamate during ischemia and reperfusion in the rat retina are not known. However, it is generally accepted that under ischemic conditions, when oxygen and nutrients deprivation causes disruption of cellular energy metabolism and oxidative stress occurs, the neurons fail to maintain membrane potential; the concomitant inefficiency of neurons and glia to remove the neurotransmitter from the extracellular space leads to accumulation of excitotoxic levels of glutamate. The increase of glutamate levels induced by ischemia is thought to be the result of a combination of vesicular release and reversal of the excitatory transporter function (Nicholls and Attwell, 1990; Phillis *et al.*, 2000; Rossi *et al.*, 2000). Four distinct EAATs have been identified in the retina: the glutamate/aspartate transporter (GLAST), in Müller cells; glutamate transporter 1 (GLT-1) and EAAT-5 in photoreceptors and bipolar

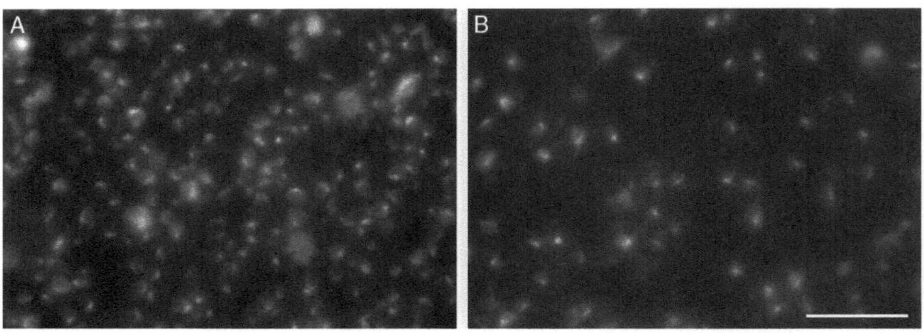

FIG. 1. Representative fluorescent photomicrographs of retinal whole mounts showing the loss of Fluorogold (FG)-labeled RGCs in ischemic retina of rat. RGCs were retrogradely labeled with the fluorescent dye FG injected, under stereotaxic guidance, bilaterally into the superior colliculus of a rat 4 days after 50 min ischemia and sacrificed after additional 4 days. Obvious reduction of FG-labeled RGCs is evident in the retina undergone ischemia/reperfusion (panel B) as compared to the contralateral, nonischemic, retina (panel A). Photomicrographs were obtained from the peripheral area of the superior quadrant of the retina. Scale bar: 50 μm.

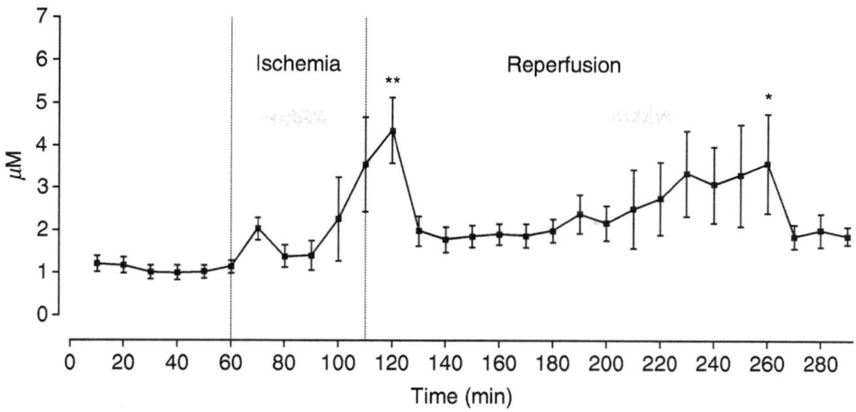

FIG. 2. Transient retinal ischemia increases intravitreal glutamate in rat. Neurochemical data from microdialysis experiments carried out in anesthetized rats to demonstrate that ischemia/reperfusion insult increases intravitreal glutamate. The extracellular level of glutamate shows a moderate increase during the first 10 min of ischemia, more evident toward the end of the ischemic period, to reach statistical significance at 10 and 150 min of reperfusion. Baseline glutamate concentrations (basal values) are the mean concentrations obtained by averaging six samples collected consecutively at 10 min intervals immediately before the onset of ischemia ($n = 6$ rats). Glutamate values (μM) are expressed as mean $\pm$ S.E.M. Statistical significance was assessed by ANOVA followed by Dunnett's test for multiple comparisons. $*P < 0.05$ and $**P < 0.001$ versus basal values.

cells; and excitatory amino acid carrier (EAAC1) expressed by horizontal, amacrine, and ganglion cells (Rauen, 2000).

Under physiological conditions, the Müller cell transporter, GLAST, is primarily responsible for the clearance of extracellular glutamate (Barnett *et al.*, 2001; Pow and Barnett, 1999; Rauen *et al.*, 1998). However, under acute retinal ischemia, the glial transporter GLAST has a reduced capability to maintain the inward transport of glutamate and the neuronal uptake by EAAT-5, GLT-1, and EAAC-1 is not sufficient to compensate for the failure of the dominant transporter GLAST leading to the accumulation of extracellular glutamate (Barnett *et al.*, 2001).

The central role played by GLAST in the prevention of glutamate neurotoxicity after ischemia is further supported by the observation that GLAST deficient mice show more severe damage after pressure-induced retinal ischemia compared to GLT-1 deficient and wild-type animals (Harada *et al.*, 1998).

The crucial role played by GLTs in the pathophysiology of glaucoma is strengthened by the observed reduction of GLAST expression (Naskar *et al.*, 2000) in conjunction with the induction of a splice variant of GLT-1 in RGCs (Sullivan *et al.*, 2006) in glaucomatous eyes.

Excitotoxic-induced Ca^{2+} overload causes a large production of free radicals overwhelming the antioxidant defenses of the tissue. Oxidative stress, a condition that takes place when generation of reactive oxygen species (ROS) exceed the antioxidant capacity of the cell, has been implicated in a large number of neurodegenerative disease and aging processes (Andersen, 2004; Potashkin and Meredith, 2006). Growing evidence suggests that oxidative stress has also a role in the RGCs death associated with retinal ischemia and glaucoma (Tezel, 2006).

It has been recently demonstrated that transient retinal ischemia leads to the formation of free radicals and lipid peroxidation, and administration of free radical scavengers or metal chelators prevents the associated RGCs death (Banin *et al.*, 2000; Celebi *et al.*, 2002; Shibuki *et al.*, 2000).

During a hypoxic-ischemic event, the decline of ATP levels induces the failure of the ionic pumps, in particular, the Na^+/K^+ ATPase that is responsible for maintaining the resting membrane potential. This event induces membrane depolarization, which opens the voltage-gated Ca^{2+} and Na^+ channels and stimulates the release of neurotransmitters, including glutamate (Nicholls and Attwell, 1990). Accumulation of $[Ca^{2+}]_i$, mainly through the opening of the NMDA receptor-associated cation channel, induces the production of free radicals by activating metabolic reactions catalyzed by Ca^{2+}-dependent enzymes such as phospholipase A_2 (Au *et al.*, 1985; Sevanian *et al.*, 1983), xanthine oxidase (McCord *et al.*, 1985), and NOS (Dawson, 1995).

Among the oxygen-derived free radicals, the species of primary concern include superoxide anion (O_2^-), hydrogen peroxide (H_2O_2), and hydroxyl radical (OH^-). The superoxide is further converted in peroxynitrite ($ONOO^-$) by reacting with nitric oxide.

Mitochondria are the main source of free radicals in the cell and, in turn, ROS can cause inhibition of complex enzymes in the electron transport chain of the mitochondria leading to the shutdown of energy production and amplifying generation of mitochondrial free radicals (Orrenius, 2007). Free radicals can then cause extensive cellular damage by causing oxidation of lipids, proteins, and DNA.

Using a proteomic approach, Tezel *et al.* (2005) have recently reported evidence of oxidative modulation of many retinal proteins in ocular hypertensive eyes. Oxidized proteins accumulate as proteolysis-resistant aggregates leading to the loss of protein function and abnormal protein clearance (Berlett and Stadtman, 1997). For instance, glutamine synthase, the enzyme responsible for glial conversion of glutamate in glutamine, has been identified as one of the proteins undergoing oxidative modification in ocular hypertensive eyes (Tezel *et al.*, 2005) leading to the accumulation of excitotoxic levels of glutamate (Levine, 1983).

Accumulation of free radicals may also damage GLT proteins (Muller *et al.*, 1998) decreasing the capacity of neurons and glia to metabolize glutamate and exacerbating the excitotoxic damage (Sandhu *et al.*, 2003; Trotti *et al.*, 1996). Free radical scavengers are useful pharmacological tools to prevent neuronal death under excitotoxic conditions (Lipton and Rosemberg, 1994). Coenzyme Q10 (CoQ10) is an essential cofactor of the mitochondrial electron transport chain endowed with potent antioxidant activity that has been shown to mediate neuroprotection in Parkinson's, Huntington's, and other neurodegenerative diseases in which impaired mitochondrial function and excessive oxidative damage has been reported (Beal, 1999, 2004; Beal and Matthews, 1997; Shults *et al.*, 2002, 2004).

Intravitreal administration of CoQ10 limits accumulation of glutamate and minimizes RGCs loss induced by transient retinal ischemia (Nucci *et al.*, 2007; Russo *et al.*, 2008b). However, pretreatment with the antioxidant vitamin E does not have significant effect on glutamate accumulation (Nucci *et al.*, 2007). This apparent discrepancy between CoQ10 and vitamin E can be rationalized on the basis of limited availability of vitamin E on the site of damage for pharmacokinetic reasons, being the latter extremely lipophilic; a more attractive hypothesis would be that mechanisms other than the antioxidant properties may account for the effect of CoQ10 on extracellular glutamate accumulation. Accordingly, it has been recently reported that CoQ10 inhibits the opening of the mitochondrial permeability transition pore (PTP) preventing the loss of the mitochondrial membrane potential that leads to apoptosis (Papucci *et al.*, 2003). Therefore, CoQ10 neuroprotection may stem from the prevention of mitochondrial energy metabolism derangement thus making it possible to sustain the function of GLTs. The latter would account for limiting accumulation of excitotoxic levels of extracellular glutamate thus preventing RGC death and loss (Nucci *et al.*, 2007; Russo *et al.*, 2008b).

Under ischemic conditions, accumulation of pathological levels of glutamate is accompanied by an early and robust activation of the proteolytic, calcium-dependent, enzyme calpain (Branca, 2004) that is maintained over 24-h period of reperfusion (Oka *et al.*, 2006; Sakamoto *et al.*, 2000) and is prevented by treatment with the NMDA receptor antagonist MK801 (Russo and Morrone, personal communication).

The occurrence of the proteolytic activity of calpain confirms that abnormal glutamate overactivates the NMDA receptor leading to intracellular calcium overload. Neuroprotection afforded by calpain inhibitors lends further support to the role of glutamate in RGC degeneration (Oka *et al.*, 2006; Sakamoto *et al.*, 2000).

Activation of calpain is usually associated with the progression of a necrotic type of cell death (Wang, 2000). However, neuronal necrosis and apoptosis occur in parallel after ischemic injury *in vitro* and *in vivo* (Charriaut-Marlangue *et al.*, 1996). Retinal ischemia causes precocious necrosis of neurons in the ganglion cell layer (GCL) and INL, whereas apoptosis appears as the delayed component of neuronal death associated with transient retinal ischemia (Joo *et al.*, 1999).

Morphological features of apoptosis following retinal ischemia, such as DNA fragmentation, nuclear condensation, and chromatin marginalization have been recently described (Nucci *et al.*, 2005) in conjuction with upregulation of Bcl-2-associated death promoter (Bad), a proapoptotic member of the Bcl-2 family (Russo *et al.*, 2008a). Treatment with neuroprotective doses of the NMDA receptor antagonist, MK801, minimizes the induction of Bad, suggesting that blockade of the excitotoxic cascade reduces the activation of the mitochondrial, proapoptotic pathway triggered by retinal ischemia (Russo *et al.*, 2008a).

The pathophysiological role of increased extracellular glutamate in the retinal damage induced by transient ischemia is further proven by the observation that systemic or intravitreal treatment with NMDA and non-NMDA glutamate receptor antagonists (i.e., MK801 and GIKI52466), as well as systemic treatment with L-NAME, an inhibitor of NOS, prevents RGC death (Adachi *et al.*, 1998; Nucci *et al.*, 2005; Russo *et al.*, 2008a; Sucher *et al.*, 1997). Treatment with the noncompetitive open-channel blocker of the NMDA receptor, for example, memantine (Chen and Lipton, 2006; Chen *et al.*, 1992), is endowed with neuroprotective effects on both acute and chronic animal models of RGC death (Hare *et al.*, 2004; Osborne, 1999; Wolde-Mussie *et al.*, 2002). Quite importantly, memantine has been recently shown to be effective in reversing the increase in the tonic level of RGC spiking activity, the reduction in RGC spike amplitude, and, in some cells, the block of spike generation caused in rabbit retina by experimental manipulations that mimic features of glutamatergic, excitotoxic, insults such as (1) application of NMDA, a selective NMDA receptor agonist; (2) application of TBOA (DL-threo-β-benzyloxyaspartic acid), an EAATs inhibitor; or (3) perfusion with magnesium-free medium (Hare and Wheeler, 2009).

At variance with MK801, memantine is suggested to possess a relatively high off-rate that limits pathological activity of the NMDA receptor while preserving normal synaptic activity (Chen and Lipton, 1997).

III. Blockade of Excitotoxicity Sustains the PI3-K/Akt Prosurvival Pathway in Retinal Ischemia

The phosphoinositide-3 kinase (PI3-K)/Akt pathway is activated as a self-defense mechanism of RGCs against injuries inflicted by intravitreal injection of NMDA (Manabe and Lipton, 2003; Nakazawa *et al.*, 2005), episcleral vein cauterization (Kanamori *et al.*, 2004; Kim and Park, 2005), optic nerve clamping (Nakazawa *et al.*, 2003), and translimbal photocoagulation (Levkovitch-Verbin *et al.*, 2007).

Conversely, the protein kinase Akt is deactivated in multiple model of cell death, including NMDA excitotoxicity (Corasaniti *et al.*, 2007; Fukunawa and Kawano, 2003), free radical exposure (Konishi *et al.*, 1999), and brain ischemia (Ouyang *et al.*, 1999).

The PI3-K pathway (Brazil and Hemmings, 2001) is physiologically activated by several neurotrophins, such as BDNF, insulin-like growth factor I (IGF-I), and

NGF (Kaplan and Miller, 2000). PI3-K generates 3′-phosphorylated phosphoinositides that, in turn, activates Akt, also known as protein kinase B (PKB), a serine/threonine kinase with prosurvival and antiapoptotic activities (Franke *et al.*, 2003).

Once activated, Akt acts on a variety of targets and promotes cell survival by inhibiting substrates involved in the apoptotic pathway including Bad, caspases-9, members of the forkhead family of transcription factors (FOXO), and glycogen synthase kinase-3β (GSK-3β) (Brazil and Hemmings, 2001).

Full activation of Akt is due to phosphorylation of Ser473 and Thr308 (Alessi *et al.*, 1996). Delayed RGC death, caused by transient retinal ischemia, is preceded, during the ischemic phase, by Akt deactivation detectable as a strong reduction of Akt phosphorylation status (Russo *et al.*, 2008a). The initial drop of Akt activity is followed, during reperfusion (up to 24 h), by a sustained PI3K-dependent activation, that peaks at 1 h of reperfusion (Russo *et al.*, 2008a) (Fig. 3).

Nakazawa *et al.* (2005) have reported a causative role for Akt deactivation in RGC death showing the mutual exclusion between TUNEL positive and Akt-activated positive cells. Therefore, the transient deactivation of Akt after retinal ischemia, that is coincident with the activation of the proapoptotic target protein GSK-3β (Russo *et al.*, 2008a), may represent one of the early events triggering the death machinery.

Increased activity of GSK-3β has been reported during neuronal degeneration (Beurel and Jope, 2006) and it has been associated with the promotion of the intrinsic mitochondrial pathway of apoptosis (Brazil and Hemmings, 2001). Pharmacological inhibition of GSK-3β reduces neuronal cell death caused by cerebral ischemia (Kelly *et al.*, 2004), suggesting that prevention of GSK-3β activation may represent a potential approach to minimize neurodegeneration in the retina.

Under the experimental paradigm described above, the initial drop of Akt activation is followed by a sustained and prolonged increase of the kinase activity, accompanied by the return to physiological level of GSK-3β. Double immunofluorescence experiments indicate that activated Akt is localized in the innermost part of the retina and, particularly, in the GCL (Russo *et al.*, 2008a) (Fig. 4).

Treatment with the PI3-K inhibitor wortmannin reduces the number of surviving RGCs after retinal ischemia/reperfusion suggesting that, Akt activation is indeed endowed with RGCs neuroprotective properties and represents a prosurvival response of the retina to the ischemic injury (Russo *et al.*, 2008a).

This hypothesis is further supported by studies reporting that intravitreous administration of IGF-I or BDNF prevents RGCs death in axotomized eye by activating Akt (Klocker *et al.*, 2000; Nakazawa *et al.*, 2002) and treatment with PI3-K inhibitors increases the loss of RGCs after optic nerve clamping (Nakazawa *et al.*, 2003) and IOP elevation (Huang *et al.*, 2008).

The neuroprotection afforded by treatment with neurotrophic factors is increased by the association with free radical scavengers. In fact, combined

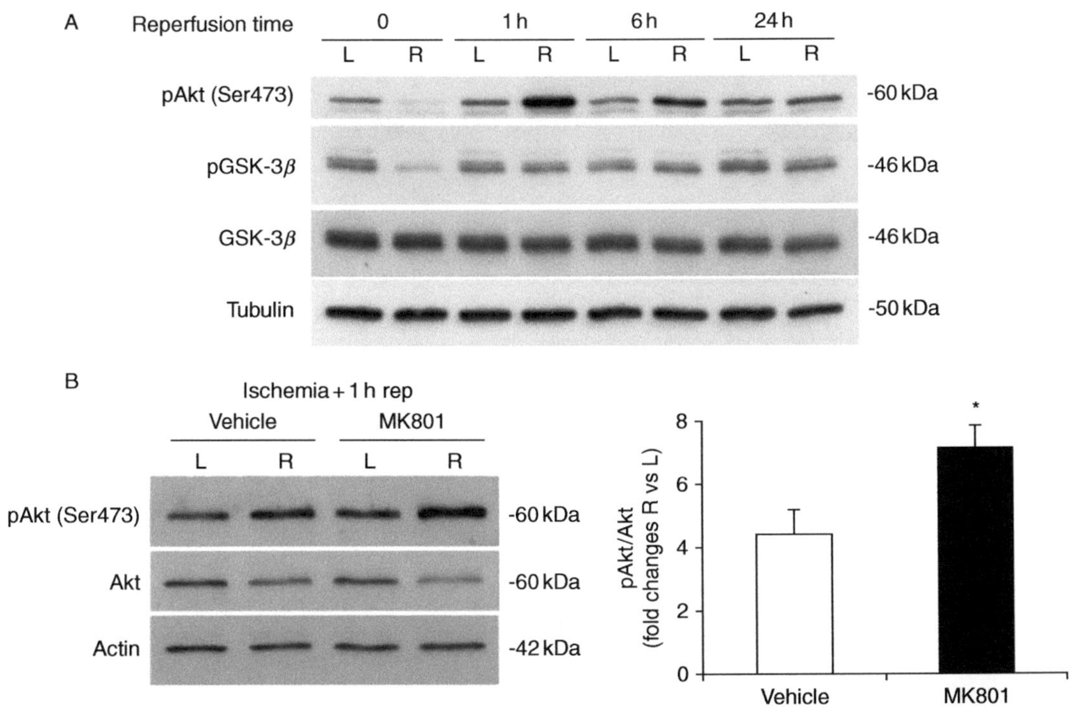

FIG. 3. Changes in Akt and GSK-3β phosphorylation induced by transient retinal ischemia and effect of intravitreal application of MK801 in rat. Animals were subjected to retinal ischemia for 50 min in the right eye (R) and reperfusion was allowed for 1, 6, or 24 h. The left eye (L) was used as control. (A) Phosphorylation of Akt on Ser473 is significantly diminished after retinal ischemia and is accompanied by a transient dephosphorylation (activation) of GSK-3β. During the reperfusion phase, Akt activation is increased within 1 h whereas GSK-3β phosphorylation status returns to basal level. (B) Intravitreal treatment with MK801 enhances the phosphorylation of Akt reported after 1 h reperfusion. Histograms show the results of the densitometric analysis of immunoreactive bands. *$P < 0.05$ versus vehicle.

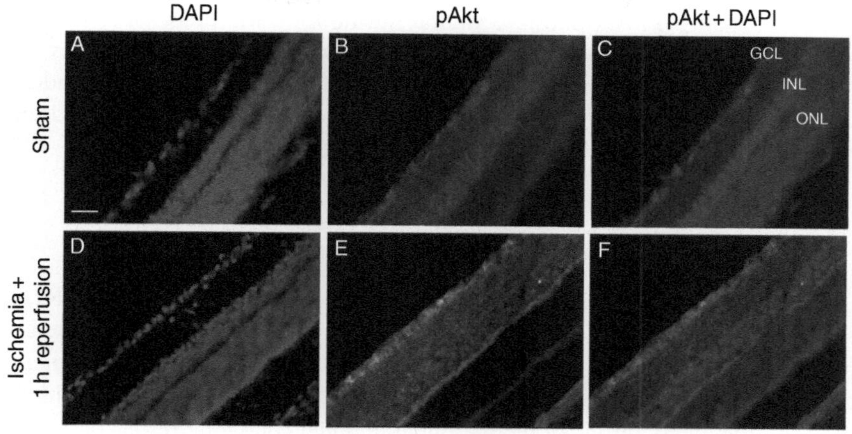

FIG. 4. Representative immunofluorescence showing the localization of pAkt in sham (A–C) and ischemic (D–F) retina after 1 h of reperfusion in rat. A strong increase of pAkt immunoreactivity is detectable in the inner segment of the retina and in particular in ganglion cells (GCL) and inner plexiform (IPL) layers. Images were obtained using identical exposure time and conditions. Scale bar: 20 μm. (See Color Insert.)

treatment of brain-derived neurotrophic factor (BDNF) with a nonspecific free radical scavenger, *N-tert*-butyl-(2-sulfophenyal)-nitrone (S-PBN), increased survival of RGCs in ocular hypertensive eyes (Ko *et al.*, 2000). Furthermore, the association between BDNF and a NOS inhibitor potentiated the neuroprotective effect of BDNF on axotomized RGCs (Klocker *et al.*, 1998). These findings suggest that trophic factors and antioxidants have synergistic effects on rescuing RGCs from excitotoxic death.

NMDA injection in the retina induces Akt dephosphorylation by activating protein phosphatases (PP), presumably PP-2A (Nakazawa *et al.*, 2005). An attempt has been made to elucidate the role of excitotoxicity in the modulation of the PI3-K/Akt under transient retinal ischemia. Interestingly, dephosphorylation of Akt induced by ischemia is not prevented by MK801 whereas the latter potentiates the increase of Akt phosphorylation seen during reperfusion (Russo *et al.*, 2008a) (Fig. 3). The latter evidence is in agreement with recent data showing that neuroprotective treatment with MK801 promotes Akt activation in the rat cortex after brain ischemia (Ahn *et al.*, 2005; Seo *et al.*, 2007).

The lack of prevention of Akt deactivation by MK801 would suggest that mechanisms other than abnormal, excitotoxic, activation of NMDA receptor may account for the observed dephosphorylation of the prosurvival kinase. It is conceivable that dephosphorylation may stem from activation of nonidentified PPs caused by the ischemic drop of ATP and consequent energy depletion (Nakazawa *et al.*, 2005; Ouyang *et al.*, 1999). On the other hand, the potentiating effect of MK801 on Akt phosphorylation during the early phase of reperfusion would identify an additional mechanism of neuroprotection afforded by this NMDA receptor antagonist. Moreover, the last finding would further underlie the role of excitotoxicity in retinal ischemia not only by triggering death-associated pathway but also limiting endogenous prosurvival responses.

IV. Conclusions

Under pathological conditions, when retinal ischemia occurs, death of RGCs results from the activation of death-associated pathways and impairment of prosurvival endogenous defenses. Excitotoxicity is believed to be the major route to damage and loss of RGCs under retinal ischemia. However, despite the favorable safety profile of memantine, the nonencouraging results of a phase 3, randomized, multicenter, placebo-controlled, double-blind clinical trial, in which more than 1000 open-angle glaucoma patients have been evaluated for assessing its efficacy for prevention of progression of the disease (FDA website: http://clinicaltrials.gov), apparently undermine the perspective for future application of the above rationale to further investigation. Interestingly, the rapid activation of the endogenous prosurvival pathway PI3-K/Akt, observed during the early phase of reperfusion, is potentiated by MK801 and this opens a new lane of investigation for the development of novel neurotherapeutics for the treatment of glaucoma and other major retinal pathologies.

Acknowledgments

Partial financial support from the Ministry of Health and University of Calabria is gratefully acknowledged.

References

Aarts, M. M., Arundine, M., and Tymianski, M. (2003). Novel concepts in excitotoxic neurodegeneration after stroke. *Expert Rev. Mol. Med.* **5,** 1–22.

Adachi, K., Kashii, S., Masai, H., Ueda, M., Morizane, C., Kaneda, K., Kume, T., Akaike, A., and Honda, Y. (1998). Mechanism of the pathogenesis of glutamate neurotoxicity in retinal ischemia. *Graefes Arch. Clin. Exp. Ophthalmol.* **236,** 766–774.

Ahn, Y. M., Seo, M. S., Kim, S. H., Kim, Y., Yoon, S. C., Juhnn, Y. S., and Kim, Y. S. (2005). Increased phosphorylation of Ser473-Akt, Ser9-GSK-3beta and Ser133-CREB in the rat frontal cortex after MK-801 intraperitoneal injection. *Int. J. Neuropsychopharmacol.* **8,** 607–613.

Alessi, D. R., Andjelkovic, M., Caudwell, B., Cron, P., Morrice, N., Cohen, P., and Hemmings, B. A. (1996). Mechanism of activation of protein kinase B by insulin and IGF-1. *EMBO J.* **15,** 6541–6551.

Andersen, J. K. (2004). Oxidative stress in neurodegeneration: Cause or consequence? *Nat. Med.* **10,** S18–S25.

Au, A. M., Chan, P. H., and Fishman, R. A. (1985). Stimulation of phospholipase A2 activity by oxygen-derived free radicals in isolated brain capillaries. *J. Cell Biochem.* **27,** 449–453.

Banin, E., Berenshtein, E., Kitrossky, N., Pe'er, J., and Chevion, M. (2000). Gallium-desferrioxamine protects the cat retina against injury after ischemia and reperfusion. *Free Radic. Biol. Med.* **28,** 315–323.

Barnett, N. L., Pow, D. V., and Bull, N. D. (2001). Differential perturbation of neuronal and glial glutamate transport systems in retinal ischaemia. *Neurochem. Int.* **39,** 291–299.

Beal, M. F. (1999). Coenzyme Q10 administration and its potential for treatment of neurodegenerative diseases. *Biofactors* **9,** 261–266.

Beal, M. F. (2004). Therapeutic effects of coenzyme Q10 in diseases. *Methods Enzymol.* **382,** 473–487.

Beal, M. F., and Matthews, R. T. (1997). Coenzyme Q10 in the central nervous system and its potential usefulness in the treatment of neurodegenerative diseases. *Mol. Aspects Med.* **18**(Suppl.), 169–179.

Berlett, B. S., and Stadtman, E. R. (1997). Protein oxidation in aging, disease, and oxidative stress. *J. Biol. Chem.* **272,** 20313–20316.

Beurel, E., and Jope, R. S. (2006). The paradoxical pro- and anti-apoptotic actions of GSK3 in the intrinsic and extrinsic apoptosis signaling pathways. *Prog. Neurobiol.* **79,** 173–189.

Branca, D. (2004). Calpain-related diseases. *Biochem. Biophys. Res. Commun.* **322,** 1098–1104.

Brazil, D. P., and Hemmings, B. A. (2001). Ten years of protein kinase B signalling: A hard Akt to follow. *Trends Biochem. Sci.* **26,** 657–664.

Buchi, E. R., Suivaizdis, I., and Fu, J. (1991). Pressure-induced retinal ischemia in rats: An experimental model for quantitative study. *Ophthalmologica* **203,** 138–147.

Camacho, A., and Massieu, L. (2006). Role of glutamate transporters in the clearance and release of glutamate during ischemia and its relation to neuronal death. *Arch. Med. Res.* **37,** 11–18.

Celebi, S., Dilsiz, N., Yilmaz, T., and Kükner, A. S. (2002). Effects of melatonin, vitamin E and octreotide on lipid peroxidation durino ischemia-reperfusion in the guinea pig retina. *Eur. J. Ophthalmol.* **12,** 77–83.

Charriaut-Marlangue, C., Margaill, I., Represa, A., Popovici, T., Plotkine, M., and Ben-Ari, Y. (1996). Apoptosis and necrosis after reversible focal ischemia: An *in situ* DNA fragmentation analysis. *J. Cereb. Blood Flow Metab.* **16,** 186–194.

Chen, H. S., and Lipton, S. A. (1997). Mechanism of memantine block of NMDA-activated channels in rat retinal ganglion cells: Uncompetitive antagonism. *J. Physiol.* **499,** 27–46.

Chen, H. S., and Lipton, S. A. (2006). The chemical biology of clinically tolerated NMDA receptor antagonists. *J. Neurochem.* **97,** 1611–1626.

Chen, H. S., Pellegrini, J. W., Aggarwal, S. K., Lei, S. Z., Warach, S., Jensen, F. E., and Lipton, S. A. (1992). Open-channel block of *N*-methyl-D-aspartate (NMDA) responses by memantine:therapeutic advantage against NMDA receptor-mediated neurotoxicity. *J. Neurosci.* **12,** 4427–4436.

Choi, D. W. (1988). Glutamate neurotoxicity and diseases of the nervous system. *Neuron* **1,** 623–634.

Choi, D. W. (1990). Methods for antagonizing glutamate neurotoxicity. *Cerebrovasc. Brain Metab. Rev.* **2,** 105–147.

Corasaniti, M. T., Maiuolo, J., Maida, S., Fratto, V., Navarra, M., Russo, R., Amantea, D., Morrone, L. A., and Bagetta, G. (2007). Cell signaling pathways in the mechanisms of neuroprotection afforded by bergamot essential oil against NMDA-induced cell death *in vitro*. *Br. J. Pharmacol.* **151,** 518–529.

Danbolt, N. C. (2001). Glutamate uptake. *Prog. Neurobiol.* **65,** 1–105.

Dawson, V. L. (1995). Nitric oxide: Role in neurotoxicity. *Clin. Exp. Pharmacol. Physiol.* **22,** 305–308.

Dirnagl, U., Iadecola, C., and Moskowitz, M. A. (1999). Pathobiology of ischaemic stroke: An integrated view. *Trends Neurosci.* **22,** 391–397.

Doble, A. (1999). The role of excitotoxicity in neurodegenerative disease: Implications for therapy. *Pharmacol. Ther.* **81,** 163–221.

Dreyer, E. B., Zurakowski, D., Schumer, R. A., Podos, S. M., and Lipton, S. A. (1996). Elevated glutamate levels in the vitreous body of humans and monkeys with glaucoma. *Arch. Ophthalmol.* **114,** 299–305.

Franke, T. F., Hornik, C. P., Segev, L., Shostak, G. A., and Sugimoto, C. (2003). PI3K/Akt and apoptosis: Size matters. *Oncogene* **22,** 8983–8998.

Fukunaga, K., and Kawano, T. (2003). Akt is a molecular target for signal transduction therapy in brain ischemic insult. *J. Pharmacol. Sci.* **92,** 317–327.

Harada, T., Harada, C., Watanabe, M., Inoue, Y., Sakagawa, T., Nakayama, N., Sasaki, S., Okuyama, S., Watase, K., Wada, K., and Tanaka, K. (1998). Functions of the two glutamate transporters GLAST and GLT-1 in the retina. *Proc. Natl. Acad. Sci. USA* **95,** 4663–4666.

Hare, W. A., and Wheeler, L. (2009). Experimental glutamatergic excitotoxicity in rabbit retinal ganglion cells: Block by memantine. *Invest. Ophthalmol. Vis. Sci.* First published on January 10, 2009 as doi:10.1167/iovs.08-2103..

Hare, W. A., WoldeMussie, E., Weinreb, R. N., Ton, H., Ruiz, G., Wijono, M., Feldmann, B., Zangwill, L., and Wheeler, L. (2004). Efficacy and safety of memantine treatment for reduction of changes associated with experimental glaucoma in monkey, II: Structural measures. *Invest. Ophthalmol. Vis. Sci.* **45,** 2640–2651.

Honkanen, R. A., Baruah, S., Zimmerman, M. B., Khanna, C. L., Weaver, Y. K., Narkiewicz, J., Waziri, R., Gehrs, K. M., Weingeist, T. A., Boldt, H. C., Folk, J. C., Russell, S. R., *et al.* (2003). Vitreous amino acid concentrations in patients with glaucoma undergoing vitrectomy. *Arch. Ophthalmol.* **121,** 183–188.

Huang, Y., Li, Z., Wang, N., van Rooijen, N., and Cui, Q. (2008). Roles of PI3K and JAK pathways in viability of retinal ganglion cells after acute elevation of intraocular pressure in rats with different autoimmune backgrounds. *BMC Neurosci.* **9,** 78.

Iijima, T., Iijima, C., Iwao, Y., and Sankawa, H. (2000). Difference in glutamate release between retina and cerebral cortex following ischemia. *Neurochem. Int.* **36,** 221–224.

Joo, C. K., Choi, J. S., Ko, H. W., Park, K. Y., Sohn, S., Chun, M. H., Oh, Y. J., and Gwag, B. J. (1999). Necrosis and apoptosis after retinal ischemia: Involvement of NMDA-mediated excitotoxicity and p53. *Invest. Ophthalmol. Vis. Sci.* **40,** 713–720.

Kanamori, A., Nakamura, M., Nakanishi, Y., Nagai, A., Mukuno, H., Yamada, Y., and Negi, A. (2004). Akt is activated via insulin/IGF-1 receptor in rat retina with episcleral vein cauterization. *Brain Res.* **1022,** 195–204.

Kaplan, D. R., and Miller, F. D. (2000). Neurotrophin signal transduction in the nervous system. *Curr. Opin. Neurobiol.* **10,** 381–391.

Kelly, S., Zhao, H., Hua Sun, G., Cheng, D., Qiao, Y., Luo, J., Martin, K., Steinberg, G. K., Harrison, S. D., and Yenari, M. A. (2004). Glycogen synthase kinase 3beta inhibitor Chir025 reduces neuronal death resulting from oxygen-glucose deprivation, glutamate excitotoxicity, and cerebral ischemia. *Exp. Neurol.* **188,** 378–386.

Kim, H. S., and Park, C. K. (2005). Retinal ganglion cell death is delayed by activation of retinal intrinsic cell survival program. *Brain Res.* **1057,** 17–28.

Klöcker, N., Cellerino, A., and Bähr, M. (1998). Free radical scavenging and inhibition of nitric oxide synthase potentiates the neurotrophic effects of brain-derived neurotrophic factor on axotomized retinal ganglion cells *in vivo*. *J. Neurosci.* **18,** 1038–1046.

Klöcker, N., Kermer, P., Weishaupt, J. H., Labes, M., Ankerhold, R., and Bahr, M. (2000). Brain-derived neurotrophic factor-mediated neuroprotection of adult rat retinal ganglion cells *in vivo* does not exclusively depend on phosphatidyl-inositol-3′-kinase/protein kinase B signaling. *J. Neurosci.* **20,** 6962–6967.

Ko, M. L., Hu, D. N., Ritch, R., and Sharma, S. C. (2000). The combined effect of brain-derived neurotrophic factor and a free radical scavenger in experimental glaucoma. *Invest. Ophthalmol. Vis. Sci.* **41,** 2967–2971.

Konishi, H., Fujiyoshi, T., Fukui, Y., Matsuzaki, H., Yamamoto, T., Ono, Y., Andjelkovic, M., Hemmings, B. A., and Kikkawa, U. (1999). Activation of protein kinase B induced by H(2)O(2) and heat shock through distinct mechanisms dependent and independent of phosphatidylinositol 3-kinase. *J. Biochem.* **126,** 1136–1143.

Levine, R. L. (1983). Oxidative modification of glutamine synthetase. I. Inactivation is due to loss of one histidine residue. *J. Biol. Chem.* **258,** 11823–11827.

Levkovitch-Verbin, H., Martin, K. R., Quigley, H. A., Baumrind, L. A., Pease, M. E., and Valenta, D. (2002). Measurement of amino acid levels in the vitreous humor of rats after chronic intraocular pressure elevation or optic nerve transection. *Glaucoma* **11,** 396–405.

Levkovitch-Verbin, H., Harizman, N., Dardik, R., Nisgav, Y., Vander, S., and Melamed, S. (2007). Regulation of cell death and survival pathways in experimental glaucoma. *Exp. Eye Res.* **85,** 250–258.

Lipton, S. A. (2006). Paradigm shift in neuroprotection by NMDA receptor blockade: Memantine and beyond. *Nat. Rev. Drug Discov.* **5,** 160–170.

Lipton, S. A., and Nicotera, P. (1998). Calcium, free radicals and excitotoxins in neuronal apoptosis. *Cell Calcium* **23,** 165–171.

Lipton, S. A., and Rosenberg, P. A. (1994). Excitatory amino acids as a final common pathway for neurologic disorders. *N. Engl. J. Med.* **330,** 613–622.

Louzada-Junior, P., Dias, J. J., Santos, W. F., Lachat, J. J., Bradford, H. F., and Coutinho-Netto, J. (1992). Glutamate release in experimental ischaemia of the retina: An approach using microdialysis. *J. Neurochem.* **59,** 358–363.

Lucas, D. R., and Newhouse, J. P. (1957). The toxic effect of sodium L-glutamate on the inner layers of the retina. *Am. Med. Assoc. Arch. Ophthalmol.* **58,** 193–201.

Lynch, D. R., and Guttmann, R. P. (2002). Excitotoxicity: Perspectives based on *N*-methyl-D-aspartate receptor subtypes. *J. Pharmacol. Exp. Ther.* **300,** 717–723.

Manabe, S., and Lipton, S. A. (2003). Divergent NMDA signals leading to proapoptotic and antiapoptotic pathways in the rat retina. *Invest. Ophthalmol. Vis. Sci.* **44,** 385–392.

McCord, J. M., Roy, R. S., and Schaffer, S. W. (1985). Free radicals and myocardial ischemia. The role of xanthine oxidase. *Adv. Myocardiol.* **5,** 183–189.

Michaelis, E. K. (1998). Molecular biology of glutamate receptors in the central nervous system and their role in excitotoxicity, oxidative stress and aging. *Prog. Neurobiol.* **54,** 369–415.

Muller, A., Maurin, L., and Bonne, C. (1998). Free radicals and glutamate uptake in the retina. *Gen. Pharmacol.* **30,** 315–318.

Nakazawa, T., Tamai, M., and Mori, N. (2002). Brain-derived neurotrophic factor prevents axotomized retinal ganglion cell death through MAPK and PI3K signaling pathways. *Invest. Ophthalmol. Vis. Sci.* **43,** 3319–3326.

Nakazawa, T., Shimura, M., Tomita, H., Akiyama, H., Yoshioka, Y., Kudou, H., and Tamai, M. (2003). Intrinsic activation of PI3K/Akt signaling pathway and its neuroprotective effect against retinal injury. *Curr. Eye Res.* **26,** 55–63.

Nakazawa, T., Shimura, M., Endo, S., Takahashi, H., Mori, N., and Tamai, M. (2005). *N*-Methyl-D-Aspartic acid suppresses Akt activity through protein phosphatase in retinal ganglion cells. *Mol. Vis.* **11,** 1173–1182.

Naskar, R., Vorwerk, C. K., and Dreyer, E. B. (2000). Concurrent downregulation of a glutamate transporter and receptor in glaucoma. *Invest. Ophthalmol. Vis. Sci.* **41,** 1940–1944.

Nicholls, D., and Attwell, D. (1990). The release and uptake of excitatory amino acids. *Trends Pharmacol. Sci.* **11,** 462–468.

Nucci, C., Tartaglione, R., Rombola, L., Morrone, L. A., Fazzi, E., and Bagetta, G. (2005). Neurochemical evidence to implicate elevated glutamate in the mechanisms of high intraocular pressure (IOP)-induced retinal ganglion cell death in rat. *Neurotoxicology* **26,** 935–941.

Nucci, C., Tartaglione, R., Cerulli, A., Mancino, R., Spanò, A., Cavaliere, F., Rombolà, L., Bagetta, G., Corasaniti, M. T., and Morrone, L. A. (2007). Retinal damage caused by high intraocular pressure-induced transient ischemia is prevented by coenzyme Q10 in rat. *Int. Rev. Neurobiol.* **82,** 397–406.

Oka, T., Tamada, Y., Nakajima, E., Shearer, T. R., and Azuma, M. (2006). Presence of calpain-induced proteolysis in retinal degeneration and dysfunction in a rat model of acute ocular hypertension. *J. Neurosci. Res.* **83,** 1342–1351.

Olney, J. W. (1969). Glutamate-induced retinal degeneration in neonatal mice. Electron microscopy of the acutely evolving lesion. *J. Neuropathol. Exp. Neurol.* **28,** 455–474.

Orrenius, S. (2007). Reactive oxygen species in mitochondria-mediated cell death. *Drug Metab. Rev.* **39,** 443–455.

Osborne, N. N. (1999). Memantine reduces alterations to the mammalian retina, *in situ*, induced by ischemia. *Vis. Neurosci.* **16,** 45–52.

Osborne, N. N., Ugarte, M., Chao, M., Childlow, G., Bae, J. H., Wood, J. P., and Nash, M. S. (1999). Neuroprotection in relation to retinal ischemia and relevance to glaucoma. *Surv. Ophthalmol.* **43** (Suppl. 1), s102–s128.

Osborne, N. N., Chidlow, G., Layton, C. J., Wood, J. P., Casson, R. J., and Melena, J. (2004). Optic nerve and neuroprotection strategies. *Eye* **18,** 1075–1084.

Ouyang, Y. B., Tan, Y., Comb, M., Liu, C. L., Martone, M. E., Siesjö, B. K., and Hu, B. R. (1999). Survival- and death-promoting events after transient cerebral ischemia: Phosphorylation of Akt, release of cytochrome C and Activation of caspase-like proteases. *J. Cereb. Blood Flow Metab.* **19,** 1126–1135.

Papucci, L., Schiavone, N., Witort, E., Donnini, M., Lapucci, A., Tempestini, A., Formigli, L., Zecchi-Orlandini, S., Orlandini, G., Carella, G., Brancato, R., and Capaccioli, S. (2003). Coenzyme Q10 prevents apoptosis by inhibiting mitochondrial depolarization independently of its free radical scavenging property. *J. Biol. Chem.* **278,** 28220–28228.

Phillis, J. W., Ren, J., and O'Regan, M. H. (2000). Transporter reversal as a mechanism of glutamate release from the ischemic rat cerebral cortex: Studies with DL-threo-b-benzyloxyaspartate. *Brain Res.* **868,** 105–112.

Potashkin, J. A., and Meredith, G. E. (2006). The role of oxidative stress in the dysregulation of gene expression and protein metabolism in neurodegenerative disease. *Antioxid. Redox Signal.* **8,** 144–151.

Pow, D. V., and Barnett, N. L. (1999). Changing patterns of spatial buffering of glutamate in developing rat retinae are mediated by the Müller cell glutamate transporter GLAST. *Cell Tissue Res.* **297,** 57–66.

Rauen, T. (2000). Diversity of glutamate transporter expression and function in the mammalian retina. *Amino Acids* **19,** 53–62.

Rauen, T., Taylor, W. R., Kuhlbrodt, K., and Wiessner, M. (1998). High-affinity glutamate transporters in the rat retina: A major role of the glial glutamate transporter GLAST-1 in transmitter clearance. *Cell Tissue Res.* **291,** 19–31.

Rossi, D. J., Oshima, T., and Attwell, D. (2000). Glutamate release in severe brain ischaemia is mainly by reversed uptake. *Nature* **403,** 316–321.

Russo, R., Cavaliere, F., Berliocchi, L., Nucci, C., Gliozzi, M., Mazzei, C., Tassorelli, C., Corasaniti, M. T., Rotiroti, D., Bagetta, G., and Morrone, L. A. (2008a). Modulation of pro-survival and death-associated pathways under retinal ischemia/reperfusion: Effects of NMDA receptor blockade. *J. Neurochem.* **107,** 1347–1357.

Russo, R., Cavaliere, F., Rombolà, L., Gliozzi, M., Cerulli, A., Nucci, C., Fazzi, E., Bagetta, G., Corasaniti, M. T., and Morrone, L. A. (2008b). Rational basis for the development of coenzyme Q10 as a neurotherapeutic agent for retinal protection. *Prog. Brain Res.* **173,** 575–582.

Russo, R., Cavaliere, F., Watanabe, C., Nucci, C., Bagetta, G., Corasaniti, M. T., Sakurada, S., and Morrone, L. A. (2008c). 17Beta-estradiol prevents retinal ganglion cell loss induced by acute rise of intraocular pressure in rat. *Prog. Brain Res.* **173,** 583–590.

Sakamoto, Y. R., Nakajima, T. R., Fukiage, C. R., Sakai, O. R., Yoshida, Y. R., Azuma, M. R., and Shearer, T. R. (2000). Involvement of calpain isoforms in ischemia-reperfusion in rat retina. *Curr. Eye Res.* **21,** 571–580.

Sandhu, J. K., Pandey, S., Ribecco-Lutkiewicz, M., Monette, R., Borowy-Borowski, H., Walzer, P. R., and Sikorska, M. (2003). Molecular mechanisms of glutamate neurotoxicity in mixed cultures of NT2-derived neurons and astrocytes: Protective effects of coenzyme Q10. *J. Neurosci. Res.* **72,** 691–703.

Seo, M. S., Kim, S. H., Ahn, Y. M., Kim, Y., Jeon, W. J., Yoon, S. C., Roh, M. S., Juhnn, Y. S., and Kim, Y. S. (2007). The effects of repeated administrations of MK-801 on ERK and GSK-3beta signalling pathways in the rat frontal cortex. *Int. J. Neuropsychopharmacol.* **10,** 359–368.

Sevanian, A., Muakkassah-Kelly, S. F., and Montestruque, S. (1983). The influence of phospholipase A2 and glutathione peroxidase on the elimination of membrane lipid peroxides. *Arch. Biochem. Biophys.* **223,** 441–452.

Shibuki, H., Katai, N., Yodoi, J., Uchida, K., and Yoshimura, N. (2000). Lipid peroxidation and peroxynitrite in retinal ischemia-reperfusion injury. *Invest. Ophthalmol. Vis. Sci.* **41,** 3607–3614.

Shults, C. W., Oakes, D., Kieburtz, K., Beal, M. F., Haas, R., Plumb, S., Juncos, J. L., Nutt, J., Shoulson, I., Carter, J., Kompoliti, K., Perlmutter, J. S., *et al.* (2002). Effects of coenzyme Q10 in early Parkinson disease: Evidence of slowing of the functional decline. *Arch. Neurol.* **59,** 1541–1550.

Shults, C. W., Beal, M. F., Song, D., and Fontaine, D. (2004). Pilot trial of high dosages of coenzyme Q10 in patients with Parkinson's disease. *Exp. Neurol.* **188,** 491–494.

Sisk, D. R., and Kuwabara, T. (1985). Histologic changes in the inner retina of albino rats following intravitral injection of monosodium L-glutamate. *Graefe's Arch. Clin. Exp. Ophthalmol.* **223,** 250–258.

Sucher, N. J., Lipton, S. A., and Dreyer, E. B. (1997). Molecular basis of glutamate toxicity in retinal ganglion cells. *Vision Res.* **37,** 3483–3493.

Sullivan, R. K., Woldemussie, E., Macnab, L., Ruiz, G., and Pow, D. V. (2006). Evoked expression of the glutamate transporter GLT-1c in retinal ganglion cells in human glaucoma and in a rat model. *Invest. Ophthalmol. Vis. Sci.* **47,** 3853–3859.

Tezel, G. (2006). Oxidative stress in glaucomatous neurodegeneration: Mechanisms and consequences. *Prog. Ret. Eye Res.* **25,** 490–513.

Tezel, G., Yang, X., and Cai, J. (2005). Proteomic identification of oxidatively modified retinal proteins in a chronic pressure-induced rat model of glaucoma. *Invest. Ophthalmol. Vis. Sci.* **46,** 3177–3187.

Trotti, D., Rossi, D., Gjesdal, O., Levy, L. M., Racagni, G., Danbolt, N. C., and Volterra, A. (1996). Peroxynitrite inhibits glutamate transporter subtypes. *J. Biol. Chem.* **271,** 5976–5979.

Wang, K. K. (2000). Calpain and caspase: Can you tell the difference? *Trends Neurosci.* **23,** 20–26.

WoldeMussie, E., Yoles, E., Schwartz, M., Ruiz, G., and Wheeler, L. A. (2002). Neuroprotective effect of memantine in different retinal injury models in rats. *J. Glaucoma* **11,** 474–480.

Yang, X. L. (2004). Characterization of receptors for glutamate and GABA in retinal neurons. *Prog. Neurobiol.* **73,** 127–150.

INDEX

D

E

F

G

N

Q

R

U

V

CONTENTS OF RECENT VOLUMES

Volume 39

Volume 40

Volume 41

Volume 42

Volume 43

Volume 44

Volume 45

Volume 47

Volume 48

Volume 49

Volume 50

Volume 51

Volume 54

Volume 55

Volume 56

Volume 57

Volume 58

Volume 59

Volume 60

Volume 61

Volume 62

Volume 63

Volume 64

Volume 65

Volume 66

Volume 67

Volume 68

Volume 69

Volume 70

Volume 71

Volume 72

Volume 73

Volume 74

Volume 75

Volume 76

Volume 77

Volume 78

Volume 79

Volume 79

Volume 81

Volume 82

Volume 83

Volume 84

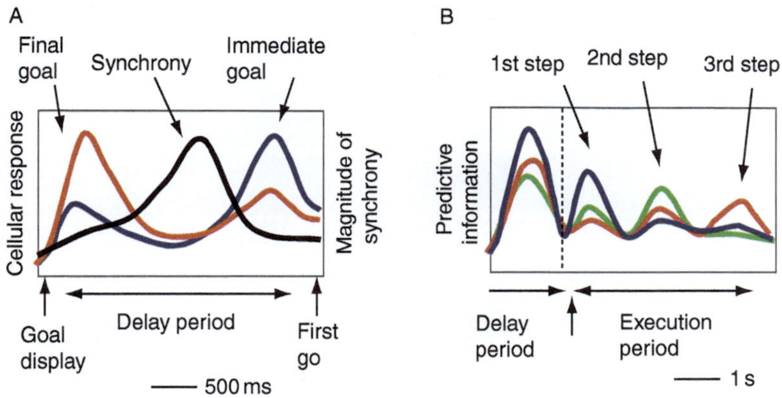

MUSHIAKE *ET AL.*, CHAPTER 1, FIG. 2. (See Legend in Text.)

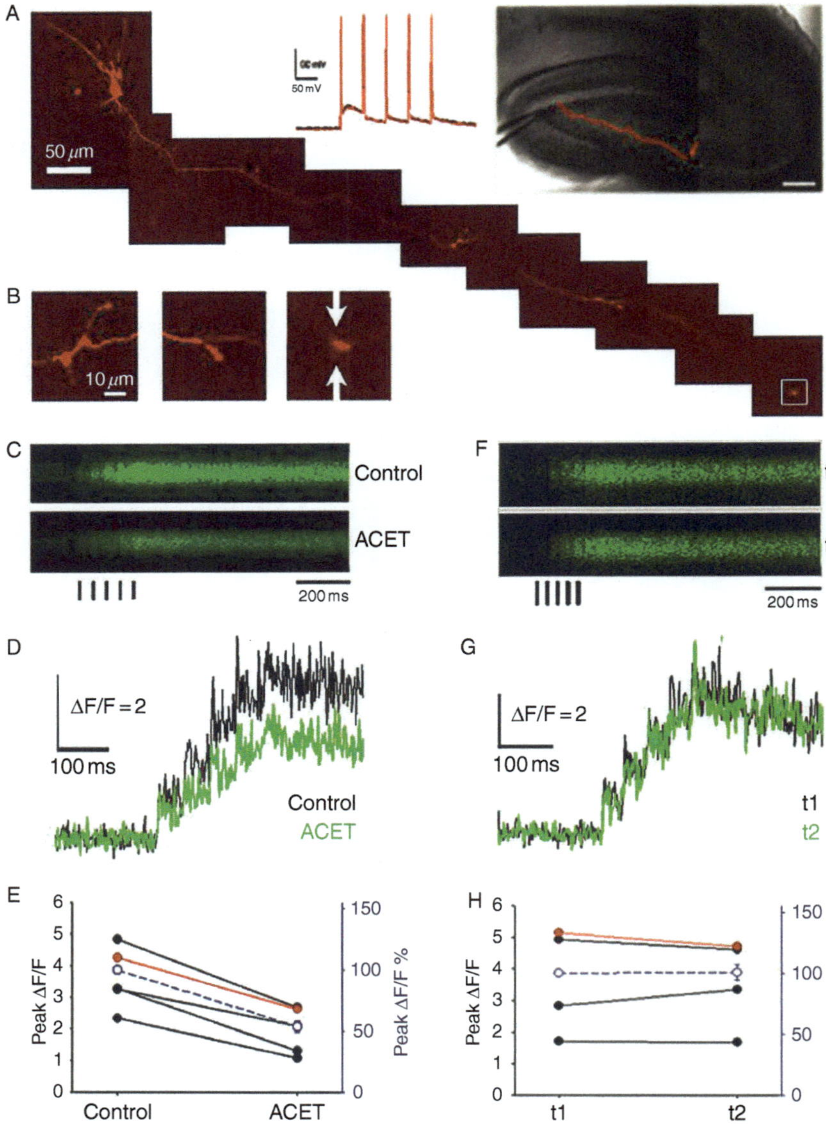

NISTICÒ *ET AL.*, CHAPTER 2, FIG. 4. (See Legend in Text.)

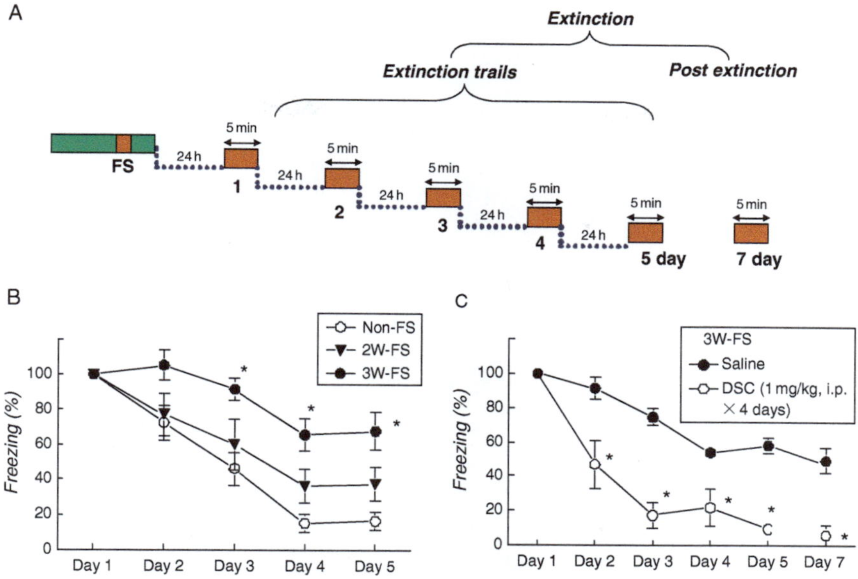

MATSUMOTO *ET AL.*, CHAPTER 7, FIG. 3. (See Legend in Text.)

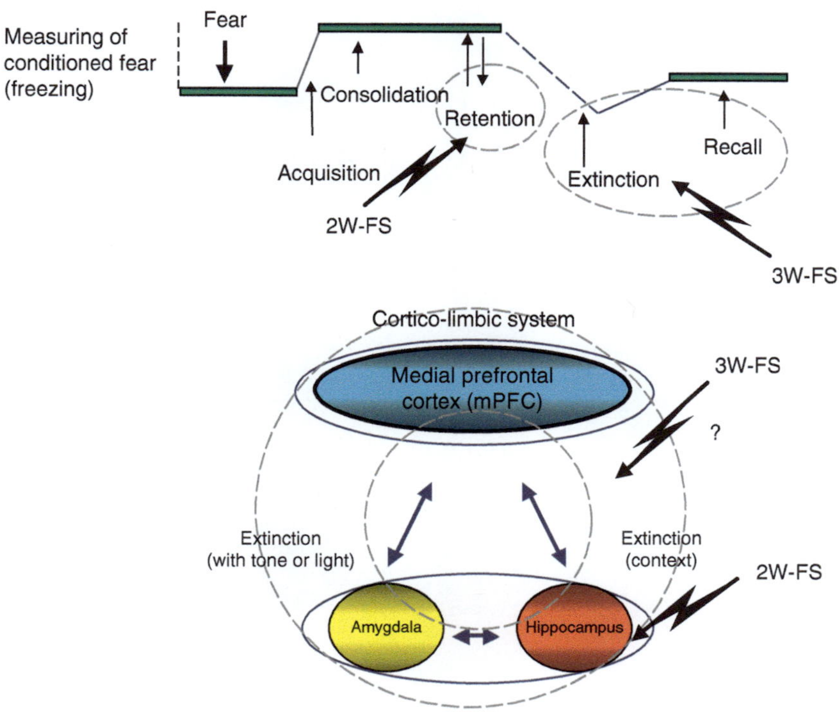

MATSUMOTO *ET AL.*, CHAPTER 7, FIG. 4. (See Legend in Text.)

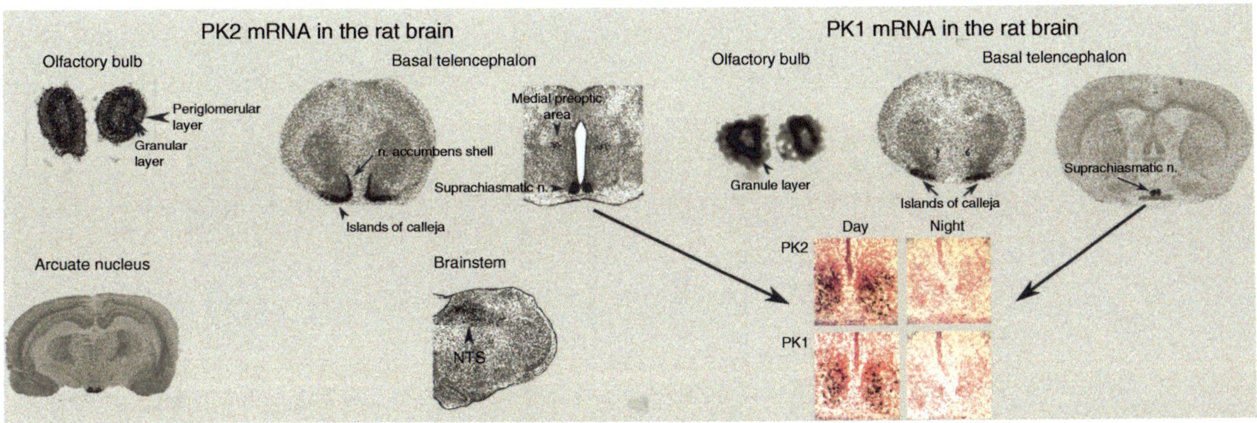

NEGRI *ET AL.*, CHAPTER 11, FIG. 1. (See Legend in Text.)

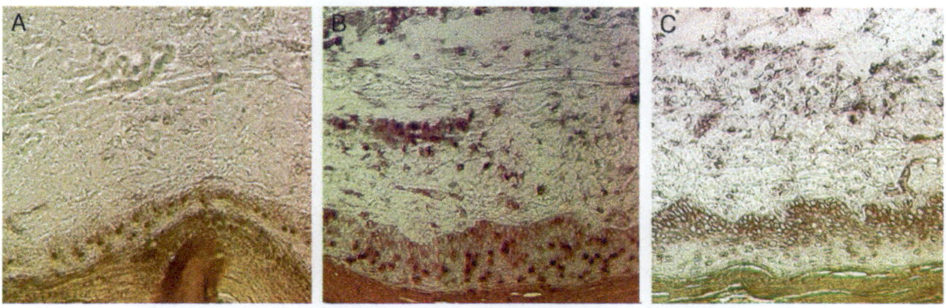

NEGRI *ET AL.*, CHAPTER 11, FIG. 4. (See Legend in Text.)

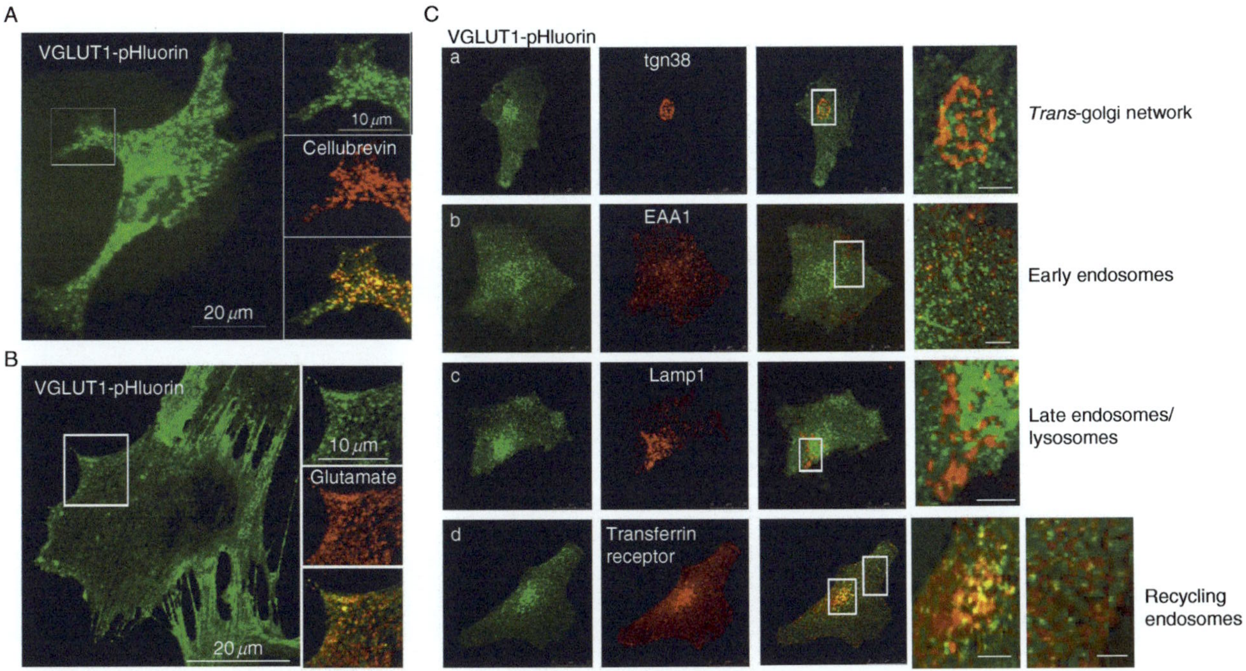

CALÌ *ET AL.*, CHAPTER 20, FIG. 1. (See Legend in Text.)

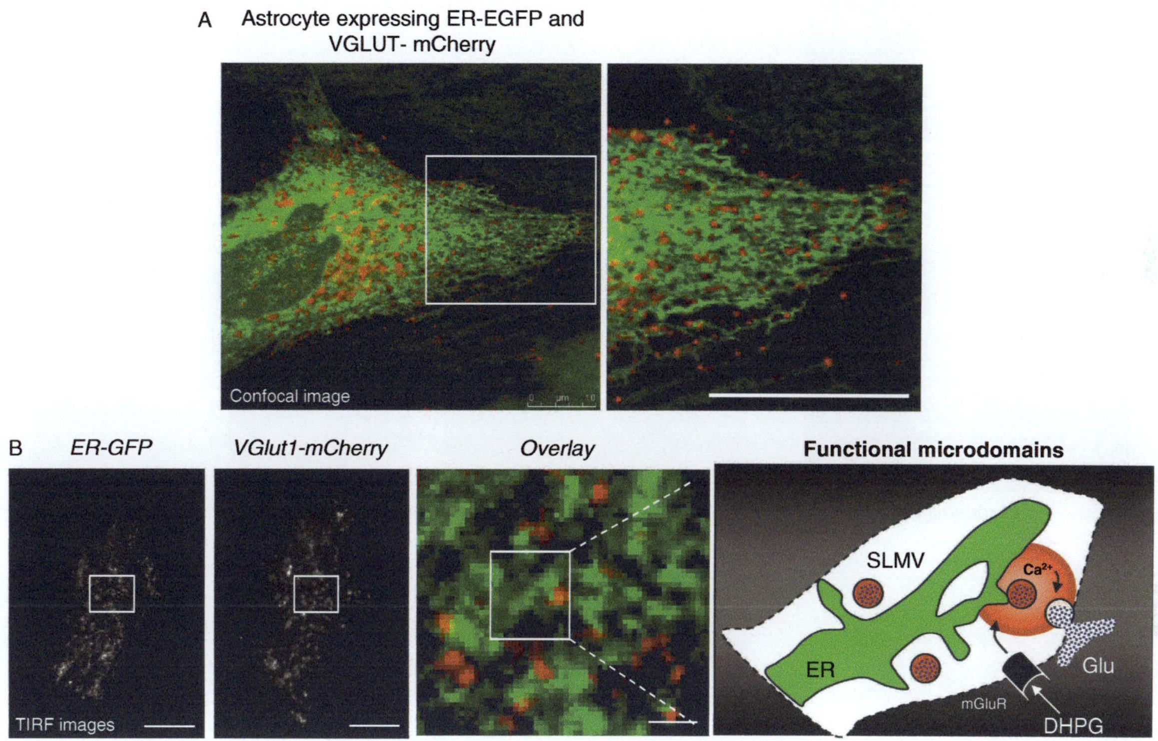

CALÌ *ET AL.*, CHAPTER 20, FIG. 3. (See Legend in Text.)

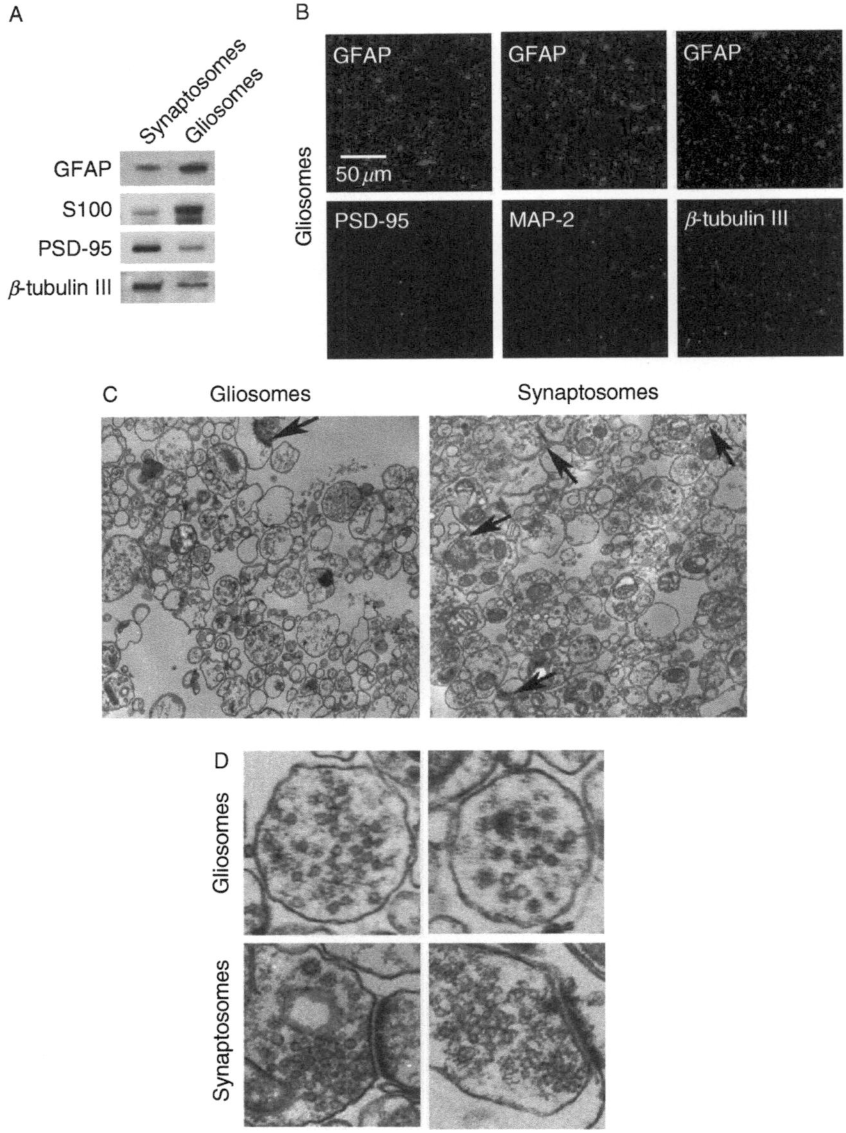

MILANESE *ET AL.*, CHAPTER 21, FIG. 1. (See Legend in Text.)

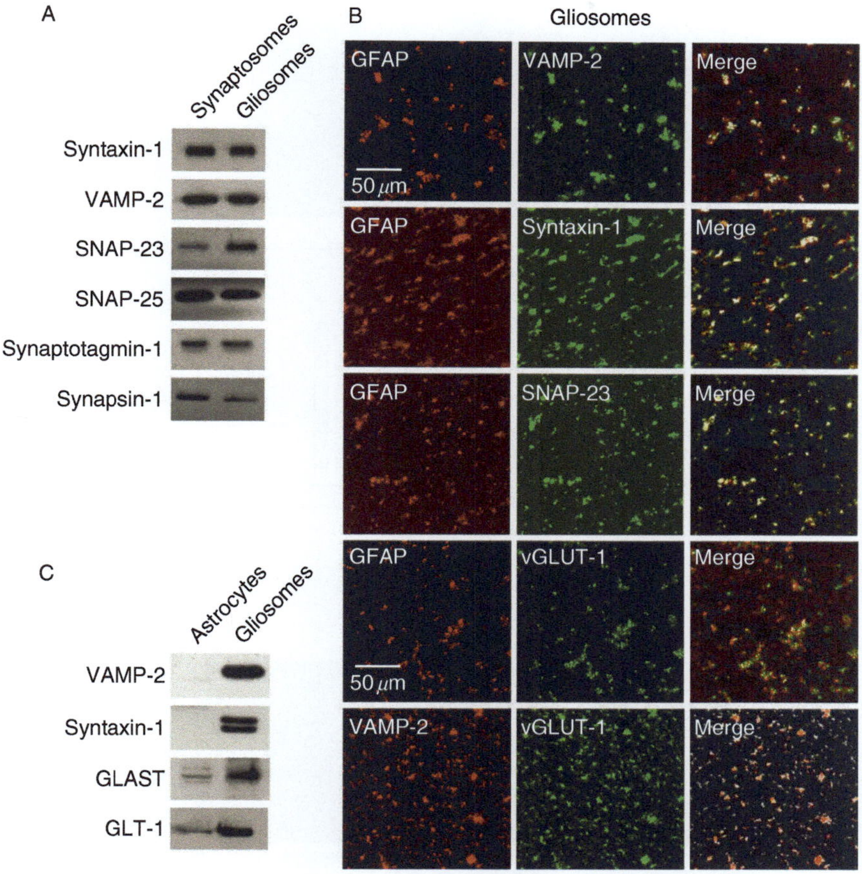

MILANESE *ET AL.*, CHAPTER 21, FIG. 2. (See Legend in Text.)

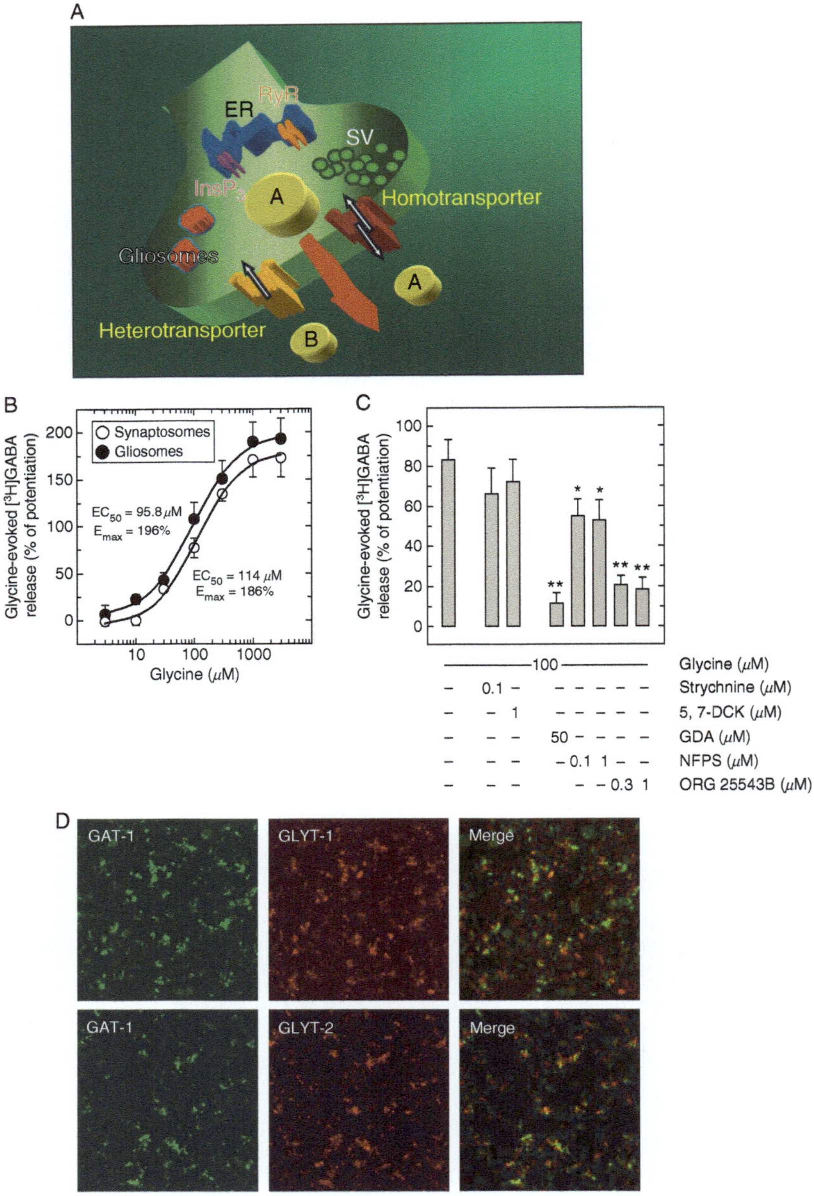

MILANESE *ET AL.*, CHAPTER 21, FIG. 6. (See Legend in Text.)

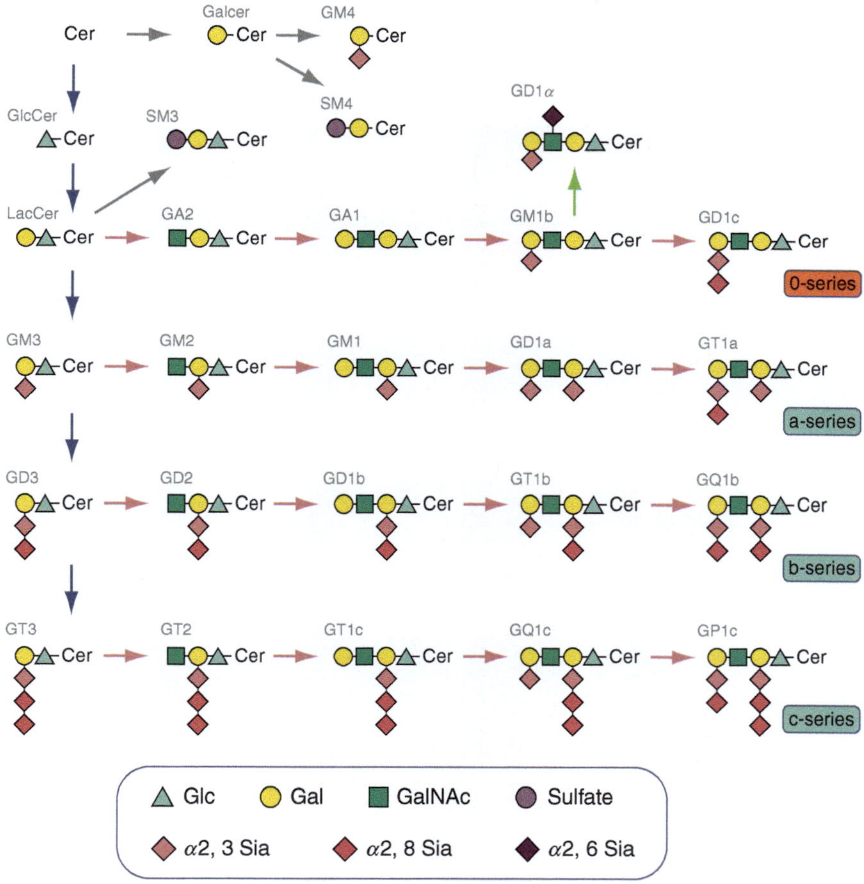

Jin-ichi Inokuchi, Chapter 22, Fig. 1. (See Legend in Text.)

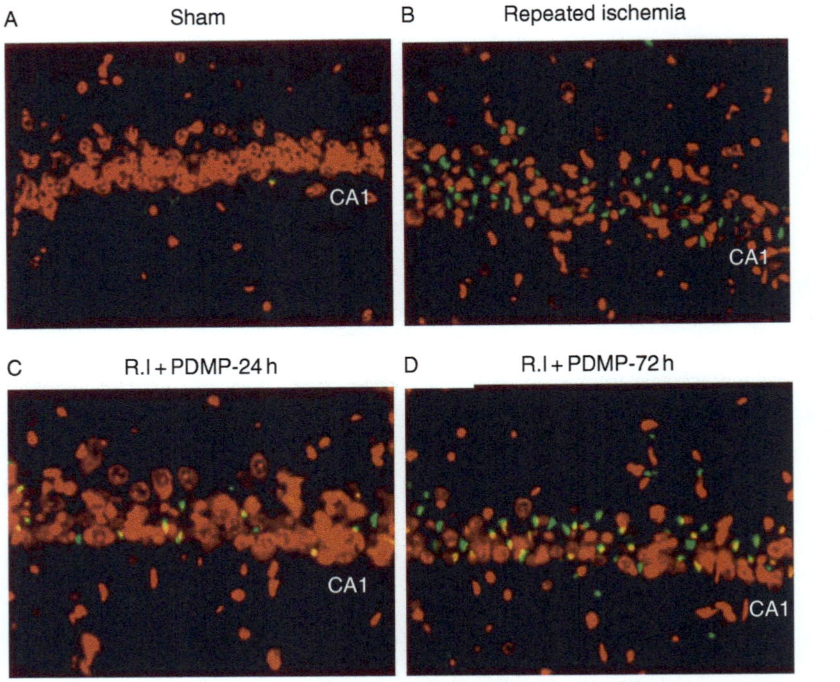

JIN-ICHI INOKUCHI, CHAPTER 22, FIG. 8. (See Legend in Text.)

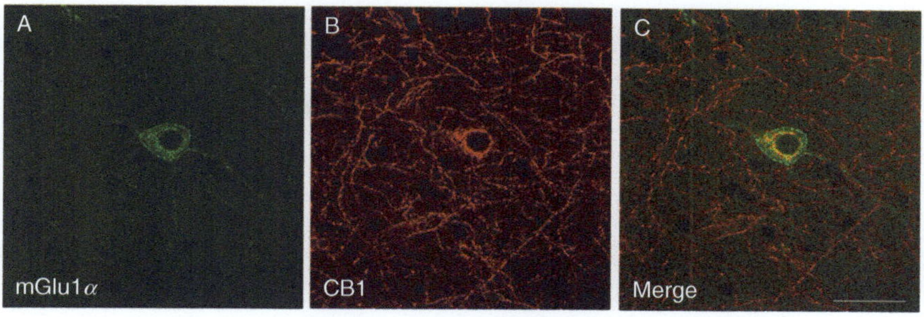

LANDUCCI *ET AL.*, CHAPTER 23, FIG. 2. (See Legend in Text.)

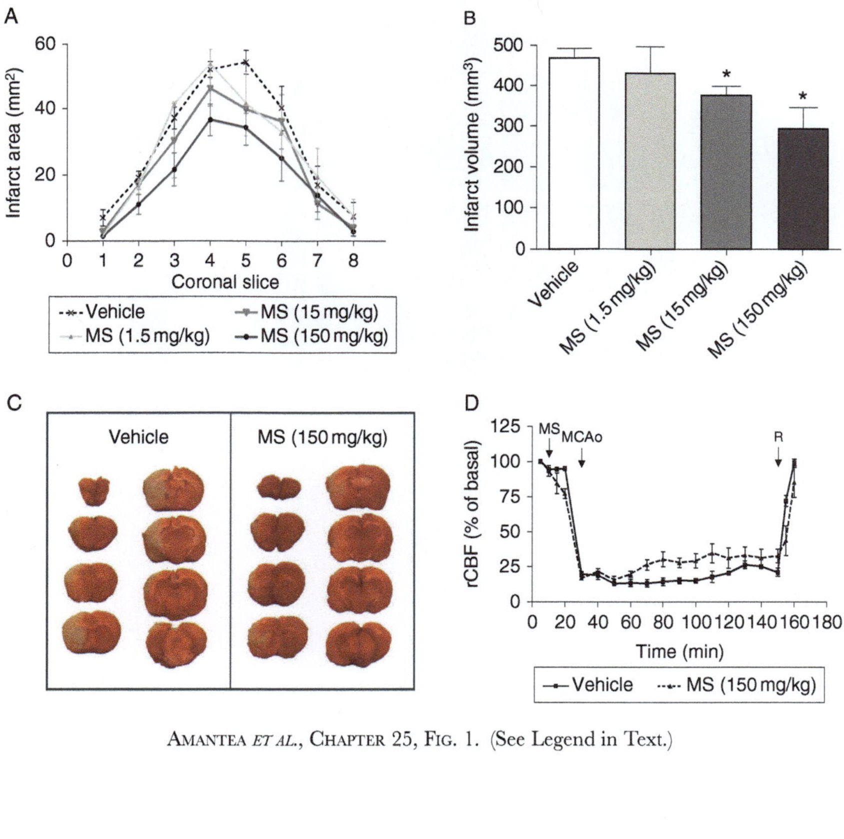

AMANTEA *ET AL.*, CHAPTER 25, FIG. 1. (See Legend in Text.)

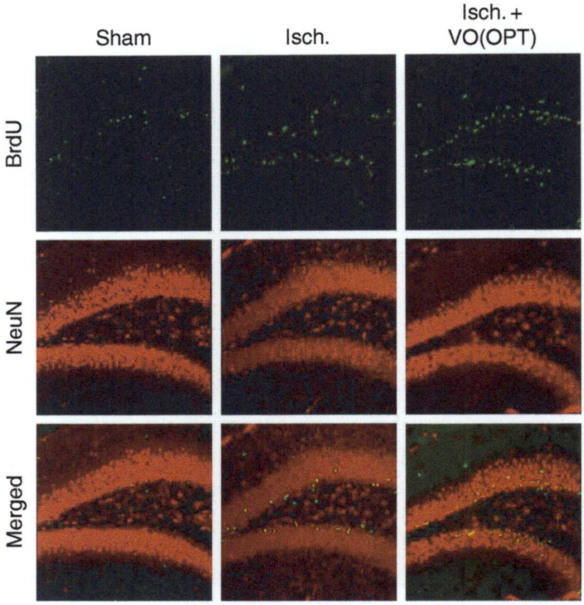

SHIODA *ET AL.*, CHAPTER 26, FIG. 1. (See Legend in Text.)

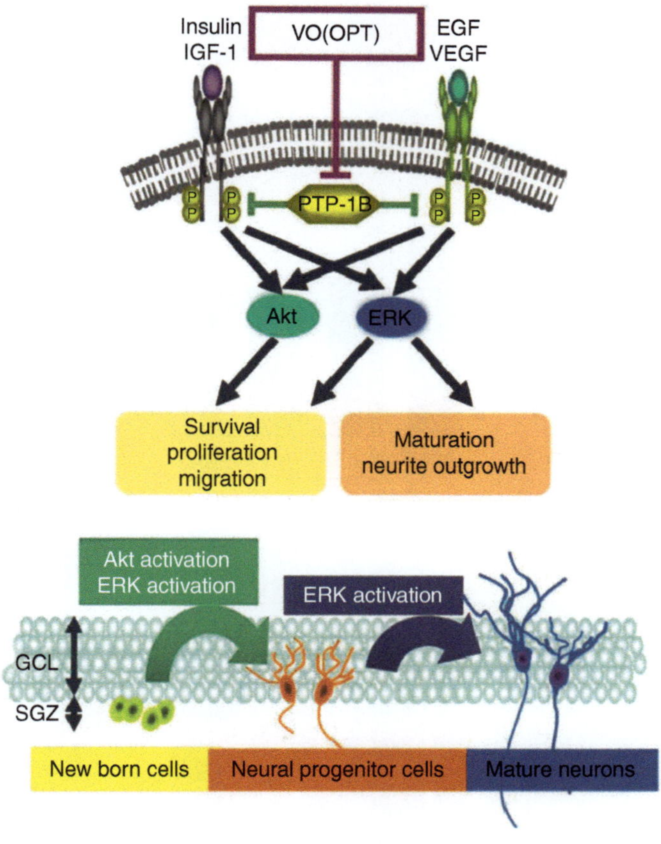

SHIODA *ET AL.*, CHAPTER 26, FIG. 2. (See Legend in Text.)

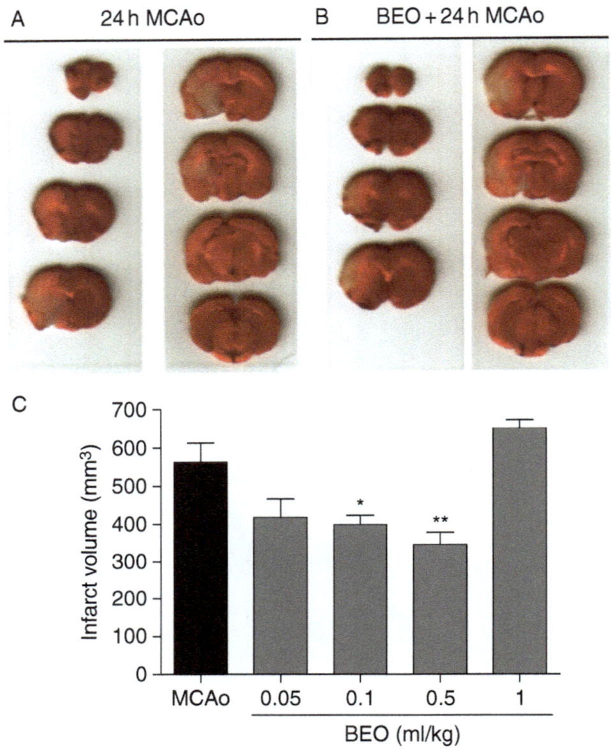

Amantea *et al.*, Chapter 27, Fig. 1. (See Legend in Text.)

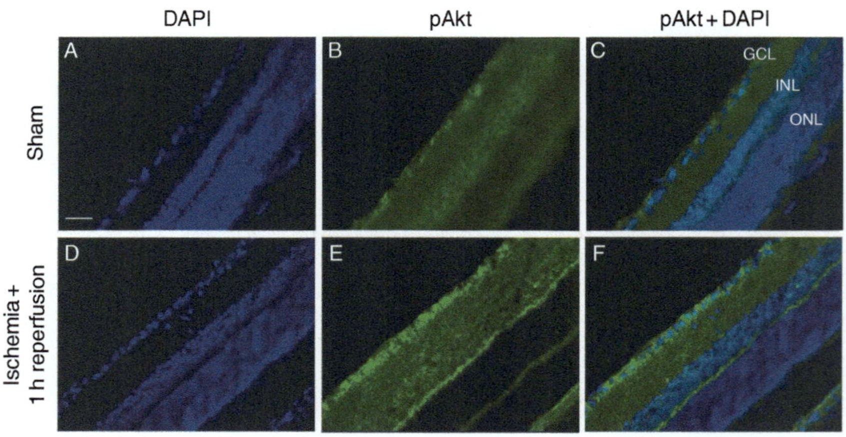

Russo *et al.*, Chapter 28, Fig. 4. (See Legend in Text.)

FSC
www.fsc.org
MIX
Papier | Fördert gute Waldnutzung
FSC® C083411

Kolibri 360 GmbH
Ferdinand-Jühlke-Straße 7
99095 Erfurt, Deutschland
produktsicherheit@kolibri360.de